Cereal Grain Quality

Cereal Grain Quality

Edited by

R.J. Henry

and

P.S. Kettlewell

CHAPMAN & HALL

London · Weinheim · New York · Tokyo · Melbourne · Madras

Published by Chapman & Hall, 2–6 Boundary Row, London SE1 8HN, UK

Chapman & Hall, 2–6 Boundary Row, London SE1 8HN, UK

Chapman & Hall GmbH, Pappelallee 3, 69469 Weinheim, Germany

Chapman & Hall USA, Fourth Floor, 115 Fifth Avenue, New York NY 10003, USA

Chapman & Hall Japan, ITP-Japan, Kyowa Building, 3F, 2-2-1 Hirakawacho, Chiyoda-ku, Tokyo 102, Japan

DA Book (Aust.) Pty Ltd, 648 Whitehorse Road, Mitcham 3132, Victoria, Australia

Chapman & Hall India, R. Seshadri, 32 Second Main Road, CIT East, Madras 600 035, India

---

First edition 1996

Typeset in 10/12pt Palatino by Cambrian Typesetters, Frimley, Surrey

Printed in Great Britain by The University Press, Cambridge

ISBN 0 412 61180 5

A Catalogue record for this book is available from the British Library

Library of Congress Catalogue Card Number: 96–85281

Printed on permanent acid-free text paper, manufactured in accordance with ANSI/NISO Z39.48-1992 (Permanence of Paper).

Errata

1. p. 293. Caption for the table should read:

   Table 10.2 Distribution of lipids in wheat (μg/grain). (Calculated from Hargin and Morrison, 1980)

2. p. 422. Figure 13.6 should be replaced with the following figure:

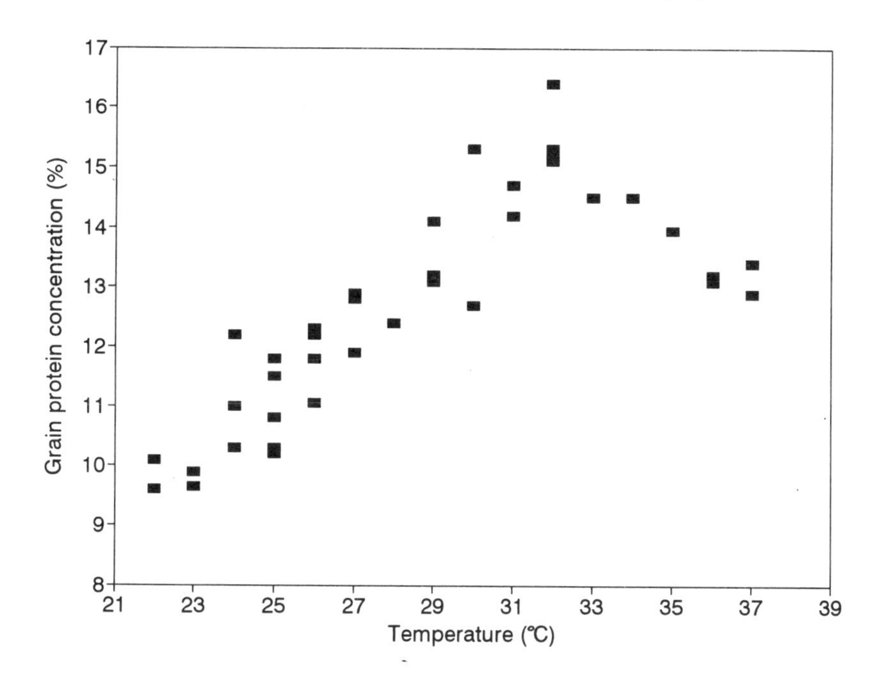

Cereal Grain Quality. Edited by R.J. Henry and P.S. Kettlewell.
Published in 1996 by Chapman & Hall, London. ISBN 0 412 61180 5.

Contents

Part Two
Chemistry and Biochemistry of Cereal Quality

Part Three
Breeding for Cereal Quality

Preface

Cereal uses range from human food and beverages to animal feeds and industrial products. It is human food and beverages which are the predominant uses covered in this book, since the nutritional quality of cereals for animal feed is described in other publications on animal nutrition, and industrial products are a relatively minor use of cereals. Cereals are the main components of human diets and are crucial to human survival. Three species, wheat, rice and maize, account for the bulk of human food. Barley is the major raw material for beer production and ranks fourth in world production. Other species such as sorghum are regionally important.

This book covers all the major cereal species: wheat, rice, maize, barley, sorghum, millet, oats, rye and triticale. Specific chapters have been devoted to a description of the major end-uses of each of the species and to definition of the qualities required for each of their end uses. The functional and nutritional quality of cereals determines their suitability for specific purposes and may limit the quality of the end-product, influencing greatly the commercial value of grain. An understanding of the factors that determine grain quality is thus important in the maintenance of efficient and sustainable agricultural and food production.

The biochemical constituents of the grain that determine quality have been described in chapters on proteins, carbohydrates and other components. An understanding of the relationships between grain composition and quality is important in selecting grain for specific uses.

The breeding of varieties of cereals to meet the quality requirements for specific end-uses has been included, with accounts of both conventional and molecular approaches using biotechnology. Plant breeding offers the potential of new qualities and even new end-uses for grain in the future.

The quality of grain is determined by the genetic potential of the cereal variety and the environment in which the grain is produced. Quality also depends upon the post-harvest storage and handling of the grain. Specific chapters cover agronomy and post-harvest management of quality.

This book should provide a useful first point of reference for almost any aspect of cereal grain quality. The literature cited will allow the reader to access more detailed information on specific aspects of cereal quality.

R.J. Henry
P.S. Kettlewell

Contributors

O. Anderson
USDA
800 Bucannan Street
Albany
CA 94710
USA

A.B. Blakeney
Yanco Agricultural Institute
Yanco NSW 2703
Australia

S.R. Eckhoff
Department of Agricultural Engineering
University of Illinois
360-C Agricultural Engineering Sciences Building
1304 West Pennsylvania Avenue
Urbana IL 61801
USA

M.J. Edney
Grain Research Laboratory
Canadian Grain Commission
1404–303 Main Street
Winnipeg
Manitoba R3C 3G8
Canada

Y. Fujino
Sonda Campus College
Department of Food Science
Minami Tsukaguchi 7-29-1
Amagaski Hyogo-Ken
Japan 661

R.J. Henry
Centre for Plant Conservation Genetics
Southern Cross University
PO Box 157
Lismore
NSW 2480
Australia

P.S. Kettlewell
Crop and Environment Research Centre
Harper Adams Agricultural College
Newport
Shropshire TF10 8NB
UK

J. Kuwata
Sond Campus College
Department of Food Science
Minami Tsukaguchi 7-29-1
Amagaski Hyogo-Ken
Japan 661

Y. Mano
Sonda Campus College
Department of Food Science
Minami Tsukaguchi 7-29-1
Amagaski Hyogo-Ken
Japan 661

J.T. Mills
Agriculture Canada
195 Dafoe Road
Winnipeg
Manitoba R3T 2M9
Canada

C.F. Morris
USDA-ARS Western Wheat Quality Laboratory
E-202 Food Science and Human Nutrition Facility E
Washington State University
Pullman
WA 99164–6394
USA

M. Ohnishi
Sonda Campus College
Department of Food Science
Minami Tsukaguchi 7-29-1
Amagaski Hyogo-Ken
Japan 661

M.R. Paulsen
Department of Agricultural Engineering
University of Illinois
338 Agricultural Engineering Services Building
1304 West Pennsylvania Avenue
Urbana
IL 61801
USA

L.W. Rooney
Cereal Laboratory
Texas A & M University
College Station
TX 77843–2474
USA

S.P. Rose
National Institute of Poultry Husbandry
Harper Adams Agricultural College
Newport
Shropshire TF10 8NB
UK

P.R. Shewry
Long Ashton Research Station
Department of Agricultural Sciences
University of Bristol
Long Ashton
Bristol BS18 9AF
UK

B.A. Stone
School of Biochemistry
La Trobe University
Bundoora
Vic 3083
Australia

F.H. Webster
Quaker Oats Co
John Stuart Research Laboratories
617 West Main Street
Barrington
Illinois 60010
USA

D. Weipert
Federal Centre for Cereal, Potato and Lipid Research
Institute for Milling and Baking Technology
Schutzenberg 12
D-32756 Detmold
Germany

C.W. Wrigley
CSIRO Grain Quality Research Laboratory
Division of Plant Industry
PO Box 7
North Ryde (Sydney)
NSW 2113
Australia

Part One

Quality Requirements of Cereal Users

1

Wheat

C.F. Morris and S.P. Rose

1.1 INTRODUCTION

1.1.1 Scope, significance and definition of quality

Wheat is arguably the most important food crop in the world. World wheat production currently stands at about 550 million metric tons (MMT) (International Wheat Council, 1994). Of this, about 100 MMT is traded each year on the international market.

This chapter will review the major classifications of consumer products and their flour quality requirements, the quantitative assessment of flour quality, milling quality and its relationship to flour functionality, the intrinsic quality of grain and grain lots, and the feed uses of wheat. Additional information appropriate to small-scale testing and cultivar development is included in Chapter 11.

First, **quality** is broadly defined as (i) any feature that distinguishes or identifies something, or (ii) the degree or grade of excellence. Although both are useful to describe wheat grain and flour quality (for example, the quality of flour is defined by its inherent physical-chemical qualities) the latter definition will be the one primarily used here. In this sense, quality may also be thought of as **suitability.**

Conceptually, quality can be thought of as being ultimately defined by the preferences of the consumer and the success of a product in the marketplace. For the consumer, quality relates to the senses: sight, sound, feel (touch and mouth feel), smell, and taste. These organoleptic characteristics are integrated with the cost to produce what can be termed **value.** Value is the cornerstone of market success and provides the fundamental basis for defining the quality requirements of wheat foods. Consider, for example, a hypothetical loaf of bread of the highest possible quality made with the finest ingredients and an elaborate processing system. The bread may be so expensive to produce that few, if any, consumers will choose to buy it. Rather, the consumer chooses to sacrifice some degree of quality so as to maximize the value of their food purchase. Consider, also, the inexpensive budget foods. Owing to the retail price and cost of production, these products may be of noticeably lower quality, yet find great consumer acceptance and success in the marketplace because of their value.

A second important consequence of a consumer definition of quality is that the quality of a particular flour is not necessarily low or high until it is judged in the context of a particular end-use. For example, a flour that exhibits very high bread baking quality will generally exhibit very poor cookie or cake baking quality. From this it follows that wheat quality is defined by an almost infinite number of different food products which contain flour, starch, gluten, bran, whole and cracked grain, etc.

One other concept is central to a discussion of wheat quality. Wheat grain is a biological entity – a living, breathing, complex collection of tissues and organs; and although the central ingredient in most wheaten foods, less than adequate allowance is often made for the inherent variability and complexity of wheat grain and flour. For comparison, consider the relatively simple ingredients sugar, salt and fat – all of which can be defined in basic physical and chemical terms.

Consequently, the cereal or food technologist is often faced with the task of predicting variation in end-product quality based on variation in wheat grain or flour. It is the identification and measurement of this variation in the context of quality that this chapter is devoted to.

1.1.2 Botanical description/classification

Botanically, wheat is a diverse family of related grasses. Over two dozen individual species have been characterized as members of the genus *Triticum*. Of these, only four (*T. monococcum* L., *T. turgidum* L., *T. timopheevii* Zhuk., and *T. aestivum* L. em Thell.) are widely cultivated. Of these, only durum (*T. turgidum* L. var. *durum*) and common (*T. aestivum* L.) wheat will be discussed here. Common wheats include varieties that may have soft or hard endosperm, red or white bran colour, winter or spring growth habit. Club wheat, primarily grown in the Pacific Northwest region of the US, is no longer considered a separate species (*T. compactum* Host.), but rather a variant within *T. aestivum* L.

Durum and common wheats are allotetraploids (AABB) and allo-hexaploids (AABBDD), respectively. The genetic constitution of wheat is important because all quality traits result from the expression of genes and their interaction with the environment. The allopolyploid nature of wheat is important because the homoeologous chromosome groups (A, B and D) are highly related and possess similar genetic traits. For example, the genes that code for the high molecular weight glutenins occur in three sets in common wheat, one each corresponding to a homoeologous group. This duplication of genes confers significant advantages in end-use quality compared to the tetraploid or diploid species. A further consequence of the genetics of wheat is that wheat is closely related to other cultivated cereals, most notably rye. As such, rye provides a unique source of traits and genes which can be incorporated into wheat. This introgression of genetic material from other species has proven particularly advantageous for the plant breeder (see Chapter 11), but may adversely affect end-use quality.

1.2 MAJOR CLASSIFICATIONS OF CONSUMER PRODUCTS AND THEIR END-USE REQUIREMENTS

1.2.1 Introduction

Wheat is used in countless food products. Since a discussion of the end-use requirements of each is beyond the scope of this book, products which share general characteristics are grouped together. This grouping is rather subjective and significant overlap among groups is common. The general characteristics used here include product formula and geometry, and method of preparation. The following groups will be reviewed: fermented (leavened) breads, flat breads and crackers, cookies, cakes and other soft wheat products, noodles, breakfast foods, starch/gluten, and pasta and other durum wheat products.

1.2.2 Fermented (leavened) breads

Leavened breads include traditional white pan, variety, hearth and sourdough breads; and sweet goods including doughnuts. Rolls are simply smaller-portioned products and may be similar to other, larger breads and sweet goods. Leavened breads are characterized by a crumb texture that is light, airy and porous, yet chewy. This unique texture confers the wide popularity and appeal of leavened breads and defines the quintessential functional characteristic of wheat gluten. One of the primary determinates of this texture is the quantity and quality of gluten protein. The gluten must be capable of holding the carbon dioxide and other gases which result from yeast and bacterial fermentation. As such, the gluten must form a continuous matrix that can stretch to accommodate the increase in volume.

Because these types of bread rely on fermentation for most of their characteristics, fermentable substrate is an important quality requirement. Fermentable substrate is produced through the addition of sucrose and glucose (dextrose) or through the production of maltose from flour starch. Since the addition of sugars is expensive, fermentation most commonly relies on the production of fermentable sugars from starch. Maltose is produced by the combined action of α- and β-amylase on starch. Starch granules damaged during the milling process are much more susceptible to enzymatic hydrolysis, and as a consequence, hard wheat flour is preferred because of its higher level of mechanically damaged starch. α-Amylase comes from three biological sources: endogenous in the flour due to preharvest sprouting, and exogenous in the form of malted barley or wheat flour, or of fungal origin (isolated from *Aspergillus oryzae*). Most formulas contain 0.1 to 0.3 parts malted wheat or barley flour per 100 parts wheat flour which is added at the flour mill (B.L. Bruinsma, *personal communication*). Malted flour may be added to achieve a desired **amylograph** or **falling number specification** (Greenwood *et al.*, 1981). Excessive α-amylase due to pre-harvest sprouting produces unacceptably poor-quality bread with low loaf volume and sticky crumb.

Most commercial breads now rely on bakers yeast of high purity and uniformity. Originally, bread doughs were fermented with wild yeast and bacteria. This traditional method is still common where bread is produced at the household or village level (see Section 1.2.3; flat breads).

A second characteristic feature of leavened breads is a well-defined crust. Fermentation produces flavour compounds and free amino acids which are involved in the Maillard reaction and give bread crusts their characteristic brown colour, crunchy texture, and unique flavour. Soft brown crusts are also common.

By far the most common type of leavened bread is white pan bread (Figure 1.1) made from straight-grade or patent flours of hard red spring and winter wheats using the sponge-and-dough and straight dough processes. Other common processes include **liquid ferment**, **Chorleywood bread process** and **'no-time'** dough methods. White pan breads require high flour water absorption, medium to strong dough mixing strength, extensible gluten, and good fermentation and mixing tolerance. Large loaf volume and fine, smooth crumb grain are desirable.

A variant of white pan bread is the sandwich, especially hamburger, bun (Figure 1.1). The hamburger bun has received international exposure due to fast-food restaurant chains like McDonald's. McDonald's alone sells over 8 billion sandwich buns per year worldwide, nearly 5 billion in the US (Palmer, 1994b). The US portion of sales consumes about 180 million kg of flour – about 1% of all the flour produced in the US (Palmer, 1994b).

Another variant of white pan bread is raisin bread (Figure 1.1) . Raisins are added to a white pan bread formula at the rate of 50% or

Figure 1.1 Upper left, moving clockwise: white pan bread, sandwich buns (hamburger and hot dog), hearth breads, sweet goods, Danish and croissant, and raisin bread.

more on a flour basis. Raisin bread owes its popularity to the unique combination of bread and fruit, and is usually toasted and enjoyed as a breakfast food. Due to the heavy load of raisins, strong gluten, high protein flour performs best. Additional yeast is formulated to counteract the tartaric acid introduced by the raisins.

Variety breads are embellished versions of white pan. These breads are prepared using whole-wheat flour and/or as many as 13 different ancillary ingredients. Among the most popular are maize, oat, barley and millet flours or meals, sunflower and sesame seeds, various tree nuts such as walnuts, honey, and cracked or whole wheat kernels, or bran. Since most of these ingredients provide no functional performance and interfere with gluten matrix formation, supplemental vital wheat gluten may be added to reduce the losses in volume and crumb structure. However, since vital gluten is an expensive ingredient, high protein short-patent flours made from hard red spring wheats are preferred.

Hearth breads are also similar to, and were the forerunners of, pan breads (Figure 1.1). Hearth breads are baked on the oven floor, or sole. As a consequence, they range from having an ovate or spherical shape to nearly a horizontal cylinder. As well as affecting the shape, the absence of baking pans exposes the entire bread surface to the oven environment. This large, exposed surface area produces the greater crust to crumb ratio characteristic of hearth breads. Perhaps the most famous hearth bread is the French baguette. The baguette is traditionally produced with a lean formula containing only flour, water, salt and yeast, and considerable hand manipulations. Enjoyed for its thick crunchy crust, the crumb of the baguette is coarse and open. The crust to crumb ratio of baguette may reach 1:1. Baguette is consumed fresh and has a shelf life of less than 24 hours due to rapid staling. Italian and Vienna hearth breads are variations which resemble the baguette. Many hearth breads which originated in Northern Europe contain rye (Chapter 7).

Hearth breads typically require less gluten strength than white pan and variety breads. Also, high flour water absorption is less critical. Hearth breads are generally prepared from straight-grade or higher extraction flours of hard red winter or blends of hard and soft red winter varieties. Sourdough hearth bread is similar to white hearth bread except for its characteristic sour, acidic flavour and extra-chewy texture. The sour flavour is obtained by a combination of yeast and bacterial fermentation which produces acetic and lactic acids as well as many minor, albeit important, constituents. A popular style of sourdough bread is the San Francisco sourdough French bread. Unlike many other hearth breads, San Francisco sourdough French bread requires higher protein, strong gluten flour, such as a patent or long patent from hard red spring wheat. Additionally, up to 2% vital wheat gluten may be added. Product geometry is similar to baguette.

Sweet goods are characterized by rich formulas high in fat and sugar and containing milk solids and eggs. These products are more varied but generally require flours similar to other leavened breads. Products include yeasted doughnuts, cinnamon rolls, coffee cakes, Danish and puff pastries, and french brioche (Figure 1.1). Sweet goods are generally denser than white pan breads and have fine crumb structures. Strong, chewy crumb is considered undesirable. Conversely, gluten strength must be sufficient to carry the large amount of ingredients and withstand the mechanical strains imposed during processing and still yield desirable product volume and density. For some products where the physical demands are particularly severe, such as where the dough is extruded as a continuous ribbon, high protein hard red spring wheat flours of 13% or more protein perform best. For many sweet good products, however, weaker hard red winter or blends of hard and soft red winter wheats yield satisfactory products.

Doughnuts are of two general types: yeast leavened and cake. The unique characteristics of doughnuts is their traditional ring shape and the method of frying in hot oil. Yeast leavened and cake doughnuts differ markedly in formulation and, as such, each has been grouped with its respective associates. Yeast leavened doughnuts are made from yeast-fermented doughs using a sponge-and-dough or straight-dough process. Doughnuts are fried at 180–195°C and take up approximately 20–30% fat, by weight. After cooling, doughnuts may be coated with powdered or glazing sugar.

Danish pastries, puff pastries and croissants are products with a flaky texture, a greater density compared to other breads, and are produced from a rich formula (Figure 1.1). These products derive their characteristic flakiness from the action of steam trapped between multiple layers of dough during baking. The dough layers are separated and lubricated by fat. The main differences between Danish pastries, puff pastries and croissants lie in the much reduced level of sweeteners and the characteristic crescent shape of croissants. Most commercial croissants are extruded and receive considerable sheeting and folding operations, therefore flour quality requirements are similar to those of Danish pastries and puff pastries.

Brioche is a French sweet good made from a rich formula containing eggs (60 parts) as the only source of liquid and a high level of fat (50 parts) (flour 100 parts). Brioche may be made using a straight-dough or sponge-and-dough method.

Steamed breads are simply breads that are steamed rather than baked. As such, they lack the brown crust that is characteristic of oven-baked breads. Steamed breads tend to be about 10–15 cm in diameter and ovate to spherical in shape. They may be plain or with filling. Fillings include savory mixtures of meat and/or vegetables, or sweet bean paste. Steamed breads are traditional wheat foods in the Peoples Republic of China, Japan, The Philippines, and other countries of the Pacific Rim. Like most wheat foods, tremendous local and regional variation exists. Two main types are consumed in China: northern- and southern-style.

Northern-style are larger and chewier, and are made from the higher protein, stronger-gluten wheats of that region. Southern-style are smaller and less dense. Steam breads are generally leavened using a 'mother' sponge – a source of sourdough bacteria and yeast. In The Philippines, the local variation is called *pan de sol*. In Japan, steamed breads may be fermented (*Saka-manju*) or chemically leavened (*Mushi-manju*) (Nagao, 1981).

1.2.3 Flat breads and crackers

Flat breads and crackers are wheat foods of various dimensions but usually no more than 6 cm and often less than 3 cm thick. Most flat breads are higher in water content than crackers, and are therefore chewy as opposed to crisp. Most crackers have no discernable crumb and crust.

The majority of flat breads are consumed in North Africa, the Middle East and the Indian subcontinent. Major types include chapati, rotti, naan, paratha, poori, balady, pita and barabri (Figure 1.2). These breads typically have high crust to crumb ratios and limited crumb. They are generally baked at very high temperatures for very short time periods (e.g. 550°C for 30 s). Consequently, they are relatively insensitive to variations in flour water absorption and α-amylase.

Flat breads are typically produced from high extraction (75–90%) or whole wheat (*atta*) flours. Due to the higher bran content of these flours, white wheat is preferred. Much less gluten strength is required compared to pan and hearth breads for two reasons. One, the limited crumb does not require much gas-holding capacity, and two, many of these breads are prepared by hand and weaker gluten is easier to mix and sheet.

Owing to their geometry and structure, tortillas, pizza crust, English muffins and crumpets, bagels, and pretzels can be considered flat breads. Tortillas, which are traditionally made from maize, are flat breads indigenous to Mexico, Central America, and the southwestern US. Tortillas have enjoyed a tremendous surge in popularity in recent years and are now commonly made from wheat, as well as maize. In 1992, US sales of tortillas exceeded $1.8 billion (Sjerven, 1994). Tortillas form the basis for several ethnic foods including tacos, enchiladas and burritos. Flour tortillas are made from relatively dry, stiff doughs prepared from high protein hard red wheats. Common sizes are 20 and 30 cm in diameter and about 60–80 g (20-cm size) (Figure 1.2).

Pizza crusts vary in texture from thin, crisp doughs that are essentially fried in a thin layer of oil, to thicker bread-like doughs of chewy texture. Since doughs must carry the physical load of toppings and resist water

Figure 1.2 Upper left, moving clockwise: naans, pita (or pocket) bread, English muffins, pretzels (bows and sticks), bagels and tortilla.

penetration, high protein, strong gluten flours are preferred. In addition, 1–2% vital wheat gluten may be added (flour basis).

English muffins and crumpets are flat breads characterized by a coarse, open grain, brown grilled top and bottom with light sides, and a flat disk geometry of about 2.5 × 10 cm (thickness × diameter) (Figure 1.2). English muffins are made from high water content formulas and the dough is over-mixed to break down the gluten structure. This combination of processing treatments produces the characteristic open crumb texture. Upon toasting, the desirable crisp texture is produced. Flour for English muffins is usually made from strong gluten, high protein hard red spring wheat of 13% or more protein. To further augment the native protein, vital wheat gluten at 1–3% may be added to improve product performance.

Bagels and pretzels are unique because they each require an intermediate processing step. Bagels are doughnut-shaped rolls, commonly 70–90 g (Figure 1.2). The word bagel comes from the Austrian-German, *bugel*, meaning stirrup (Petrofsky, 1986). Bagels were invented over 300 years ago to honour Jan Sobieski, king of Poland and famed equestrian. Bagel doughs are retarded for up to 20 hours and then boiled just prior to baking. Bagel doughs are produced at low water absorption (50–55%, flour basis). High protein, strong gluten flours typical of good quality hard red winter wheat give best results (Petrofsky, 1986). Bagels may be plain, egg or 'wheat', lean or rich. Egg bagels contain 2.5–10% whole egg (flour basis). Wheat bagels are made by replacing about one-third of the white flour with whole wheat flour. Lean bagels typically contain only flour, water, yeast, salt and less than 1% sugar; while rich bagels contain about 3% each, sugar and vegetable oil.

Two types of pretzels are popular. The first tend to be small, commonly about 10 cm, and may be in the characteristic twisted shape or other configurations such as short, straight sticks or rings (Figure 1.2). These products are baked and then dried to about 2–3% moisture content and have a crunchy texture and a relatively long shelf-life. The second type, soft pretzels, tend to be much larger, up to 30 cm with the baked dough piece diameter about 2–4 cm. Moisture content is higher and the texture resembles chewy bread. In addition to the characteristic twisted shape, the feature that sets pretzels apart is the use of caustic lye (usually 1.25% $NaOH$). The dough pieces are immersed briefly in hot lye solution, baked and then sprinkled with coarse granular salt. Flour for pretzels is generally a higher ash soft red winter wheat or a blend of soft and some hard (Loving and Brenneis, 1981). US sales of pretzels in 1993 exceeded $1 billion (Anon., 1994b).

Crackers are characterized by thin product geometry (often about 0.5 cm), low water content and crisp, crunchy texture. Crackers are popular in North America, Europe and Australia. Cracker sales in the US exceeded $2.8 billion in 1993 (Palmer, 1994a). Products include soda crackers, cream crackers, water biscuits, graham crackers, sprayed crackers, and savory crackers. In this section, only crackers produced using fermentation are discussed (soda crackers, cream crackers, water biscuits). In these products, gluten is developed and is referred to as

hard dough in the UK (Greenwood *et al.*, 1981; Hoseney *et al.*, 1988; Thacker, 1994). Chemically-leavened products, whether produced by doughs or batters, are discussed under Section 1.2.4, soft wheat products. Hard dough sweet biscuits, often called semi-sweet biscuits, such as Marie, are chemically leavened and likewise appear under Section 1.2.4.

Soda crackers, or saltines, are the most definitive cracker product (Figure 1.3). They are prepared by fermenting a relatively stiff, dry sponge dough (25–35 parts water to 70 parts flour) for 18–24 hours with additional flour (30 parts) at dough-up. Saltine formulas typically contain diastatic malt and shortening. Fermentation is accomplished by the action of yeast and lactic acid-producing bacteria. Sodium bicarbonate (hence the name soda cracker) is added at dough-up to neutralize the drop in pH which occurs during fermentation. Generally, crackers are produced from stronger, medium protein soft red winter wheat flour. If

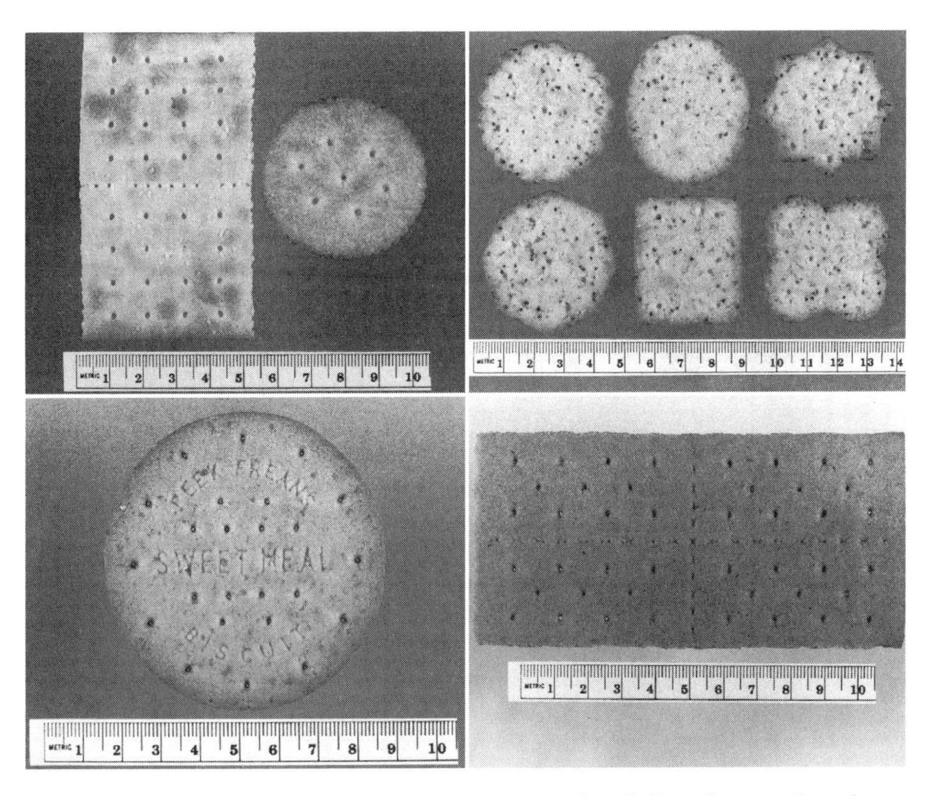

Figure 1.3 Upper left, moving clockwise: soda cracker (left) and sprayed cracker (right), savory snack crackers, graham cracker and sweetmeal cracker.

more gluten strength is required, a blend using some hard red winter is used. Usually, if two flours are used, the stronger is used for the sponge. The final dough must be extensible and sheet easily. Enough gluten strength must remain, however, to produce the necessary oven spring or increase in product volume during baking. Owing to the protracted fermentation, diastatic activity of the dough is important. Bakers rely on naturally-occurring diastatic activity associated with limited levels of sprouting or add diastatic malt to sound flours as described earlier. Diastatic specifications generally vary in the range of 500–600 BU.

Cream crackers of the UK and former colonies are similar to US soda crackers. Fermentation of cream crackers is relatively shorter, lasting from 4–24 h. Cream crackers rely on well-developed gluten and are made from flours of 10–11.5% protein (Wade, 1972). UK breadmaking wheats of 9–10.5% are also used (Thacker, 1994).

1.2.4 Cookies, cakes and other baked goods made from soft wheat

Soft wheat is used in a myriad of diverse food products, often as the major, but commonly ranking second or third in ingredients. Soft wheat is used in these applications because (i) the stronger gluten and higher protein levels of hard wheat reduce product quality, (ii) soft wheat flours have lower levels of starch damage and consequently lower water absorption and viscosity, and (iii) soft wheat flours generally have a finer texture, or smaller particle size distribution. In the case of hard wheat flour, the inherent viscoelasticity of gluten is the primary reason for its use. In the case of soft wheat flour, strong gluten is generally detrimental and starch functionality is of much greater importance. The major food uses of soft wheat flour are chemically-leavened crackers, pie crust, cookies (biscuits), American-style biscuits, scones, moon cake, products made from batters (sugar wafer cookies, ice cream cones, pancakes and waffles), cakes, tempura, breadings and soup thickeners.

Chemically-leavened crackers are similar to fermented crackers except for their method of leavening, which is usually sodium bicarbonate with an acidifying salt. Since these crackers do not go through gluten development and fermentation, there is no particular requirement for blending hard wheats with higher levels of stronger gluten. Some chemically-leavened crackers are made from very lean formulas, however most have higher contents of fat or sugar (or both), and inclusion of savoury flavourings, such as garlic, onion, or cheese. Popular variations include sprayed crackers, savory snack crackers, Marie biscuits, Sweet Meal and digestive biscuits, and graham crackers (Figure 1.3). As fat and especially sugar content rises, the differentiation between crackers and cookies becomes obscure. Currently in the US, only traditional saltines are produced using fermentation. The increasingly diverse snack cracker industry uses chemical leavening.

Pie crust is another sheeted product that does not go through gluten development. As such, pie crust is usually made from soft wheat pastry flour to ensure a tender texture. Fat levels may vary from 35–80% (flour basis). Water, added in limited amount, and salt are the only other ingredients. Pies may be of the typical dish style or a smaller fried variant (Downs, 1971; Loving and Brenneis, 1981). Tarts may also be considered variants of pie-type dough. Flour for pie dough should be from sound, unsprouted soft wheat. According to Loving and Brenneis (1981) flours should be coarser granulation of 8.0–9.5% protein; ash content is not critical and may be as high as 0.65%.

Cookies (biscuits) are food products with generally limited three-dimensional structure, high contents of sugar and fats, and generally low moisture contents. Most are chemically leavened to reduce product density and impart a desirable texture. According to Kulp (1994), little if any starch gelatinization occurs in cookies. This lack of gelatinization is in marked contrast to most other wheat foods.

Owing to the plethora of minor variations and styles, cookies are most easily classified according to the method of production which, in turn, governs the water content and viscosity of doughs. The major types (from lowest to highest water content) are rotary moulded, cutting machine, wire-cut and deposit. Wafer cookies, which are made from a batter, are covered later in this section.

In the UK and former colonies, chemically-leavened cookie doughs are classified as short-dough sweet biscuits (sweet biscuits, short doughs, or short sweet doughs) and hard dough semi-sweet biscuits (Thacker, 1994). Marie and Rich Tea are popular examples of hard dough semi-sweet biscuits. In these products, gluten is partially developed. Short-dough sweet biscuits may be classified according to the US system, i.e. rotary moulded, cutting machine, wire-cut and deposit (Thacker, 1994).

Rotary moulded cookies (Figure 1.4) are made from relatively dry, stiff doughs. Water in the formula may run about 8% on a flour weight basis. The dough is pushed into dies cut on the surface of a smooth cylinder, excess is scraped off, and the cookie is released onto a canvas conveyor and moved to the oven. The dies produce three-dimensional raised designs, wording or logos on the top of the cookie. These cookies are commonly used to make sandwich cookies, where, as the name implies, a cream filling is deposited between two cookies. Rotary-moulded cookies are made from slightly stronger flours with minimal spread. These characteristics ensure that the moulded design will remain through baking and that the constant diameter and resistance to breaking will accommodate the sandwiching process.

Figure 1.4 Upper left, moving clockwise: rotary moulded cookies, wire-cut chocolate chip, deposit (Danish butter), and cutting machine and moulded cookies.

Cutting machine cookies (Figure 1.4) are produced from a sheeted dough. This process feature is unique and requires that the dough possesses sufficient cohesiveness to accommodate the sheeting process, yet not be excessively strong to detrimentally affect product texture and performance. Various shapes are cut from the dough sheet and baked. Scrap dough remaining after the cutting process is returned and incorporated into subsequent doughs. An example of the use of cutting machine cookies is in the production of ice cream sandwiches.

The wire-cut cookie (Figure 1.4) is perhaps the most common commercial cookie in the US. A popular variation includes chocolate chips. As the name implies, wire-cut cookies are produced by passing a taught wire through the extruded dough stream. The wire cuts the dough into individual cookies which drop onto a belt and are conveyed to the oven. Flour for wire-cut cookies must produce dough that is sufficiently cohesive, yet extrudes and separates cleanly upon cutting. A variation of the wire-cut type is referred to as bar-type. Bar-type cookies are extruded as a continuous bar of dough and cut into individual units after being deposited onto the conveyor belt (before or after baking).

The last major class of cookies is referred to as deposit cookies (Figure 1.4). Deposit cookies are made from higher moisture doughs, some approaching the viscosity of cake batters. A popular variation of the deposit cookie is the vanilla wafer. Deposit cookies are produced by extruding the dough through nozzles where reciprocating cutters release the dough onto the conveyor, or oven band. A popular variation is the Danish butter cookie.

Some miscellaneous products fall between cookies and cakes. These products are chemically leavened, are produced from doughs, as opposed to batters, but are lower in enriching ingredients and higher in moisture than typical cookies. These products include American-style biscuits, scones, and baked Chinese buns (*yit bien*, moon cake). American-style biscuits are popular in southeastern US. They are commonly prepared from 'self-raising' flour – flour which has had leavening salts (e.g. anhydrous monocalcium phosphate and sodium bicarbonate) and salt (NaCl) added prior to packaging and sale (Loving and Brenneis, 1981). Biscuit flour is typically low-extraction (0.33–0.36% ash) short patent milled from soft wheat. Low protein hard wheat flour may be blended for greater strength.

Baked Chinese buns, *yit bien* or moon cake, is a popular product enjoyed during the Festival of the Moon according to the Chinese calendar. Buns typically range from about 7–12 cm in diameter and contain a semi-sweet filling (Nagao, 1981). The surface is embossed with decorative designs and brushed with egg to produce a glossy, rich brown appearance. Flours with moderate protein content and strength are preferred, generally a blend of 80% confectionery flour and 20% hard wheat flour, or a short-patent all-purpose flour (Nagao, 1981).

As formula water increases, the dough viscosity drops to a point where it can flow, thus becoming a batter. Batter systems offer several advantages: no gluten development, high ingredient and air cell carrying capacity, generally a moist and tender final product, and ease of processing, e.g. batter can be pumped. Batter products include some low moisture, crisp products like wafer cookies and ice cream cones; griddle cakes like pancakes and waffles; a vast array of higher volume cakes, and coatings such as tempura.

Sugar wafer cookies (Figure 1.5) are produced from high-water formulations (1.4–1.8 × four weight) by injecting the batter between metal plates. After baking, the wafers are cut, and sandwiched with a sugar-fat filling. The resulting product is light and crisp in texture. Two types of ice cream cones are common: sugar cones and moulded (or wafer) cones (Figure 1.5). Both are prepared from batters. Sugar cones may be satisfactorily made from lower grade flours of soft wheat. They are baked between textured metal plates, producing a flat disk of about 0.5 cm in thickness. While still hot and pliable, the disk is rolled into a cone shape where it cools and hardens. Moulded cones are produced by injecting the batter into a die and baking, similar to wafer cookies. The texture is nearly identical to wafer cookies. For all these products, flour water absorption and starch pasting quality are important for batter viscosity, product colour and texture, and ease of processing. Higher quality soft wheat flours free of α-amylase activity perform best.

Pancakes are relatively thin (1–2 cm), round (10–20 cm) cakes prepared by cooking a batter on a hot griddle or pan (Figure 1.5). Oil on the griddle surface fries the outer surface of the pancake to a golden brown. Pancakes are chemically leavened with baking powder and/or soda. Formulations typically contain much less sweetener than regular cakes. Waffles (Figure 1.5) are prepared from similar batters but are

Figure 1.5 Upper left, moving clockwise: sugar wafer cookie, ice cream cones (wafer, large sugar and smaller sugar, left to right), cake doughnuts, and pancake (left) and waffle (right).

cooked in a two-sided griddle which imparts the characteristic three-dimensional structure and crisp texture desired by the consumer. Eating texture and appearance are primary quality considerations. Due to the thin nature of pancakes, stable gas cell formation is critical. If cells coalesce, they easily migrate the short distance to the cake surface and escape, increasing product density. Pancakes and waffles are made from higher protein soft or a combination of soft and hard wheat flours.

Whereas pancakes are fried in a thin layer of oil, doughnuts are completely immersed in hot cooking oil. As mentioned above, doughnuts are of two general types: yeast leavened and cake. Although both are cooked in a similar fashion, the two types differ markedly in formulation. Cake doughnuts (Figure 1.5) are made from a formula much like a standard layer cake which uses both chemical leavening and incorporated air to obtain the desired low product density. Flour protein is generally somewhat stronger than that used for layer cakes. Stronger-gluten soft wheats or mill streams may be selected, or some hard wheat flour may be added to obtain the desired performance.

Cakes, although distinct and unique foods, share attributes of both breads and cookies. Like bread, cakes are baked foams and use many of the same ingredients as bread, but usually at substantially higher proportions, like cookies. Unlike bread, which relies principally on fermentation, cakes derive their light, porous texture through the use of leavening agents like sodium bicarbonate and/or through the entrapment of minute air cells. Like cookies, cakes exhibit little gluten development; gluten development is detrimental to cake quality and imparts a tough texture. In contrast to cookies which are generally less than 2–3% moisture, cakes are formulated to produce tender, moist, appealing textures. Cakes, depending on type, may have relatively high amounts of sugar, eggs and fat in their formula, often exceeding the quantity of flour. In high ratio cakes, as the name implies, sugar-to-flour ratios may range from 1 to as high as 1.2. In the case of angel food cake, sugar and egg whites may each exceed flour by a ratio of 2.75.

Cakes are made from batters with substantially lower viscosity than cookies or bread doughs. Consequently, the stable entrapment of minute air cells is critical to cake volume and texture. The formation of emulsions with formula fat or egg lipids is crucial for the entrapment of air cells and for preventing the establishment of a continuous gluten network. In many cakes, the structure-forming role of eggs, whole or only whites, is substantial. Cakes can also be highly-flavoured with high proportions of chocolate, fruit, and other ingredients.

Some of the main classifications of cakes are high-ratio, pound, chocolate (devil's food, fudge, chocolate layer, and milk chocolate), yellow and white layer, sponge, Swiss roll, chiffon, angel food, cup cakes and muffins, and Japanese *castilla* (Figure 1.6).

Figure 1.6 Upper left, moving clockwise: chocolate (left) and white (right) layer cake, pound cake, poppy seed muffin, and angel food cake.

Flour requirements for cakes vary somewhat, depending on the style and quality of cake. The quality of most cakes is dramatically better when made from soft wheat flours of low protein. Some cakes, such as angel food, are especially responsive to flour quality and are best made from low extraction, premium soft wheat flours (e.g., 0.23% ash and 4.7% protein (Dubois, 1961)). A more normal range of specifications for an average cake flour would be 8.5% protein and 0.36% ash. Often the functional performance of cake flours is enhanced by reducing the mean particle size through pin milling. Low starch damage, low viscosity, and low amylase activity are also generally considered desirable traits (Miller *et al.*, 1967).

In the US, cakes flours may still be chlorinated, although the practice is decreasing in response to consumer concerns. In many countries around the world, chlorine is not used. The improving effect of chlorination has long been recognized and has been the subject of considerable research. The exact mechanism of chlorine action has not been established, although it appears to act on protein, starch and lipid, and lowers pH.

Tempura is a Japanese batter used to coat seafood and vegetables prior to deep-fat frying (Nagao, 1981). Soft wheat flour of low protein and weak gluten performs best. Short-patent flours milled from US

Western White are most commonly used for tempura in Japan (Nagao, 1981). In the US, a similar batter containing spices is popular for coating spirally-cut potatoes prior to frying.

In addition to batters, dry breadings are commonly used to coat meat, especially poultry and ground meat patties, prior to baking, pan or deep-fat frying. Breadings may be made from moderate-extraction soft or hard wheat flours, or ground bread or cracker crumbs (Loving and Brenneis, 1981). Colour of the final product is most critical, followed by texture.

Wheat flour is commonly used to thicken processed, canned soups. Hard or soft wheat flours may be used. Primary considerations are thickening ability of the starch. As such, wheat must be sound with no sprout damage. Further, flours exhibiting high starch pasting (gelatinization and gelation) viscosity are preferred. Starch pasting quality of the flour may be assessed using the Brabender Viscoamylograph or the Newport Rapid Visco Analyzer (see Section 1.3).

1.2.5 Noodles made from common (hexaploid) wheat

In this chapter, differentiation is made between **noodles** prepared from common, or hexaploid, wheat flour and **pasta** prepared from durum semolina (see Sections 1.1.2 and 1.2.8). Noodles may be generally defined as boiled 'strings' of unleavened wheat dough. However, this definition has many exceptions and caveats due to the tremendous complexity and diversity of noodle products. Most noodles are produced and consumed in the orient, and as such, their use runs parallel to bread consumption in the West. Noodles are the traditional form in which wheat is consumed and predates recorded history. In addition to rice, noodles form the foundation of the carbohydrate-based diets in this part of the world.

Although tremendously diverse, the vast majority of noodles are prepared from very simple formulas: flour, water and salts. Because of this, wheat flour quality is particularly critical in determining noodle quality and consumer acceptance. For the purposes of classification and discussion, noodles may be grouped according to formulation (primarily type of salts), noodle geometry, method of preparation, post-preparation processing, method of packaging, and manner in which consumed (Figure 1.7 and Table 1.1).

Based on formulation, noodles may be classified as white salted, alkaline, soba, or egg. White salted noodles are produced from the simplest formula and contain only flour (100 parts), water (28–40 parts, 34 standard), and salt (NaCl, 2–3 parts) (Nagao, 1981). The three main features of white salted (udon) noodle quality are colour, appearance, and mouth feel. Noodle colour should be clear and bright, not dull. The

Table 1.1 Noodle classification and flour use in Japan (FY1992) (thousands of tonnes)

Fresh noodle		718.3			
Udon					
wet				40.4	
boiled				217.3	
Chinese (chuka-men)					
wet				216.8	
boiled				61.9	
steamed				87.6	
dough sheets				25.0	
soba					
wet				27.2	
boiled				42.1	
buckwheat flour					25.7
Dry noodle		271.9			
thick					
standard (udon)			61.7		
flat (hira-men)				11.7	
thin					
thin (hiya-mugi)			37.3		
very thin (so-men)				42.7	
handmade very thin (te-nobe so-men)				75.4	
soba				35.9	
buckwheat flour					13.5
Chinese (chuka-men)				7.2	
Instant fried noodle	322.6				
packed					
Chinese (chuka-men)					
fried				153.7	
non-fried				23.0	
udon and soba					
fried				4.6	
non-fried				1.3	
buckwheat flour					1.6
snack type				140.0	
Frozen noodle		39.8			
Udon					
wet				0.9	
boiled				25.8	
Chinese (chuka-men)					
wet				0.8	
boiled				6.4	
steamed				0.3	
dough sheets				1.7	

Table 1.1 *Continued*

soba				
wet			0.3	
boiled			3.6	
buckwheat flour				1.4
Pasta		139.8		
Total noodle production	1452.6			

Adapted from information supplied by the Japanese Food Agency, May 1994.

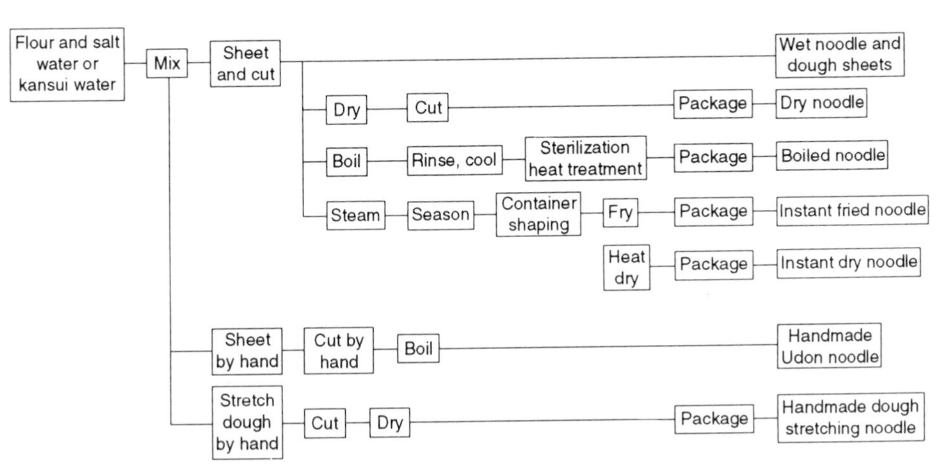

Figure 1.7 Noodle processing in Japan. (Adapted from information supplied by the Japanese Food Agency, May 1994).

noodle should appear smooth with a glossy surface and should have sharp, well-defined edges rather than rounded. Mouth feel and biting characteristics should be soft and elastic. Wheat and flour quality characteristics that contribute to these desired quality traits include good milling characteristics with ease of bran separation, bright endosperm and flour colour, weak but extensible gluten, relatively high amylo-viscosity, low amylase activity, and lower amylose content. Australian Standard White (ASW) wheat is a preferred wheat for producing white salted noodle flours. Particularly in Japan, Western Australia ASW segregation is considered the best wheat for traditional udon noodle and is blended with US Western White and domestic wheat, typically soft red winter. The lower starch damage, finer granulation and mellower gluten associated with soft wheats is preferred. Western Australia (WA)

ASW apparently possesses superior starch pasting properties. Until recently, WA ASW was almost exclusively the variety Gamenya.

Alkaline noodles, as the name implies, are prepared using alkaline salts. A typical formulation would include 1–3% NaCl and 1–3% alkaline salts. Alkaline salts, often referred to as *kan sui* (can soo-ee, or can-swee) may be composed of combinations of potassium and sodium salts of carbonate (K_2CO_3 and Na_2CO_3) and/or sodium hydroxide. Like white salted noodle, the main quality attributes of alkaline noodle are colour, appearance and texture. Colour should range from creamy white to yellow. Appearance should be bright, not grey or dull, with as few specks as possible. The colour should be stable and not deteriorate during storage. Texture should be elastic rather than weak and soft, and should not soften or deteriorate during storage. Milling quality, endosperm colour, gluten characteristics and starch pasting quality all contribute to alkaline noodle quality. Alkaline noodles are typically prepared from higher protein, stronger gluten hard wheats, similar to those used for bread. Lower protein hard wheats or a blend of hard and soft wheats may be used depending on regional preferences for chewiness and other texture characteristics. A top-quality alkaline noodle flour in Japan is specified as semi-strong, 0.34% ash, and 11.1% protein. Typical flour specifications for Japan and South Korea are presented in Tables 1.2 and 1.3.

Two popular variations on the basic, simple formula include buckwheat and egg noodles. Buckwheat noodles, call soba, are made from wheat flour and buckwheat (*Fagopyrum esculentum* Moench.), about 3 parts wheat flour to 1 part buckwheat flour (Table 1.1). Egg noodles, as the name implies, are prepared using eggs. Egg noodles may be prepared using whole fresh eggs as the only source of formula liquid. In the US, dry egg noodles contain at least 5.5% egg solids and are typically 0.5 to 1.5 cm wide (Hoseney, 1986).

Table 1.2 Flour quality specifications by noodle type in Japan

Noodle type	*Ash*	*Protein*	*Flour type**
Udon	0.35–0.40	8.0–9.5	Medium
Binding flour for soba	–	9.0–15.0	
Chinese, top quality	0.33–0.37	10.5–11.5	Semi-strong
normal grade	0.37–0.40	10.5–11.5	
instant	0.40–0.45	10.0–11.5	Semi-strong
steamed	0.38–0.43	9.0–11.0	

*See Tables 1.4 and 1.5 for flour classification in Japan.

Adapted from information supplied by the Japanese Food Agency, May 1994.

Table 1.3 Flour quality specifications by noodle type in Korea

Type of noodle	*Ash*	*Protein*	*Absorption**	*Wheat type*†
Common bag	0.50–0.54	10.3–10.6	57–59	HRW:WW (50:50)
High-quality bag	0.40–0.44	9.4–10.6	58–63	HRW:WW, AH:ASW (20:80)
Udon-type bag	0.39–0.42	8.5–8.7	56–60	ASW, HRW:WW
Cup and bowl	0.40–0.45	11.0–11.5	62–63	HRW:DNS (70:30)

*Optimum water absorption as determined by the farinograph.
†HRW, US Hard Red Winter; WW, US Western White; AH, Australian Hard; ASW, Australian Standard White; DNS, US Dark Northern Spring
Adapted from Kim, 1993.

Figure 1.8 Instant noodles. Left, cup-style noodle in polystyrene foam cup with cut-away view. Hot water is added directly to the cup, which, after steeping, functions as the serving container. Right, bag and bowl-style noodles showing seasoning packet (lower right).

In addition to formulation, noodles may be categorized and discussed based on processing. Figure 1.7 presents a classification of noodles in Japan based on processing. Instant noodles, in particular, are worthy of separate discussion (Figure 1.8). In Korea, instant noodles are by far the predominant and most popular style of noodle. Korean instant noodles may be categorized based on price (common or high-quality), packaging method (bag, cup or bowl), and noodle geometry (thin or thick, udon-type). Flour quality requirements follow these categories (Table 1.3 and Kim, 1993).

As in the West, convenience is an important force driving consumer spending. In addition to the various styles of instant noodles described above, two additional avenues for providing convenience in the Japanese market are frozen and long-life noodle. Frozen noodle has the advantage of 'fixing' quality at the time of freezing, thus providing stability during and extending storage, and providing a convenience food that is quick to prepare. Long-life alkaline noodle is boiled or steamed, packaged in non-permeable film and then steam-sterilized. These noodles have a very long shelf-life (5 months) and do not require refrigeration.

In addition to thin noodle strands, several other boiled and/or fried noodle-like products are prepared from sheeted doughs. Examples include won-ton wraps, Peking ravioli, and egg and spring roll wraps. In the US, chow mein noodles are prepared from higher extraction soft red winter wheat flours. The dough is sheeted, cut, steamed and fried. Although similar to ramen instant fried noodles, chow mein noodles are relatively dark and have a blistered, airy surface and crunchy texture (Loving and Brenneis, 1981).

1.2.6 Breakfast foods

Although many rolls, breads, noodles etc. are eaten at breakfast time, a separate group of wheat foods are considered to be specifically 'breakfast foods'. The subject will not be covered extensively here. Hoseney (1986), and Loving and Brenneis (1981) group cereal-based breakfast foods into those requiring cooking and those 'ready to eat'. In a sense, these breakfast foods date back to the earliest consumption of wheat. Porridge and gruel led eventually to controlled fermentation as a means of producing alcohol-containing beverages/foods, and to doughs which were also fermented and leavened by natural microflora before baking.

Common types of breakfast cereals involving wheat are farina (pieces of endosperm from hard hexaploid wheat, similar to durum semolina), whole kernel flakes, shredded biscuits, cereal granules (granules of ground, dense, baked bread), and puffed whole kernel wheat. Most of the ready to eat varieties have been previously boiled, steamed or extruded, or a combination of these. For most products, grain should be free of sprout damage. White grain of soft texture promotes easy flaking and maintains better colour after roasting.

1.2.7 Starch/gluten

Wheat starch and gluten each have unique compositions and properties. As such, it is useful to isolate each in relatively pure form. Separation is accomplished through one of three general methods: Martin and batter processes, and differential centrifugation. All three typically employ roller-milled flour, usually harder, higher protein, stronger gluten wheats. The Martin process mixes flour and water into a dough and then washes out the starch and water solubles. Often, NaCl is included to promote gluten agglomeration. The batter process uses initially greater volumes of water to disperse the flour into a batter. Small aggregates of agglomerated gluten are recovered by sieving. In both the Martin and batter processes, starch is recovered through centrifugation. Differential centrifugation separates non-aggregated gluten from starch, the main advantage being reduced wash water and hence, plant effluent. Once separated, most gluten is dried and sold as a free-flowing light yellow powder.

In the US and other countries with large quantities of maize, wheat starch is of minor importance, gluten being the primary reason for starch–gluten separation. In countries such as Australia, however, starch is an important co-product. After isolation, starch is usually further processed into sugar syrups, ethanol and specialty modified starches. In this regard, wheat starch use parallels that of maize.

The primary reason for starch–gluten isolation is to obtain vital gluten. In the US, the annual use of vital wheat gluten has doubled from approximately 59 million kg in 1986, to 118 million kg in 1994 (Anon., 1994a). **Vital** refers to the fact that the isolated gluten retains its functional viscoelastic properties. The main use of vital gluten is in yeast leavened bread products, especially variety breads where the load of non-functional ingredients is relatively high. Up to 4% or more vital gluten may be added to these products. Vital gluten interacts with endogenous gluten to help maintain product volume and internal structure. During isolation, controlled drying is key to maintaining the vital nature of gluten. Excessive heat denatures gluten and renders it non-vital. Non-vital gluten, though of little value in baking, does find application in pet foods and other uses. For further reference the reader is directed to Pomeranz (1988), Pyler (1988) and Simmonds (1989).

1.2.8 Pasta and other durum wheat products

Alimentary paste, or **pasta**, is comprised of extruded dough pieces of various shapes and sizes, usually dried during production, and later boiled and consumed. Baroni (1988) groups all pasta products into four categories: long, short, Bologna-type and 'nests and skeins' (Figure 1.9). Long pasta – generally thought of as noodles – may be further divided into round and flat geometries. In this chapter, a distinction between **noodles** prepared from hexaploid wheat flour, soft or hard (see Section 1.2.5) and **pasta** prepared from durum (*Triticum turgidum* var. *durum*) semolina is made. Short pasta is what many in the West would consider macaroni: elbows, shells etc. Bologna-type and 'nests and skeins' are not nearly as common in the West as are the long and short types. For additional information on these types, refer to Baroni (1988).

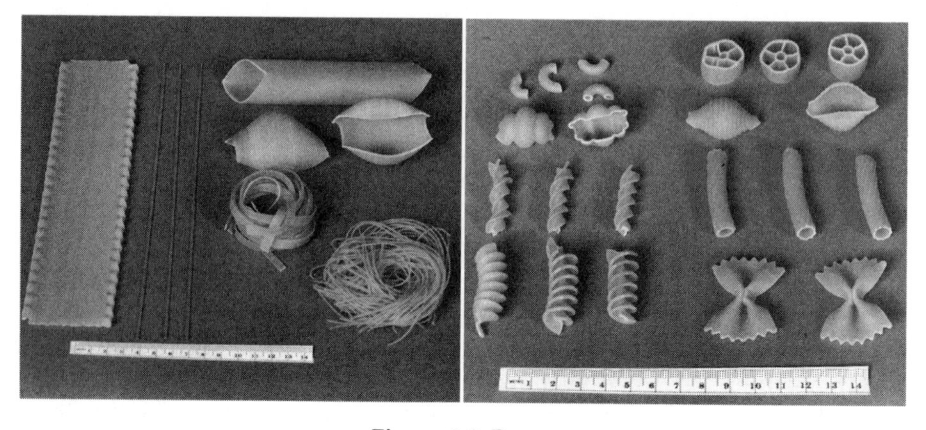

Figure 1.9 Pasta.

Although Italian in origin (Baroni, 1988; Bizzarri and Morelli, 1988), pastas such as spaghetti, macaroni, lasagna, and fettuccine are now a common part of culture and language in the US and worldwide. In addition to the pastas shown in Figure 1.9, those containing eggs or powdered vegetables such as carrot and spinach are popular. The powdered vegetables create appealing orange and green coloured pasta which are then often marketed in an equally-proportioned mixture with standard semolina-only pasta. Some pasta products contain fillings, such as tortellini and ravioli with cheese or meat. A small portion of the total pasta market comprises fresh pasta (high moisture content, ready to boil) and canned pasta (fully cooked, usually with sauce, ready to heat and serve).

As mentioned earlier, pasta is an extruded product. Semolina is first mixed with water in an approximate ratio of 30:100 (water:semolina) into a stiff dough. The dough is forced at high pressure through a die to produce the desired shape and size. The pasta is then collected and dried. Mixing, kneading and drying are critical steps to successful pasta manufacture. Baroni (1988) describes the commercial manufacture of pasta products.

Semolina is the primary product of durum wheat milling (see Section 1.4) and is simply larger-sized pieces of endosperm, free of adhering bran. Semolina size ranges and requirements are discussed by Bizzarri and Morelli (1988), but are of the order of 130 to 550 μm. Joppa and Williams (1988) define semolina as that milling fraction retained on a US no. 100 sieve, flour defined as the fraction passing through. In addition to the particle size of semolina, colour and vitreousness and protein quantity and quality are key traits. Higher levels (>13%) of strong gluten are preferred, although grain below 10% protein may be utilized for some products (Bizzarri and Morelli, 1988). The protein must form a continuous matrix to entrap the starch granules so that the pasta surface does not become sticky during cooking. In most cases, the addition of hexaploid wheat to pasta is considered undesirable and is prevented by law in some countries. Consequently, several methods have been devised to detect hexaploid wheat adulteration.

Other non-pasta durum wheat products include primarily bread, couscous and bulgur. Durum bread is prepared from durum flour rather than semolina. The salient differences between typical hexaploid and durum wheat flours are coarser particle size, higher starch damage, higher water absorption, strong gluten with low extensibility, high dough stability, and yellow colour of durum flour (Quaglia, 1988). Due to the high damaged starch and water absorption of durum flour, durum bread stales much more slowly and consequently has an extended shelf-life, compared to hexaploid wheat bread. Quaglia (1988) indicates that yellow colour, taste and shelf-life are the main quality traits of durum bread. The best colour is obtained by including a high proportion of amber durum in the milling grist.

Couscous and bulgur are precooked, unleavened foods traditionally consumed in North Africa and the Middle East. Couscous is prepared from semolina by hydrating, mixing, steaming, drying and size-fractionating the resultant particles. Couscous is then rehydrated with oil and meat and/or vegetable sauce when eaten. Bulgur, on the other hand, is prepared from whole or cracked kernels by soaking in water, parboiling, drying and grinding. Prior to eating, bulgur is boiled or steamed for 15–20 min (Quaglia, 1988).

1.3 ASSESSMENT OF FLOUR QUALITY

1.3.1 Introduction

The primary reason for assessing flour quality is to predict commercial end-product quality. In this sense, flour quality must be thought of in terms of **suitability** (see Section 1.1.1). Approaches to the assessment of flour quality may be grouped into two categories: laboratory end-product tests and component tests. Both have their own merits and are described below. End-product tests tend to produce a summation of quality – the sum total of all the components of quality as well as their interaction, if any. For this reason, end-product tests are generally considered the best predictors of commercial end-product quality. On the downside, end-product tests are generally labour- and capital-intensive, requiring more personnel, time, flour, and equipment compared to component tests.

Component tests tend to assess one or more fundamental property, or component, of flour, such that end-product quality may be predicted. The advantage of component tests is that they are generally conservative of resources, quick, and amenable to large numbers of samples. Component tests also have the advantage of more precisely identifying why a particular flour may have better or poorer quality. The main shortcoming of component tests is their limited ability to predict commercial end-product quality. Of particular concern is what statistically is referred to as Type II error: a flour is predicted to perform adequately, but fails to do so. Small-scale component tests used primarily in cultivar development are described in Chapter 11.

1.3.2 End-product tests

End-product tests typically use a scaled-down, semi- or non-automated procedure that mimics the large-scale industrial process. Product formula may exactly match that of the commercial product or may be a standardized test product such as the sugar-snap cookie procedure of the American Association of Cereal Chemists (AACC, 1985). Standardized end-product tests approved by the AACC include pan bread (100 g 'pup' and pound loaves, straight and sponge-and-dough methods), yeasted sweet goods, cookies (sugar snap and wire-cut), American biscuits, pie, layer and angel food cakes, and pasta. Laboratory-scale end-product tests have also been developed for steamed bread (Faridi and Rubenthaler, 1983a), flat breads (Faridi *et al.*, 1981; Faridi and Rubenthaler, 1983b), bagels (Bath and Hoseney, 1994), saltine crackers (Pizzinatto and Hoseney, 1980), Japanese sponge cake (Nagao *et al.*, 1976), and Japanese udon (Nagao *et al.*, 1976; Toyokawa *et al.*, 1989) and Chinese noodles (Miskelly and Moss, 1985; Moss *et al.*, 1987).

Quality is assessed on the characteristics of the prepared product, for example, the volume of bread loaves and internal appearance and texture of crumb, the diameter of cookies, and the cooking loss of noodles and pasta. Often the inherent subjectivity of end-product quality assessment is a disadvantage of these tests.

1.3.3 Component tests

The component test approach to the assessment of flour quality is founded on the premise that end-product quality can be predicted from various attributes of the flour *a priori*. Beyond predicting end-product quality, there is the practical necessity of ascertaining the processing properties of flour so that plant production runs smoothly and efficiently. As often happens, end-product quality and processing quality are closely linked. For example, the mixing requirement of a bread flour impacts the time that the dough must be mixed (often as a

450–900 kg dough piece) as well as the volume and crumb grain texture of the finished loaf. A flour with a lower mixing requirement must be accommodated by shortening the processing time or it will become over-mixed. Even if mixed to optimum, the resulting bread may be inferior. If over-mixed, problems such as release from the mixer will occur, and the bread will certainly be inferior.

It is the inherent, fundamental properties of wheat flour that make it the single most important and diverse food in the world's diet. It is also these fundamental properties that are characterized, and by doing so, a prediction of the processing and end-product quality of a flour is made. In this context, the fundamental properties, or components, of wheat flour may be grouped as: protein, starch, water relations and colour.

Protein involves the quantity and quality of protein. As discussed earlier, it is the viscoelastic nature of gluten that allows gases produced during fermentation to be trapped in the dough, thereby increasing the volume and producing the appealing texture of bread. Likewise, it is these same gluten proteins that can impart a tough or undesirably chewy texture to many cakes, cookies and pastries.

Starch can undergo profound physical changes. These changes primarily relate to the events known as gelatinization and gelation. During the gelatinization process, starch can interact with several times its own weight in water and, in doing so, forms a gel structure. This gel structure is critical to the structure, texture and quality of many foods. In many instances such as soup thickeners, the concomitant increase in viscosity associated with gelatinization is the main purpose for including wheat flour as an ingredient. Conversely, the hydrolysis of starch provides a ready source of fermentable carbohydrates for yeast fermentation. Starch hydrolysis, however, is not always desirable, especially if it results from preharvest sprouting – germination of grains in the field prior to harvest. Preharvest sprouting occurs when non-dormant grain imbibes rain water or dew and the biochemical machinery normally associated with seedling establishment commences.

The third fundamental aspect of flour functionality relates to water relations. Even though gluten hydration during mixing and starch gelatinization are major contributors in this respect, water relations are often examined independently because, (i) there are additional, often poorly characterized sources of water uptake, and (ii) often the simple sum total expression of these various components is of immediate interest to the food processor.

Finally, colour is an important consideration because food must appeal to the eye. Colour in the sense of end-product quality may be considered in the context of the production or maintenance of desirable colour or the prevention or absence of undesirable colour. Two examples are the golden brown of bread crust which results from

Maillard reaction, and the bright yellow of alkaline chinese noodles. If crust is too dark or noodles too dull, the product may not appeal to the consumer.

With this introduction, each of the four categories of fundamental flour properties will be dealt with from the standpoint of technological quantitation. By no means is this intended to be an exhaustive treatment of the subject. Likewise, various instruments may have considerable overlap in terms of the properties which they characterize.

The quantity of protein is often important in predicting end-use quality. There are two main approaches: determination of elemental nitrogen with an empirical conversion to protein, and an empirically-derived prediction of protein based on spectroscopy using specific wavelengths of light in the near-infrared (NIR) region. The techniques include Kjeldahl and combustion methods of elemental nitrogen determination and NIR reflectance and transmission spectroscopy (AACC, 1985; AOAC, 1995).

The quantitative assessment of protein quality, generally in reference to gluten quality, is substantially more challenging and often empirical. Standard methods exist for the determination of wet and dry gluten (AACC, 1985). Flour is mixed with water into a dough and the starch and soluble fractions are washed away using large volumes of water. Usually, NaCl is included to promote gluten agglomeration. The wet gluten is weighed, dried and weighed again. A semi-automated procedure has been developed (Perten, 1990). Two other approaches to gluten quality include recording dough mixers and machines that characterize the elasticity and extensibility of doughs. In the first category are the Mixograph and Farinograph. In the second category are the Extensigraph and Alveograph (AACC, 1985).

Recording dough mixers produce a curve from which can be derived assessments of gluten hydration, development, and stability. Additionally, an assessment of optimum water absorption by flour can be made. The extensibility of doughs, and therefore gluten, is assessed by the extensigraph by uniaxially stretching a dough piece to the point of failure (dough piece tears). The Alveograph biaxially stretches a dough piece by blowing a bubble using air pressure. The bubble volume is likewise increased to the point of failure.

Ash is an old but extensively used method of assessing milling extraction and therefore such things as protein quality, and colour. Central endosperm is low in ash while bran is high in ash. Flour from the central endosperm generally contains more functional protein (better protein quality) as opposed to flour from near the bran or outer portion of the kernel. Similarly, bran often contributes undesirable colours.

Wheat flour is about 75–80% starch on a dry weight basis (Bauer and

Alexander, 1979; McCleary *et al.*, 1994a, 1994b). Starch content may be estimated by subtracting known contents of protein, moisture, fat, crude fibre, ash and pentosans (Bauer and Alexander, 1979), though often these extensive data are not available for flours. Alternatively, the starch content of flour can be directly quantitated using specific starch degrading enzymes (Bauer and Alexander, 1979; AACC, 1985; McCleary *et al.*, 1994a, 1994b).

Usually of much greater interest than the quantity of starch is the functional performance of starch – starch quality, primarily the gelatinization and gelation properties. Two common methods available for assessing starch quality are the Brabender ViscoAmylograph (or Amylograph) (AACC, 1985) and the Newport Scientific Rapid Visco Analyzer (RVA) (Ross *et al.*, 1987). Both work on starch–water or flour–water slurries and record viscosity as a function of resistance to stirring during gelatinization, or gelatinization and gelation.

Often the functional performance of starch has less to do with starch quality *per se* than it does with the presence and activity of starch-degrading enzymes. These enzymes may be endogenous – the result of preharvest sprouting, or exogenous – in the form of malt flour such as that added to pan bread formulas (see Section 1.2.2). The Amylograph and RVA can assess existing starch damage and can also assess potential *in situ* starch damage which might occur during processing. Similarly, they can assess the potential effects of added malt. The falling number test (AACC, 1985) is based on viscometry during gelatinization. A weighted plunger is allowed to fall through a gelatinizing sample of ground grain or flour. Primarily aimed at assessing sprout damage, the method also incorporates inherent differences in starch gelatinization properties. Occasionally, the activity of a specific hydrolytic enzyme is of interest. Methods are available for the assessment of α-amylase (Barnes and Blakeney, 1974; AACC, 1985; McCleary and Sheehan, 1989) and *β*-amylase (McCleary and Codd, 1989) activities.

Starch quality and performance are also affected by mechanical starch damage. As noted previously, damaged starch absorbs more water and is more susceptible to enzymatic attack. Although the level of starch damage is highly influenced by grain hardness (Section 1.4), milling procedures have a major impact and different millstreams will differ markedly in damaged starch. Quantitative measures of starch damage rely on the susceptibility of granules to α-amylase attack or ability to bind I_2/KI (AACC, 1985; Gibson *et al.*, 1991, 1993).

Water relations are critical to end-product quality and processing efficiency. Assessment of water relations can be made from dough mixers and starch quality (see above). Two methods assess flour hydration in batter systems by adjusting the pH to alkaline or acid ranges. Alkaline water retention capacity (AWRC) measures the amount of alkaline solution retained by a flour after mixing and centrifugation (AACC, 1985). Generally, low AWRC is associated with good soft wheat quality. Many soft wheat products are prepared from alkaline formulas due to the alkaline nature of eggs. The viscosity of acidulated flours is used to assess effects of hydrated, swollen proteins and other flour constituents. Commonly referred to as MacMichael viscosity, based on the viscometer of the same name, the lack of availability of MacMichael equipment has lead to a replacement method using a viscometer produced by Brookfield Engineering Laboratories, Inc. (Gaines, 1990). Although Gaines (1990) suggests that no direct relationship exists between acid viscosity and end-product quality, low acid viscosity is a trait historically associated with high-quality soft white club wheats from the US Pacific Northwest. The low acid viscosity of these club wheats may simply reflect the very weak and low hydration nature of club wheat storage proteins. In terms of pH, acid viscosity may be most applicable to saltine cracker doughs.

Food must be appealing to the eye as well as the other senses. As such, colour is an important quality consideration. Although visual assessment of colour is often satisfactory (e.g. Pekar colour slick for flour, AACC, 1985), the development of hand-held reflectance colourimeters has made triaxial quantitation of colour common.

The importance of inherent colour systems associated with wheat and flour varies depending on end-product. For example, the whiteness of a soft wheat flour may be crucial to the appeal of an angel food cake whereas it may matter little in a chocolate devils food cake. Two products where colour is particularly important are noodles and pasta. Noodles must be bright, not dull and should be creamy white at neutral pH and yellow at alkaline pH. Off-colours are unacceptable and generally result from chemical or enzymatic changes. Pasta, on the other hand, requires high levels of carotenes and xanthophylls to impart a bright, deep yellow appearance. Loss of colour during boiling should be minimized. Several standard colour tests are described in the *Approved Methods of the American Association of Cereal Chemists* (AACC, 1985).

1.4 MILLING QUALITY

The wheat flour miller is in a central and critical position in the utilization of wheat as a food. The miller takes a raw material of biological origin and converts it to a food ingredient for countless numbers of diverse food products. Much of the day-to-day success of the baker is a direct result of the miller's technical ability and expertise. In this regard, the overriding aspect of flour quality is consistency of performance. This endeavor is by no means simple or straightforward. Wheat is a biological entity and as such can vary dramatically over geographical locations, over years and importantly among different cultivars. Often this variation is subtle and complex. It is part of the miller's task to remove this variation and thereby produce a consistent product.

This section describes some of the basic features of wheat flour milling. In contrast to feed milling, the flour miller must constantly consider the functional, end-use quality of the food product for which the flour is intended.

In a practical sense, the miller's goal is to separate the botanical parts of the wheat kernel – germ and bran from endosperm – and then to reduce the endosperm to flour-sized particles. This separation is accomplished by exploiting the differences in inherent material properties among the various botanical parts. Bran tends to be leathery and resistant to breakage when hydrated, while germ tends to be more plastic and therefore deformable. Additionally, bran, and likewise pieces of endosperm with adhering bran, is less dense and has greater buoyancy in an air stream compared to pure endosperm. A limited number of physical and mechanical means are employed in the modern flour mill and aim to exploit these differences in material properties. These include the steel roller mill, sifters and purifiers.

The roller mill comprises pairs of long cylindrical horizontal steel rolls rotating counter to one another. Rolls have two purposes: to dislodge endosperm from the adhering bran and to reduce larger pieces of endosperm to flour-sized particles. Rolls may be classified according to these two purposes. Break rolls break open the kernel and scrape away chunks of endosperm from the adhering bran using compressive and shearing forces. They are usually corrugated and operate at a 2.5:1 differential of rotation, usually 500–550 rpm, fast roll (Bass, 1988). Reduction rolls crush the endosperm chunks and reduce them to flour-sized particles (approx. 100 μm). They are usually smooth and operate at a 5:4 differential and a rotational speed similar to break rolls (Bass, 1988).

Sifters effect particle separation based on size using screens of various aperture rotating in an horizontal plane. After each roll pass, stock is sifted to separate three to four types of particles. These are flour – small endosperm particles preferably free of bran, middlings – larger pieces of endosperm with varying amounts of bran (preferably bran-free), shorts – smaller pieces of bran with some endosperm, and bran – large pieces of the outer tissues of the kernel with varying amounts of endosperm. In North America, flour is generally defined as stock passing through a 112 μm opening (Bass, 1988). A dressed flour is expected to pass completely through a 132 μm sieve. As noted in Section 1.2.4, cake flour quality is often enhanced by reducing the particle size. In this regard, a cake flour may mostly pass through a 93 μm aperture (Bass, 1988). After sifting, large pieces of bran are sent to additional break rolls for more scraping and endosperm recovery.

Particles of similar, intermediate size (usually referred to as break middlings) are sent to machines called purifiers. Purifiers exploit differences in density and drag of particles subjected to an air stream. The purifier consists of two to three levels of gently sloping sieves; the entire purifier oscillates to agitate and stratify the stock on the sieves into layers (Bass, 1988). The stock is separated into pure endosperm, composite pieces (endosperm with adhering bran) and bran. In keeping with the general milling technique, each separated fraction is directed to a break roll, reduction roll, or feed stock, as appropriate. Purification systems are especially important in durum milling where bran-free semolina is the primary goal.

In addition to these basic conceptual and practical aspects of flour milling, several other topics are worthy of mention. They include grain storage and cleaning, sanitation, conditioning, flour extraction, flour treatment, and wheat hardness. Most of these topics will not be discussed in depth here. Grain storage is extensively covered in Sauer (1992) and by Bass (1988). Issues related to sanitation are covered by Mailhot and Patton (1988) and Mills and Pederson (1990).

The aim of cleaning is to remove all materials which might adversely affect flour quality or damage the milling equipment. Such materials include, but are not limited to, stones, dirt, ergot, metals, straw and chaff, non-wheat seeds, insect-damaged wheat, and shrivelled or broken wheat kernels. Bass (1988) provides a discussion of the various means employed to clean wheat prior to milling.

Conditioning, or tempering, is the process whereby the miller changes the material properties of the various parts of the kernel by adding water to the grain. Conditioning improves the milling process in two important ways. Bran becomes more leathery when hydrated and so tends to resist fragmentation. Consequently, it can be separated from flour and middlings by sieving, due to its larger size. A second important role that conditioning plays is that the material strength of the endosperm is reduced. This reduction in strength means less power is required to fracture the endosperm. On the other hand, too high a moisture content produces more flaking rather than fracturing during reduction, and also causes poor sifting (bolting) properties (Ford and Kingswood, 1981). To a large extent, the optimum moisture content for conditioning is based on the milling response of the endosperm. Soft, hard and durum wheats are tempered to different moisture contents, the harder wheats receiving more water. Typically, these values are as low as 12% for soft wheat and up to 17.5% for durum (Bizzarri and Morelli, 1988). Conditioning is conducted at moisture contents above those which are safe for grain storage (Sauer, 1992). However, resting times are usually only 8–20 h. Typically, temper water is added in multiple steps and resting time is longer for harder wheats.

Flour extraction, or flour extraction rate, refers to the quantity of flour produced relative to various unit measures of beginning or end products. Bass (1988) provides five commonly used methods of calculating flour extraction: (i) wheat 'as received', (ii) clean dry wheat, (iii) clean tempered wheat, (iv) total products off the mill, and (v) total mill output (includes screenings). Since a higher extraction rate usually means more flour per unit wheat, extraction rate relates to the profitability and efficiency of the mill. Naturally, flour extraction can theoretically range from zero to 100% (whole wheat flour). But typically, a flour in commerce might run about 70–80% extraction (total products). As noted in the previous section, ash can be used as a relative indicator of flour extraction (and quality). In other words, low ash flour represents relatively purer central endosperm with less bran contamination. This basic relationship between ash and milling performance is represented in what are referred to as 'cumulative ash curves'.

Figure 1.10 shows a typical ash curve from a three break, five reduction experimental pilot mill. The milling system employs no purifiers. To construct the curve, each flour stream is analysed for ash content and the results are plotted beginning with the lowest ash stream, sequentially moving to streams with greater and greater ash

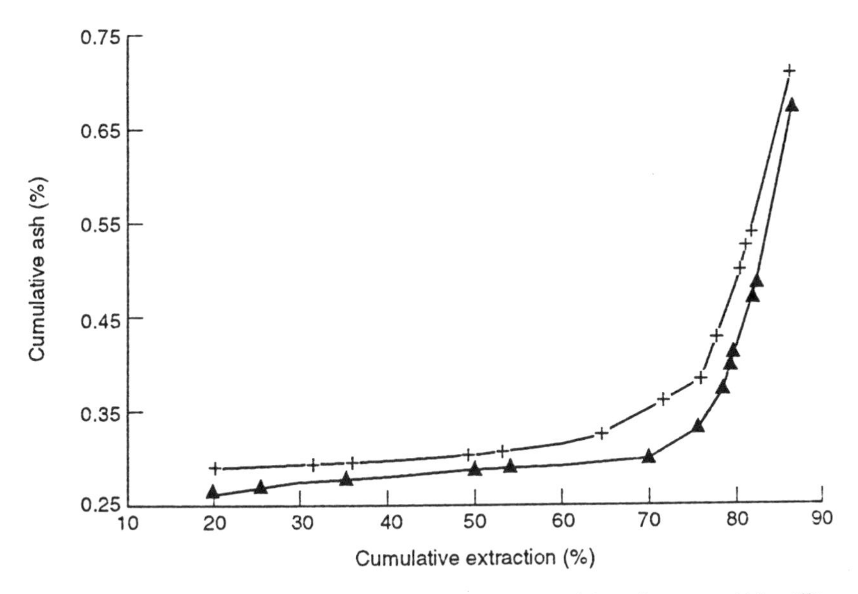

Figure 1.10 Cumulative ash curves for a better (triangle) and poorer (+) milling wheat.

contents. On the abscissa is plotted the accumulative yield (extraction) of that particular stream. Ash and yield are cumulative until theoretically reaching 100% extraction and an ash content equal to the whole grain. These curves are valuable in that they provide an indication of the milling quality of a lot of wheat and the performance of the milling process itself. Likewise, the extraction rate expected given a certain ash specification may also be estimated. In similar fashion, cumulative protein curves may also be constructed.

Flours of different extraction rate and ash content have different names in the milling vernacular. Examples from lower to higher ash include short patent, standard patent, straight grade, first, second, third clears etc. The current commercial trading price of over 15 individual classes of flour are quoted weekly in *Milling and Baking News* (Anon., 1995b). Additionally, prices of 'family flour' (retail parcels), semolina, germ, and millfeed are listed.

Table 1.4 lists the major flour classes of Japan based on ash content and end-use. In practice, a single flour milling company may produce well over 100 different flours. This large variety of flours is produced through three basic approaches: blending different grain lots, blending different flour streams and post-milling flour treatments (chemical or mechanical). Table 1.5 lists the typical wheats used alone or in blends to produce the various flours listed in Table 1.4. Blending wheats usually involves selecting grain lots with different protein levels to reach an intermediate protein. Similarly, grain lots with different hardness or other traits may be blended for other considerations. Morris (1992) has examined the effects of blending soft with hard white wheats on end-use quality. In the US, **Western White Wheat** is, by law, a blend of soft white club and common wheat.

Currently in North America, the practice of blending flours by selecting streams of the same or different mill run is more common than blending grain lots. Not only does flour blending provide much greater flexibility to the miller in providing an assortment of flours for numerous end-uses, but it also permits the milling of individual, more uniform grain lots. Uniformity of grain lots usually means better conditioning response and greater milling performance. Bass (1988) describes the process of stream selection, or 'flour dividing', as a means of producing different flours. Ford and Kingswood (1981) and Bass (1988) provide comparative stream, flour, and other milling terms for English, French, Spanish, German, and American systems/languages.

Post-milling flour treatments may be classified as being chemical or mechanical. Chemical treatments include supplementation with nutrients and vitamins; treatment with oxidizing, bleaching or maturing agents; chlorination, and addition of malt or enzymes, leavening agents and reducing agents (Ford and Kingswood, 1981; Bass, 1988; Hoseney *et*

Table 1.4 Major flour classes and uses, and approximate protein range (below use), in Japan

Flour grade	*Approximate ash content*	*Flour type*					
		Strong	*Semi-strong**	*Semi-strong**	*Medium*	*Soft**	*Soft**
Premium	<0.4	Bread 11.5–12.5	Bread 11.0–12.0	Chinese noodle 10.5–11.5	Japanese noodle 8.0–9.0	Japanese confections 7.5–8.5	Cake flour[‡] 6.5–8.0
First	0.4	[†]Bread 11.7–12.7	Bread 11.2–12.2	Instant noodle 10.5–11.5	Japanese noodle 8.5–9.5	Japanese confections 8.0–9.0	Confections 7.8–8.5
Second	0.5	Bread 12.0–13.0	Bread 11.5–12.5	Bread	All-purpose 9.5–10.5	Japanese confections 9.0–10.0	Confections and All-purpose 8.0–9.0
Third	1.0	Gluten/starch	Gluten/starch	–	–	–	None
Clear	2–3	Plywood, feed	Plywood, feed	–	Plywood, feed	–	Plywood, feed

[†]Most common bread flour.
*Extraction methods differ depending on particular end-product (e.g. bread vs. Chinese noodle).
[‡]Cake flours of 0.35, 0.37 ash.

Table 1.5 Major flour classes in Japan and the wheats used to produce them

Strong
No. 1 Canada Western Red Spring (13.5% protein)
US Dark Northern Spring (14.0% protein)
*US Hard Red Winter (13.0% protein)
Semi-strong
No. 1 Canada Western Red Spring (13.5% protein)
US Dark Northern Spring (14.0% protein)
US Hard Red Winter (13.0% protein)
US Hard Red Winter (11.5% protein)
*US Hard Red Winter (ordinary protein)
Medium
*US Hard Red Winter (ordinary protein)
Australian Standard White (Western Australia)
Japanese Soft
*US Western White
Soft
*Japanese Soft
US Western White

*Denotes use is optional, variable, sometimes blended with other wheats listed.
Adapted from information supplied by the Japanese Food Agency, May 1994.

al., 1988; Pyler, 1988). In some countries, certain chemical treatments are prohibited by law. Mechanical treatments include pin milling and air classification, as well as physical treatments such as storage (ageing) and heat treatment (Ford and Kingswood, 1981; Bass, 1988; Hoseney *et al.*, 1988; Pyler, 1988). Pin milling is used to reduce the particle size of flours by disrupting flour aggregates. Air classification exploits differences in size, mass, and density to separate particles in the range of 15–80 μm (Bass, 1988). Separation is effected in a cyclone-type apparatus where air drag and centrifugal force oppose one another. The finest fraction (<17 μm) is enriched in protein bodies and may be 25% protein. A fraction of 17–35 or 40 μm is mostly free starch granules and may be 5% protein. This fraction is particularly useful for making cake flours. The fraction larger than 40 μm is similar to the parent flour (Bass, 1988). Storage relies on natural oxidation with atmospheric oxygen to age flours. Chemical oxidizing and maturing agents aim to accomplish the same end in much less time. Heat treatment also ages and matures flour. Seguchi has studied extensively the effects of chlorination and heat treatment in flour ageing and lipophilization of starch (see Seguchi, 1993, and references therein).

A final important consideration is wheat grain hardness. As noted in previous sections, soft, hard and durum wheats are typically used for different end products. Indeed, it is the inherent characteristics associated with each type that makes them suitable for particular end uses, as well as unsuitable for others. Although modern plant breeders have included many additional, different traits into these three hardness classes of wheat, the single most important aspect of their utilization remains their grain hardness.

From a milling standpoint, the softer wheats require less temper water and shorter conditioning times, fracture more easily on the mill, generate more 'break' flour in the head of the mill, require greater sifting area but fewer purifiers, more break rolls but fewer reduction rolls, produce a finer flour with poorer bolting and flow characteristics, and produce flours with less starch damage. Hardness is discussed further in the following section.

1.5 INTRINSIC QUALITY OF GRAIN AND GRAIN LOTS

Grain hardness is arguably the single most important aspect of wheat utilization. Reviews of grain hardness are provided by Pomeranz and Williams (1990) and Anjum and Walker (1991). Important as grain hardness is to wheat utilization, the fundamental biological basis for the trait is poorly understood. A major source of variation for grain hardness has been assigned to the short arm of chromosome 5D (Mattern *et al.*, 1973; Law *et al.*, 1978). Consequently, durum wheats which lack the D genome are very hard, while hexaploids may be soft or hard, depending on the allelic state. Bettge *et al.* (1995) have shown that the expression of this gene follows additive gene action. Although the actual mechanism of the expression of this gene has not been elucidated, a family of 15 kDa lipophilic proteins and two classes of bound polar lipids are always associated with isolated soft wheat starch (Greenwell and Schofield, 1986; Jolly *et al.*, 1993; Morris *et al.*, 1994; Greenblatt *et al.*, 1995). The presence of these 15 kDa proteins can be used to characterize the genotypic state (i.e. soft, hard, or durum) of individual wheat grains (Bettge *et al.*, 1995). Generally, of much greater importance to the miller or end-user is the phenotypic hardness of a given grain lot. A long-used physical method, referred to as **particle size index** (PSI), relies on the fact that upon grinding, soft wheats produce finer meals and flours (AACC, 1985). By sifting and weighing, wheats may be classified as soft or hard, or assigned a quantitative numerical value. More recent developments in the assessment of grain hardness include a near infra-red spectrophotometric (NIR) method (AACC, 1985) and a single kernel crushing device (Martin *et al.*, 1993).

In addition to hardness, kernel mass and morphology, and protein content, there are other intrinsic traits that affect end-use quality. Generally, millers prefer uniformly large, well-filled (plump) kernels.

Uniformity ensures the best milling response for any given mill stand, sifter or purifier setting. Large plump kernels maximize the volume to surface ratio, or in other words, endosperm to bran. Kernel morphology has also been used as a means of classifying grain in market channels. For example, the US market classes Hard Red Spring, Hard Red Winter, Soft White Common, and Soft White Club have historically each had particular kernel shapes associated with each.

The measurement of kernel mass is typically expressed in terms of thousand kernel weight, while kernel morphology is approximated using bulk density (hectolitre or bushel weight). The single kernel crushing device (Martin *et al.*, 1993) provides measures of kernel weight and outer dimension, as well as hardness and moisture.

Protein quantity, along with protein quality, is the second most important determinant of end-use quality; the quality of protein is primarily controlled by genetics (Chapter 8). The quantity of protein, however, is highly influenced by the environment including cultural practices, most notably nitrogen fertilizer. A range of 7–17% protein is possible within one cultivar. The methods of determining grain protein are similar to those employed for flour (see Section 1.3.3). Whole-grain NIR methods are particularly convenient and obviate the need for grinding.

Moisture content determines the storability of grain, the relative concentration of other kernel constituents (e.g. protein), and the amount of additional water needed during tempering. For safe storage, wheat must be about 12–13% moisture, or drier to prevent the growth of various moulds. In this regard, moisture, time and temperature all interact to affect grain storage. Sauer (1992) presents a lengthy discussion of moisture and its measurement, fungi and other aspects of grain storage.

A final intrinsic quality trait of grain is soundness, or sprouting. Preharvest sprouting occurs when rains and high humidity coincide with grain maturation and delayed harvest. The inferior quality of sprouted grain relates to the presence of carbohydrases, proteases and other hydrolytic enzymes normally associated with germination (see Section 1.3.3). The two main methods of assessing sprout damage are by visual inspection of kernels (noting pericarp rupture over the embryo or other signs of germination) and the falling number assay (AACC, 1985). Other methods used for flour or starch are amenable, though not commonly used for the analysis of grain. The worldwide importance of sprouting in wheat and other cereals is exemplified by the holding of international symposia on the subject (Walker-Simmons and Ried, 1993).

Finally, the quality of grain and grain lots may be reduced by the presence or prior activity of moulds, insects and rodents, seeds of other species, non-millable material, stones etc. (see Section 1.4). The presence of pesticide residues is increasingly an important issue.

1.6 CLASSES AND GRADES OF WHEAT

To promote the orderly marketing of wheat, most developed countries have a system of classes and grades. Classes and grades aim to give both the buyer and the seller an estimate of the potential quality of a grain lot. In this regard, wheat with lower test weight, sprouted kernels, mixed classes (e.g. soft and hard) etc., is discounted in the market place due to its higher risk and lower value. A detailed discussion of the various systems of classifying, grading and marketing wheat is beyond the scope of this book. For reference, the reader is directed to the works of Simmonds (1989) for Australia, Heilman and Wilson (1988) for durum, Halverson and Zeleny (1988) for descriptions of US, Canadian, Australian and EEC standards, and Hill (1990) for a thorough treatment of the historical development of grain standards in the US for all major crops.

1.7 FEED USES OF WHEAT (contributed by S.Paul Rose)

1.7.1 Introduction

One fifth of the world's total annual wheat production is used in animal feeds (International Wheat Council, 1992). Wheat accounts for around 15% of the total world cereal usage in animal feeds compared to over 50% for maize. The importance of wheat as an animal feed varies markedly around the world. Europe accounts for three-quarters of the wheat that is fed to animals in the world each year (Lucbert, 1990; International Wheat Council, 1992), mostly in poultry and pig feeds. Conversely, South American, Asian and African countries use less than 5% of their total wheat production as animal feed.

Wheat, like all cereals, provides a concentrated source of energy in animal feed and it contributes to the animals' protein requirements. All farm livestock have a large dietary requirement for energy and protein. All cereals are deficient in most minerals required for growth or reproduction, although potassium and phosphorus are exceptions. The potassium in wheat is almost completely available to pigs and poultry (Combs and Miller, 1985), but only around half of its phosphorus is available to monogastric animals (Beers and Jongbloed, 1993). Cereals contain small amounts of vitamin E, choline and some water-soluble vitamins. All cereals, except maize, have low levels of the essential fatty acid, linoleic acid.

1.7.2 Energy

Energy is the single most expensive component of farm animal feeds. Forages are used as a low cost energy source for ruminants but these feeds have a low energy availability for non-ruminants. Growing wheat pasture may be used as a forage for ruminants in some agricultural systems (Ralston *et al.*, 1990). Maize and wheat grains are the most concentrated source of metabolizable energy for poultry and pigs (Table 1.6) and comprise the greater part of the feeds for these animals. Barley may also be used, particularly in pig feeds. The cereal that gives the lowest cost per unit of metabolizable energy will be used (Pye, 1987). For example, a 1992 survey of the UK feed compounding industry showed that broiler chicken and laying hen rations contained 57% and 54% of wheat respectively. No other cereals were used in these feeds.

The concentration of available energy in a wheat sample is the single most important nutrient that influences its quality for use in pig and poultry rations (Pye, 1987). Energy availability is determined as metabolizable energy (ME) or sometimes digestible energy (DE) with pigs. ME is the gross energy content of a feed given to an animal with the gross energy of the animal's faeces and digesta deducted. A technique called true metabolizable energy (TME) is also used for poultry; TME determinations correct ME for the endogenous energy loss from the digestive tract (McNab and Blair, 1988). There are large variations in the determined ME values of different wheat samples. Studies since 1970 of large numbers of wheat samples have generally given ranges of poultry MEs of 13.1–16.6 MJ/kg of dry matter (Wiseman, 1990; McNab, 1991). Poultry TMEs and pig MEs have a similar range but are approximately 0.5 MJ/kg higher (Smith *et al.*, 1988). There are two major exceptions to this range of values.

Mollah *et al.* (1983) and Rogel *et al.* (1987) both examined samples of

Table 1.6 Energy availability and protein quality of cereals for farm livestock

	Wheat	*Barley*	*Maize*	*Oats*	*Rye*
Metabolizable energy (MJ/kg)					
Sheep and cattle[1]	13.6	12.8	13.8	12.0	14.0
Pigs[2]	13.9	12.7	14.3	11.5	12.6
Poultry[3]	13.1	11.1	14.1	10.7	11.0
Crude protein (g/kg)[2]	114	115	85	118	120
Amino acid balance (g/kg crude protein)[2]					
Lysine	32	34	29	34	34
Methionine + cystine	51	32	47	31	30

Superscripts indicate the data source: 1, MAFF (1990); 2, NRC (1988); 3, NRC (1984). Data is expressed on an air-dry basis.

wheat grown in Australia and found that some of the samples had poultry MEs of less than 11.0 MJ/kg. There are no reliable predictors of the MEs of wheat samples. Starch accounts for over 80% of the available energy in wheat, yet the total starch content of a wheat sample is not well correlated to its ME (Rogel *et al.*, 1987). It is unclear whether the digestibility of the starch is an indicator of ME. Wheat starch is completely digested by adult birds (Longstaff and McNab, 1986) and growing pigs (Fuller *et al.*, 1989).

Rogel *et al.* (1991) showed there was a significant correlation (r^2 = 0.85) between starch digestibility and the ME of the wheat for young broiler chickens. This study included a large number of very low ME wheat samples that were grown at the end of a severe drought. There was a much more variable relationship ($p < 0.05$) between these two parameters ($r^2 = 0.35$) when only the wheats with MEs greater than 13.1 MJ/kg were selected from this study. Nicol *et al.* (1993) found a similar poor correlation when studying low ME UK wheat samples.

Starch is the only plant polysaccharide known to be hydrolysed by the digestive enzymes of pigs and poultry. Wheat contains about 11% non-starch polysaccharides (Englyst, 1989). There is about 2% cellulose and two pentosans, arabinoxylans and xylans, account for 3.3% and 4.8% respectively. Choct and Annison (1990) showed that more than 5% total pentosan in low ME Australian wheats had an antinutritive effect in poultry. The high pentosan levels reduced the digestibility of fat, protein and starch (Choct and Annison, 1992). However, there was no evidence of a correlation between pentosan levels and ME in low ME wheat samples from the UK (Nicol *et al.*, 1993), in Canadian wheat samples (Coates *et al.*, 1977) or in high ME Australian wheats (Annison, 1990).

There are no consistent differences in ME between wheat varieties, although Waldron *et al.* (1993) observed differences in the growth of broiler chickens between two different wheat varieties. There are no good correlations with ME for specific weight (McNab, 1991), 1000 grain weight (Coates *et al.*, 1977) or wheat hardness (Rogel *et al.*, 1987). McNab (1991) and Rose *et al.* (1993) observed that the Hagberg falling number of a sample was positively correlated with the ME of different wheat samples. However, there was only a poor correlation between α-amylase activity and ME (Rose, unpublished data). Wheat contains α-amylase inhibitors that are active against the digestive enzymes of pigs and poultry (Silano, 1987). Wheat varieties have different inhibitor activities (Warchalewski *et al.*, 1989). The inhibitors are highly heat labile (Snow and O'Dea, 1981) and their nutritional significance has not been established.

Anderson and Bell (1983) examined wheat cultivars over two growing seasons. They concluded that environmental conditions that affect crop growth and seed maturity have important effects on the energy availability in wheat samples. Wheat samples grown in low rainfall seasons have lower starch and higher protein and fibre contents (Nik-Khah *et al.*, 1972). Drought conditions during crop growth have been suggested as the reason for some low ME Australian wheats (Annison, 1990). However, preharvest sprouting (Gatel and Bourdon, 1989), frost damage (Anderson and Bell, 1983) or the soil type or soil series (Nik-Khah *et al.*, 1972) do not have large effects on energy availability.

1.7.3 Protein

Cereals may provide over one-third of the total protein in practical pig and poultry rations. The amino acid balance of cereal protein is poor. It is particularly deficient in lysine. An ideally balanced protein should contain 55–70 g of lysine/kg crude protein for animal growth whereas cereal proteins contain only 29–34 g/kg (see Table 1.1). A lysine-rich protein concentrate, such as soya bean meal, is invariably needed to provide a balanced protein supply in cereal-based feeds.

The digestibility of the amino acids in wheat protein is low. For example, the digestibility of lysine for cockerels was only 0.81 (McNab, 1991) and Sauer *et al.* (1981) found a range of lysine digestibilities for pigs of 0.72–0.78 in Canadian wheats. Fuller *et al.* (1989) observed that the amino acid digestibility was greater in high protein winter wheat varieties than in low protein varieties. Wheat contains a proteinase inhibitor that is particularly active against trypsin. The importance of its antinutritive effect is not established, but it is unlikely to be a major influence on wheat amino acid availability (Boisen, 1983)

Variation in the total crude protein content of cereals does not markedly affect their economic value as feedstuffs for pigs or poultry (Pye, 1987). The improved protein digestibility in high protein varieties may be a factor that favours their use in nutrient dense pig and poultry feeds. Increasing the protein content by application of nitrogen to the growing crop does not change the availability of the limiting amino acids (Fuller *et al.*, 1989; McNab, 1991).

1.7.4 Effects on productive performance

ME and digestible protein concentrations are the two best measures of feed wheat quality, but there is a poor correlation between these two parameters and the growth of *ad libitum* fed pigs (Bell and Anderson, 1984) and poultry (Rose *et al.*, 1993). Factors that affect the voluntary feed intakes of animals may be important in determining this aspect of productive performance. The α-amylase in digestive enzymes cannot penetrate within some gelatinized starch granules (Würsch *et al.*, 1986). Differences in the rate of amylase attack may affect the rate of digestion of the starch and the speed of movement of digesta through the digestive tract. These factors may directly influence gut fill and so affect voluntary feed intakes.

Microbial or fungal contamination of wheat is also a factor that affects voluntary feed intakes. For example, contamination of wheat crops by *Fusarium graminearium* can contaminate the grains with vomitoxin and zearalenone (Neish and Cohen, 1981). There is an inverse linear relationship between vomitoxin contamination and voluntary feed intakes in growing pigs (Friend *et al.*, 1982). Zearalenone affects the perfomance of reproducing animals (Diekman and Green, 1992). The contamination of wheat grains by smut spores of *Tilletia caries* and *Tilletia contraversa* also reduces the voluntary feed intakes of growing pigs (Westermann *et al.*, 1988).

Microbial and fungal contamination of wheat is difficult to detect by simple tests and so is often only monitored visually when the quality of wheat is assessed for animal feeds. Visual inspection of a wheat sample is probably only able to detect high levels of bacterial or fungal contamination that may be potentially toxic. Subclinical levels may not be detected but may still affect voluntary feed intakes. Large seasonal differences in contamination are likely to occur.

1.7.5 Determination of the nutritional quality of wheat in practice

The proximate nutrient composition of a wheat sample remains the primary chemical test of quality by the animal feed compounding industry (Figure 1.11). Individual lots of wheat are not held at feed compounding mills long enough for detailed chemical tests to be performed. Dry matter is estimated by drying at 100–105°C in an air-oven for at least 4 h (AOAC, 1990), or estimation by NIR reflectance (Osborne and Fearn, 1983). The total ash content of the sample is determined by incineration in a muffle furnace at 450°C (Egan *et al.*, 1987). Ash can also be estimated by near infra-red reflectance techniques. Dry matter and ash give an approximate indication of the organic matter that remains available for digestion and is closely correlated to the gross energy of a sample (McNab, 1991). Samples of wheat may be segregated according to their crude protein contents. High protein wheats may be economically more valuable in nutrient dense feeds for growing animals.

Measurements of ME and protein digestibility are made using live animals. TME methods are the most rapid but they still take over one week for results to be obtained. There are no reliable physical or

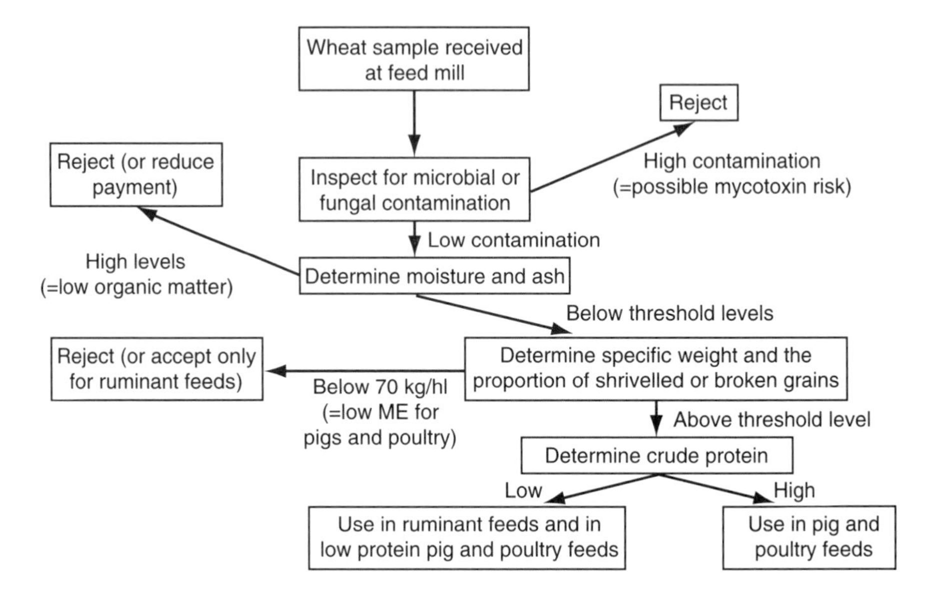

Figure 1.11 A practical system of rapid quality testing of feed wheat by animal feed compounders.

chemical predictors of the ME or protein digestibility of an individual cereal sample. McNab (1991) showed that if wheat samples had specific weights of less than 70 kg/hl there was a significant correlation between specific weight and the ME. This information could be used to establish a threshold level for detecting very low ME wheat samples.

REFERENCES

AACC (1985) *Approved Methods of the American Association of Cereal Chemists*, 8th edn, American Association of Cereal Chemists, St. Paul, Minnesota, USA.

Anderson, D.M. and Bell, J.M. (1983) The digestibility by pigs of dry matter, energy, protein and amino acids in wheat cultivars. II. Fifteen cultivars grown in two years, compared with Bonanza and Fergus barleys, and 3CW-grade hard red spring wheat. *Canadian Journal of Plant Science*, **63**, 393–406.

Anjum, F.M. and Walker, C.E. (1991) Review on the significance of starch and protein to wheat kernel hardness. *Journal of the Science of Food and Agriculture*, **56**, 1–13.

Annison, G. (1990) Polysaccharide composition of Australian wheats and the digestibility of their starches in broiler chicken diets. *Australian Journal of Experimental Agriculture*, **30**, 183–6.

Anon. (1994a) Manildra Milling building new wheat starch, gluten plant. *Milling & Baking News*, **72**(52), 13.

Anon. (1994b) Pretzels are taking on a new look. *Milling & Baking News*, **73**(10), 34–40.

AOAC (1990) *Official methods of analysis of the Association of Official Analytical*

Chemists. 15th edn, Association of Official Analytical Chemists, Washington, DC.

AOAC (1995) Protein in cereal grains, method 992–23, in *Official Methods of Analysis of the AOAC*, 16th edn, Association of Official Analytical Chemists, Washington, DC.

Barnes, W.C. and Blakeney, A.B. (1974) Determination of cereal alpha amylase using a commercially available dye-labelled substrate. *Starch*, **26**, 193–7.

Baroni, D. (1988) Manufacture of pasta products, in *Durum Wheat: Chemistry and Technology* (eds G. Fabriani and C. Lintas), American Association of Cereal Chemists, St. Paul, Minnesota, USA, pp. 191–216.

Bass, E.J. (1988) Wheat flour milling, in *Wheat: Chemistry and Technology*, 3rd edn, Vol. 2 (ed. Y. Pomeranz), American Association of Cereal Chemists, St. Paul, Minnesota, USA, pp. 1–68.

Bath, D.E. and Hoseney, R.C. (1994) A laboratory-scale bagel-making procedure. *Cereal Chemistry*, **71**, 403–8.

Bauer, M.C. and Alexander, R.J. (1979) Enzymatic procedure for determination of starch in cereal products. *Cereal Chemistry*, **56**, 364–6.

Beers, S. and Jongbloed, A.W. (1993) Phosphorus digestibility and requirement of pigs. *Feed Mix*, **1**, 28–32.

Bell, J.M. and Anderson, D.M. (1984) Comparisons of wheat cultivars as energy and protein sources in diets for growing and finishing pigs. *Canadian Journal of Animal Science*, **64**, 957–70.

Bettge, A.D., Morris, C.F. and Greenblatt, G.A. (1995) Assessing genotypic softness in single wheat kernels using starch granule-associated friabilin as a biochemical marker. *Euphytica*, **86**, 65–72.

Boisen, S. (1983) Protease inhibitors in cereals. Occurrence, properties, physiological role, and nutritional influence. *Acta Agricultura Scandinavica*, **33**, 369–81.

Bizzarri, O. and Morelli, A. (1988) Milling durum wheat, in *Durum Wheat: Chemistry and Technology* (eds G. Fabriani and C. Lintas), American Association of Cereal Chemists, St. Paul, Minnesota, USA, pp. 161–89.

Chavez, E.R. (1984) Vomitoxin-contaminated wheat in pig diets: Pregnant and lactating gilts and weaners. *Canadian Journal of Animal Science*, **64**, 717–23.

Choct, M. and Annison, G. (1990) Anti-nutritive activity of wheat pentosans in broiler diets. *British Poultry Science*, **31**, 811–21.

Choct M. and Annison, G. (1992) The inhibition of nutrient digestion by wheat pentosans. *British Journal of Nutrition*, **67**, 123–32.

Coates, B.J., Slinger, S.J., Summers, J.D. and Bayley, H.S. (1977) Metabolizable energy values and chemical and physical characteristics of wheat and barley. *Canadian Journal of Animal Science*, **57**, 195–207.

Combs, N.R. and Miller, E.R. (1985) Determination of potassium availability in K_2CO_3, $KHCO_3$, corn and soybean meal for the young pig. *Journal of Animal Science*, **60**, 715–19.

Diekman, M.A. and Green, M.L. (1992) Mycotoxins and reproduction in domestic livestock. *Journal of Animal Science*, **70**, 1615–27.

Downs, D.E. (1971) Basic aspects of fried pie production. *Bakers Digest*, **45**(3), 62–8.

Dubois, D.K. (1961) Achieving flexibility in variety cake production. *Proceedings of the American Society of Bakery Engineers*, pp. 274–82.

Egan, H., Kirk, R.S, and Sawyer, R. (1987) *Pearson's chemical analysis of foods*, 8th edn, Longman Scientific and Technical, Harlow.

Englyst, H. (1989) Classification and measurement of plant polysaccharides. *Animal Feed Science and Technology*, **23**, 27–42.

Faridi, H.A and Rubenthaler, G.L. (1983a) Laboratory method for producing Chinese steamed bread and effects of formula, steaming and storage on bread starch gelatinization and freshness, in *Proceedings of the Sixth International Wheat Genetics Symposium* (ed. S. Sakamoto), Plant Germplasm Institute, Kyoto University, Kyoto, Japan, pp. 863–7.

Faridi, H.A. and Rubenthaler, G.L. (1983b) Experimental baking techniques for evaluating Pacific Northwest wheats in North African breads. *Cereal Chemistry*, **60**, 74–9.

Faridi, H.A., Finney, P.L. and Rubenthaler, G.L. (1981) Micro baking evaluation of some US wheat classes for suitability in Iranian breads. *Cereal Chemistry*, **58**, 428–32.

Ford, M. and Kingswood, K. (1981) Milling in the European Economic Community, in *Soft Wheat: Production, Breeding, Milling, and Uses* (eds W.T. Yamazaki and C.T. Greenwood), American Association of Cereal Chemists, St. Paul, Minnesota, USA, pp. 129–67.

Friend, D.W., Trenholm, H.L., Elliot, J.I., Thompson, B.K. and Hartin, K.E. (1982) Effect of feeding vomitoxin-contaminated wheat to pigs. *Canadian Journal of Animal Science*, **62**, 1211–22.

Fuller, M.F., Cadenhead, A., Brown, D.S., Brewer, A.C., Carver, M. and Robinson, R. (1989) Varietal differences in the nutritive value of cereal grains for pigs. *Journal of Agricultural Science, Cambridge*, **113**, 149–63.

Gaines, C.S. (1990) Use of a spindle-type viscometer (Brookfield) to measure the apparent viscosity of acidulated flour-water suspensions. *Cereal Foods World*, **35**, 741–7.

Gatel, F. and Bourdon, D. (1989) Effects of preharvest sprouting on the feeding value of wheat for pigs. *Pig News and Information*, **10**, 159–60.

Gibson, T.S., Al Qalla, H. and McCleary, B.V. (1991) An improved enzymatic method for the measurement of starch damage in wheat flour. *Journal of Cereal Science*, **15**, 15–27.

Gibson, T.S., Kaldor, C.J. and McCleary, B.V. (1993) Collaborative evaluation of an enzymatic starch damage assay kit and comparison with other methods. *Cereal Chemistry*, **70**, 47–51.

Greenblatt, G.A., Bettge, A.D. and Morris, C.F. (1995) The relationship between endosperm texture, and the occurrence of friabilin and bound polar lipids on wheat starch. *Cereal Chemistry*, **72**, 172–6.

Greenwell, P. and Schofield, J.D. (1986) A starch granule protein associated with endosperm softness in wheat. *Cereal Chemistry*, **63**, 379–80.

Greenwood, C.T., Guinet, R. and Seibel, W. (1981) Soft wheat uses in Europe, in *Soft Wheat: Production, Breeding, Milling, and Uses* (eds W.T. Yamazaki and C.T. Greenwood), American Association of Cereal Chemists, St. Paul, Minnesota, USA, pp. 209–66.

Halverson, J. and Zeleny, L. (1988) Criteria of wheat quality, in *Wheat: Chemistry and Technology*, 3rd edn, Vol. 1 (ed. Y. Pomeranz), American Association of Cereal Chemists, St. Paul, Minnesota, USA, pp. 15–45.

Heilman, R.G. and Wilson, W.W. (1988) Durum marketing, in *Durum Wheat: Chemistry and Technology* (eds G. Fabriani and C. Lintas), American Association of Cereal Chemists, St. Paul, Minnesota, USA, pp. 303–16.

Hill, L.D. (1990) *Grain Grades and Standards*, University of Illinois Press, Urbana, Illinois, USA.

Hoseney, R.C. (1986) *Principles of Cereal Science and Technology*, American Association of Cereal Chemists, St. Paul, Minnesota, USA.

Hoseney, R.C., Wade, P. and Finley, J.W. (1988) Soft wheat products, in *Wheat:*

Chemistry and Technology, 3rd edn, Vol. 2 (ed. Y. Pomeranz), American Association of Cereal Chemists, St. Paul, Minnesota, USA, pp. 407–56.

International Wheat Council (1992) *World Grain Statistics*. International Wheat Council, London.

International Wheat Council (1994) *Grain Market Report 229 – October 1994*, International Wheat Council, London, p. SUMMARY-1.

Jolly, C.J., Rahman, S., Kortt, A.A. and Higgins, T.J.V. (1993) Characterisation of the wheat Mr 15000 'grain-softness protein' and analysis of the relationship between its accumulation in the whole seed and grain softness. *Theoretical and Applied Genetics*, **86**, 589–97.

Joppa, L.R. and Williams, N.D. (1988) Genetics and breeding of durum wheat in the United States, in *Durum Wheat: Chemistry and Technology* (eds G. Fabriani and C. Lintas), American Association of Cereal Chemists, St. Paul, Minnesota, USA, pp. 47–68.

Kim, S.-K. (1993) *Instant Noodles: An Amazing Growth Story*, Department of Food Science and Nutrition, Dankook Univ., Seoul, South Korea.

Kulp, K. (1994) Functionality of ingredients in cookie systems, in *Cookie Chemistry and Technology* (ed. K. Kulp), American Institute of Baking, Manhattan, Kansas, USA, pp. 210–79.

Law, C.N., Young, C.F., Brown, J.W.S., Snape, J.W. and Worland, A.J. (1978) The study of grain protein control in wheat using whole chromosome substitution lines, in *Seed Protein Improvement by Nuclear Techniques*, International Atomic Energy Agency, Vienna, Austria, pp. 483–502.

Longstaff, M. and McNab, J.M. (1986) Influence of site and variety on starch, hemicellulose and cellulose composition of wheats and their digestibilities by adult cockerels. *British Poultry Science*, **27**, 435–49.

Loving, H.J. and Brenneis, L.J. (1981) Soft wheat uses in the United States, in *Soft Wheat: Production, Breeding, Milling, and Uses* (eds W.T. Yamazaki and C.T. Greenwood), American Association of Cereal Chemists, St. Paul, Minnesota, USA, pp. 169–207.

Lucbert, J. (1990) L'alimentation animale peut rester le premier debouche du ble en France. *Cultivar-Paris*, **282**, 50–1.

McCleary, B.V. and Codd, R. (1989) Measurement of β-amylase in cereal flours and commercial enzyme preparations. *Journal of Cereal Science*, **9**, 17–33.

McCleary, B.V. and Sheehan, H. (1989) Measurement of cereal α-amylase: A new assay procedure. *Journal of Cereal Science*, **6**, 237–51.

McCleary, B.V., Gibson, T.S., Solah, V. and Mugford, D.C. (1994a) Total starch measurement in cereal products: Interlaboratory evaluation of a rapid enzymatic test procedure. *Cereal Chemistry*, **71**, 501–5.

McCleary, B.V., Solah, V. and Gibson, T.S. (1994b) Quantitative measurement of total starch in cereal flours and products. *Journal of Cereal Science*, **20**, 51–8.

McNab, J. (1991) Factors affecting the nutritive value of wheat for poultry. *HGCA project Report No.43*. Home Grown Cereals Authority, London.

McNab, J.M. and Blair, J.C. (1988) Modified assay far true and apparent metabolisable energy based on tube feeding. *British Poultry Science*, **29**, 697–707.

MAFF (1990) *Feed Composition. UK Tables of feed composition and nutritive value for ruminants*, 2nd edn, Chalcombe Publications, Marlow, UK.

Mailhot, W.C. and Patton, J.C. (1988) Criteria of flour quality, in *Wheat: Chemistry and Technology*, 3rd edn, Vol. 2 (ed. Y. Pomeranz), American Association of Cereal Chemists, St. Paul, Minnesota, USA, pp. 69–90.

Martin, C.R., Rousser, R. and Brabec, D.L. (1993) Development of a single-kernel wheat characterization system. *Transactions of the American Society of Agricultural Engineers*, **36**, 1399–404.

Mattern, P.J., Morris, R., Schmidt, J.W. and Johnson, V.A. (1973) Location of genes for kernel properties in the wheat variety 'Cheyenne' using chromosome substitution lines, in *Proceedings of the 4th International Wheat Genetics Symposium*, Columbia, Missouri, USA, pp. 703–7.

Miller, B.S., Trimbo, H.B. and Powell, K.R. (1967) Effects of flour granulation and starch damage on the cake making quality of soft wheat flour. *Cereal Science Today*, **12**, 245–7; 250–2.

Milling and Baking News (1995) Sosland Publishing Co., Kansas City, Missouri, USA (ISSN 0091–4843).

Mills, R. and Pederson, J.R. (1990) *A Flour Mill Sanitation Manual*, Eagen Press, St. Paul, Minnesota, USA.

Miskelly, D.M. and Moss, H.J. (1985) Flour quality requirements for Chinese noodle manufacture. *Journal of Cereal Science*, **3**, 379–87.

Mollah, Y., Bryden, W.L., Wallis, I.R., Balnave, D. and Annison, E.F. (1983) Studies on low metabolisable energy wheats for poultry using conventional and rapid assay procedures and the effects of processing. *British Poultry Science*, **24**, 81–9.

Morris, C.F. (1992) Impact of blending hard and soft white wheats on milling and baking quality. *Cereal Foods World*, **37**, 643–8.

Morris, C.F., Greenblatt, G.A., Bettge, A.D. and Malkawi, H.I. (1994) Isolation and characterization of multiple forms of friabilin. *Journal of Cereal Science*, **21**, 167–74.

Moss, R., Gore, P.J. and Murray, I.C. (1987) The influence of ingredients and processing variables on the quality and microstructure of hokkien, cantonese and instant noodles. *Food Microstructure*, **6**, 63–74.

Nagao, S. (1981) Soft wheat uses in the Orient, in *Soft Wheat: Production, Breeding, Milling, and Uses* (eds W.T. Yamazaki and C.T. Greenwood), American Association of Cereal Chemists, St. Paul, Minnesota, USA, pp. 267–304.

Nagao, S., Imai, S., Sato, T., Kaneko, Y. and Otsubo, H. (1976) Quality characteristics of soft wheats and their use in Japan. I. Methods of assessing wheat suitability for Japanese products. *Cereal Chemistry*, **53**, 988–97.

National Research Council (1984) *Nutrient Requirements of Poultry*, 8th edn, National Academy Press, Washington, DC.

National Research Council (1988) *Nutrient Requirements of Swine*, 9th edn, National Academy Press, Washington, DC.

Neish, G.A. and Cohen, H. (1981) Vomitoxin and zearalenone production by *Fusarium graminearum* from winter wheat and barley in Ontario. *Canadian Journal of Plant Science*, **61**, 811–15.

Nicol, N.T., Wiseman, J. and Norton, G. (1993) Factors determining the nutritional value of wheat varieties for poultry. *Carbohydrate Polymers*, **21**, 211–15.

Nik-Khah, A., Hoppner, K.H. Sosulski, F.W., Owen, B.D. and Wu, K.K. (1972) Variation in proximate fractions and B-vitamins in Saskatchewan feed grains. *Canadian Journal of Animal Science*, **52**, 407–17.

Osborne, B.G. and Fearn, T. (1983) Collaborative evaluation of near infra red reflectance analysis for the determination of protein, moisture and hardness in wheat. *Journal of the Science of Food and Agriculture*, **34**, 1011–17.

Palmer, E. (1994a) Cracker makers go wherever consumers are snacking. *Milling & Baking News*, **73**(26), 24–8.

Palmer, E. (1994b) McDonald's picks up the pace of world-wide expansion. *Milling & Baking News*, **73**, 26–30.
Perten, H. (1990) Rapid measurement of wet gluten quality by the gluten index. *Cereal Foods World*, **35**, 401–2.
Petrofsky, R. (1986) Bagel production and technology. *American Institute of Baking Research Department Technical Bulletin*, **8**(11), 1–5.
Pizzinatto, A. and Hoseney, R.C. (1980) A laboratory method for saltine crackers. *Cereal Chemistry*, **57**, 249–52.
Pomeranz, Y. (ed.) (1988) *Wheat: Chemistry and Technology*, 3rd edn, American Association of Cereal Chemists, St. Paul, Minnesota, USA.
Pomeranz, Y. and Williams, P.C. (1990) Wheat hardness: Its genetic, structural, and biochemical background, measurement, and significance, in *Advances in Cereal Science and Technology*, Vol. X (ed. Y. Pomeranz), American Association of Cereal Chemists, St. Paul, Minnesota, USA, pp. 471–544.
Pye, R.E. (1987) Maximising the use of cereals in animal feeds. The effects of quality and price. *The Feed Compounder*, **7**, 10–13.
Pyler, E.R. (1988) *Baking Science & Technology*, 3rd edn, Sosland Publishing Co., Merriam, Kansas, USA.
Quaglia, G.B. (1988) Other durum wheat products, in *Durum Wheat: Chemistry and Technology* (eds G. Fabriani and C. Lintas), American Association of Cereal Chemists, St. Paul, Minnesota, USA, pp. 263–82.
Ralston, R.E., Knight, T.O., Coble, K.H. and Lippke, L.A. (1990) The wheat and stocker cattle analyzer: A microcomputer decision aid for evaluating wheat production and stocker cattle grazing decisions. *Southern Journal of Agricultural Economics*, **22**, 185–93.
Rogel, A.M., Annison, E F., Bryden, W L and Balnave, D. (1987) The digestion of wheat starch in broiler chickens. *Australian Journal of Agricultural Research*, **38**, 639–49.
Rose, S.P., Kettlewell, P.S., Reynolds, S.M. and Watts, R.M. (1993) The nutritive value of different wheat varieties for poultry. *Proceedings of the Nutrition Society*, **52,** 206A.
Ross, A.S., Walker, C.E., Booth, R.I., Orth, R.A. and Wrigley, C.W. (1987) The Rapid Visco-Analyzer: A new technique for the estimation of sprout damage. *Cereal Foods World*, **32**, 827–9.
Sauer, D.B. (ed.) (1992) *Storage of Cereal Grains and Their Products*, 4th edn, American Association of Cereal Chemists, St. Paul, Minnesota, USA.
Sauer, W.C. Kennelly, J.J. Aheme, F.X. and Cichon, R.M. (1981) Availabilities of amino acids in barley and wheat for growing pigs. *Canadian Journal of Animal Science*, **61**, 793–802.
Seguchi, M. (1993) Effect of wheat flour aging on starch-granule surface proteins. *Cereal Chemistry*, **70**, 362–4.
Silano, V. (1987) Alpha-amylase inhibitors, in *Enzymes and their role in cereal technology* (eds J.E. Kruger, D. Lineback and C.E. Stauffer), American Association of Cereal Chemists, St Paul, Minnesota, USA, pp. 141–99.
Simmonds, D.H. (1989) *Wheat and Wheat Quality in Australia*, CSIRO, Australia.
Sjerven, J. (1994) Nation's consumers take a 'fresh' look at tortillas. *Milling & Baking News*, **73**(13), 29–32.
Smith, W.C., Moughan, P.J. and Pearson, G. (1988) A comparison of bioavailable energy values of ground cereal grains measured with adult cockerels and growing pigs. *Animal Feed Science and Technology*, **19**, 105–10.
Snow, P. and O'Dea, K. (1981) Factors affecting the rate of hydrolysis of starch in food. *American Journal of Clinical Nutrition*, **34**, 2721–7.

Thacker, D. (1994) British markets, in *Cookie Chemistry and Technology* (ed. K. Kulp), American Institute of Baking, Manhattan, Kansas, USA, pp. 36–50.
Toyokawa, H., Rubenthaler, G.L., Powers, J.R. and Schanus, E.G. (1989) Japanese noodle qualities. I. Flour components. *Cereal Chemistry*, **66**, 382–6.
Wade, P. (1972) Technology of biscuit manufacture: Investigation of the role of fermentation in the manufacture of cream crackers. *Journal of the Science of Food and Agriculture*, **23**, 1021–34.
Waldron, L.A., Rose, S P and Kettlewell, P.S. (1993) Difference in productive performance of broiler chickens fed two wheat varieties. *Aspects of Applied Biology, Cereal Quality III*, **36**, 485–9.
Walker-Simmons, M.K. and Ried, J.L. (eds) (1993) *Pre-Harvest Sprouting in Cereals 1992*, American Association of Cereal Chemists, St. Paul, Minnesota, USA.
Warchalewski, J.R., Madaj, D. and Skupin, J. (1989) The varietal differences in some biological activities of proteins extracted from flours of wheat seeds harvested in 1986. *Die Nahrung*, **33**, 805–21.
Westermann, H.D., Bamikol, H., Fiedler, E., Rang, H. and Thalmann, A. (1988) Gesundheitliche Risiken bei Verfutterung von Brandweizen (Weizenstein-brand und Zwergbrand). 2. Mitteilung: Analytik, Standorterhebung 1984, Futterungsversuch (Mastschweine). Landsirtsch, Forschung, **41**, 169–76.
Wiseman, J. (1990) Quality requirements of wheat for poultry feed. *Aspects of Applied Biology, Cereal Quality II*, **25**, 41–52.
Würsch, P., DelVedovo, S. and Koellrevtter, B. (1986) Cell structure and starch nature as key determinants of the digestion rate of starch in legume. *American Journal of Clinical Nutrition*, **43**, 25–9.

2

Rice

A.B. Blakeney

2.1 INTRODUCTION

Rice (*Oryza sativa L.*) is the staple food of over half the world's people. Worldwide, it is grown on 150 million hectares, more than 10% of total arable land. Total world production exceeds 500 million tonnes of rough rice (paddy). Ninety five per cent of the world's rice is grown by less developed countries, mostly in Asia (IRRI, 1995).

Asian farmers plant 90% of the world's harvested rice area and account for 92% of global rice production. In the humid and subhumid tropics, rice is the primary source of human energy. In Bangladesh, Cambodia, Indonesia, Laos, Burma, Thailand, and Vietnam, rice provides 55–80% of the calories consumed. With the exception of the highest income countries, per capita rice consumption has remained stable in Asia over the past 30 years (Duff, 1991). In most African and Latin American countries, rice is less important.

Rice production and consumption are often associated with low income and poverty. Of the 23 countries in the world that produce more than one million tons, almost half have a per capita income of less than US$500. These are the countries categorized by the World Bank as 'least developed'. Rice is one of their cheapest sources of food energy and protein. Most rice is consumed as white polished grain. Despite the dramatic food value losses resulting from milling, brown rice is unpopular because it requires more fuel for cooking, it may cause digestive disturbances and the oil in the bran tends to turn rancid and reduce storage life (IRRI, 1993).

Globally, the rice-consuming population is growing at 2% a year (Table 2.1). In humid and subhumid Asia where rice is the primary staple food, population is expected to increase by 18% during the 1990s, and by 58% over the next 35 years. Recent projections indicate a world rice food need of about 758 million tons in 2025 – 70% more rice than is

Table 2.1 World rice production and consumption. (Modified from IRRI, 1995)

Major rice consuming and producing countries	*Population by years*		*Rough rice (1993)*			*Milled rice consumption per capita*		*Milled rice 1992*	
	Estimated 1991	*Projected 2000*	*Production*	*Area*	*Yield*	*Estimated 1990*	*Projected 2000*	*Imports*	*Exports*
	(millions)		*(000 t)*	*(000 ha)*	*(t/ha)*	*(kg/year)*		*(000 t)*	
Asia	3,157	3,653	482,549	131,665	3.7	85	91	5,944	10,748
Bangladesh	111	128	28,000	10,900	2.0	155	163	39	0
Cambodia	9	10	2,500	1,800	1.4	167	172	81	0
China	1,150	1,316	187,211	33,403	6.0	94	106	107	1,034
India	866	1,006	111,011	41,200	2.7	66	70	43	560
Indonesia	181	209	47,885	10,932	4.4	138	152	610	42
Japan	124	128	9,793	2,139	4.6	62	51	18	*
Korea, DPR	22	25	2,940	1,000	2.9	125	149	50	5
Korea, Republic	43	47	6,597	1,135	5.8	92	77	1	0.4
Laos	4	6	1,251	539	2.3	190	186	40	0
Malaysia	18	22	2,100	665	3.2	79	444	0.1	
Myanmar	43	51	17,434	5,794	3.0	190	198	0	205
Nepal	19	24	3,100	1,240	2.5	102	92	12	0
Pakistan	116	147	5,927	2.207	2.7	19	24	*	1,512
Philippines	63	74	9,530	3,450	2.8	99	90	1	35
Sri Lanka	17	19	2,450	790	3.1	94	111	237	1
Thailand	57	64	19,090	8,970	2.1	128	117	0	5,151
Vietnam	68	82	22,300	6,466	3.4	146	155	2	1,950

Latin America	445	524	18,294	6,468	2.8	26	27	2,011	782
Brazil	151	178	10,193	4,431	2.3	43	44	480	1
Colombia	33	38	1,650	386	4.3	32	40	80	1
Cuba	11	12	186	78	2.4	48	52	300	0
Dominican Republic	7	8	530	105	5.0	49	50	18	0
Ecuador	11	13	814	205	4.0	38	27	3	0
Guyana	1	1	300	90	3.3	67	146	0	115
Peru	22	27	950	174	5.5	36	36	418	0
Surinam	0.5	1	260	69	3.8	89	91	0	77
Uruguay	3	3	700	126	5.6	9	10	*	328
Africa	645	831	14,802	7,145	2.1	14	16	3,760	202
Côte d'Ivoire	12	17	675	506	1.3	24	45	380	0
Egypt	54	62	4,159	538	7.7	28	30	0.1	188
Guinea	6	8	733	850	0.9	61	71	247	0
Liberia	3	3	71	65	1.1	98	105	128	0
Madagascar	12	15	2,550	1,200	2.1	104	121	60	1
Sierra Leone	4	5	486	382	1.3	95	118	107	0
Nigeria	99	127	3,400	1,750	1.9	12	14	270	0
Australia	17	20	858	106	8.1	5	4	32	519
USA	253	273	7,081	1,146	6.2	6	5	175	2,165
Rest of world	833	867	3,829	987	3.9	6	5	3,313	1,357
World	5,350	6,168	527,413	147,517	3.6	55	59	15,235	15,772

consumed today. In South Asia, where poverty is extensive, the food need for rice is expected to double over the next 40 years. Production needs can be expected to be even higher, to provide stocks, seed, and non-food uses (IRRI, 1995). The International Rice Research Institute collects and collates a book of world rice statistics (IRRI, 1988).

Perhaps due to this traditional distribution, rice grain quality has been studied less extensively and by fewer scientists. The major reviews are the AACC book *Rice Chemistry and Technology* edited by B.O. Juliano (1985), and *Rice: Utilisation* and *Rice: Production* edited by Luh (1991a,b). The *International Bibliography of Rice Research* produced by the International Rice Research Institute is an excellent way of searching for other rice literature; unfortunately it is not yet available in an electronic form.

2.2 RICE TRADE AND THE INTERNATIONAL MARKET

In most of Asia, rice is grown on small, 1–3 hectare farms. A typical Asian farmer plants rice primarily to meet family needs; less than half goes to market, and even that is mostly sold locally.

International rice trade accounts for only 4% of world production, and is mostly in quality rice (Table 2.1). For example, Basmati, the high-quality, scented rice produced in Pakistan and north west India, commands an international market price four times higher than the domestic price of the coarse local rice low-income people eat. Major exporters are Thailand (36% of the world market), the United States (19%), Vietnam (10%) and Pakistan (7%); Australia, China, India, and Uruguay each account for 1–3% of the market. Other world rice markets are for specific quality or processing types (Efferson, 1985).

Most countries cannot depend on imports to meet the food needs of their people. The world market is small and few reserves are held. For example, if China wanted to buy 10% of its domestic consumption, the demand for rice in the world market would increase by more than 80%, and that would dramatically affect international prices. Within south east Asian countries the quality of rice has been shown to directly influence price. It is possible to calculate the value of each increment (Unnevehr *et al.*, 1985).

An important political objective in most rice-dependent countries is self-sufficiency in rice production, in order to maintain stable prices (in particular for rapidly increasing numbers of urban consumers). But rice harvests can fluctuate widely, especially in the less favourable environments. Variable natural conditions cause year-to-year shortages and surpluses, and that means wide variation in the amount farmers send to market. This makes domestic prices highly unstable. Price controls through maintenance of large stocks can benefit urban consumers, but often keep farm prices below a profitable level.

Figure 2.1 Structure of the (paddy) rice grain. 1, Brown rice (caryopsis); 2, palea; 3, lemma; 4, rachilla; 5, sterile lemmas; 6, rudimentary glumes and part of the pedicel.

2.3 GRAIN MORPHOLOGY

The rice grain, commonly called a seed, consists of the true fruit or brown rice (caryopsis) and the hull, which encloses the brown rice (Figure 2.1). Brown rice consists mainly of the embryo and endosperm. The surface (pericarp) contains several thin layers of differentiated tissues that enclose the embryo and endosperm (Figure 2.2). The palea, lemmas, and rachilla constitute the hull of indica rices. In japonica rices, however, the hull usually includes rudimentary glumes and perhaps a portion of the pedicel. The dry weight of a single grain weighs between 10 and 45 mg. Grain length, width, and thickness vary widely among varieties (Table 2.2). Hulls average about 20% of total grain weight. Grain dimensions are an important appearance quality of rice and rice entering trade is usually classified on its grain size and uniformity.

2.4 GRAIN: QUALITY AND QUALITY CONTROL

At the varietal level, rice grain quality comprises grain appearance, milling quality and cooking quality. In all these areas quality control is a monitoring operation with feedback only through the breeding programme or, in the case of some milling quality problems, through the agronomic or processing practices. The most comprehensive survey of

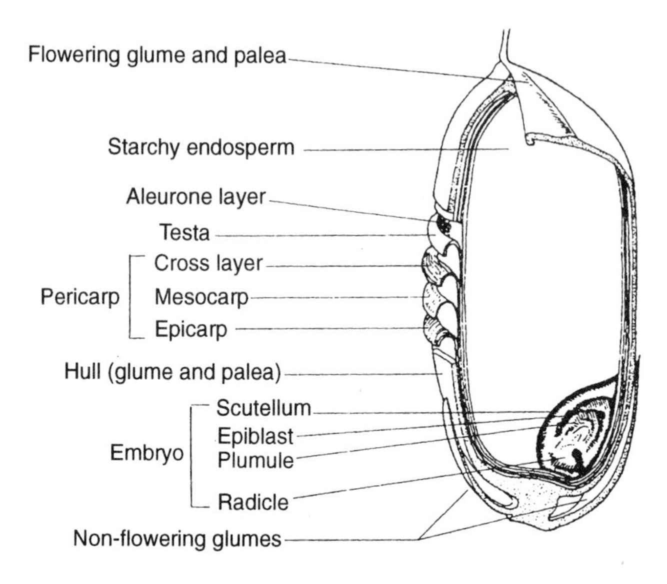

Figure 2.2 Internal structure of a rice grain.

methods used in equating rice grain quality is that of Juliano and Villareal (1993) who collated information supplied by rice research from all major rice producing countries.

2.4.1 Grain appearance

The quality of commercial rice samples is usually assessed in terms of moisture content, cleanliness, freedom from empty hulls, straw and other seeds, colour and uniformity. When rice growers deliver rice for sale all these parameters are usually assessed. Samples may be taken for later test milling and cooking tests. Rice is often harvested at moisture contents above 20% so it is crucial to know the moisture content of a delivered load so that the microbial stability and potential milling quality can be preserved by appropriate drying and storage. It has usually been assessed with conductance or capacitance moisture meters but whole grain near infrared transmittance instruments have recently been introduced in Australia and Japan (Blakeney *et al.*, 1994).

Unlike other cereals that are ground into flours, rice is consumed as whole grain. Consumers in different regions of the world have definite expectations as to how their rice should look. Grain size, shape, colour, gloss, translucency and uniformity all need to be considered when evaluating rice grain appearance. Appearance measurements must also

Table 2.2 Size and shape classification of milled rice

Length (mm)			*Shape (L/B)*			*Weight (mg)*		
Class	*Abbrvn*	*Specification*	*Class*	*Abbrvn*	*Specification*	*Class*	*Abbrvn*	*Specification*
Extra long	EL	Over 7.0	Slender	s	Over 3.0	Giant	G	Over 23
Long	L	6.0–7.0	Quasislender	q	2.4–3.0	Big	B	18.1–23
Medium	M	5.0–5.99	Bold	b	2.0–2.39	Small	S	12–18
Short	S	Less than 5.0	Round	r	Less than 2.0	Tiny	T	Less than 12

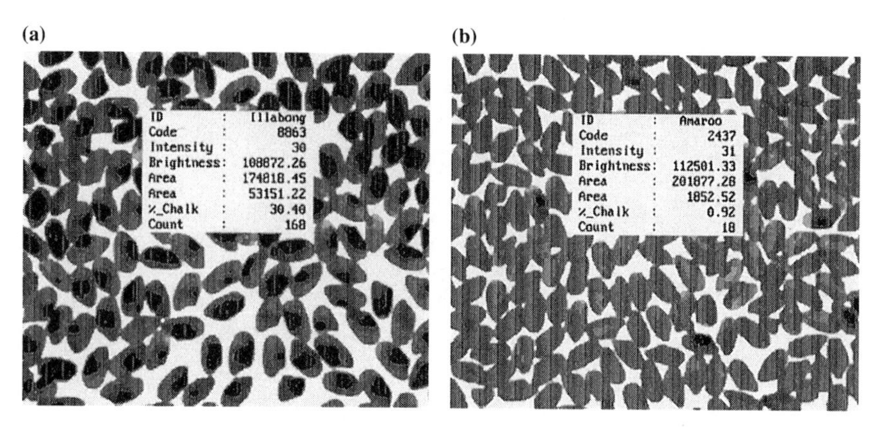

Figure 2.3 Image analysis of chalk in two rice varieties, analysed measuring total area of grain, total area of chalk and reports of the percentage chalk. (a) Amaroo Australian medium grain variety with a low (0.44%) chalk content. (b) Illabong Australian Arborio type variety with high (31.10%) chalk).

be made on both brown and white rice. Rice appearance has traditionally been evaluated subjectively by eye, or with the aid of a microscope for measuring grain dimensions (Ikehashi and Khush, 1979). Objective methods for colour (Blakeney *et al.*, 1994b) using a reflectance spectrophotometer and translucency and size and shape using image analysis are now available (Figure 2.3; Reece and Blakeney, 1993). Milled rice is classified on the basis of grain size and shape; Bhattacharya and Sowbhagya (1980) reviewed the previous classifications and proposed those shown in Table 2.2.

2.4.2 Milling quality

Individual varieties vary in length, width, thickness and weight. Table 2.3 illustrates the wide range of these attributes. Rice milling consists of removal of hulls, bran and germ with a minimum breakage of endosperm. The milling quality of rice is based primarily on the yield of whole grain rice obtained, the economic value of rice being dependent on this factor. Broken grain is usually worth only about half as much as whole grain. Grain breakage during processing is a serious problem in all rice growing areas. Yields of whole grain vary with both variety and environment. The two major causes of breakage of grain during milling are checking and chalky grain.

Checking, or suncracking results when harvesting and threshing is delayed or when too rapid drying of the crop takes place. The cause

Table 2.3 Grain characteristics of a wide range of rice varieties. (Adapted from Yoshida, 1981)

Variety	*Country of origin*	*Wt (mg/grain) dry wt basis*	*Length (mm)*	*Width (mm)*	*Thickness (mm)*	*Hull (%)*
Khao Lo	Laos	44.4	11.30	3.20	2.53	18.6
Cseljaj	Hungary	36.6	9.43	3.67	2.52	21.5
Ku 70-1	Thailand	34.7	8.91	3.93	2.32	19.6
Hiderishirazu	Japan	26.4	7.98	3.66	2.21	20.9
Rikuto Norin 21	Japan	23.1	6.94	3.48	2.19	20.6
Bergreis	Austria	21.8	6.81	3.25	2.21	19.3
Ai Yeh Lu	China	20.8	6.82	3.08	2.05	20.8
IR747B2–6	IRRI	16.4	7.61	2.46	1.81	20.4
Bangarsal	India	14.7	7.73	2.03	1.72	20.1
Bomdia	Portuguese Guinea	11.6	5.89	2.45	1.65	25.4
Kalajira	Bangladesh	11.9	5.81	2.43	1.89	21.8

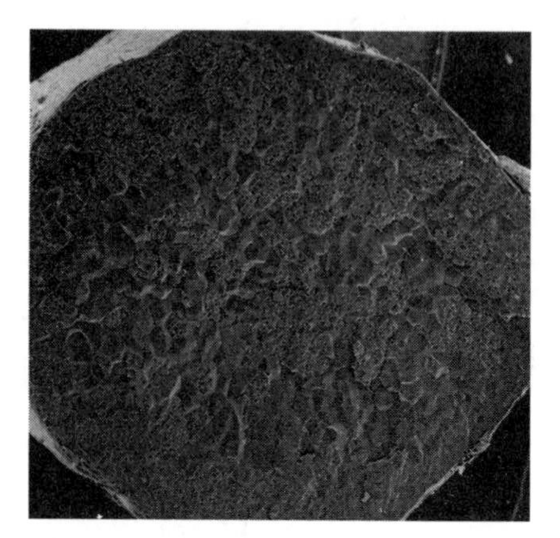
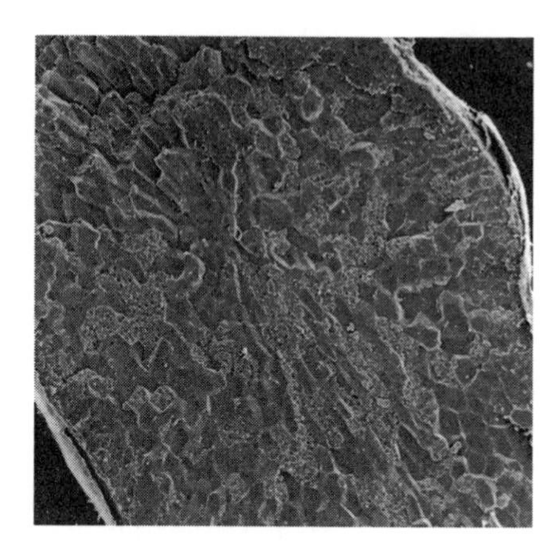

Figure 2.4 Natural suncracked surface of the rice variety Kulu. The endosperm cells are compact and arranged in a relatively random pattern. They resist pressures within the grain so that Kulu suffers less from suncracking than other varieties, such as Caloro. The cells in Caloro are in a radical pattern, and cracks form along the lines of the cell walls. (× 27).

appears to be mechanical, fluctuations in temperature and moisture content causing the outer portion of the grain to expand more quickly than the centre, resulting in the formation of cracks along the line of endosperm cell walls (Figure 2.4). A similar phenomenon has been reported for wheat. The importance of these factors has been recognized for many years and local conditions and varieties thoroughly investigated. Change in agronomic practice, to harvest early at higher moisture levels, together with controlled drying during storage have overcome much of this problem.

Chalky and immature grains also break easily on milling. Chalky portions of the grain are layers of cells with loose packing of starch granules (Figure 2.5). The formation of chalky grain appears to be both genetic and environmental; immature grains are generally chalky. The report that the starch from the chalky area of IR 8 had a gelatinization temperature 2°C higher than the rest of the endosperm, together with the work of Briones and coworkers (1968) which shows a fall in gelatinization temperature from flowering to maturity, would seem to confirm this.

Other factors that contribute to milling quality include moisture content (Siebenmorgen, 1994) and infestation, together with grain shape (Srinivas and Bhashyan, 1985) and hardness. The type and design of milling equipment may also influence milling results as could the design of harvesting equipment (Satake, 1994).

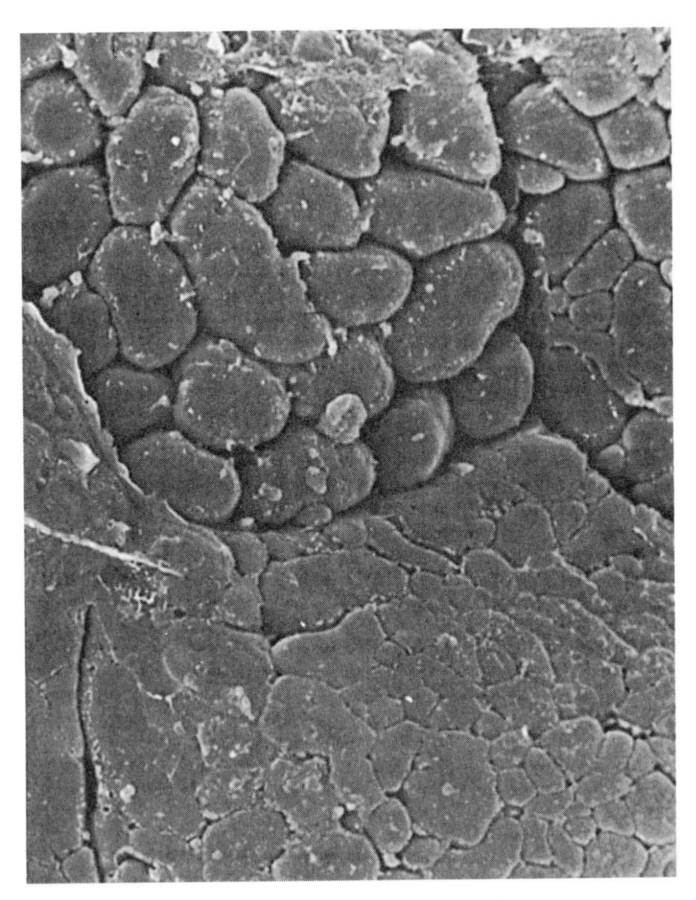

Figure 2.5 A sectioned rice grain showing where chalky belly starts. The two cells at the top of the photo are loosely packed with lots of air between the amyloplasts giving the chalky appearance. The grain then changes suddenly to densely packed amyloplasts within the cells, normally found in translucent grain. (× 4000).

Parboiling

Parboiling, a method of pre-cooking rice grain before milling, is well known to reduce rice breakage during milling. Parboiling originated in India and has been refined into a modern industrial process. Paddy rice is steeped in water, steamed, and dried prior to milling (Garibaldi, 1974). Improvement is attributed to greater hardness and to the sealing of internal cracks and chalkiness in the parboiled grain. Adequate penetration and distribution of water throughout the rice grain is critical to the quality of parboiled rice. Failure to bring about the complete gelatinization of the starch throughout the endosperm results in grain containing opaque spots indicating the presence of ungelatinized starch granules. Grains containing these spots are susceptible to breakage during milling. Due to its pregelatinized nature, the cooking quality of parboiled rice is different, though related, to the raw rice from which it is derived. Parboiling also reduces the amount of protein extractable with water and leaches B group vitamins from the aleurone layers to the endosperm, increasing the nutrient value of the milled product. Parboiled rice has a storage life of 2–3 years.

2.4.3 Cooking quality

The preferred texture, flavour and other cooking characteristics of rice vary with ethnic group and geographical region. In evaluating cooking quality an attempt is made to relate chemical and physical properties of the grain to the desired subjective characteristics of the cooked rice.

Rice varieties differ greatly in cooking and processing qualities. Cooking characteristics are generally associated with grain size and shape. Usually short and medium grain varieties become somewhat sticky on cooking. Although this type is preferred in some north Asian countries there is a distinct preference in Australia and most western countries for the drier, flaky quality usually found in long grain varieties. The short grain varieties are used for making puffed rice and are also used successfully for precooked canned rice and in dry quick cooking rice products in the US. Both types are used in the manufacture of dry breakfast cereals and both are used in the parboiling process. Length of grain is not, however, a completely reliable indicator of the cooking quality of rice, because several varieties give cooked products different from the type expected on the basis of grain length classification.

Gelatinization temperature and amylose content of the component starch are the two most important physiochemical properties relating to cooking quality in rice. Starch comprises approximately 90% of the total dry substances of polished rice. The starch in the endosperm of rice exists in compound granules, unlike other cereals except oats. These granules are formed by the following process. Many small spherical granula are initiated in the proplastid stroma, and by assuming a polyhedral shape with growth, soon fill the plastid. Fusions between granula often occur so that each may not be completely surrounded by a stroma layer. At the molecular level, starch (a glucose polymer) is composed of two principle fractions: a linear fraction, amylose, and a branched fraction, amylopectin. Amylose gets its linear structure from the α-(1→4) bonds that link glucose units. The branched structure of amylopectin results from the presence of α-(1→6) bonds, in addition to

Figure 2.6 Waxy, low and high amylose rices illustrating their different cooked grain bulk densities.

the α-(→4) bonds. The molecular structure of starch is covered in more detail in Chapter 9.

Cooking rice consists mainly of gelatinizing the starch. Gelatinization temperature is related to cooking time, granula size, molecular size of the starch subfraction polymers and their relative amounts. Varieties have been classified on their gelatinization temperatures. Amylose content has been shown by numerous workers to be very important to the cooking and processing qualities of rice. Correlation between amylose content and eating quality preferences has been reported. Cooked rice with high amylose content is flaky and dry, while rice with low amylose is sticky and moist when cooked (Figure 2.6). The determination of amylose content and gelatinization temperature and the relationship between these two factors have been the major indirect means of selecting for cooking quality in rice breeding programmes.

Amylose determination

Halick and Keneaster (1956) used an empirical starch–iodine blue test to differentiate between long grain samples. Hogan and Planck (1958) pointed out that, under the particular conditions used, in which the amylose responsible for the development of the blue colour was leached from ground rice at 77°C, the test may actually have been a measure of ruptured starch granules. These objections were overcome by modification of the test using extractions at 99.5°C, but even this modified test is of limited value for determining amylose in milled rice in the 20–30% range. The original test has, however, been found to have a highly significant correlation of 0.860 with amylose content in samples with a gelatinization temperature below 77°C and is useful for detecting varieties with high gelatinization temperatures (Juliano, 1964).

Rao and coworkers (1952) reported that amylose content determined by potentiometric iodine titration was closely related to 'swelling number' of the 22 samples of Indian rice they examined. Williams and coworkers (1958), using the same method of assay, examined a number of varieties of rice grown in southern USA and found a trend toward higher amylose content in the long grain varieties. Juliano and coworkers (1964) concluded that generally the higher the amylose content, the higher the gelatinization temperature but that these are exceptions and that a cause–effect relationship does not seem to exist. Evidence of the lack of any direct relationship between amylose content and gelatinization temperature together with an exception to the grain shape classification was presented by Halick and Kelly (1959). These workers pointed out that two atypical long grain varieties, Century Patna 231 and Toro, have approximately the same amylose content, yet differ widely in gelatinization characteristics. The gelatinization temperatures, determined as the temperature at which an increase in viscosity of a 20% (w/v) slurry of ground rice was first recorded by the amylograph, differed greatly and reflected their cooking qualities.

The method used by Halick and Kelly (1919) for determining gelatinization temperature on the amylograph using a high concentration of ground rice agreed well with gelatinization values obtained by the orthodox method of determining microscopically the loss of birefringence or granule swelling. However, after the work of Sanstedt and Abbott (1964) on other cereal starches the concentration of the slurry may need further investigation to show true first swelling. In the international survey of methods used for rice quality conducted by Juliano (1982), four laboratories used this approach to gelatinization temperature measurement.

Gelatinization temperature is influenced by environment. Beachell and Stansel (1963) noted that cooler temperatures, especially during ripening, produced starch with lower gelatinization temperatures. Kihara and Kajikawa noted that high temperature during ripening gave a slower rate of alkali digestion of the endosperm which indicated a high gelatinization temperature range. The maturity of the sample also seems to be involved.

A method closely related to determining gelatinization temperature was used by Little and Hilder (1960) who showed a high correlation between the microscopically determined extent of alteration of starch granules, in ground rice heated in water at 62°C for 30 minutes, and the taste panel scores for cohesiveness for cooked samples of the same lots of rice.

A positive correlation between water uptake, related to the degree of gelatinization, and taste panel scores for cohesiveness was reported by Batcher and coworkers (1963). Hogan and Planck (1958) reported that when hydration was carried out at 70°C for 20–30 minutes rather than at near boiling temperature, differences between short, medium and long grain types were readily observed. Furthermore, it was possible to detect long grain varieties that were not true to type. The water absorption values of the hydrated grains agreed with the cohesive properties and cooking characteristics of the grain. High water absorption values obtained under these conditions indicated poor cooking quality and stickiness; long grained rices yielding dry fluffy cooked products were characterized by low water absorption at 70°C. However, Batcher and coworkers (1963) found that high amylose rice generally absorbs more water and expands more during cooking at 99°C in excess water, than low amylose rice. Differences in the above findings are probably due to the different cooking temperatures employed acting through gelatinization temperature.

Amylograph studies are undertaken to simulate cooking and study changes occurring during pasting, cooking and cooling of aqueous starch systems. Halick and Kelly (1959) reported that amylograph curves of 10% w/v ground rice slurries yielded useful information about the cooking properties of rice samples. Amylose content was correlated with pasting characteristics of rice as measured by the amylograph. The drop in viscosity after cooking for 20 minutes was also highly correlated with amylose content indicating that resistance to disintegration of rice during cooking is also indicative of high amylose content. Set-back viscosity, which is the difference in viscosity on cooling to 50°C and peak viscosity, is a measure of the retrogradation of amylose during cooling of the paste. This and the viscosity on cooling to 50°C have both been correlated to amylose content.

Webb (1975), after using the amylograph test for several years in the evaluation of breeding selections, concluded that transition temperature, peak viscosity and resistance to thinning during 10 minutes holding at 95°C in the amylograph furnished useful information. This was of even greater value if considered collectively with the amylose content and the gelatinization temperature of the sample. A similar outlook was recommended by Beachell and Stansel in selecting rice of specific cooking characteristics in a breeding programme. The methodology for using the amylograph for rice was reviewed by Suzuki (1979).

The use of amylograph type post-viscometry was made significantly easier with the introduction of the Rapid Viscoanalyser (RVA) (Newport Scientific, Warriwood, Australia) (Welsh *et al.*, 190; Blakeney *et al.*, 1991; Kohlwey, 1994). This instrument has the advantage of requiring only a 2.5 g sample and taking only 12 minutes per test.

In the Australian rice breeding programme we now use RVA paste viscosity testing at all stages from F_4 grain onward to estimate cooking quality and class. From the traces we measure paste gel temperature, peak viscosity, breakdown after peak, setback and viscosity on cooling to 50°C. Although these measurements are useful, the major use of RVA traces is for direct comparison with class standard varieties. Figure 2.7 shows two Australian standard long grains (Pelde and Doongara), an Indian Indica (Dular) and a Californian Waxy variety (Calmochi 202). These three figures will illustrate the genetic range of pasting patterns that can be encountered. Figure 2.8 illustrates the influence environment can have on the pasting properties of a single variety, all the traces in this figure are of the premium Japanese variety Koshihikari grown at seven locations in Japan. It should be noticed that rice from the most favoured area, Niigata prefecture, can easily be identified. Where the trace of a new line very nearly matches that of the class standard, it can be progressed with the confidence that it will have similar cooking qualities to the standard. The results are computer compatible and a trial of 10 or 20 lines can be superimposed in colour for visual appraisal. A black and white presentation in which closely related lines are compared to standard varieties is shown in Figure 2.9. This does not do justice to the discrimination which can be achieved on a colour screen or with a colour printer. By using two RVAs, one technician can evaluate over 50

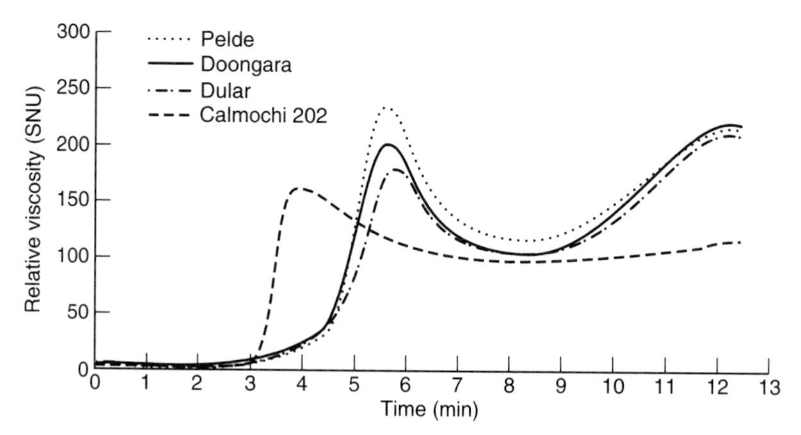

Figure 2.7 RVA pasting curve of a waxy (Calmochi 202) and three long-grained rices known to vary in cooked grain texture.

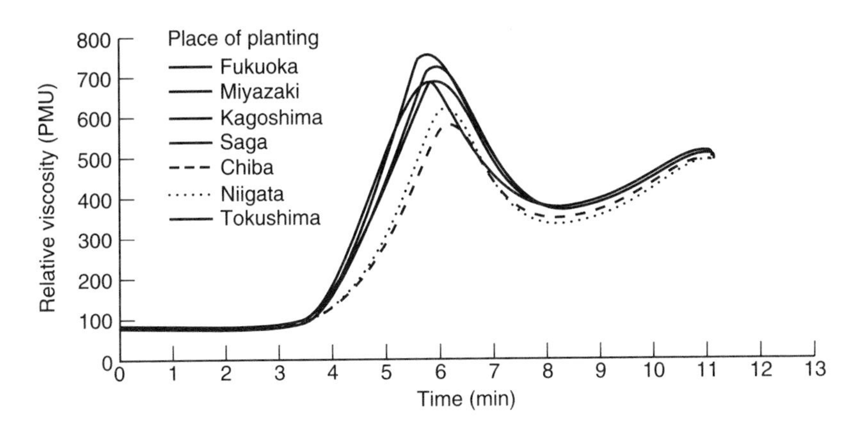

Figure 2.8 The premium Japanese variety Koshihikari grown at seven locations throughout Japan.

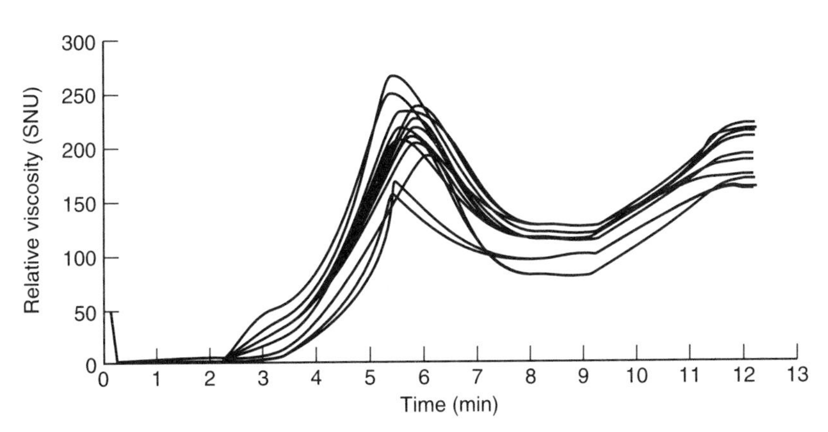

Figure 2.9 Rices from a breeding programme superimposed to allow classification and selection.

lines per day and quality selection can easily be completed between harvest and sowing.

The possible significance of the differences in the molecular weight of the amylose fraction and the internal and external chain lengths of the amylopectin fraction from different varieties is not yet apparent and requires further study.

Other quality factors

Although there seems to be little doubt that the properties of the rice starch and the changes it undergoes during cooking are of major importance in determining the characteristics of the cooked grain the possible influence of other factors has not been overlooked. Dawson and coworkers suggested that proteins, fats, cell wall carbohydrates and minerals may influence cooking quality. They also suggested that the alkali test, which has been used to indicate cohesiveness, may involve protein, fat and mineral components of rice as well as starch.

Protein

Protein has two roles in rice quality. Firstly, it is related to the nutritive value of rice and secondly, protein content has been correlated with cooking and eating quality. Protein content of the world rice collection has been surveyed and shown to vary from 5–17% with a mean of 10.5% and a significant correlation between total protein and lysine levels (IRRI, 1967). Varieties vary significantly in protein content depending on the conditions under which they are grown. High protein lines would contribute to the nutritive value of rice as a food. The role of rice in human nutrition was recently reviewed by Juliano (1993) for FAO.

Protein content can influence milled rice colour and the rate at which rice endosperm yellows with age. In a study on Australian farm samples white rice yellowness index was found to be well correlated with brown rice colour (Figure 2.10). The yellowness index used is ASTM D1925 which combines the brightness (L*) and yellowness (b*) parameters of colour (Blakeney *et al.*, 1994).

Juliano and coworkers (1965) have positively correlated the colour of cooked rice with protein content. Protein content also affects the

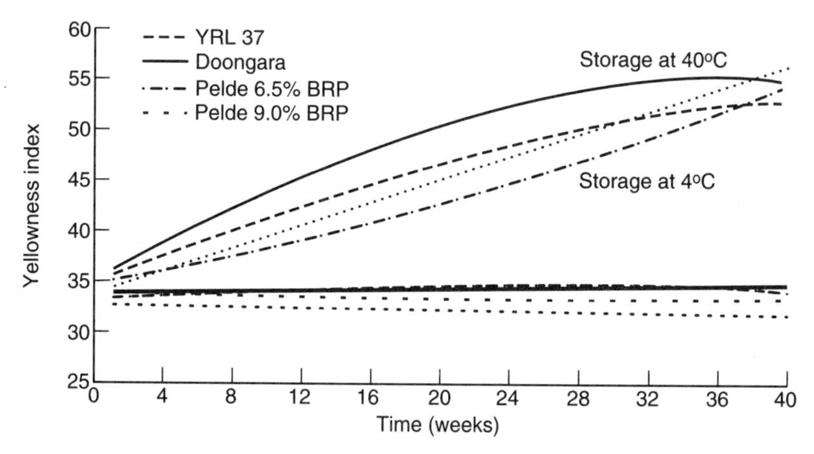

Figure 2.10 Effect of protein content and storage temperature on grain yellowness.

tenderness and cohesiveness of cooked rice: within the one variety, cooked rice of higher protein content was found to be significantly less cohesive. Protein content has also been reported to affect the physical hardness of the grain and so may be a factor in milling quality.

An example of the way non-starch ingredients can affect the properties of cooked rice is reported by Brokaw (1962) who demonstrated reduced stickiness in rice grain treated with monoglycerides prior to cooking or cooked in water in which monoglycerides had been dispersed. It was suggested that the action occurred because of complexing of the amylose leached from the swollen granules during cooking. However, a possible additional explanation may lie in the known effect of monoglycerides in delaying the swelling of starch granules.

A study of the histology and histochemistry of raw and cooked rice grains by Little and Dawson (1966) led them to suggest that suppression of swelling of the starch in rice grains to a small fraction of its capacity and the different patterns of disruption of the structure of the grain during cooking may be attributed largely to delaying or limiting effects of the cell walls or protein. The effectiveness of the cell walls in this respect might be related to thickness, composition, and distance apart and that of protein to its composition and concentration. Differences in microscopic appearance between varieties were noted but observations were not sufficient to establish a relationship between structure, or non-starch components, and cooking characteristics.

Cooking time and methods

The time required for rice to cook is related to the rate of penetration of water into the grain. Desikachar and Subrahmanyan (1961) reported that cracks which developed in rice grains soaked in water appear to be responsible for their greatly reduced cooking time. Rice cooking procedures (Juliano, 1982) and cooked rice evaluation procedures (del Mundo, 1979) vary widely and are dependent on the cuisine and culture in which the rice is being used. It is, however, generally agreed that texture is the major component of cooked rice quality. Texture can be assessed by taste panels or measured using Kramer, Ottawa or back-extrusion cells on force testing instruments (Blakeney, 1979; Juliano *et al.*, 1984).

Rice as a food ingredient

Rice is widely used in breakfast and snack foods in the West as well as being an ingredient in traditional foods. There are several excellent sources of information on this usage including Juliano (1985, 1993) and

Luh (1991b). Sheng (1995) has recently reviewed the use of rice as an ingredient in crackers, snack dips, breakfast foods and baked goods in the USA.

Rice quality data

Several collections of rice quality data exist, notably the United States Department of Agriculture survey of foreign and domestic rices (USDA, 1965). This is now a little dated but contains a good review of quality analysis methods at the time; variety data is unfortunately not included. The IRRI physicochemical data bulletin on rice grain (Juliano and Villareal, 1993) has been updated several times and represents the best source of regional and varietal rice quality information.

REFERENCES

Batcher, O.M., Staley, M.G. and Deary, P.A. (1963) Palatability of Foreign and Domestic Rices Cooked by Different Methods I. *Rice Journal*, **66**, 19–27.

Batcher, O.M., Staley, M.G. and Deary, P.A. (1963) Palatability of Foreign and Domestic Rices Cooked by Different Methods II. *Rice Journal*, **66**, 13–15.

Beachell, H.M. and Stansel, J.W. (1963) Selecting Rice for Specific Cooking Characteristics in a Breeding Program, in 'Proceedings af the 10th Pacific Science Congress, Symposium on Rice Problems', *International Rice Commission Newsletter Special Issue*, pp. 25–40.

Bhattacharya, K.R. and Sowbhagya, C.M. (1980) Size and Shape Classification of Rice, Il, *Riso*, **29**, 181–5.

Blakeney, A.B. (1979) Instron measurement of cooked-rice texture, in *Proceedings of the Workshop on Chemical Aspects of Rice Grain Quality*, International Rice Research Institute, Los Baños, pp. 343–54.

Blakeney, A.B., Welsh, L.A. and Bannon, D.R. (1991) Rice Quality Analysis Using a Computer Controlled RVA, in *Cereals International* (eds D.J. Martin and C.W. Wrigley), Cereal Chemistry Division, Royal Australian Chemical Institute, Parkville.

Blakeney, A.B., Welsh, L.A. and Reece, J.R. (1994) Recent Developments in Rice Quality Analysis, in *Temperate Rice – Achievments and Potential*, Proceedings of the Temperate Rice Conference, Yanco 1994, Vol. 1, (eds E. Humphreys, E.A. Murray, W.S. Clampett and L.G. Lewin), Temperate Rice Conference Organising Committee, Grifith, NSW, Australia.

Briones, V.P., Magbanua, L.G. and Julianoa, B.O. (1968) Changes in Physiochemical Properties of Starch in Developing Rice Grain. *Cereal Chemistry*, **B45,** 351–7.

Brokaw, G.Y. (1962) Distilled Monoglycerides for Food Foaming and for Starch Complexing, *Ca. Food Ind.*, **33**, 36–9.

Dawson, E.H., Batcher, O.M. and Little, R.R. (1960) Cooking Quality of Rice, *Rice Journal*, **63(5)**, 16–22.

Duff, B. (1991) Trends and Patterns in Asian Rice Consumption, in *Rice Grain Marketing and Quality Issues*, Selected papers from the International Rice Research Conference 27–31 August 1990. International Rice Research Institute, Manila, pp. 1–22.

Efferson, J.N. (1985) Rice Quality in World Markets, in *Rice Grain Quality and Marketing*, Papers presented at the International Rice Research Conference 1–5 June 1985. International Rice Research Institute, Manila, pp. 1–13.

Garibaldi, F. (1974) *Rice Parboiling*, Food and Agriculture Organization of the United Nations, Rome.

Halick, J.V. and Kelly, V.J. (1959) Gelatinization and Pasting Characteristics of Rice Varieties as Related to Cooking Behaviour. *Cereal Chemistry*, **36**, 91–8.

Halick, J.V. and Keneaster, K.K. (1956) The Use of a Starch–iodine-blue Test as a Quality Indicator of White Milled Rice. *Cereal Chemistry*, **33**, 315–19.

Ikehashi, H. and Khush, G.S. (1979) Methodology of assessing appearance of the rice grain, including chalkiness and whiteness, in *Proceedings of the Workshop on Chemical Aspects of Rice Grain Quality*, International Rice Research Institute, Los Baños, pp. 223–9.

IRRI (1967) *The International Rice Research Institute Annual Report 1967*, Cereal Chemistry Section, International Rice Research Institute, Los Baños.

IRRI (1988) *World Rice Statistics 1987*, International Rice Research Institute.

IRRI (1993) *IRRI Rice Almanac*, International Rice Research Institute, Manila.

IRRI (1995) *IRRI Rice Facts*, International Rice Research Institute, Los Baños.

Juliano, B.O. (1982) An International Survey of Methods Used for Evaluation of the Cooking and Eating Qualities of Milled Rice, *IRRI Research Paper Series Number 77*, International Rice Research Institute, Manila.

Juliano, B.O. (1985) *Rice: Chemistry and Technology*, American Association of Cereal Chemists, St Paul, Minnesota.

Juliano, B.O. (1993) *Rice in Human Nutrition*, FAO Food and Nutrition Series No. 26, Food and Agriculture Organization of the United Nations, Rome.

Juliano, B.O. and Villareal C.P. (1993) *Grain Quality Evaluation of World Rices*, International Rice Research Institute, Manila.

Juliano, B.O., Bautista, G. Lugay, J.S. and Reyes, A.C. (1964) Rice Quality. Studies on the Physiochemical Properties of Rice. *J. Agr. Food Chem.*, **12**, 131–8.

Juliano, B.O., Onate, L. and Del Mondo, A (1965) The Relation of Starch Composition, Protein Content and Gelatinization Temperature to Cooking and Eating Qualities of Rice, *Food Technology*, **19**, 116–21.

Juliano, B.O., Perez, C.M., Alyoshin, E.P., Romanov, V.B., Blakeney, A.B., Welsh, L.A., Choudhury, N.H., Delgado, L., Iwasaki, T., Shibuya, N., Mossman, A.P., Siwi, B., Damardjati, D.S., Suzuki, H. and Kimura, H. (1984) International cooperative test on texture of cooked rice. *Journal of Texture Studies*, **15**, 357–76.

Kohlwey, D.E. (1994) New Methods for the Evaluation of Rice Quality and Related Terminology, in *Rice Science and Technology* (eds W.E. Marshall and J.I. Wadsworth) Marcel Dekker, New York, pp. 113–37.

Little, R.R. and Dawson, E.H. (1960) Histology and Histochemistry of Raw and Cooked Rice Kernels. *Food Research*, **35**, 111–15.

Luh B.S. (1991a) *Rice: Production*, Vol. 1, 2nd edn, Van Nostrand Reinhold, New York.

Luh, B.S. (1991b) *Rice: Utilization*, Vol. 2, 2nd edn, Van Nostrand Reinhold, New York.

del Mundo, A.M. (1979) Sensory Assessment of Cooked Milled Rice, in *Proceedings of the Workshop on Chemical Aspects of Rice Grain Quality*, International Rice Research Institute, Los Baños, pp. 313–26.

Ong, M.H., Jumel, K., Tokarczuk, P.F., Blanshard, J.M.V. and Harding, S.E. (1994) Simultaneous Determinations of the Molecular Weight Distributions of Amyloses and the Fine Structures of Amylopectins of Native Starches. *Carbohydrate Research*, **260**, 99–117.

Reece, J.R. and Blakeney, A.B. (1993) Image Analysis for the Assessment of Chalk in Milled Rice, in *Proceedings of the 43rd Australian Cereal Chemistry Conference* (ed. C.W. Wrigley), Cereal Chemistry Division of the Royal Australian Chemical Institute, North Melbourne. pp. 21–4.

Reece, J.R., Blakeney, A.B., Welsh, L.A., Ronalds, J.A. and Sharman, J.P. (1994) Paddy Rice Grain Moisture Analysis by Near Infrared Transmission, in *Proceedings of the 44th Australian Cereal Chemistry Conference* (eds J.F. Panozzo and P.G. Downiw), Cereal Chemistry Division of the Royal Australian Chemical Institute, North Melbourne. pp. 118–21.

Satake, R.S. (1994) New Methods and Equipment for Processing Rice, in *Rice Science and Technology* (eds W.E. Marshall and J.I. Wadsworth), Marcel Dekker, New York. pp. 229–62.

Sheng, D.Y. (1995) Rice-based ingredients in cereals and snacks. *Cereal Food World*, **40**, 538–40.

Siebenmorgan, T.J. (1994) Role of Moisture Content in Affecting Head Rice Yield, in *Rice Science and Technology* (eds W.E. Marshall and J.I. Wadsworth), Marcel Dekker, New York. pp. 341–80.

Srinivas, T. and Bhashyan, M.K. (1985) Effect of Variety and Environment on Milling Quality of Rice, in *Rice Grain Quality and Marketing*, Papers presented at the International Rice Research Conference 1–5 June 1985. International Rice Research Institute, Manila, pp. 49–59.

Suzuki, H. (1979) Amylography and Alkali Viscography of Rice, in *Proceedings of the Workshop on Chemical Aspects of Rice Grain Quality*, International Rice Research Institute, Los Baños, pp. 261–82.

Unnevehr, L.J., Juliano, B.O. and Perez, C.M. (1985) Consumer Demand for Rice Grain Quality in Southeast Asia, in *Rice Grain Quality and Marketing*, Papers presented at the International Rice Research Conference 1–5 June 1985. International Rice Research Institute, Manila, pp. 14–23.

USDA (1965) Quality evaluation studies of foreign and domestic rices. *USDA Tech. Bull. No. 1331*, 186 pp.

Webb, B.D. (1975) Cooking, Processing and Milling Qualities of Rice, in *Six Decades of Rice Research in Texas*, Texas Agricultural Experiment Station, Research Monograph No. 4, Beaumont, Texas, pp. 97–106.

Yoshida, S. (1981) *Fundamentals of Rice Crop Science*, International Rice Research Institute, Los Baños.

3

Maize

S.R. Eckhoff and M.R. Paulsen

3.1 MAIZE AS A STARCH CROP

Maize is known in different parts of the world as *Zea mays*, maize or corn. It is indigenous to the Americas with its origin believed to be in Central Mexico (Benson and Pierce, 1987; Johnson, 1991) although it is currently produced on every continent except Antarctica. It is a highly domesticated plant (evidence suggests that it was domesticated prior to 5000 BC; Benson and Pierce, 1987) with little remaining ability to propagate in the wild. It probably has the widest range of genetic diversity of any of the major cereal grains.

Maize is grown primarily because it is easy to cultivate, has a large yield, can be stored easily and is high in starch, which can readily be metabolized into energy. Maize is not a complete food product for most livestock and humans because it is deficient in lysine, an essential amino acid. When lysine is available in the diet from other sources, maize is a good source of protein and energy and can constitute a high percentage of the diet. In some African and Central and South American cultures, maize is a vital part of the daily human diet.

Commercial maize is an annual grass which has an erect stalk that usually produces one ear per plant, although it can set more than one ear depending upon genetics and environmental factors. The ear consists of a central pithy section (cob), which acts to transport nutrients to the kernels during development, surrounded radially by individual kernels (300–1000 per ear; Benson and Pierce, 1987). When mature, the kernels can be removed from the cob by hand or through the use of a mechanical sheller. The cob can be used as an energy source, as animal bedding or as a feed stock for conversion into industrial chemicals (Bagby and Widstrom, 1987). The kernels can be further processed into food or animal feed.

3.2 STRUCTURE AND COMPOSITION

Maize kernels are composed of approximately 73% starch, 10% protein, 5% oil and the remainder fibre, vitamins and minerals (Table 3.1). There are three major structural parts to the maize kernel: pericarp, endosperm and germ (Figure 3.1).

The pericarp is the outer protective coating of the kernel composed of primarily cellulose and hemicellulose, which resists water uptake by the

Table 3.1 Corn kernel composition by parts (% dry basis). (Data from Earle *et al.* (1946) and Watson (1984))

		Proportion of whole kernel	*Starch Protein*	*Oil*	*Ash*	*Sugar*	
Kernel	Mean	100	71.5	10.3	4.8	1.4	2.0
	range	–	67.8–74.0	8–11.5	3.9–5.8	1.3–1.5	1.6–2.2
Endosperm	Mean	81.9	86.4	9.4	0.8	0.3	0.6
	range	80.3–83.5	83.9–88.9	6.7–11	0.7–1.1	0.2–0.5	0.5–0.8
Germ	Mean	11.9	8.2	18.8	34.5	10.1	10.8
	range	10.5–13.1	5.1–10.0	17.3–20	31–39	9.4–11.3	10–12.5
Pericarp	Mean	5.3	7.3	3.7	1.0	0.8	0.3
	range	4.4–6.2	3.5–10.4	2.9–3.9	0.7–1.2	0.3–1.0	0.2–0.5
Tip cap	Mean	0.8	5.3	9.6	3.8	1.6	1.6
	range	0.8–1.1	–	9.1–11	3.7–3.8	1.4–2.0	–

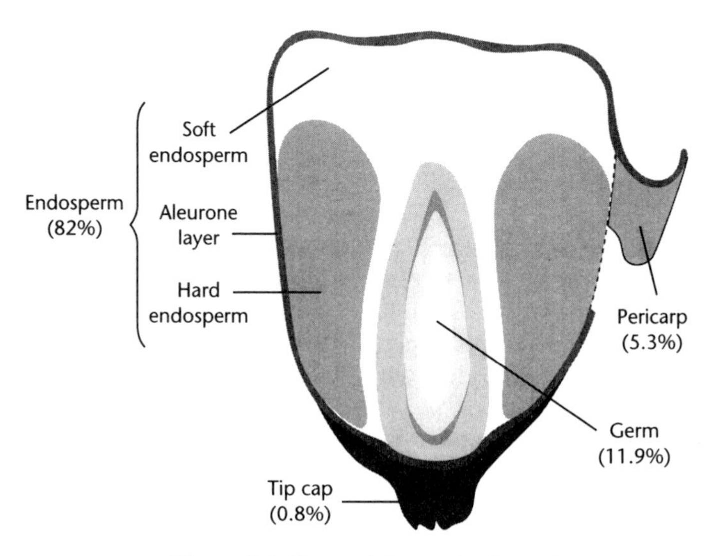

Figure 3.1 Parts of the maize kernel.

kernel as well as protects the kernel against microbial infection and insect infestation. The tip cap is a distinct part of the pericarp which plays a major role in seed germination and in the processing of the kernel. As the attachment point for the maize kernel to the cob, the tip cap area acted as the conduit for the transport of components into the kernel from the maize plant. A black layer of dense cells, known as the hilar layer, is laid down by the seed after maturation of the kernel in an attempt to seal this path into the kernel. The tip cap is the weakest area of the pericarp and is easily broken during the early stages of steeping, tempering or water uptake during germination to provide easier access into the internal structure of the kernel.

The germ is the living part of the kernel, containing the necessary genetic information for the propagation of the maize plant. The germ structure is complex and has been detailed by Wolf *et al.* (1952). From the perspective of processing of the maize kernel, the germ is important for two reasons. The first is that it is a concentrated source of oil. The second is that the germ has a higher rate of moisture absorption than other kernel components and acts as a passage way into the endosperm during water absorption (Ruan *et al.*, 1992)

The endosperm constitutes the major portion of the maize kernel and is composed primarily of starch granules encased in a protein matrix. Starch granules are approximately spherical in shape with the size of the granules ranging from 5–35 μm in diameter (Watson, 1987). There are two types of endosperm, hard (also called vitreous, horny or translucent) and soft (also called floury or opaque). The two types of endosperm are similar in composition although hard endosperm generally has a higher protein content. Structurally the soft endosperm is weaker and less dense. This region of the kernel collapses during dry-down because it cannot support the weight of the structure once the water is removed. Totally opaque kernels appear wrinkled because of this internal structural collapse.

The endosperm protein matrix is composed of alkali soluble prolamine protein known as glutelin. Embedded in the glutelin protein matrix are small spherical protein bodies composed of the protein zein. Zein is an alcohol soluble protein which has industrial applications as a coating for pharmaceutical capsules. These zein bodies apparently add little to the structural integrity of the endosperm but can be packed so densely into the hard endosperm protein matrix that they cause indentations on the surface of the starch granules.

3.3 TYPES OF MAIZE

The commercially important cultivars can be divided into five types: Dent, Floury, Flint, Popcorn and Sweet Corn.

Dent maize

Dent maize is the most widely grown type of maize constituting the majority of corn grown in the United States. Dent corn is a cross between flint and floury maize, which was found to have superior traits by settlers in the early 1800s (Benson and Pearce, 1987), and can have varying proportions of flinty and floury endosperm. Dent corn gets its name from a characteristic dent in the crown area of the kernel which occurs during maturation and drying. The dent results from the collapse of the protein matrix structure in the floury endosperm region, which is generally in the centre of the kernel. This region is too weak to support itself during dry down and collapses, pulling the pericarp at the crown toward the centre of the kernel. Dent maize hybrids with a small percentage of floury endosperm can exhibit little dent while hybrids with a high percentage of floury endosperm will look shrivelled and immature.

Floury maize

Floury maize is one of the oldest types of maize and is characterized by the lack of any hard or vitreous endosperm (Zuber and Darrah, 1987). The kernels are generally large flat kernels. During dry down the endosperm uniformly shrinks and as a result no denting of the kernel occurs. This type of maize is grown primarily in portions of South America.

Flint maize

Flint maize has a thick, hard vitreous endosperm which surrounds a small centre of soft or floury endosperm (Zuber and Darrah, 1987). During dry down there is no dent formed in the kernel due to the predominance of hard endosperm. Flint maize is grown extensively in Argentina and to a lesser extent in other South American countries and Europe.

Popcorn maize

Popcorn maize is an early variation of flint maize which has been selected over the years for the ability of the kernels to pop during rapid cooking. The individual kernels pop because of the build-up of internal pressure during rapid heating. Moisture in the hard endosperm vaporizes but is unable to escape due to the low diffusivity of the thick dense protein matrix encapsulating the starch granules (Hoseney *et al.*, 1983). The vapour congregates at a pin hole size open area located at the centre of each starch granule. When the total pressure in the kernel exceeds the mechanical strength of the kernel, there is a pop or release of water vapour from the kernel. Each individual starch granule explodes, expanding the hot protein matrix which surrounds the granule like a balloon. Evaporative cooling freezes the protein matrix in this expanded position with the gelatinized starch granule splattered on the interior surface of the protein matrix. Expansion of the kernel occurs to a bulk density of over 48 cm^3/g for premium quality popcorn. Hard endosperm dent or flint maize will also pop to a limited extent depending upon the diffusive character of the endosperm.

Sweet maize

Sweet maize is a genetic variant maize which inhibits the conversion of sugar into starch during kernel filling. The result is that there is an accumulation of sugar in the endosperm as the kernels develop. Ears harvested prior to denting (approximately 20 days post-pollination; Zuber and Darrah, 1987) and boiled have a sweet pleasing taste. Commercially, the sweet maize kernels can be cut off the ear and canned for preservation and consumption at a later date.

There are three classes of modifications to the dominant maize genetics which result in sweet maize: standard sweet maize, augmented-sugary sweet maize and other mutants as described by Marshall (1987). Standard sweet maize results from the selection of the sugary mutant gene (*su*) and results in the accumulation of sucrose and phytoglycogen in the endosperm. Standard sweet maize hybrids have lower total sugar content (approximately 16% sucrose) and are less sweet as compared to the other classes of sweet maize. Augmented-sugary sweet maize hybrids result from the use of genetic modifications in conjunction with the sugary gene to enhance the sugar content, including the use of the sugary *extender* gene (*se*). The other mutants class does not use the *sugary* (*su*) mutant but other mutants such as *shrunken-2* (sh_2) *brittle-2* (bt_2) and *brittle-1* (bt_1), to increase the sucrose content of the endosperm. These single mutants have higher sucrose content (approximately 35%) and have other characteristics which are desirable for commercial production. It is also possible to combine multiple mutants, such as *amylose extender* (*ae*), *dull* (*du*) and *waxy* (*wx*), to produce high sugar sweet maize.

3.4 REGIONS OF PRODUCTION

Worldwide production of maize is approximately 500 million metric tons (Anon., 1994). Nearly 50% of the world's production of maize occurs in North America with over 40% of the world's production occurring in the

United States corn belt. The corn belt consists of part or all of the 12 states which include or touch Illinois, Indiana and Iowa. Other major production areas include Northeast China, Eastern Europe (Baltic States), Brazil, Argentina, Thailand, France, Italy, Turkey and South Africa.

3.5 MAJOR MAIZE USES

3.5.1 Animal feed

The major use of the 500 million metric tonnes annual worldwide corn crop is as an animal feed, either directly or as part of a pre-processed feed. As the volume of corn processed by wet or dry milling continues to increase, more maize-based animal feed will result from the co-products of the milling processes.

Maize is not a complete animal feed because of the lack of the essential amino acid lysine. Lysine can be produced by microbial fermentation and as a supplement can reduce total feed cost by supplanting higher valued lysine sources such as soybean meal. The production of high lysine corn can also be a good means of providing the necessary lysine. Opaque 2 high lysine corn can produce 30–100% more lysine than normal dent hybrids (Mertz, 1970; Wilson, 1987), but the extremely soft endosperm of the corn makes it more susceptible to mechanical damage during handling and more susceptible to insect and microbial infection during storage. Quality protein maize (QPM), high-lysine corn with a hard endosperm, has been developed for use in human food and for animal feed (Mertz, 1992) and has better mechanical handling characteristics. The co-products of maize wet milling, gluten meal, gluten feed, germ meal and heavy steepwater all have considerably higher levels of lysine and other essential amino acids than does maize alone (Wilson, 1987).

Carotenoids, which are concentrated in the endosperm of yellow dent maize, are desired by poultry processors for their ability to provide a consistent appealing yellow colour to egg yolks and broiler skin. Maize is the only cereal grain which contains enough carotenoids to be useful for this purpose (Wright, 1987). White maize has almost no carotenoids. Corn gluten meal (60% protein, wet weight basis; w.b.) from the wet milling of maize is the preferred means of providing carotenoids to poultry diets in the United States. During wet milling, the carotenoids of yellow maize are concentrated approximately 12 to 1 into the gluten meal fraction since the carotenoids are retained with the endosperm protein (Wright, 1987).

Maize can be fed in whole form to most types of livestock but feed efficency increases if the maize is pre-processed to reduce particle size.

Pre-processing techniques include hammer milling, steam flaking and roller milling and they are often performed to allow for the more efficient blending of various feed components to form a complete diet. The blended feed can be given in granular form or can be pelletized for better utilization. For more information on feed processing and utilization see Wright (1987) and McEllhiney (1994).

Maize used for animal feed should be relatively free from mycotoxins (the acceptable level depending upon the type of toxin and the target animal), should have low levels of insect and microbial infection and should not contain high levels of damaged kernels. For many livestock feeders, the current international grading systems can be acceptable if mycotoxin concerns are addressed. However, as feed manufacturers become more sophisticated in their feed formulations, they are demanding more consistent quality and compositional characteristics. Some livestock feeders already have specifications on protein, oil and starch. Factors such as breakage susceptibility and stress cracks are also of interest because they indicate the intrinsic mechanical quality of the kernels. Broken maize can be difficult to handle properly through many feed processing steps.

3.5.2 Wet milling

Wet milling is the second largest use of maize with an estimated 50–60 million metric tons milled annually worldwide, with 33 million metric tons milled in the United States alone (Anon., 1994). Wet milling is performed primarily for the purpose of the isolation and recovery of starch which is used to produce food grade modified and unmodified starches, dextrose syrups, fructose syrups, ethanol and other chemicals via fermentation.

The process of wet milling (Figure 3.2) is capital and energy intensive but is economically viable due to the value of the starch and co-products. Because the co-products are more homogeneous in composition than can be achieved by other maize milling processes, they command a premium price. Wet milling can be divided into five sections: steeping, germ recovery, fibre recovery, protein recovery and starch washing.

Steeping is the heart of the process in that if the maize is not properly steeped it is difficult to get the desired quality and quantity of starch and co-products. Steeping is a biochemical, chemical and mechanical process whereby the kernel is prepared for separation. The primary objectives of steeping are to hydrate the kernel to 45–55% moisture (w.b.), induce differential swelling between the pericarp, germ and endosperm to loosen the connective tissue and to weaken the endosperm protein

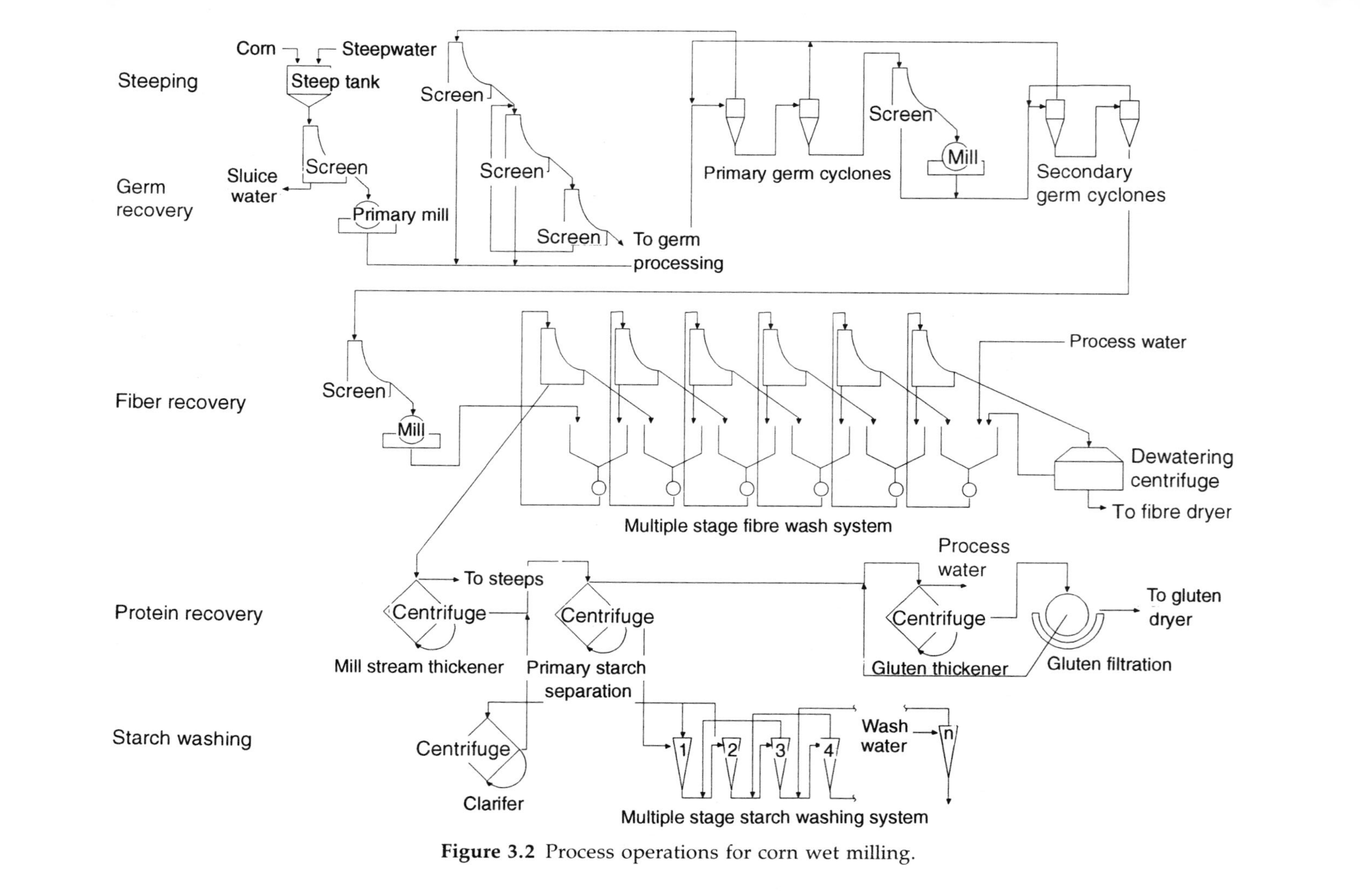

Figure 3.2 Process operations for corn wet milling.

matrix surrounding the starch granules by action of sulfur dioxide and endogenous protease.

Maize is steeped commercially for 20–48 hours in large stainless steel tanks, through which is sequentially pumped a steeping solution of sulfur dioxide and water. The steepwater is usually process water from the mill stream thickener containing solubles washed from the starch and co-products to which sulfur dioxide is added. During the steeping process sulfur dioxide is absorbed from the steepwater into the kernels where it diffuses into the interior areas of the endosperm. This sulfur dioxide activates proteolytic enzymes (Wahl, 1970; Eckhoff and Tso, 1991a) and breaks disulfide bonds in the protein matrix (Wahl, 1971). The steepwater, at 50–53°C, is pumped counter-currently through the tanks such that the oldest maize is exposed to the highest level of sulfur dioxide (usually 1500–2000 ppm) and the newest maize is exposed to the lowest level of sulfur dioxide (30–200 ppm).

The level of sulfur dioxide in the steepwater covering the new maize (this is the oldest steepwater) is too low to inhibit bacterial growth. The naturally occurring *Lactobacillus* sp. organisms tend to favour the steeping environment and the first 6–12 hours of steeping includes a fermentation which produces lactic acid to a level of 1.5–3.0% in the steepwater. This lactic acid increases the softness of the kernel and increases starch yield by improving the separation of starch from fibre (Eckhoff and Tso, 1991b). Du *et al.* (1996) found that other organic and inorganic acids can be substituted for lactic acid to achieve similar effects. The production of lactic acid stops when the level of sulfur dioxide in the steepwater gets to an inhibitory level, usually after 8–16 hours of steeping.

Once the steepwater passes through each of the steeptanks, it contains approximately 7–10% soluble solids of which 35–50% will be soluble protein. This product is known as light steepwater. Approximately half of these solids are carried into the steeptanks by the process water used to make the steepwater. The other half are either naturally occurring solubles in the maize kernel or are created during steeping due to the action of the sulfur dioxide. The light steepwater is usually evaporated in multiple effect vertical tube evaporators to a solids level of approximately 50% which is known as heavy steepwater. At this concentration the steepwater forms a hydrated complex which does not allow for further concentration via evaporation. Heavy steepwater can be used as a nutrient source for pharmaceutical fermentations or can be added to the fibre as part of the gluten feed fraction. When added to dry fibre, the hydrated complexes in the heavy steepwater are disrupted and the remaining water can be removed.

After the maize is properly steeped, it is removed from the tanks, dewatered over a stationary screen and lightly milled in a disc attrition mill to break the kernel apart without damaging the germ. The most common degerminating mills consist of two intermeshing knobby plates with one stationary and the other rotating. The resulting slurry consists of released starch, gritty endosperm pieces, germ attached to endosperm and/or pericarp, free germ and pericarp pieces with attached endosperm. This slurry is pumped through a hydrocyclone (15–23 cm diameter) which separates the free germ from the mash by use of the density difference between the oil laden germ and the endosperm and pericarp pieces. After steeping, the germ will contain 45–55% oil due to the leaching out of the soluble protein, sugars and minerals naturally occurring in the germ. A second degermination mill, set slightly tighter, and hydrocyclone recovery system is often used to increase total recovery of the germ. The recovered germ is washed, dewatered in a germ press and then dried to a moisture content of approximately 3% (w.b.) prior to oil recovery via expelling or expelling/extraction. The germ meal remaining after oil recovery is high in protein (approximately 22%, dry weight basis; d.b.) and can be added to the fibre fraction to produce gluten feed or sold as a separate feed product.

The degermed mash is finely ground using double disc attrition mills with the discs rotating in opposite directions or larger single disc mills. The objective is to impart mechanical energy into the mash to tear open endosperm cells and enhance starch release from the protein matrix.

After fine grinding, the mash is passed over a series of 50–75 μm pressure-fed screens and counter-currently washed with process water. The screens act to shear attached starch from the fibre as well as to wash unbound starch off the fibre. A six-stage washing system is common. The washed fibre is dewatered in a press and dried prior to the addition of germ meal and heavy steepwater.

The liquid which passes through the screens contains starch and protein. Separation of starch and protein occurs using a series of disc-nozzle centrifuges and starch washing hydrocyclones. A four-centrifuge system for protein/starch separation is generally used, although three-centrifuge systems have been used successfully. The first centrifuge in a four-centrifuge system passed through is known as a mill stream thickener. The mill stream thickener is used to thicken up the slurry for the primary starch centrifuge and to reduce the volume of water sent to the downstream processes. The water recovered from the mill stream thickener has a lot of solubles and is sent to the steeps for addition of sulfur dioxide and use as steepwater.

The primary centrifuge takes the concentrated starch–protein slurry and makes a split such that the overflow is protein rich containing approximately 70% (d.b.) protein and the underflow is starch rich containing 98–98.5% (d.b.) starch. The overflow is further concentrated in a third centrifuge known as a gluten thickener and dewatered using belt filters or decanting centrifuges. The process water from the gluten thickener is used either as germ or fibre wash water. The protein fraction discharging from the belt filters is approximately 42% moisture (w.b.) and is dried using a ring or flash drier to approximately 10% (w.b.). This product, known as gluten meal, is a highly valued poultry feed because of its intense yellow colour and high protein content.

The underflow from the primary centrifuge is sent to the inlet of the starch washing system, which is a series of 10 mm hydrocyclones connected so as to allow for counter-current washing with fresh water. Most of the hydrocyclones operate at a 60/40 underflow/overflow ratio. The overflow is routed back to the inlet of the previous hydrocyclone stage while the underflow is sent to the next stage. Each 10 mm hydrocyclone has a capacity of approximately 4 l/min and each stage of washing consists of hundreds of hydrocyclones in parallel, housed in a common shell to increase total capacity. It takes 12–14 stages of hydrocyclones to wash out residual protein and to decrease the protein level in the starch to 0.3–0.5% at an inlet fresh water wash rate of 0.9–1.4 l/kg of maize.

The protein in the overflow fraction is gradually washed upstream and discharges from the starch washing system as the overflow of the first stage hydrocyclone. This liquid fraction is sent to the fourth centrifuge, known as a clarifier. The clarifier thickens the protein-rich overflow from the starch washing system and sends the thickened underflow back to the primary centrifuge to try and recover the protein in the gluten meal fraction. The overflow water from the clarifier is low in solubles and is used to wash the germ or fibre products.

Starch slurry exiting the last stage of the starch washing system can be sent for chemical modification, dried to produce pearl starch or sent for liquefaction and saccharification for production into glucose or fructose syrups. Representative commercial facility yields are shown in Table 3.2. For more information see Watson (1984), Blanchard (1992), Johnson (1991) and May (1987).

Wet millers desire maize which is low in mycotoxins, has a large amount of starch and which can be easily processed. Mycotoxin levels are of particular concern in a wet milling facility because mycotoxins tend to concentrate in the feed fractions resulting from a wet mill (corn germ meal, corn gluten meal and corn gluten feed). The amount of concentration in each fraction depends upon the specific mycotoxin (Bennett and Anderson, 1978; Romer, 1984).

Test weight is important to a wet miller because steeping and first grind are volume limited processes. Wet milling of low test weight maize decreases total plant throughput even though low test weight by itself does not effect starch yield (Eckhoff and Denhart, 1994). Test weights appreciably above 56 lb/bu are not desirable because they

Table 3.2 Representative industrial wet milling yields

Fraction	*Yields (% d.b.)*		
	Blanchard (1992)	*May (1987)*	*Johnson (1991)*
Starch	67.5	66.0	67.5
Steepwater	6.5	7.0	7.5
Fibre	12.5	13.0	11.5
Germ	7.3	7.9	7.5
Gluten meal	5.4	5.7	5.8
Loss	0.8	0.4	0.2

indicate the presence of more hard endosperm. Hard endosperm is denser than soft endosperm and inhibits the penetration of sulfur dioxide, subsequently resulting in the need for longer steep times. Totally soft (opaque) maize can be adequately wet milled in less than 12 hours (Fox and Eckhoff, 1993) while flint type maize containing totally hard endosperm may require nearly 60 hours to wet mill.

Starch content is not equivalent to starch yield. High starch content is required for high starch yield but high starch content alone will not ensure high starch yield. There are a variety of factors which can inhibit starch recovery including light or small starch granules, irregular shaped granules and binding between the starch and protein matrix. Figure 3.3 shows the range of starch yields from 131 samples representative of the hybrids grown in the US corn belt.

Mechanical damage (any damage which results in exposure of the endosperm) increases the soluble level in the steepwater (Wang, 1994). These increased solubles are lost into the feed fraction and can cause evaporator fouling problems as well as foaming. Broken maize has the same problems as mechanically damaged maize but it also poses a problem in that it tends to segregate during filling of the steeptanks and can result in non-uniform steeping of the product and plugging of the steeptank screens.

The procedures used to dry the maize are very important in determining the wet milling quality. Kernel drying temperatures above the gelatinization temperature of starch (approximately 60°C) can result in starch gelatinization, protein denaturation and loss of endogenous protease activity (Kerpischi, 1988; Eckhoff and Tso, 1991a). The initial moisture content (Vojnovich, 1975) and the rate of drying (Lasseran, 1973) have also been shown to affect starch recovery. A rough estimate is that starch yield will decrease 1–2% for every 10°C above 60°C.

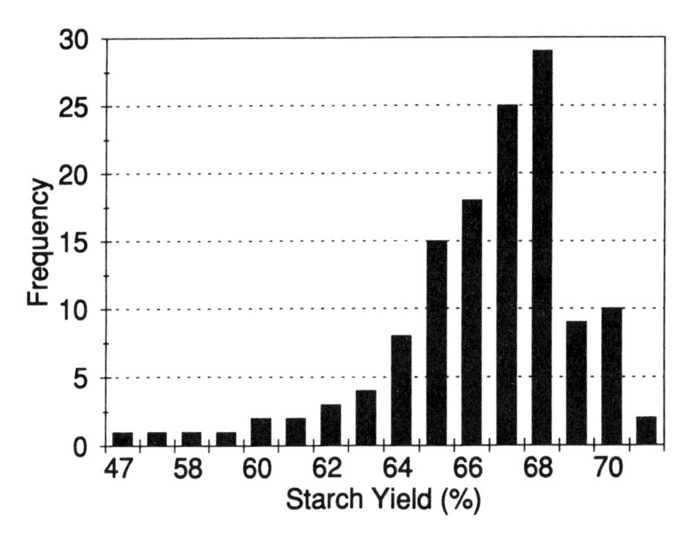

Figure 3.3 Starch yield distribution for 131 commercial yellow dent hybrids grown in the US cornbelt. (Reprinted from Eckhoff, 1995, with permission.)

Excessive drying also causes problems in wet milling beyond the loss of starch. High temperature dried maize hydrates more slowly and to a lower equilibrium value. The maize is also harder to grind, has more brittle germs (resulting in lower oil recovery) and the resulting starch will have lower peak amylograph values and may be more difficult to modify chemically.

Low germination of the kernels can result from age, high temperature drying, or microbial or insect infestation. Loss of germination results in some irreversible enzyme changes within the kernel which makes the kernel harder to steep and mill resulting in lower starch and oil recovery. The loss of germination with age appears to be related to the natural cycling of temperature and moisture which occurs at most ambient storage conditions. Experience has shown that constant temperature refrigeration of laboratory samples results in consistent starch yields throughout an 18-month storage period while industrial experience indicates that starch yields will decrease with storage time. A kernel age related condition known as 'new crop phenomenon' is probably caused by increased levels of naturally occurring endogenous proteases shortly after harvest. Millers find that maize recently harvested can create problems in the plant due to excessive foaming in the steep tanks and in downstream processes. The elevated levels of protease creates more water soluble protein during steeping causing foaming. Usually within 2 months of harvest, the new crop phenomenon disappears, due to natural loss of this protease activity. Intensity of the problems associated with the new crop phenomenon vary year to year.

3.5.3 Dry milling

Maize dry milling is a term that is often used to refer to one of three different processes. The first process is the oldest and simplest process for milling maize. This process, non-degerminated dry milling, grinds the whole kernel into a flour consistency using either a stone grinder or a metal roller mill. The presence of ground germ in the flour decreases storage life due to oil rancidity, although the germ tends to increase the flavour of the flour.

The non-degerminated dry milling process requires no tempering or separation of maize components. The dry maize is fed directly into the mill where it is ground to a flour. The ground material can be sifted and reground if necessary to achieve the desired consistency. This process is used in some cultures due to the lack of equipment for degerminating dry milling but is often used due to taste preference as the oil imparts a more distinctive maize flavour to the products. This process is used as preparation for some of the ethnic foods discussed later in this chapter.

The second process which is often called dry milling is the dry grind ethanol process (Maisch, 1987). In the dry grind ethanol process, whole kernels are ground using a hammer mill for the primary purpose of size reduction. The ground maize is then cooked, saccharified and put into a fermentor for conversion of the starch to ethanol. The grinding of the maize increases total surface area and increases the conversion rate.

The third process, degerminating dry milling, is a process in which the maize is initially tempered to facilitate separation of the germ, pericarp and endosperm pieces. The major steps of the dry milling process (Figure 3.4) are tempering, degermination, drying and separation.

During tempering, dry maize (12–15% moisture, w.b.) is rapidly increased in moisture to 18–24% (w.b.) by using hot or cold water or low-pressure steam. The pericarp and germ preferentially absorb the moisture during the first 15–30 minutes of tempering, which toughens them so that they do not shatter during the subsequent mechanical impact forces of degermination. Because of the moisture absorption, the germ and pericarp swell, creating shearing stresses on the connecting tissue holding the pericarp to the endosperm, pericarp to the germ and germ to the endosperm. This differential swelling enhances the separation between components in dry milling. Long tempering times, as used in wheat milling, result in more uniform distribution of moisture throughout the kernel and subsequently poorer separation.

Degermination of the tempered maize occurs by using a Beall type degerminator, an Entoleter, granulators, disc mills, roller mills (Brekke,

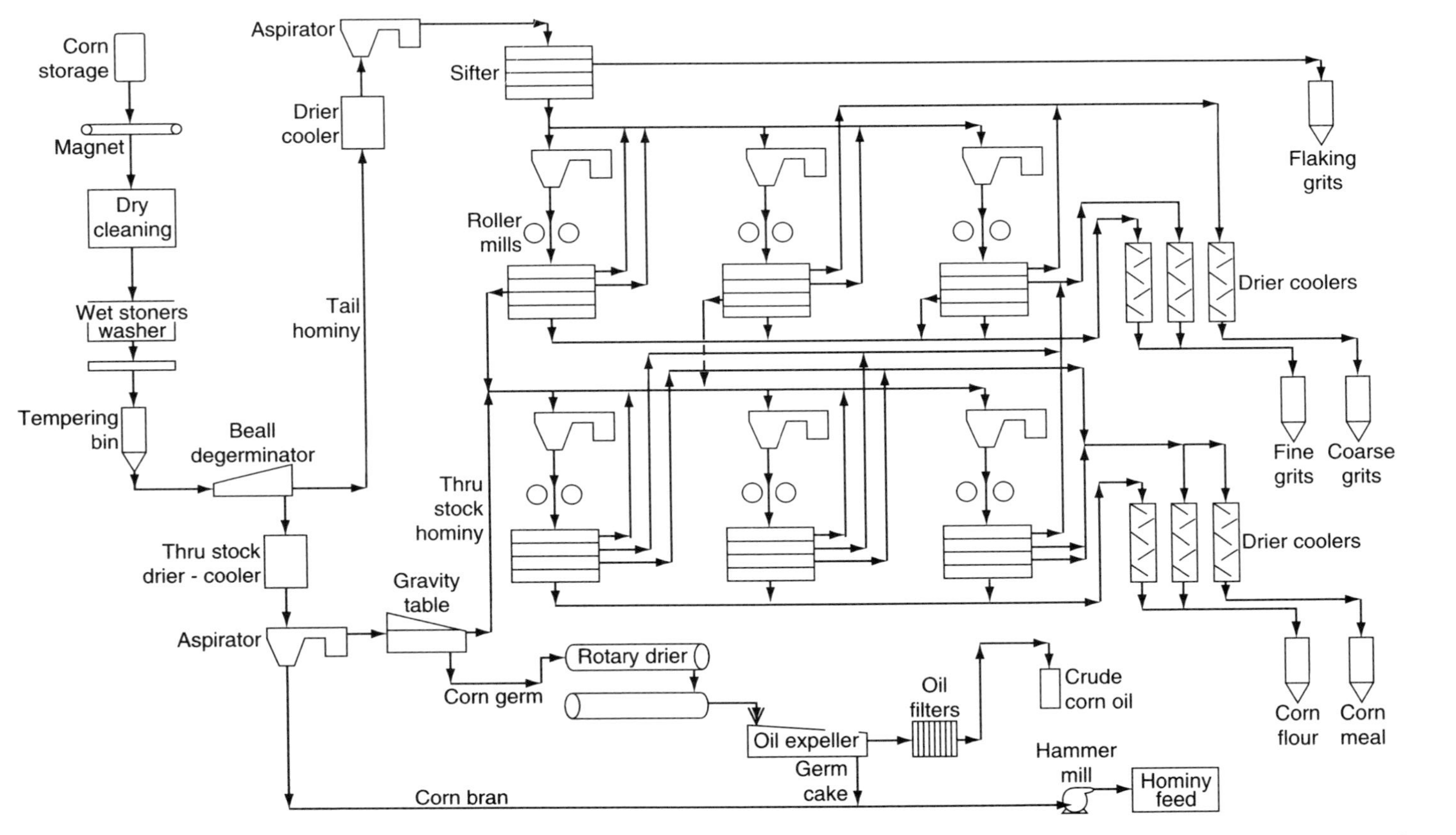

Figure 3.4 Production flow chart for a typical corn tempering–degerminating system. (Reprinted from Alexander, 1987, with permission.)

1970) or a decorticator. The Beall type degerminator is preferred for the making of large grits because of the combined shearing and impact action which occurs in the machine, the ability to change the machine clearance and the separation of the product into two distinctly different streams.

Once the maize has been passed through some type of degerminating equipment, the remainder of the process is spent on separating the components into as pure fractions as possible. The tempering and degermination steps are by far the most important in dry milling because poor degermination and/or tempering cannot be compensated for by the downstream separation processes. Figure 3.4 shows a representative process flow for separation using roller mills, aspirators, gravity tables and sifters. Drying is performed at various steps through the process in order to enhance separation characteristics. The selection of a specific process flow depends upon the available equipment and the desired product mix.

There are five main classification of products resulting from a dry mill: germ, hominy feed, grits, meal and flour. The germ is a fraction constituting 8–15% of the maize resulting from either separation on a gravity table or by roller milling the undried through stock on a Beall degerminator. The germ can either be added directly to the hominy feed fraction or the oil in the germ fraction can be recovered by either expelling and/or solvent extraction, depending upon economics. Hominy feed is the low valued fraction constituting 25–40% of the total dry material weight. It is composed of the pericarp (fibre) fraction, spent germ (oil removed), broken corn from the cleaners and high fat fine material which cannot be readily marketed. The fibre fraction is recovered by the use of aspiration.

The grits, meal and flour are all pieces of endosperm, varying in size from large (grits) to small (flour). Table 3.3 lists a classification of products by sieve size, but the industry can produce a wide range of products depending upon the requirements of the end-user. The major requirements imposed by the end-user usually involve particle size distribution and fat (oil) content. One widely used product not shown in Table 3.3 is maize cones, which result from roller milling of grits to produce a small size meal product (−40+80; through a 40 mesh screen, retained on an 80 mesh screen). Because the product is produced by the reduction of hard endosperm, it has different characteristics to meal resulting from soft endosperm. Cones are used in a variety of extruded maize products. Table 3.3 also shows representative product yields reported from several sources.

Quality plays a major role in the cost-effectiveness of the maize dry milling process. The most valuable pieces of the kernel to the dry miller are the large endosperm pieces. The large pieces can be reduced in size

Table 3.3 Representative production yields from degerminated dry milling

		% Yield		
Product endosperm‡	*Sieve Size**	*Regular Yellow Dent*†	*Regular*	*Hard*
Flaking grits	–3.5 + 6	12	6.0	14.7
Coarse grits	–10 + 14	15	–	–
Regular grits	–14 + 28	23	29.0	20.8
Coarse meal	–28 + 50	3	–	–
Fine meal	–50 + 75	4	16.0	17.2
Flour	–75 + Pan	4	6.0	9.9
Oil	NA	1	–	–
Hominy feed	NA	35	43.0	37.4

*Brekke (1970).
†Brekke (1970).
‡Eckhoff (1987); product classifications are approximate.
NA, Not Applicable.

to produce any of the endosperm products desired but high fat/high fibre flour is undesirable and has market value primarily related to animal feed prices. Damage to the kernel which results in smaller endosperm pieces has a negative effect on profitability.

Soft endosperm has little structural integrity and in dry milling it primarily ends up as undesirable high fat degerminator flour. Most of this is not easily recoverable and is often lost to the hominy feed fraction. Dry millers desire hard endosperm dent maize to achieve maximum recovery of the primary products of low-fat grits and meal. The hard endosperm is less likely to break apart during handling or during degermination resulting in larger endosperm pieces.

Some US and international maize dry millers purchase maize based upon endosperm hardness either by use of test weight premiums, producer direct contract purchasing of specific hybrids or by other testing methods. There is considerable hybrid variability as to the acceptability of the maize for dry milling although it is generally accepted that the higher the percentage of hard endosperm, the better the milling characteristics. Flint maize dry mills differently from hard endosperm dent maize primarily because of the small kernel size, a round kernel shape and different moisture diffusion characteristics. The smaller kernel size results in smaller grits and the round kernel shape makes degermination more difficult.

Dry millers are very sensitive to microbial deterioration of the maize due to the potential for mycotoxins and the increased levels of microbial cells. Mouldy maize increases the mold level throughout the processing plant potentially resulting in sanitation problems. The higher microbial levels also decrease the potential storage life of the final product (Mills and Pederson, 1990; Bothast *et al.*, 1974). According to Bothast *et al.* (1974), the average microbial count in dry milled maize in the US is less than 5000 counts/g with levels above 10 000 counts/g considered unacceptable. Mycotoxins in dry milled maize distribute throughout the various products based upon the specific toxin. However, unlike wet milling where most of the toxin concentrates in the animal feed fractions rather than in the starch, the prime products of dry milling can retain significant levels of various mycotoxins.

Drying temperature is also a major concern to dry millers since rapid drying results in the formation of stress cracks. These stress cracks weaken the endosperm structure and affect the number and size of resulting endosperm particles during degermination. Some single stress cracks can occur naturally in the field in hard endosperm maize but dual or multiple stress cracks result from an excessive rate of drying.

3.5.4 Masa production

Masa is the dough material resulting from the fine grinding of alkali (lime) cooked maize which can be sheeted, cut and baked or fried into tortillas, taco shells, tortilla chip or maize chips. Alkali cooked maize is the major source of nutrition in Mexico and other Central and South American countries.

Production of masa results from mixing maize with lime (approximately 0.5% by weight of maize) with water in a water:maize ratio of 1.5–3.0:1 (Rooney and Serna-Saldivar, 1987). The mixture is cooked at near boiling temperatures (85–95°C) for 20–60 minutes, depending upon various process and kernel characteristics. The cooked maize is allowed to cool overnight and the cooking liquor, known as nejayote, is decanted off and discarded. The cooked whole maize, known as the nixtamal, is fully hydrated and the pericarp has been partially degraded by the lime cooking. Residual lime and any detached pericarp is washed off the nixtamal using fresh water and the water is discarded. The washed nixtamal is ground to a moderately fine consistency using either lava or aluminium oxide stone grinders. Stone grinders impart specific desirable characteristics to the resulting masa (Rooney and Serna-Saldivar, 1987; Serna-Saldivar *et al.*, 1990b).

The cooking of the masa results in softening of the kernel, chemical loosening of the pericarp, partial solubilization of the glutelin endosperm protein matrix and partial gelatinization of the starch. The gelatinization which occurs causes the masa to become sticky which helps hold together the gritty parts (ungelatinized pieces) of the endosperm. Overcooking results in too much gelatinization of starch and the masa loses the desired gritty texture. Excessive loss of solubles and difficulty forming sheets also occurs from overcooking. If the maize is undercooked, the kernels are difficult to grind, the masa is too gritty and the masa will not hold together during forming of the tortilla (Serna-Saldivar *et al.*, 1990a).

Probably the most important quality factors to masa producers are stress cracks, pericarp damage, broken maize and endosperm hardness. Kernel to kernel variability in endosperm hardness (caused by blending of hybrids of different endosperm hardness) or variability within a kernel (moderately hard dent corn) results in non-uniform cooking. Some kernels (or parts of kernels) will be undercooked while others will be overcooked. Masa producers prefer the kernel hardness to be consistent and not a blend of kernel characteristics. They prefer totally hard endosperm dent corn because the cooking characteristics are more uniform and because the handling and mechanical characteristics of the maize are improved. It is possible to produce a good quality masa from opaque or totally soft endosperm corn by decreasing the cooking time.

Pericarp damage, stress cracks and broken maize change the diffusional pathways into the kernels and result in accelerated cooking of those kernels. Broken kernels also leads to excessive solubilization of solids. All other factors equal, hard endosperm corn decreases the amount of broken corn resulting from mechanical handling. Stress cracks are generally the result of a fast drying rate.

Ease of pericarp removal is a characteristic upon which some masa producers select their preferred hybrids. There is considerable variability between hybrids in the ease of removal of the pericarp during the washing of the maize. Residual pericarp is always present in masa but too much pericarp affects the sheeting and forming operations (Rooney and Serna-Saldivar, 1987).

General sanitation of the maize used in masa production (insect and microbial infestation) is important. Insect infestation or microbial infection in incoming maize affects the overall plant sanitation, reduces the storage life of the maize and shelf-life of the product, increases the risk of mycotoxin production and decreases product palatability.

3.5.5 Popcorn

Popcorn is used as a snack food and can be popped and coated to provide a range of flavoured products. In the United States, the majority of popcorn is popped at home by the ultimate consumer and is consumed warm with the addition of salt and/or butter. Home size mechanical poppers, hot air or hot oil, are in wide use although the trend is to pop the maize in a microwave using prepackaged portions. Microwave popcorn can be packaged with a variety of flavourings which have enhanced the worldwide acceptance of popcorn as a nutritious snack food.

Popcorn pops due to the pressure of water vapour trying to escape from the kernel during rapid heating. The endosperm of popcorn is dense and inhibits the diffusion of the vapour resulting in a build-up of pressure within each starch granule (Hoseney *et al.*, 1983). Maximum expansion depends upon several genetic and quality factors. The shape of the kernel needs to be rounder than normally occurs with dent maize in order to maximize the diffusional path lengths for vapour escape. Song and Eckhoff (1994a) and Song *et al.* (1991) found that larger kernels had higher expansion values and less unpopped kernels (UPK).

The pericarp also plays a major role in the complete expansion of the kernels since it acts as a diffusional barrier requiring water vapour to diffuse all the way down to the tip cap. Single popcorn kernels dropped into hot oil will gyrate or move in the oil due to the propulsion by the water vapour escaping out of the tip cap. Any pericarp damage decreases the expansion volume but the pericarp itself is not necessary for a kernel to pop. The pericarp can be mechanically abraded away and the kernel will still pop to a limited extent.

Factors which influence the effective popping of kernels include the moisture content and the rate of heat transfer. Optimal moisture content depends upon kernel size but is in the range of 12.8–13.5% (w.b.) (Song and Eckhoff, 1994a). There is also natural variability in the popping characteristics of the kernels dependent upon their location on the ear (Song and Eckhoff, 1994b). The precise optimal moisture will be influenced by genetics and the type of popping procedure (hot air vs. oil popping vs. microwave). If there is too little moisture, there is insufficient vapour to allow for effective popping. Higher moisture contents do not cause as significant a decrease in popping volume as occurs at lower moisture but at higher moisture, storability becomes a problem due to potential microbial damage. If the rate of heat transfer to the kernel is too slow, the water vapour has a longer period of time to diffuse out of the kernel, and there is an insufficient pressure build up in the starch granules.

Since popcorn is consumed directly, the quality factors of foreign material, mould infestation and insect infestation are important and should be minimized, with zero defects the target. A hard endosperm texture is required to create the low water vapour diffusivity and thus high test weight (>60 lb/bu) is common and desirable. Since the popcorn is consumed directly as human food, the corn should be tested for all mycotoxins of concern.

3.5.6 Other food uses

Maize is widely used in Asia, Africa, Central America and parts of the former Soviet Union for direct food consumption. According to Rooney and Serna-Saldivar (1987), the maize can be made into bread (arepa, tortilla, corn bread, hoe cake or blintzes) and can be either fermented or unleavened. It can be made into either unfermented or fermented porridges (atole, ogi, kenkey, ugali, ugi, edo pap, maizena, posho or asidah), steamed products (tamales, couscous, Chinese breads, dumplings and chengu), alcoholic beverages (koda, chicha, Kaffir beer or maize beer), non-alcoholic beverages (mahewu, magou or chica dulce) or snacks (empanada, chips, tostadas or fritters).

There are some general quality characteristics important for food use maize regardless of the specific product being produced. The maize should be free from visual microbial or insect deterioration, should consist of intact sound kernels with minimal pericarp damage and should be essentially free of microbial toxins. Microbial or insect deterioration decreases the palatability of the resulting products, creates sanitation problems and decreases the shelf-life of the product. Pericarp damage or broken maize are undesirable because most of the cooking or processing steps in the production of products for human food use are diffusion-limited. Kernels with pericarp damage or broken kernels have different diffusional and cooking characteristics resulting in less product uniformity and in some cases undesirable product consistency.

Microbial toxins result from improper storage conditions and/or from environmental factors during maize production. There are a variety of such mycotoxins (Frayssinet, 1988; Wilson and Abramson, 1992) which can result in sickness or death if consumed. Testing for these mycotoxins is important to ensure the safety of the food chain. Additional quality factors for specific food use products may also be important but they probably relate to characteristics similar to those needed for wet milling, dry milling or alkali processing, depending upon the specifics of the process used. For example, the quality factors important for the commercial production of arepa in Venezuela are similar to those needed for commercial dry milling.

Sweet maize is another direct food use of maize and is consumed either directly cooked on the ear or cut off the ear and frozen or canned. On the ear the maize is boiled in hot water until the kernels are softened and hot. The ears are buttered and salted to taste and the kernels are eaten directly off the ear. Sweet maize for canning is mechanically harvested, the kernels cut off the ear, canned and heat-treated for preservation. Similarly, sweet maize can be cut off the ear, packaged in plastic bags and frozen for preservation. Quality factors for sweet maize are similar to those for general food use maize.

3.6 LABORATORY METHODS FOR QUALITY MEASUREMENT

3.6.1 Grading standards

US maize grading standards

Maize (*Zea mays* L.) or corn as it is commonly called in the United States, is routinely inspected and graded to provide orderly marketing and an indication of its intrinsic value. The US Grain Grading Standards were established in 1916 by the Federal Grain Inspection Service (FGIS). They will be discussed initially as a means of providing background for maize grading standards of other countries which have many similarities. Over the years, the purposes of the US Grain Standards have evolved to: (i) define uniform and descriptive terms to facilitate trade; (ii) provide information to aid in determining grain storability; (iii) offer end-users the best possible information from which to determine end-product yield and quality; (iv) create a means for the market to establish improvement incentives.

There are four factors in the US maize grading standards: broken maize and foreign material (BCFM), test weight, heat-damaged kernels and total damaged kernels. These factors determine grades 1 through 5, as shown in Table 3.4. Any objectionable odour, certain weed seeds, or other foreign substances, can cause the maize to be graded the lowest possible grade: sample grade.

The BCFM grading factor is the percentage of fine maize particles and other material that will pass through a 4.76 mm round-hole sieve plus all matter other than maize, such as small pieces of cob or stalk, that remain on top of the 4.76 mm sieve after sieving (Anon., 1990c).

Test weight is a grade factor that indicates bulk density. It is determined by filling a quart cup level full from a funnel hopper placed 2 inches above the top of the cup. The contents are weighed, providing grams/quart which is converted to kg/hectolitre or to lbs/bu. Heat-damaged kernels represent the percentage of maize kernels by weight which are materially discoloured and damaged by heat. Total damage includes heat-damaged kernels and damage due to heat generated by fungal metabolism. Total damage also includes kernels and pieces of maize kernels that are badly ground-damaged, weather-damaged, damaged by fungi, insect-bored, frost-damaged, germ-damaged, sprout-damaged, or otherwise materially damaged (Anon., 1990c).

The lowest of any of these four factors determines the maximum grade given.

Maize grading standards of other countries

Argentina, South Africa and People's Republic of China are major maize exporting countries, as well as the United States. The Argentine maize

Table 3.4 Grades and grade requirements for maize (Anon., 1990c)

Grade	*Maximum limits of:*			
	Minimum test weight per bushel (lb)	*Damaged kernels*		*Broken corn and foreign material (%)*
		Heat damaged kernels (%)	*Total (%)*	
US No. 1	56.0	0.1	3.0	2.0
US No. 2	54.0	0.2	5.0	3.0
US No. 3	52.0	0.5	7.0	4.0
US No. 4	49.0	1.0	10.0	5.0
US No. 5	46.0	3.0	15.0	7.0
US Sample grade				

US Sample grade is maize that:
(a) Does not meet the requirements for the grades US Nos. 1, 2, 3, 4, or 5; or
(b) Contains eight or more stones which have an aggregate weight in excess of 0.20% of the sample weight, two or more pieces of glass, three or more crotalaria seeds (*Crotalaria* spp.), two or more castor beans (*Ricinus communis* L.), four or more particles of an unknown foreign substance(s) or a commonly recognized harmful or toxic substance(s), eight or more cockleburs (*Xanthium* spp.) or similar seeds singly or in combination, or animal filth in excess of 0.20% in 1000 g; or
(c) Has a musty, sour, or commercially objectionable foreign odour; or
(d) Is heating or otherwise of distinctly low quality.

Table 3.5 Grading factors and maximum allowable percentages for Argentine maize (Paulsen and Hill, 1985; Bender *et al.*, 1992)

Grade	*Damaged kernels (%)*	*Broken kernels (%)*	*Foreign material (%)*	*Minimum test weight (kg/hl)*
1	3	2	1	75
2	5	3	1.5	72
3	8	5	2	69
4*	12	9	2	–

*Grade 4 is added only in years when damage levels exceed Grade 3.

grading standards consider four factors as shown in Table 3.5. Test weight was recently added in 1991. Tables 3.6 and 3.7 show grading standards for South Africa and China, respectively.

3.6.2 Moisture content

Moisture content is important for determining whether maize will store safely and in marketing to determine the amount of dry matter actually

Table 3.6 Grading factors and maximum allowable percentages for South African Yellow Maize (Hill, 1992)

Grade	*Defective kernels* (%)*	*Foreign material (%)*	*Other colour kernels (%)*	*Pinked labels (%)*	*Deviations[†] (%)*
1	9	0.30	2	12	7
2	20	0.50	5	12	13
3	30	0.75	5	12	25

*Defective kernels are those kernels and pieces of kernels that pass through a 6.35 mm round-hole sieve plus any damaged kernels from fungi, insects, punctures, etc.
[†]Deviations where the value given represents the maximum sum of defective kernels, foreign material, and other coloured kernels allowed for that grade.

Table 3.7 Maize grading factors for People's Republic of China, PRC Standard UDC 633:15 (1986). All classes of maize use Grade 2 as an intermediate standard

Grade	*Purity minimum* (%)*	*Impurity[†] (%)*	*Moisture content[‡] (%)*	*Colour*
1	97			
2	94	1	14/18	Normal
3	91			

*Sources of impurities are immature kernels, insect-damage, sprout-damage, mould-damage, heat-damage, broken kernels with damage to endosperm or germ.
[†]Impurities are similar to foreign material and represent materials which pass through a 3.0 mm round-hole sieve, inorganic impurities such as soil, pellets, organic impurities such as other grains.
[‡]Maximum moisture content is generally 14%, but for some northeastern and Mongolia geographical areas it is 18%.

bought or sold. Initially, the measurement of moisture in grain would appear to be an easy task. Moisture measurement methods fall into three categories. Those categories are: (i) basic reference methods used to determine the absolute moisture in the sample; (ii) practical reference methods, usually oven tests, which are calibrated to basic reference methods; (iii) rapid methods which are usually electronic meters or thermal methods. A review of these methods is provided in Multon (1991).

Moisture content in grain is usually expressed on a wet basis, MC_{wb}:

$$MC_{wb}\% = 100\ W_w/(DM + W_w)$$

where W_w is the mass of water contained in the sample; *DM* is the mass of dry matter in the sample.

Multon (1991) stated that moisture is actually adsorbed and is contained in cell tissues. A practical definition of moisture is 'the quantity of water lost by a substance when it is brought to true equilibrium with a water vapour pressure of zero, in conditions where possible perturbing reactions are avoided'. This means moisture is considered to be zero when exposed to 0% equilibrium relative humidity for an extended period of time. The complicating factor is that water can still be extracted from this 'dry' grain if starch was oxidized which would give off H_2O as a byproduct of the reaction.

The basic reference methods typically involve either a P_2O_5 method or the Karl Fischer method. The P_2O_5 method creates a 0% equilibrium relative humidity condition by heating a ground sample in a vacuum oven at 50°C in the presence of a strong desiccant (P_2O_5). The drying time takes several days to a month, but the repeatability is about 0.05% moisture (Multon, 1991).

The Karl Fischer method involves grinding a sample in anhydrous methanol in a sealed container for a specified time. The water–alcohol extract is titrated with Karl Fischer reagent. The amount of Karl Fischer reagent used is a measure of the amount of water present. With this method, moisture is defined as that water removed by alcohol extraction. An advantage of the Karl Fischer method is that it can be run in about a half hour.

For the practical reference methods, oven drying is usually used. The common air oven method used in the US for maize is the AACC Method 44-15A which uses a 103 ± 1°C oven for 72 hours. Fifteen grams of unground samples are dried in tared aluminium containers. After drying, the containers are removed, lidded and placed in a desiccator to cool to room temperature. The containers are weighed to the nearest 0.1 mg. Replicate determinations should check within 0.2% moisture (Anon., 1990a, 1992).

The ICC oven reference method uses ground maize at 130°C for 4 hours. The repeatability of the method is between 0.05 and 0.1% (Multon, 1991).

The advantage of the practical reference methods is that they are simpler and can be calibrated to the basic reference methods. The AACC air oven method is typically within 0.8% of the P_2O_5 and Karl Fischer methods.

For the rapid methods of moisture measurement, most moisture meters use the dielectric principle (capacitance), resistance, a combination of dielectrics and resistance, or infrared energy absorbance. All of these methods are calibrated to oven reference methods. Since the water molecule is made up of two positively charged hydrogen atoms and one negatively charged oxygen atom, the molecule is said to be dipolar. Thus, in the presence of an electric field between two parallel plates, the molecules will try to line up with the plus dipole toward the negative voltage side and the negative dipole toward the positive side (Funk, 1991). The dielectric constant of a material is a measure of the ability of that material to store energy when exposed to an electric field. The dielectric constant of water is 80, which is about 20 times greater than that of other constituents in maize, making water relatively easy to measure.

Some of the commonly used moisture meters in the US are the Motomoco 919, approved by the Federal Grain Inspection Service of USDA, the Dickey-john GAC meters, and Steinlite meters which use the dielectric principle. In addition, moisture adjustments are made for variations in grain temperature and bulk density of the samples tested. Some of these meters also use conductance measurements. There are at least two single kernel meters, by Shizuoka Seiki and Seedburo, which use the resistance principle (Bonifacio-Maghirang *et al.*, 1994).

For the infrared energy absorbance method, NIR (near infrared reflectance) or NIT (near infrared transmittance) instruments are used. By measuring light reflected from or transmitted through a sample that absorbs selective wavelengths of light (1640 or 1910 nm for water) in proportion to the quantity of water present, moisture can be determined.

3.6.3 Maize hardness

Maize hardness is defined as the amount of vitreous endosperm in the kernel relative to the amount of floury endosperm present in the maize kernel. Hardness is affected primarily by variety and to a lesser extent by soil fertility and growing conditions. Flint type varieties are generally much harder than dent varieties. In general, high soil nitrogen and drought-like growing conditions tend to increase kernel hardness somewhat.

The reason hardness is important is that for dry milling, harder endosperm kernels result in higher percentages of prime product, i.e. large grits suitable for flaking grits (Paulsen and Hill, 1984). In wet milling, too hard endosperm tends to require excessively long steeping times and too soft endosperm breaks too easily, with the undesirable result of having more starch in the steep water.

There are numerous tests which have been used to detect hardness. Among the quickest are those based on density differences: hard endosperm maize is higher in density than soft maize and the density tests will be discussed in the following section. Other tests that can be used are visual examination, the Stenvert hardness tester and NIR.

Visual inspection gives a subjective evaluation of hardness which can be done with whole kernels or by cutting transversely or longitudinally. Techniques using machine vision for detecting kernel hardness have been developed (Liao *et al.*, 1991). In the Stenvert hardness test, a Stenvert hammer-mill grinder is used to grind 20 g of maize and the time is measured to fill up a collection tube to a certain volume (17 ml). Hard endosperm maize takes longer to grind and has higher percentages of coarse particles than soft endosperm maize. Near Infrared Reflectance (NIR) uses ground samples of maize. Hard endosperm maize grinds into coarser particle sizes than soft maize, resulting in less reflectance with NIR.

3.6.4 Density

Density of maize is an indication of hardness and of maturity. It is measured both as bulk density (test weight) or as true density of individual kernels. Variations of the true density test involve floaters tests. True density is a very important test for use by both wet and dry corn millers.

Test weight is affected by many factors, including moisture content, frost damage, maturity, growing conditions, drying conditions, fine material, kernel damage, and variety. Test weight is determined by weighing the volume of maize contained in a level full quart cup. The mass in g/quart is converted to lb/bu or kg/hl (56 lb/bu = 72.08 kg/hl). Thus, all of the above factors which can affect void spaces between kernels as well as kernel density complicate the true meaning of test weight.

True density determines only the g/cm^3 of the kernel itself, which more accurately represents maize hardness characteristics. True density may be determined using an alcohol column filled with a pre-weighed 100 g of sample. The mass of the maize divided by volume displaced by the 100 g of maize represents the true density. True densities at about 15% moisture content range from about 1.19 to 1.34 g/cm^3. Maize with densities above 1.26 to 1.28 g/cm^3 are generally considered to be reaching into the hard maize category. Another method for determining true density is to use a helium-air displacement pycnometer where the volume of gas displaced by the sample is compared to gas displaced compared to a known volume.

Another measure of hardness include the floaters test in which the number of kernels floating in a 1.275 specific gravity solution of sodium nitrate or other suitable chemical is determined. The higher the percentage of floaters, the softer the endosperm. The percentage of floaters is compared to a chart which also accounts for the natural tendency of having more floaters as moisture content increases (Wichser, 1961).

3.6.5 Whole kernels

The percentage of kernels that are whole and fully intact can vary widely. Depending on the country, grading standards may not account for non-whole kernels either as damage or as fine material if the non-whole kernel is large enough to be retained by the sieve used for determining fine material. Paulsen *et al.* (1989) found average percentages of whole kernels in two export shipments to Japan ranged from 68–81% for US No. 3 yellow maize.

Percentages of whole kernels are determined by weighing 50 g of a random sample and hand sorting for kernels that are completely intact with no pericarp damage. Whole kernel percentages are decreased by combine harvesting when maize is wet (28% moisture and higher), and by the severity of mechanical handling after harvest.

3.6.6 Pericarp damage

Pericarp damage to maize kernels is the primary type of damage that causes kernels to be non-whole. Pericarp damage consists of cracks, cuts, abrasions, and chips or pieces of missing endosperm. A leading source of pericarp damage is mechanical harvesting when the crop is wet, over about 28% moisture content (Paulsen and Nave, 1980). Less than 10% pericarp damage is generally considered very good. Pericarp damage provides an opening for easy fungal invasion and insect infestation, as well as increased breakage during subsequent handlings.

Pericarp damage can be detected using 0.1% (FCF) fast green dye or 0.1% brilliant blue R-250 dye and distilled water. Samples of 50 g of whole kernels are placed into fast green dye for 10 min or, if brilliant blue dye is used, immersed for 30 seconds. The sample is removed, washed with tap water for 30 seconds, placed on paper towels to remove excess water, and allowed to dry. The sample is next inspected for severe, minor, or negligible damage. Kernels where dye penetrates into floury endosperm are considered to be severely damaged. Those with minor cuts, scratches or abrasions are considered to have minor damage. The category of negligible damage has no dye absorption or very minor absorption near the tipcap of the kernel.

3.6.7 Stress cracks

Stress cracks in maize kernels are tiny fissures inside a kernel beginning near the centre and extending outward through the vitreous endosperm but they do not extend all the way outward to the pericarp. Stress cracks are caused primarily by large moisture gradients between the centre and the outer periphery of the kernel, but they are largely associated with high temperature drying. The more points of moisture removed in one pass of drying also increases moisture gradients and the likelihood of stress cracking.

The importance of stress crack determination is that maize lots with high stress crack levels are highly susceptible to breakage during handling. Furthermore, if the maize is intended for use in dry milling, a high percentage of multiple cracks will cause the yield of large flaking grits to be decreased. If the maize is used for wet milling, high stress crack levels provide a reasonably high probability that the maize was heated sufficiently during drying to cause protein denaturation which in turn makes separation of protein from starch more difficult, reduces starch recovery by 2–3%, and reduces wet milling process efficiencies.

Stress cracks are determined by candling 50–100 whole kernels over a light table. The germ side is turned down and light is passed up through the kernels. Stress cracks are visually inspected. The percentages of kernels with one, two, or more than two stress cracks is determined. Commercial export maize can easily have 50–60% stress cracks, while low temperature dried maize is well below 20%. Efforts are underway to automate the stress crack inspection procedure using machine vision (Yie *et al.*, 1993).

3.6.8 Breakage susceptibility tests

Breakage susceptibility is defined as the potential for kernel fragmentation or breakage when subjected to impact forces (Anon., 1990b). There are two breakage susceptibility testers that have been used widely and many others with similar operating principles.

Stein breakage tester

One of the oldest breakage testers is the Stein CK-2 tester followed by the model CK-2M. These testers use two rotating blades inside a metal cup to provide small impact and abrasion to a 100 g pre-sieved sample of grain. After 2 minutes the sample is removed and re-sieved on a 12/64 inch round hole sieve to determine the percentage of breakage generated by the tester.

Wisconsin breakage tester

A second major breakage tester was first built at the University of Wisconsin. The WBT has an impeller rotating at 1800 rpm that centrifugally impacts kernels against the inside of a vertical wall. A pre-sieved 200 g sample is passed through the impeller, collected and re-sieved usually on a 12/64 inch round hole sieve to determine the percentage of breakage generated.

The effect of the two machines on maize is somewhat different. The Stein primarily abrades away at the surface of the kernel, so soft endosperm maize tends to have higher breakage than hard endosperm. Breakage appears as fine pieces of kernels and as powder. The WBT provides one large impact, so hard endosperm maize tends to shatter and breakage will appear as large pieces of kernels. Consequently, the Stein has been said to predict breakage during handling; whereas, the WBT tends to be better correlated to stress cracks (Paulsen *et al.*, 1992). The complicating factor in this generalization is that breakage from both testers is affected greatly by sample moisture content and to a lesser extent by sample temperature (Eckhoff, 1992b).

3.6.9 Heat damage

Heat damage is a term which has several meanings. In most maize grading systems, heat damage refers to visual discoloration of the kernel due to exposure to high temperatures. The high temperature exposure may be due to excessive drying air temperature or microbial heating. Visual heat damage occurs after considerable damage has already occurred to the kernel and lack of it is not a reliable indicator that the kernel has not been subjected to excessive heat. Visual heat damage definitely indicates low quality but the absence of visual heat damage does not always indicate good quality.

Between 35°C (95°F) and 41°C (105°F) kernel germination starts to drop. Around 60°C (140°F) protein in the endosperm begins to denature causing difficulty in separating protein and starch in wet milling. At about the same temperature, starch in the kernel will begin to gelatinize, if sufficient moisture exists (usually above 28%, w.b.). Although it is possible to dry the corn to where it is not possible to recover the starch due to excessive gelatinization, usually the starch granules will only partially swell. These swollen granules are then too light to be recovered by use of hydrocyclones.

There are two tests which have been used as a measure of heat damage: the turbidity test and the ethanol solubility tests. Both tests are able to identify corn which has a moderate to excessive amount of heat damage and are currently used in some countries where high-temperature drying from elevated harvest moisture is the norm. US wet millers have not found the procedures satisfactory for identifying heat damage in US corn.

Stress crack analysis (method described above) is often used to indicate heat damage since high drying rates will cause increased stress cracks in the hard endosperm. However, stress cracks are an indirect means of determining potential heat damage. Mistry *et al.* (1993) showed that although high-humidity high-temperature drying could reduce the occurrence of stress cracks, starch yield and the peak Brabender consistency of the starch decreased.

Ethanol soluble protein

High temperatures will denature the protein in the endosperm. Changes in the solubility of the ethanol soluble protein, zein, can be used as an indication of heat damage. The procedure (Godon and Petit, 1971) involves the measurement of the protein content of supernatant resulting from extraction at 4°C for 15 h of a ground maize sample with a solution of 85% ethanol and 1.25% acetic acid.

Turbidity test

High temperatures will also denature the water-soluble and salt-soluble proteins found primarily in the germ. The procedure reported by the Institut Technique des Céréales et des Fourrages (Anon., 1987) measures the light transmittance at 550 nm of liquid extracted from ground maize using a 5% solution of NaCl. The higher the transmittance, the lower quality the maize is for wet milling purposes.

3.6.10 Germination tests

Tetrazolium test

The tetrazolium test, a viability test recognized by the AOSA, is a predictor of whether seeds are viable or dead on the basis of relative respiration rate when hydrated. At least 200 randomly selected seeds, in replicates of 100 seeds or less, are used for the test. Seeds are immersed in a 1% solution of 2,3,5-triphenyl tetrazolium chloride. Cells that are living stain red, while dead tissue does not stain. Dehydrogenase enzymes, which are responsible for reduction processes in living tissues, reduce the chemical imbibed by the seed to a red-coloured, stable, non-diffusible substance called formazan. Maize endosperm is dead tissue and does not stain but the embryo should stain red if it is living. Seeds with damage in critical areas such as the plumule, more than half of the radicle area, or the centre portion of the scutellum will generally not germinate (Delouche *et al.*, 1962).

Warm germination test

The warm germination test is the most commonly conducted germination test. It consists of placing 100 seeds on wetted paper towels, which are rolled, and placed in an environment chamber at 25°C (77°F) and about 95% relative humidity for 4 days. Normal seedlings are counted and removed. The remainder are allowed to germinate for 3 more days before counting again.

Cold test

The cold test for germination involves placing 100 kernels in a 50% soil/50% sand mixture. The soil is wetted to 70% water capacity and maintained at 10°C (50°F) and 95% relative humidity in darkness for seven days. After the cold treatment, the seeds are kept for seven days at 25°C with moderate lighting to prevent etiolation. Germination and evaluation of seedlings into normal and abnormal seedlings are obtained on the fourteenth day. The cold test more closely correlates to field emergence than does the warm germination test.

3.6.11 Starch, oil, protein

Near infrared reflectance (NIR) can be used to obtain protein and oil in maize or soybeans. NIR operates on the principle that selected wavelengths of light are absorbed in proportion to the quantity of oil and protein present. NIR equipment is very reliable. However, grinding procedures and calibrations to chemistry values are most important. Protein and oil percentages are important measures of value.

It is always essential that a completely random and representative sample be obtained so that a sample of approximately 1050 g is representative of an entire truckload, railcar, barge, or sublot on an ocean vessel. The most accurate sampling method is the mechanical diverter sampler. With this method, grain on a moving belt is sampled across the width of the belt at timed intervals. Other sampling methods use a compartmentalized probe, Ellis cup, or pelican sampler.

REFERENCES

Anon. (1987) *Mode of operation of the turbidity test*, Institut Technique des Céréales et des Fourrages, Station Experimentale, Boigneville, France.

Anon. (1990a) *Approved Methods of the AACC*, Method 44–15A, American Association of Cereal Chemists, St. Paul, MN.

Anon. (1990b) *Approved Methods of the AACC*, Method 55–20, American Association of Cereal Chemists, St. Paul, MN.

Anon. (1990c) *Grain Inspection Handbook – Book II, Grain Grading Procedures*, Chapter 4 – Corn, Federal Grain Inspection Service, Washington DC.

Anon. (1992) *ASAE Standards*, Standard no. S352.2, Moisture measurement-unground grain and seeds, American Society of Agricultural Engineers, St. Joseph, MI.

Anon. (1994) *1994 Corn Annual*, Corn Refiners Association, Washington DC.
Alexander, R.J. (1987) Corn dry milling: processes, products, and applications, in: *Corn Chemistry and Technology* (eds S.A. Watson and P.E. Ramstadt), American Association of Cereal Chemists, St. Paul, MN.
Bagby, M.O. and Widstrom, N.M. (1987) Biomass uses and conversions, in: *Corn Chemistry and Technology* (eds S.A. Watson and P.E. Ramstadt), American Association of Cereal Chemists. St. Paul, MN.
Bender, K.A., Hill, L.D. and Valdes, C.(1992) A comparison of grain grades among countries, Report No. AE-4685, Department of Agricultural Economics, Illinois Agricultural Experiment Station, University of Illinois, Urbana-Champaign, IL.
Bennett, G.A. and Anderson, R.A. (1978) Distribution of aflatoxin and/or zearalenone in wet-milled corn products: a review. *Journal of Agricultural and Food Chemistry*, **26**(5), 1055–60.
Benson, G.O. and Pearce, R.B. (1987) Corn perspective and culture, in *Corn Chemistry and Technology* (eds S.A. Watson and P.E. Ramstadt), American Association of Cereal Chemists. St. Paul, MN.
Blanchard, P.H. (1992) *Technology of Corn Wet Milling and Associated Processes*, Elsevier Publishing, New York, NY.
Bonifacio-Maghirang, E., Paulsen, M.R., Hill, L.A., Bender, K. and Wei, A. (1994) Single kernel moisture variation and fungal growth of blended corn. ASAE Paper No. 94–6054, American Society of Agricultural Engineers, St. Joseph, MI.
Bothast, R.J., Rogers, R.F. and Hesseltine, C.W. (1974) Microbiology of corn and dry milled corn products. *Cereal Chemistry* **51**, 829–38.
Brekke, O.L. (1970) Corn dry milling industry, in *Corn Culture, Processing, Products* (ed. G.E. Inglett), AVI Publishing, Westport, CT.
Delouche, J.C., Still, T.W., Raspet, M. and Lienhard, M.(1962) *The tetrazolium test for seed viability*, Mississippi Agricultural Experiment Station Technical Bulletin No. 51, Mississippi State University.
Du, L., Li, B., Lopes-Filho, J.F., Daniels, C. and Eckhoff, S.R. (1996) Effect of selected organic and inorganic acids on corn wet milling yields. *Cereal Chemistry*, **73**(1), 96–8.
Earle, F.R., Curtis, J.J. and Hubbard, J.E. (1946) Composition of the component parts of the corn kernel. *Cereal Chemistry*, **23**, 504–11.
Eckhoff, S.R. (1987) *Report on the increased yield potential and economic value of hard endosperm corn*, US Feed Grains Council, Washington DC.
Eckhoff, S.R. (1992a) Converting corn into food and industrial products. *Illinois Research*, **34**(1/2), 19–23.
Eckhoff, S.R. (1992b) Evaluating grain for potential production of fine material – breakage susceptibility testing, in *Fine Material in Grain* (ed. R. Stroshine), North Central Regional Publication No. 332. OARDC Special Circular 141. Ohio Agricultural Research and Development Centre, Wooster, OH.
Eckhoff, S.R. (1995) The future of commodity corn, *Wet Milling Notes*. No. 11, March 1995, Department of Agricultural Engineering, University of Illinois, Urbana, IL.
Eckhoff, S.R. and Denhart, R.L. (1994) Wet milling of low test weight corn. *Wet Milling Notes*. No. 10, July 1994, Department of Agricultural Engineering, University of Illinois, Urbana, IL.
Eckhoff, S.R. and Tso, C.C. (1991a) Starch recovery from steeped corn grits as affected by drying temperature and added commercial protease. *Cereal Chemistry*, **68**(3), 319–20.

Eckhoff, S.R. and Tso, C.C. (1991b) Wet milling of corn using gaseous SO_2 addition before steeping and the effect of lactic acid on steeping. *Cereal Chemistry*, **68**(3), 248–51.

Fox, E.J. and Eckhoff, S.R. (1993) Wet milling of soft-endosperm, high-lysine corn using short steep times. *Cereal Chemistry*, **70**(4), 402–4.

Frayssinet, C. (1988) Classification and chemical characteristics of fungal toxins in grains and seeds, in *Preservation and Storage of Grains, Seeds and Their By-Products* (ed. J.J. Multon). Lavoisier Publishing, New York, NY.

Funk, D.B. (1991) Uniformity in dielectric grain moisture measurement, in *Uniformity by 2000, An International Workshop of Corn and Soybean Quality* (ed. L.D. Hill) Scherer Communications, Urbana, IL.

Godon, B. and Petit, L. (1971) Le mais grain: prestockage, sechage et qualite, v. proprietes des proteines. *Annales de Zootechnie*, **20**, 641–4.

Hill, L.D. (1982) Price and value relationships explained for high-moisture c/ grain. *Feedstuffs*, **54**, 37–42.

Hoseney, R.C., Zeleznak, K. and Abdelrahman, A. (1983) Mechanism of popcorn popping. *Journal of Cereal Science*, **1**, 43.

Johnson, L.A. (1991) Corn: production, processing and utilization, in *Handbook of Cereal Science and Technology*, (eds K.J. Lorenz and K. Kulp), Marcel Dekker, New York, NY.

Kerpisch, M.R. (1988) Effect of variety and drying temperature on proteolytic enzyme activity of yellow dent corn, M.S. Thesis, Department of Food Science and Industry, Kansas State University, Manhattan, KS.

Lasseran, J.C. (1973) Incidences of drying and storing conditions of corn (maize) on its quality for starch industry. *Die Staerke*, **25**, 257–88.

Liao, K., Reid, J.F., Paulsen, M.R. and Shaw, E.E. (1991) Corn hardness classification by color segmentation, ASAE Paper No. 91–3504, American Society of Agricultural Engineers, St. Joseph, MI.

McEllhiney, R.R. (1994) *Feed Manufacturing Technology IV*, American Feed Industry Association, Arlington, VA.

Maisch, W.F. (1987) Fermentation processes and products, in *Corn Chemistry and Technology* (eds S.A. Watson and P.E. Ramstadt), American Association of Cereal Chemists, St. Paul, MN.

Marshall, S.W. (1987) Sweet corn, in *Corn Chemistry and Technology* (eds S.A. Watson and P.E. Ramstadt), American Association of Cereal Chemists, St. Paul, MN.

May, J.B. (1987) Wet milling: process and products, in *Corn Chemistry and Technology* (eds S.A. Watson and P.E. Ramstadt), American Association of Cereal Chemists, St. Paul, MN.

Mertz, E.T. (1970) Nutritive value of corn and its products, in *Corn Culture, Processing, Products* (ed. G.E. Inglett), AVI Publishing, Westport, CT.

Mertz, E.T. (1992) *Quality Protein Maize*, American Association of Cereal Chemists, St. Paul, MN.

Mills, R. and Pederson, J. (1990) *A Flour Mill Sanitations Manual*, Eagan Press, St. Paul, MN.

Mistry, A.H., Wu, X., Eckhoff, S.R. and Litchfield, J.B. (1993) Wet-milling characteristics of high-temperature, high-humidity maize. *Cereal Chemistry*, **70**(3), 360–1.

Multon, J.L. (1991) Basics of moisture measurement in grain, in *Uniformity by 2000, An International Workshop of Corn and Soybean Quality*, (ed. L.D. Hill), Scherer Communications, Urbana, IL.

Paulsen, M. R. and Hill, L.D. (1984) Corn quality factors affecting dry milling performance. *Journal of Agricultural Engineering Research*, **31**, 255–63.

Paulsen, M.R. and Hill, L.D. (1985) Quality attributes of Argentine corn. *Applied Engineering in Agriculture*, **1**(1), 42–6.
Paulsen, M.R. and Nave, W.R. (1980) Corn damage from conventional and rotary combines. *Transactions of ASAE*, **23**(5), 1110–16.
Paulsen, M.R., Hill, L.D., Shove, G.C. and Kuhn, T.J. (1989) Corn breakage in overseas shipments to Japan. *Transactions of ASAE*, **33**(3), 1007–14.
Paulsen, M.R., Darrah, L.L. and Stroshine, R.L. (1992) Genotypic differences in breakage susceptibility of corn and soybeans, in *Fine Material in Grain* (ed. R.L. Stroshine), North Central Regional Publication No. 332. OARDC Special Circular No. 141, Ohio Agricultural Research and Development Center, Wooster, OH.
Romer, T. (1984) Mycotoxins in corn and corn milling products. *Cereal Foods World*, **29**(8), 459–61.
Rooney, L.W. and Serna-Saldivar, S.O. (1987) in *Corn Chemistry and Technology* (eds S.A. Watson and P.E. Ramstadt), American Association of Cereal Chemists, St. Paul, MN.
Ruan, R., Litchfield, J.B. and Eckhoff, S.R. (1992) Simultaneous and non-destructive measurement of transient moisture profiles and structural changes in corn kernels during steeping using microscopic nuclear magnetic resonance imaging. *Cereal Chemistry*, **69**(6), 600–6.
Serna-Saldivar, S.O., Gomez, M.H., Islas-Rubio, A.R., Bockholt, A.J. and Rooney, L.W. (1990a) The alkaline processing properties of quality protein maize, in *Quality Protein Maize* (ed. E.T. Mertz), American Association of Cereal Chemists, St. Paul, MN.
Serna-Saldivar, S.O., Gomez, M.H. and Rooney, L.W. (1990b) Technology, chemistry, and nutritional value of alkaline-cooked corn products, in *Advances in Cereal Science and Technology*, Vol. 10 (ed. Y. Ponereanz), American Association of Cereal Chemists, St.Paul, MN.
Song, A. and Eckhoff, S.R. (1994a) Optimum popping moisture content for popcorn kernels of different sizes. *Cereal Chemistry*, **71**(5), 458–60.
Song, A. and Eckhoff, S.R. (1994b) Individual kernel moisture content of preshelled and shelled popcorn and equilibrium isotherms of popcorn kernels of different sizes. *Cereal Chemistry*, **71**(5), 461–3.
Song, A., Eckhoff, S.R., Paulsen, M.R. and Litchfield, J.B. (1991) Effects of kernel size and genotype on popcorn popping volume and number of unpopped kernels. *Cereal Chemistry*, **68**(5), 464–7.
Vojnovich, C., Anderson, R.A. and Griffin, Jr., E.L. (1975) Wet-milling properties of corn after field shelling and artificial drying. *Cereal Foods World*, **20**, 333–5.
Wahl, G. (1970) Biochemical-technological studies on wet-processing of maize, part 3. milieu conditions in the maize grain during the steeping process. *Die Staerke*, **22**, 77–83.
Wahl, G. (1971) Biochemical-technological studies on wet-processing of maize, part 6. examination of the proteolytic and the lipolytic enzyme system of maize. *Die Staerke*, **23**(4), 145–48.
Wang, D. (1994) Effect of broken and pericarp damaged corn on water absorption and steepwater characteristics. M.S. Thesis. Department of Agricultural Engineering, University of Illinois, Urbana, IL.
Watson, S.A. (1984) Corn and sorghum starches: production, in *Starch Chemistry and Technology* (eds R.L. Whistler, J.N. BeMiller and E.F. Paschall), Academic Press, Orlando, FL.
Watson, S.A. (1987) Structure and composition, in *Corn Chemistry and Technology* (eds S.A. Watson and P.E. Ramstadt), American Association of Cereal

Chemists, St. Paul, MN.
Wichser, W.R. (1961) The world of corn processing. *American Miller and Processor*, **89**(1), 29–31.
Wilson, C.M. (1987) Proteins of the kernel, in *Corn Chemistry and Technology* (eds S.A. Watson and P.E. Ramstadt), American Association of Cereal Chemists, St. Paul, MN.
Wilson, D.M. and Abramson, D. (1992) Mycotoxins, in *Storage of Cereal Grains and Their Products* (ed. D.B. Sauer), American Association of Cereal Chemists, St. Paul, MN.
Wolf, M.J., Buzan, C.L., MacMasters, M.M. and Rist, C.E. (1952) Structure of the mature corn kernel. IV. microscopic structure of the germ of dent corn. *Cereal Chemistry*, **29**, 362–82.
Wright, K.M. (1987) Nutritional properties and feeding value of corn and its by-products, in *Corn Chemistry and Technology* (eds S.A. Watson and P.E. Ramstadt), American Association of Cereal Chemists, St. Paul, MN.
Yie, T.J., Liao, K., Paulsen, M.R., Reid, J.F., Maghirang, E.B. (1993) Corn kernel stress crack detection by machine vision, ASAE Paper No. 93–3526, American Society of Agricultural Engineers, St. Joseph, MI.
Zuber, M.S. and Darrah, L.L. (1987) Breeding, genetics, and seed corn production, in *Corn Chemistry and Technology* (eds S.A. Watson and P.E. Ramstadt), American Association of Cereal Chemists, St. Paul, MN.

4

Barley

M.J. Edney

4.1 INTRODUCTION

Barley is the most widely adapted of the world's cereal crops. It is very tolerant of a cool, short growing season and is thus grown farther north than any other major cereal, even being found above the Arctic Circle. Barley is often the only cereal that will mature in the short growing seasons that are found at high altitudes such as in Tibet and Ethiopia. It is grown in temperate climates as well as in tropical climates where it can be second cropped during the cool season. It tolerates drought, alkaline and salt conditions better than any other cereal grain. This adaptability has led to widespread acceptance of barley in many parts of the world.

History indicates that barley has been cultivated longer than any other cereal grain. This early cultivation was due to the ease with which it responded to cultivation. Barley was first cultivated in the Fertile Crescent of the Middle East, possibly as early as 10 000 BC (Harlan, 1968), and it may even have been cultivated as early as 16 000 BC in Egypt (Wendorf *et al.*, 1979). At either location, barley would have been used mainly, if not only, for human food. Straw and barley lost in the field may have been used by wild animals, but there were no domestic animals for several more thousand years, the earliest time at which barley would have been used intentionally as feed (Harlan, 1968). The history of barley as an alcoholic beverage has followed that of food barley and is believed to have begun in either the Fertile Crescent or Egypt around 3000–5000 BC (Katz, 1979).

Today barley is grown throughout the world (Table 4.1). World production had increased steadily for over 30 years, although in recent years production has declined. This is likely due to falling barley prices caused by political changes. Only in Australia has production continued to increase, a result of increasing market opportunities in nearby Southeast Asia.

Table 4.1 Average barley production, in thousands of tonnes, for various areas of the world over the past 34 years (Gudmunds and Webb, 1994)

	*Europe**	*USSR or FSU-15†*	*North America*	*Asia**	*Oceania*	*Africa*	*South America*	*World total*
1960–64	31 910	17 834	12 755	16 766	1195	3409	535	84 404
1965–69	43 439	26 101	15 378	17 379	1480	3309	543	107 628
1970–74	54 193	40 022	19 482	14 957	2682	3375	564	135 273
1975–79	64 093	49 020	19 395	16 444	3485	4187	635	157 260
1980–84	70 489	39 242	22 968	17 640	4142	4107	541	159 129
1985–89	71 412	46 133	23 577	18 207	4343	5597	680	169 949
1990	71 691	52 535	23 083	19 838	4543	5236	751	177 677
1991	72 784	41 520	22 157	19 049	4912	7381	708	168 511
1992	59 176	53 010	21 339	19 671	5887	4896	877	164 856
1993	55 991	50 595	22 464	21 849	7020	3850	678	165 447

*Excluding USSR or FSU-15 countries.
†FSU-15, the 15 countries of the former Soviet Union.

4.2 BARLEY END-USES

Barley has three distinct end-uses: alcoholic beverages, human foods and feed (Figure 4.1). However, in contrast to earlier days when barley was used mostly for food, feed is now the dominant use with close to 75% of the world's barley being used for feed in 1993. The percentage does vary between continents with Europe, FSU-15 (the 15 countries of the former Soviet Union), North America and Oceania having the highest percentages of feed barley use. Africa and Asia, areas with the greatest use of barley for human food, also show increasing percentages of barley being used as feed (Table 4.2). The use of barley in South America has increased over the past 30 years but, in contrast to other continents, the percentage used for feed has decreased to insignificant amounts. The main use of barley in South America is in the form of malt for beer.

The world's use of barley malt for alcoholic beverages has increased significantly since the Second World War, but this trend has now started to reverse itself, especially in the West where alcohol consumption has become less popular and beer consumption, in particular, has declined. A corresponding drop in the amount of barley used for malt is apparent in a number of countries. However, beer consumption continues to increase in some areas of the world such as Japan and China. Although the sharpness of the increase has recently declined in Japan, forecasts are for continued increase in demand for malt and malting barley in China and other countries in Southeast Asia well into the new century.

In contrast to malting and feed barley, the world use of barley for food has declined constantly since the 1960s (Figure 4.1). The decline

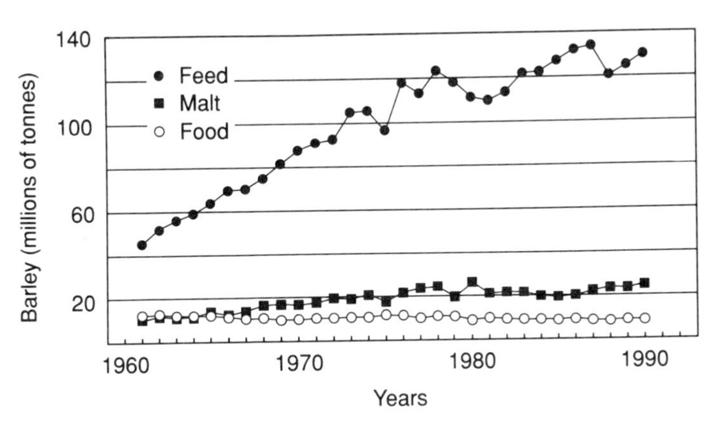

Figure 4.1 World trends in the end-use of barley from 1960 to 1990 (Anon., 1994; Gudmunds and Webb, 1994).

Table 4.2 Percentage of barley used for feed by various areas of the world over the past 25 years (Gudmunds and Webb, 1994)

Year	*Europe**	*USSR or FSU-15*[†]	*North America*	*Asia** *(% feed use)*	*Oceania*	*Africa*	*South America*	*World total*
1969	79	80	72	40 (24)[‡]	52	0	19	75
1970	78	84	74	38 (21)	66	0	15	76
1971	78	87	71	38 (24)	65	0	16	76
1972	76	83	72	39 (26)	84	2	8	74
1973	77	81	72	40 (25)	82	2	9	75
1974	77	84	68	37 (23)	80	2	6	75
1975	78	80	68	42 (23)	72	16	6	74
1976	78	83	67	43 (22)	72	15	7	76
1977	77	81	71	45 (22)	67	18	6	75
1978	77	82	69	48 (25)	64	29	6	76
1979	78	85	67	50 (29)	38	30	6	77
1980	78	73	64	60 (37)	42	34	5	73
1981	76	81	66	61 (36)	57	38	5	75
1982	77	80	70	62 (38)	47	37	8	75
1983	77	82	71	69 (41)	42	45	8	77
1984	78	82	72	69 (43)	65	47	8	77
1985	79	83	72	72 (47)	46	50	4	78
1986	78	84	73	73 (44)	53	54	4	79
1987	77	77	71	73 (46)	48	62	5	76
1988	77	79	68	71 (46)	54	58	5	76
1989	77	82	68	70 (44)	60	60	4	77
1990	76	79	71	68 (43)	59	56	3	75
1991	73	70	72	72 (44)	69	60	4	73
1992	73	74	70	69 (40)	77	65	4	74
1993	74	74	71	70 (41)	86	64	4	74

*Excluding USSR or FSU-15 countries.
[†]FSU-15 = the 15 countries of the former Soviet Union.
[‡]Bracketed figure excludes Middle East.

becomes more dramatic the farther one goes back in history. At one time, barley was the staple food of a large proportion of the world's inhabitants. The decreased use of barley for food resulted from increased availability of different foods and access to grains with better quality characteristics for preparing food. For example, barley was used as bread by the Romans and as late as the 15th century in England, but as the advantages of using wheat for bread were discovered and as wheat became more available, it replaced barley as the preferred grain for bread. Increased yields of rice and the corresponding increase in its availability have also contributed to barley's decline as a food. In more recent times, especially in the developing countries, a shift from vegetarian diet to a more omnivorous diet has also reduced demand for food barley, but contributed to the increased demand for feed barley.

Despite these declines, over 5% of the world's barley continues to be consumed as food. The area with the highest use of barley for food is in the northern countries of Africa, where 65 kg/person is consumed annually (Bhatty, 1993). Barley remains an important food in remote mountainous regions of the world such as Tibet, Ethiopia and Caucasus, where limited access and short growing seasons restrict the supply of wheat. Koreans and the Chinese also consume significant amounts of barley, although this has declined over the last 20 years. Smaller amounts are consumed in Japan and in the Western World. Barley workers are predicting significant increases in food barley consumption, especially in the Western World and Japan, as people respond to recent health claims concerning barley (see Section 4.10).

The increased use of barley as feed is related to its wide adaptability, which has allowed barley to be competitively priced with other coarse grains. Barley's adaptability has allowed it to be grown under a wide range of environmental conditions where other crops fail to grow. Its early maturity allows it to grow at high latitudes and altitudes where land is often inexpensive. The high yields that barley can achieve in these areas also contribute to its low price. Price is a key component of feed grain use.

In contrast, barley's popularity for malt is related to some of its unique quality attributes including a husk, which protects the growing kernel during malting and which is essential for successful lautering, the process in brewing beer by which spent grains are separated from the extract. The ability to produce large quantities of enzymes during germination is a key attribute that has enhanced the use of barley in the brewing and the distilling industries. Another advantage of barley is the firmness of steeped barley, which allows for handling of the barley during the malting process (Burger and LaBerge, 1985). Finally, barley and barley malt contribute unique flavours to alcoholic beverages which has added to barley's reputation. Some of these characteristics are inherent in barley, but others, such as the ability to produce enzymes, are probably the result of selection pressures over time.

4.3 TYPES OF BARLEY

There are two major types of barley, two-rowed and six-rowed. This diversity may have also contributed to the wide range of end-uses for barley. The basis for their difference is the lateral florets in the spikelets of these two barley types. In the two-rowed spikelets, the lateral florets are sterile, which results in one central kernel per spikelet. Since there are spikelets on both sides of the rachis, the single kernels in each spikelet result in the two-rowed appearance. In contrast, six-rowed barley has three fertile florets per spikelet and thus three kernels per spikelet. This gives the six-rowed appearance. These morphological differences can also affect barley quality. For example, in six-rowed barley the lateral kernels tend to be crushed and twisted during grain filling. The result is a greater variability in kernel size compared to two-rowed kernels, and slight differences in composition (Table 4.3). These differences mean that the two types of barley often have to be handled differently during end-use processing.

It is believed that all domesticated barley originated from two-rowed barley. Six-rowed barley was then selected from the original two-rowed type because of better shatter resistance, which was a necessary trait if barley was to be harvested successfully and then planted during the following growing season. Archaeological findings also indicate that two-rowed barley, through history, has tended to be used in dry areas whereas six-rowed cultivars have been grown in wet climates or under irrigation (Harlan, 1968). In some parts of the world, for example, the United States of America, there is still such a tendency, although in

Table 4.3 A comparison of some quality characteristics for two- and six-rowed barley, derived from three varieties for each type, all having been grown at 11 locations in western Canada over three summers

	*Plump kernels (%)**	*Protein (% dry matter)*	*Neutral-detergent fibre (% dry matter)*
Two-rowed	79.0 (7.3)[†]	12.6 (0.8)	13.2 (0.7)
Six-rowed	74.9 (9.4)	12.4 (0.5)	15.2 (1.3)

*Kernels remaining on a 6/64″ × 3/4″ slotted sieve.
[†]Mean (standard deviation).

Europe this is not the case. At present, both types of barley are grown and used in the production of alcoholic beverages, food or as feed.

However, the uses made of two- and six-rowed barley do vary around the world. For example, in the past the predominant malting barley in North America has been six-rowed barley while the rest of the world has relied more on two-rowed malting cultivars, with some exceptions. Two-rowed cultivars tend to malt better, hence their popularity. However, two-rowed cultivars introduced to the east coast of North America performed poorly. In contrast, six-rowed varieties introduced to California grew well. The end effect has been a dominance of six-rowed malting barley in North America. Six-rowed barley is grown throughout the world for feed as it tends to yield better than the two-rowed cultivars.

In general, two-rowed cultivars malt more slowly and produce lower enzyme levels, but the malts yield greater amounts of extract. Six-rowed malt has been favoured for adjunct brewing and grain whisky distilling because of the additional enzymes it contains. The higher enzyme levels are necessary for breaking down the adjunct starch. However, newer cultivars of two-rowed malting barley also show good enzyme potential, while newer six-rowed cultivars are also producing high levels of extract (Table 4.4). Breeding has also resulted in the development of two-rowed malting cultivars that can be grown successfully in North America. The USA continues to use large quantities of six-rowed barley for malting and brewing, but more two-rowed barley is being used, especially in Canada. In the rest of the world, two-rowed malting barley remains dominant.

Two-rowed barley is reputed to have higher feeding value than six-rowed barley which is known to be thinner, a result of the crushing of

Table 4.4 A comparison of some malting quality characteristics for two- and six-rowed barley, derived from two cultivars for each type, all having been grown at three locations over three summers. All the cultivars and locations were Canadian

	Fine-extract (% dry matter)	*70° Coarse/fine difference* (% dry matter)*	*Diastatic power (° Linter)*	*α-Amylase (dextrinizing units)*
Two-rowed	79.0 (1.8)[†]	5.5 (2.6)	150 (15.3)	51.2 (5.0)
Six-rowed	77.9 (1.1)	3.9 (1.7)	114 (9.2)	44.0 (3.6)

*Similar to ASBC fine/coarse difference, except coarse mashed at 70°C for 2 h. All other analyses are ASBC (Anon., 1992a).
[†]Mean (standard deviation).

the lateral kernels during grain filling. This can also result in lower test weights, higher concentrations of fibre and lower concentrations of starch in six-rowed barley. However, the differences may not have a significant effect on the feeding value since reports in the literature remain controversial with respect to the better feeding value of two- versus six-rowed barley (see Bhatty, 1993, for review).

Hull-less barley is another distinct barley type and it exists in both two- and six-rowed forms. Hull-less barley has a loose hull that is removed and left in the field during harvesting. The lack of hull makes the barley more digestible and thus more desirable as a food. Consequently, it was an important barley type when barley was first cultivated in the Fertile Crescent. However, as wheat became the cereal food of choice, hull-less barley production decreased while covered barley production for feed increased. Hull-less barley continued and continues to be grown at high altitudes where barley remains a food staple due to the poor supply of wheat.

Hull-less barley has begun to have a resurgence in the past 10–20 years. The interest, though, has been as a feed and not as a food. The lack of hull increases the energy content of barley for monogastrics to levels almost equal to those seen in feed wheat (Rossnagel *et al.*, 1981). Hull-less barley, more recently, has also been promoted as a food because of new health implications associated with barley (see Section 4.10).

Some of the other barley types that exist are distinguished by specific biochemical components. Examples of these types include high-lysine, high-amylose and waxy barley. High-lysine barley can have 30–40% more lysine in its protein than a standard barley (Bhatty, 1993). High-amylose barley has an altered type of starch resulting in a greater concentration of amylose (approx. 40%) and less amylopectin than standard barley which has 20–25% amylose. In contrast, the starch in waxy barley is nearly all amylopectin (*ca.* 95%) (MacGregor and Fincher, 1993).

The high-amylose and waxy barley types offer exciting possibilities for both nutritionists, who report differences in the digestibilities of the different starches, and industrialists, who are interested in isolating specialized barley starches. These barleys are the result of natural mutants and have possibly been in existence for thousands of years. They have been studied academically as abnormalities in order to better understand plant physiology and in an attempt to find specialized uses for them. There has been some commercial success with the processing of waxy barley (McIntosh *et al.*, 1993), although on a limited basis. However, in today's spirit of entrepreneurship, there is heightened interest in developing unique markets for waxy barley. These specialized barleys are not new, but few cultivars of these types have existed with acceptable agronomic properties. The fate of the newer cultivars with acceptable agronomic characteristics rests on the ability of research workers to develop specialized markets for them.

High-lysine barley has received considerable attention because of important nutritional implications. The first limiting essential amino acid in barley is known to be lysine and, therefore, any cultivars containing higher lysine would be important in both food and feed barley. Both natural and induced high-lysine mutants have been reported in the literature. Examples include the natural mutant Hiproly (Munck *et al.*, 1970), and induced mutants such as Risø 1508 (Ingversen *et al.*, 1973) and Notch II (Bansal *et al.*, 1977). Despite the potential nutritional advantages of high-lysine barley, poor agronomics and grain filling have always been associated with these barleys. Therefore, no commercial cultivars of high-lysine barley have been released. Work continues on trying to develop acceptable cultivars with high lysine and some success has recently been reported (Bang-Olsen *et al.*, 1991).

4.4 BARLEY FOR ALCOHOLIC BEVERAGES

Both malted and unmalted barleys (barley adjunct) are used in the production of alcoholic beverages. Malted barley is predominantly used in the production of beer, but it is also an important ingredient in the distilling industry and, to a limited degree, in the food industry. Exact figures on the breakdown of malt use are difficult to obtain for proprietary reasons, but in the USA and Canada, close to 95% of the malt goes into the brewing of beer, while the rest is divided between the distilling and food industries (Burger and LaBerge, 1985). Barley adjunct is also used in both the brewing and the distilling industries, although to a much lesser degree than malted barley.

Barley use as an adjunct is the result of several characteristics. First, adjuncts are generally used as inexpensive sources of carbohydrates as they are much lower priced than is malt, and barley is one of the cheapest cereal grains. Another important characteristic of barley is its low gelatinization temperature. Adjunct starch must be gelatinized if malt amylases are to break the starch down into maltose and other sugars during mashing. Therefore, the starch has to be gelatinized before the enzymes are inactivated. The gelatinization temperature of barley starch is below the inactivation temperature of many of the malt enzymes. Thus barley can be mashed with the malt and its starch will still be degraded by the malt enzymes. In contrast, other cereals, such as rice and corn, have gelatinization temperatures that are higher than the inactivation temperatures. These adjuncts, therefore, need to be heated to their gelatinization temperature in a separate cereal cooker, cooled and then added to the malt mash. Barley adjunct is thus cheaper and simpler to use than some other grain adjuncts. The low oil content of barley is another benefit because high oil content in adjuncts can cause flavour deterioration in the final beer (Briggs *et al.*, 1981). The flavour and tradition of barley are also important considerations for choosing it as an adjunct.

Barley is used in the distilling industry for reasons similar to those in the brewing industry. Barley malt is capable of producing high levels of starch degrading enzymes that are essential in the distilling industry to maximize extract from other cereal grains. The use of barley for whisky is also traditional. Several world famous whiskies have used barley malt and barley adjunct for centuries and they are now considered an integral part of the process.

4.5 MALTING BARLEY QUALITY

Nearly all traits that are indicative of a good malt are dependent on the quality of the initial barley. Malting conditions also have a significant effect on malting quality, but such conditions cannot compensate for a poor barley.

The quality of a malting barley is determined by both an inherent capacity and by the environment under which it is grown. The inherent capacity is a result of breeding malting quality potential into a barley. This potential remains consistent within a cultivar, which explains why certain cultivars gain a reputation for consistently producing quality malt. The inherent aspect of barley quality can be exploited by growing cultivars of good malting ability.

A malting barley cultivar will only achieve its full potential with the appropriate environmental conditions, but these are not as easily controlled. Weather probably has the single largest effect on the malting quality of a barley and is beyond human control. Farming practices also affect barley quality and they are more easily manipulated. For example, amounts and timing of fertilization can have a marked effect on the protein levels in a barley. Barley with good malting quality is best achieved by growing the appropriate malting barley cultivars for a particular geographical area under the conditions most suitable for that area.

Selecting barley with malting quality is the next consideration. Malting barley quality will most likely be found in cultivars bred for a particular area. Samples of those cultivars should thus be examined for quality. The appearance of a barley is the simplest and first consideration in determining the malting potential of a sample. The barley should be uniform, plump, and have a good colour. If a sample has a good appearance, then more in-depth tests, such as protein and germination tests, should be performed. The results of the visual analysis and the protein and germination tests will give a good indication of a barley's malting potential (Bamforth and Barclay, 1993).

4.5.1 Barley homogeneity

A key component to a good quality malt is to start with a sample of homogeneous or uniform barley. Homogeneity is important with respect to all the barley traits that will be discussed. When a barley sample contains kernels with differing quality (i.e., a heterogeneous barley), the kernels will modify at different rates and the result will be under-modification in some of the individual kernels. The overall quality of the malt will thus be lowered and it can be difficult to process further in the brewery. On the other hand, it is possible to produce a trouble-free beer from a poor malt, provided it is homogeneous, as brewing conditions can often be adjusted to compensate for the poor quality (Narziß *et al.*, 1990; Bamforth, 1994). However, brewing conditions cannot be adjusted for a heterogeneous malt, where a multitude of conditions may be required to handle the varying quality present in the single sample.

A heterogeneous malt will produce several problems in a brewery. Problems include a low extract because of the encapsulation of endosperm cells and high viscosity that high levels of β-glucan can produce. Heterogeneous malt could cause lautering problems, plugged filters during beer filtration and beer hazes, all the result of excess β-glucan (Bamforth, 1982; Narziß *et al.*, 1989b). Malting a barley with a mixture of kernel qualities can also lead to poor fermentation in the brewery. Fermentation problems, in turn, can lead to taste defects in the final beer (Narziß, 1976). None of these problems are tolerated by breweries and, therefore, malt must be made from homogeneous barley.

Identifying a poor, heterogeneous sample of barley can be difficult. Differences in the physical quality of a barley sample can be ascertained by a visual examination or physical testing. For example, determining the homogeneity with respect to the plumpness of a barley sample can be performed through sieving. In contrast, determining heterogeneity with respect to protein content is very difficult, as individual kernels are not analysed. A related problem is the separation of the admixture to give a homogeneous barley sample. This can be achieved with plump and thin kernels, where separation can be performed using screens. In contrast, the separation of kernels with differing biochemical quality, protein, for example, would be impossible. The most practical solution is to select sound, uniform samples of malting barley cultivars that are known for homogeneity.

4.5.2 Whole, plump barley

Whole, plump barley is preferred for malting. Plump kernels contain a lower amount of protein and this will give a higher extract (Burger and LaBerge, 1985). Malting only plump kernels will also contribute to malt homogeneity, as thin kernels malt at a quicker rate than do plump kernels. The small and intermediate kernels can be separated from the plump kernels quite easily in the malthouse. However, the amount of thin and intermediate material must be kept to a minimum for economic reasons: separated, thin kernels, purchased at malting barley prices, will command only feed barley prices.

Broken and peeled kernels are another concern in malting barley. Such kernels often fail to germinate or germinate poorly because of a lost or damaged embryo, the living portion of the kernel. Any ungerminated or partially germinated kernels will result in a heterogeneous malt and the associated problems (Burger and LaBerge, 1985). The peeling of kernels in the finished malt can also cause problems when the hulls fall off during handling. Free hulls can collect together due to their light weight and static charge, resulting in high concentrations of hulls in a mash and consequent poor extract and taste defects in the final beer. Thus, barley intended for malting should have tightly adhering hulls.

4.5.3 Barley brightness

Malting barley should have a nice clean, bright colour. This will indicate good ripening and harvest conditions, suggesting a low moisture content in the barley – another important consideration for malting barley (Narziß, 1976). In contrast, a stained barley is often an indication of a microbial infection, which can lead to problems during malting and in the finished beer. Microbes can produce mycotoxins that are serious food safety concerns. The presence of microbes on a barley sample can also lead to water sensitivity (see Section 4.5.6) and thus upset malting schedules due to poor or slow germinations. Finally, microbial contamination on the barley can produce beer of poor quality including off-flavours and gushing, a phenomenon where beer is propelled at a high force from a bottle after opening (Burger and LaBerge, 1985). Fortunately, these problems can usually be avoided by selecting bright, unstained barley with no evidence of fungal contamination.

However, not all stained barley is necessarily contaminated. A barley may be simply discoloured due to weathering, which occurs when lodged barley, or barley left standing or lying in the field to dry, experiences inclement weather resulting in discoloured barley. Such barley can still produce quality malt, which can be valuable when good malting barley is in short supply. Therefore, the ability to distinguish simple discoloration from fungal contamination can be of economic importance.

4.5.4 Varietal purity

It is very difficult to produce a good quality malt from a barley that contains a mixture of barley cultivars. Different cultivars behave differently in the malthouse, taking up water and germinating at different rates. As well, different cultivars are likely to differ in their physical and chemical characteristics. Any of these differences could result in a heterogeneous malt and the associated problems (see Section 4.5.1).

Problems are best avoided by malting pure barley cultivars. However, cultivars with similar pedigrees often malt in a like manner and can, therefore, be malted together provided they have comparable quality. Batches of barley for malting should be grown in the same year and, if possible, in the same geographical area. This will keep problems, due to differences in malting performance, to a minimum.

4.5.5 Barley protein

A high protein barley often produces poor quality malt as a significant negative correlation can exist between barley protein and the amount of extract produced from a malt made from that barley (Figure 4.2). Such a barley can have less starch as a negative relationship exists between barley protein and starch contents (Briggs *et al.*, 1981). Lower levels of starch reduce the amount of extract in the resulting malt. Lower extract

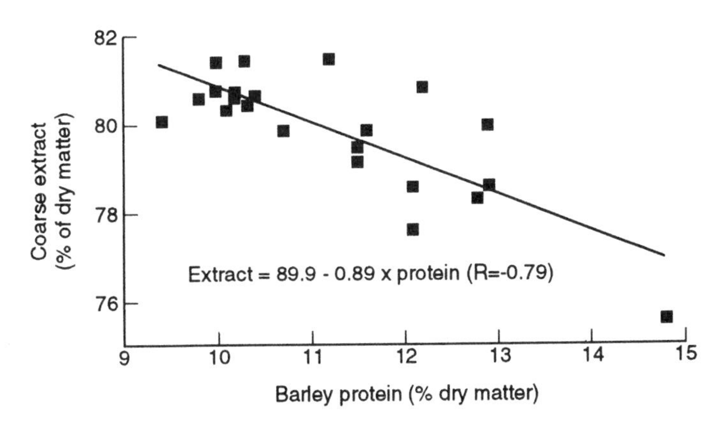

Figure 4.2 Relationship between barley protein and the coarse extract (Anon., 1987) in the resulting malt.

is an important economic concern due to the high cost of malt and, therefore, high protein is often a downgrading factor for malting barley.

In North America, the hot, dry weather normally experienced during the growing season tends to cause higher protein levels in barley. As a result, North American barley is often selected with protein levels as high as 12% (dry matter basis) for two-rowed and 13.5% for six-rowed barley. In contrast, European and Australian barley is seldom selected for malting if the protein level is over 11%, although higher levels may be required for some of the speciality malts such as those with a high degree of colour.

High protein malt can be diluted in the brewery with adjunct. However, producing low protein barley continues to be an objective of all those involved in the malting barley industry. Breeders are developing cultivars with lower protein levels, barley producers are being more careful with the fertilization of malting barley, and barley selectors select those barleys with the lowest protein levels.

Caution must be observed, though, as protein levels can also be too low. Protein levels are correlated with enzyme levels, so that barley with a very low protein level can result in malt with inadequate levels of some enzymes. The need for adequate enzyme concentrations is especially important in the brewing of beer with adjuncts. A certain amount of malt protein is also necessary for other reasons when brewing quality beer. Free amino nitrogen, predominantly amino acids, is necessary for the nutrition of the yeast. Large, soluble proteins also play important roles in defining beer colour and foam characteristics of a beer. Adequate amounts of the right size and type of malt protein are dependent on the action of proteases during malting and brewing, but the initial barley must also have an adequate level of protein.

4.5.6 Germination

The production of malt from a barley is dependent on the barley germinating. In the malting process, barley is germinated through a series of controlled processes. First, barley moisture must be increased to approximately 45% through steeping, which entails immersing the barley in water, often with intermittent air rests. The barley is then allowed to germinate under controlled moisture and temperature conditions. During germination the endosperm must become modified which is the systematic degradation of the endosperm through the action of newly synthesized or activated enzymes. The different enzymes must degrade β-glucan and protein, and begin starch degradation. Enzymes also cause the barley to respire. In order to keep respiration losses to a minimum, germination is brought to a halt through kilning after four to ten days of germination. Kilning is the controlled application of heat to the green malt. Kilning arrests the action of the enzymes and thus restricts further losses to respiration. However, temperatures must be carefully controlled in order to preserve enzyme activities in the finished malt as they are required for distilling and in the production of beer. The development of malt flavour and colour, necessary components of quality beer, also occurs during kilning.

Barley does not always germinate. Ungerminated kernels in a malt sample can lead to significant problems in the brewery such as those listed above (see Section 4.5.1). An acceptable barley sample should have a germination level of at least 95%. Barley that shows a poorer percentage should be avoided. However, the measurement of germination is time-consuming and costly. It is usually only measured in a sample that is acceptable with regard to all the other parameters mentioned.

A sample may not germinate for a number of reasons. It may not be viable or living because of improper grain drying, loss of the embryo, or it may have pre-germinated in the field and then died during storage. It may also be dormant. Dormant kernels are in a natural resting stage that, under normal conditions, will end with time. Plants rely on dormancy to prevent seeds from sprouting during inclement weather in the autumn. Dormancy is often broken by exposure to cold, which in nature, allows the seeds to germinate in the spring when they have the best chance of survival. Barley cultivars have often been bred for dormancy in order to prevent pre-germination in the field. However, such cultivars can also cause problems when they fail to germinate in the maltings early in the new malting season. Dormancy problems can often increase when damp weather occurs at harvest.

A barley needs to be treated carefully in order to overcome dormancy. Treatment should begin as soon as possible with drying, if necessary, of the freshly harvested barley. The barley should not be stored at low temperatures prior to drying or dormancy will be increased. After drying, the barley should be stored warm (35°C–40°C) for 3–10 days. The barley should then be cooled and stored at a low temperature. These steps should eliminate dormancy in a barley sample (Narziß, 1976), although they may not be practical, especially at farms.

Water sensitivity can also prevent a barley sample from germinating properly. Water sensitivity is observed in some samples of barley when they are left immersed in water too long during steeping. It has been suggested that rather than being water sensitive, the kernels have been oxygen starved (MacLeod, 1979). Some of the methods of overcoming water sensitivity support this theory; for example, steeping in a 0.1% peroxide solution or increasing the number or the length of air rests during steeping would both improve the oxygen supply to the barley.

Other methods of eliminating water sensitivity, such as heating the barley at 40–50°C for 1–2 weeks or treating the barley with a pesticide, could be destroying microbial organisms that may have been causing water sensitivity by depleting all the available oxygen (Narziß, 1976).

Water sensitivity is often present in only a small number of the kernels in a barley sample, leading to a very heterogeneous malt with a number of unmodified kernels being present. The result can be a very poor malt sample with all the problems outlined above (see Section 4.5.1). Water-sensitive barley, therefore, must be avoided or treated in a manner to overcome the sensitivity.

4.6 MALT QUALITY FOR BREWING

The quality desired in a finished malt will vary depending on its final use. In general, malt should have a high extract, the proper proportion of soluble protein and a good complement of starch degrading enzymes. It should also have low levels of a number of undesirable compounds, such as β-glucan, polyphenols and dimethyl sulfide precursor. When a finished malt has good quality with respect to the above mentioned parameters, then it should produce a trouble-free beer.

Different types of malt have different compositions, aromas, flavours and colours. Such malts are used to give special character and colour to the final beer. They are obtained by altering malting conditions. The quality required in the initial barley is relatively constant for the different malts, although slightly higher levels of protein are required in barley used for producing darker malts. The additional protein provides groups required for the Maillard reaction. This non-enzymic browning reaction, involving carbohydrates and amino groups, provides nearly all the colour components in dark malts (Narziß, 1976).

4.6.1 Enzymes

Enzymes are essential for the malting process, but they are also important components in the finished malt. During malting, enzymes are required for the degradation of endosperm cell walls, protein and starch. Some of the enzymes are already present in the barley and they are simply activated during malting, while others have to be synthesized *de novo*. During mashing, the first step in the brewing process, the malt enzymes are reactivated and they complete the breakdown of endosperm components.

The first enzymes of importance in malting are those that degrade the endosperm cell walls. The endosperm is the major storage organ of the barley kernel. It contains most of a kernel's starch and protein, which serve as nutrients for the embryo during germination. Endosperm cell walls can restrict access to the contents of endosperm cells which contain the starch and protein. The cell walls consist of β-glucan (70%) and arabinoxylans (25%), as well as some protein and other minor amounts of carbohydrates. One of the most important events in malting is the breakdown of these cell walls by a family of enzymes that are produced or activated during early stages of malting. Most of these enzymes hydrolyse the β-glucan in the cell walls, although some arabinoxylan- and protein-degrading enzymes may also be involved. The breakdown of the cell walls opens up the cells to attack by starch- and protein-degrading enzymes. This breakdown of endosperm structure, or modification as it is often referred to, must be complete throughout the endosperm if maximum extract is to be realized during the brewing process (Bamforth and Barclay, 1993).

Proteases (protein-degrading enzymes) are needed to produce free amino nitrogen and soluble proteins (see Section 4.5.5). Proteases, which consist of a large family of endo- and exo-acting enzymes, are most active during the malting of barley as they are heat labile and, therefore, rapidly inactivated at typical mashing temperatures.

Starch-degrading enzymes are very important: they hydrolyse starch into simpler sugars. Most of the sugars are metabolized by yeast into alcohol; however, some of the larger sugars, termed dextrins, are not fermentable, but they are important in terms of mouth-feel in the finished beer. The starch degrading enzymes, which consist of α-amylase, β-amylase and limit dextrinase, as well as a number of minor enzymes (see Section 4.9.2, Standard malt analysis), perform most of their hydrolysis during mashing.

4.6.2 Extract

Extract is the most important component of malt. It is the material that can be solubilized during mashing, and it determines how much beer can be made from a given weight of malt. Extract is predominantly soluble sugars and starch breakdown products but also consists of soluble protein and many other compounds; therefore, the amount of extracted material is largely dependent on how much starch and protein there is in a barley. It will also depend on how soluble that material is after malting and mashing. Solubility will depend on the amounts of protein- and starch-degrading enzymes in a malt, as well as the ease of access these enzymes have to their appropriate substrates. Access will be significantly increased by adequate degradation of endosperm cell walls (see Section 4.6.1).

The size of starch and protein within the cells may also affect the extract. Research suggests that small and large starch granules can both be solubilized, to different degrees. Small granules are thought to be less digested during mashing than large granules and this can lead to problems during brewing (Bamforth and Barclay, 1993).

4.6.3 β-Glucan

Inefficient separation of extract from spent grains can also restrict the amount of extract. The non-starch polysaccharide, β-glucan, is often a major factor in this separation. Besides acting as a barrier to enzymes in the intact endosperm (see Section 4.6.1), high amounts of β-glucan can lead to high viscosities during mashing, as it is a large, water-soluble molecule when released from cell walls. An extract with a high viscosity can lead to slow lautering which can reduce malt extract. Insoluble β-glucan can also plug the filter bed and bring the filtering process to a stop, leading to further loss of malt extract.

The easiest way to avoid these problems is to degrade the β-glucan during the malting process. This requires a barley with the ability to modify completely, which is only possible with the appropriate β-glucanases and an endosperm texture that allows enzymes access to the β-glucan (Palmer, 1975a). Intact β-glucan in the finished malt can be degraded during mashing, but if the β-glucan is not initially soluble, it will often be released only after starch gelatinizes. This occurs at temperatures where the β-glucanase enzymes are inactive (Erdal and Gjertsen, 1967). Therefore, malts with significant amounts of insoluble β-glucan can lead to the β-glucan problems mentioned above. β-Glucan that remains in solution after mashing and lautering ends up in the finished beer where it can lead to beer filtration problems (Narziß *et al.*, 1989a), as well as beer hazes.

4.6.4 Other undesirable barley/malt components

Barley contains other components that can reduce the quality of malt. These include polyphenols and dimethyl sulfide precursors. Polyphenols cause problems by binding and precipitating proteins in the finished beer, thereby causing beer haze. The majority of polyphenolic substances in finished beer come from hops, although a significant amount is derived from the malt. The level of polyphenols in a finished malt can be affected by the quality of the initial barley, as barleys with low protein levels have been shown to have higher levels of polyphenols than high protein barley. The level in the final malt is also influenced by the final kilning temperature, with higher temperatures producing higher levels of polyphenols in the finished malt (Narziß, 1976). Barley cultivars with no anthocyanogens, a type of polyphenol, have been bred. Beer produced from such barley has shown a significant improvement in quality with respect to haze formation. Unfortunately, these barley cultivars show poor agronomic properties, but breeding programmes continue to address this problem (Von Wettstein *et al.*, 1985).

Dimethyl sulfide (DMS) can be a problem in finished beer because it produces a distinctive, vegetable-like flavour. The precursors of DMS are formed in the malt during germination. Levels of DMS can be minimized by altering malting and brewing conditions such as kilning (Briggs *et al.*, 1981) and wort boiling (Narziß, 1985). However, it is possible that barley cultivars could be developed that would produce reduced levels of the DMS precursor.

The levels of polyphenols or DMS precursors are not closely monitored in testing commercial malting barley for quality. The biochemistry of these compounds within the barley kernel and the relationship between levels in the barley and in the finished product are poorly understood and continue to be studied. In the meantime, good malting and brewing techniques will continue to be used to keep problems associated with these compounds to a minimum.

4.7 MALT QUALITY FOR DISTILLING

Barley malt is used in the production of both malt and grain whiskies. Malt whisky is prepared only from barley malt with no adjuncts. The malt for malt whisky, therefore, should contain high levels of starch and only enough enzyme to degrade the endogenous starch in the malt. Barley required for such malt should be plump and have low levels of protein (Bathgate and Cook, 1989).

Grain whisky contains limited amounts of barley malt, which is used largely as a source of enzymes, while cereal grains are the main source of starch. The high levels of enzymes in the malt are needed to degrade the starch derived from the cereal adjuncts. The barley used to make the malt for grain whisky should have higher levels of protein because it must have the potential to produce the increased levels of enzymes (Bathgate and Cook, 1989).

The other quality parameters required in a barley for whisky malt, be it for malt whisky or grain whisky, are similar to those required in a barley for brewing malt.

4.8 BARLEY ADJUNCT QUALITY

Barley is predominantly used as an adjunct because it provides extract for beer or whisky at a much cheaper cost than does malt. Therefore, the primary quality concern of a barley adjunct is that it provides maximum amount of extract. This is most easily achieved by using plump, low protein and high test weight barley. However, the use of barley as an adjunct can cause processing problems due to the barley β-glucan. A portion of the adjunct's β-glucan is solubilized at mash temperatures in excess of 60–65°C (Erdal and Gjertsen, 1967), temperatures at which the malt's β-glucanases are no longer active. The result is high levels of β-glucan in the final mash that can lead to problems outlined above (see Section 4.6.3).

Reports indicate that the processing of barley adjunct prior to mashing can reduce the size of β-glucan molecules. Processing, which can be in the form of heat or infrared radiation, is used to gelatinize the starch. However, even with such processing, it appears that industrial enzymes, containing β-glucanase activity, are necessary if β-glucan problems are to be consistently avoided when using barley adjunct in brewing (Byrne and Letters, 1992).

Another concern with barley adjunct is the presence of endogenous protease inhibitors. These anti-nutrients can inhibit malt proteases and this can lead to inadequate protein hydrolysis during mashing. Problems such as poor yeast nutrition, because of a deficiency in free amino nitrogen, and poor beer foam can result (Enari *et al.*, 1964). However, these inhibitors can also be beneficial as they can be used to prevent excess proteolysis in high protein malt. Brewing with barley adjunct can thus be used to inhibit proteolysis and prevent problems such as beer hazes (Enari *et al.*, 1964).

4.9 QUALITY ANALYSIS OF MALTING BARLEY AND ITS MALT

4.9.1 Barley

Appearance

Determining the quality of malting and adjunct barley is not simple. Much of the assessment is based on the appearance of the sample. An accurate visual assessment can only be made by experienced professionals. They are trained to look for the proper colour, smell and general appearance of a sample. They have to be able to differentiate poor colour caused by harmful microbial contamination from that caused by weathering. Selection is somewhat subjective, but with proper training and the use of visual barley standards, quality barley can be successfully selected.

Standard methods

The physical and chemical methods used to measure barley characteristics are more objective. Standard methods exist for nearly all the barley parameters of interest. Most of the methods have official status, meaning they have been tested collaboratively for accuracy and reproducibility by organizations such as the American Society of Brewing Chemists (Anon., 1992a), European Brewing Convention (Anon., 1987), Institute of Brewing (Anon., 1982), American Association of Cereal Chemists (Anon., 1976) and the Association of Official Analytical Chemists (Anon., 1990). Each organization produces a list of standard methods in which exact details on the execution, calculations and expected reproducibility for each official method can be found. Not all methods of interest, however, are official methods, as they may not yet have been tested collaboratively or they could still be at the development stage. Some methods may be too specific for the particular organization, or methods may not be available for testing for proprietary reasons.

Barley methods of interest include those for proper sampling, moisture content, protein content, germination, kernel sizing and varietal purity. Proper sampling cannot be over-emphasized and if done poorly, meaningful results for all the other analyses will not be possible. Official methods for sampling and the measurement of moisture and kernel size are relatively straightforward. Moisture is most often determined using slow oven-drying techniques, although quicker methods based on near-infrared technology exist. Kernel size is usually expressed as percentage of plump kernels, which is measured by shaking barley samples on standard sized screens.

Methods for germination, protein and varietal purity are not as straightforward. Tests for germination are especially complicated as viability, dormancy and water sensitivity are being measured. In addition, the different organizations – ASBC, EBC and the Institute of Brewing – each recognize different methods. Therefore, when analysing for germination, care should be taken to test for the right factor and to specify the method used. In general, viability is measured using a staining procedure or by germinating in hydrogen peroxide. Dormancy and water sensitivity are measured by germinating barley in varying amounts of water, which can take up to five days.

The Kjeldahl method of protein determination has been the standard method for many years. Near-infrared reflectance and transmittance methods have been, however, developed recently and given official status (Anon., 1992a). Even more recently, the Dumas combustion method for measuring nitrogen has been investigated (Foster, 1992; Foster, 1993; Buckee, 1994). This method is quicker, less labour intensive and it produces no acidic wastes; it has been shown to be adequate for measuring nitrogen in barley and barley malt samples; however, repeatability has been a problem when measuring nitrogen in liquid samples such as wort and beer.

Determining varietal purity can be difficult. A visual examination is often sufficient, especially in countries where varieties are released for malting only if they can be distinguished visually. However, the best method for distinguishing barley varieties at present is based on separating barley hordein proteins using polyacrylamide gel electrophoresis (Anon., 1987). Each variety tends to have a distinct pattern, similar to a fingerprint, when its protein bands are separated. However, the method is time-consuming and experience with band pattern recognition is necessary. High performance liquid chromatography methods have also been developed for separating hordeins, but these methods are expensive (Marchylo and Kruger, 1984). More specific methods, based on DNA, are in the process of being developed (Ko *et al.*, 1994).

Prediction of malting quality

The potential malting quality of a barley sample is always of interest. It is very hard to predict the malting quality of a barley by examining only the barley. However, producing a malt from the barley and then analysing the malt is very time-consuming and can take up to two weeks. Therefore, methods have been developed for predicting the malting quality of a barley. These methods include prediction equations, which give a prediction of the malt extract to be expected based on the protein content, thousand kernel weight and a varietal constant for the barley. Malt potential is also predicted by performing an extract on the unmalted barley using enzyme supplementation and this method does have official status (Anon., 1987, 1992a). Several other methods for predicting the potential malting quality of a barley have also been proposed; including milling energy (Allison *et al.*, 1976), degree of sedimentation (Palmer, 1975b) and near-infrared technology (Henry, 1985).

4.9.2 Malt

Micro-malting

The most common method for testing the potential malting quality of a barley sample remains the malting of the sample. Micro- and pilot-maltings have been developed to simulate malthouse conditions and they require from 100 g to over 10 kg of barley. The quality of the malt produced can then be analysed using standard methods such as those used in malthouses as part of their routine quality control.

Several micro-systems have been developed. These give relatively good estimates of a barley's malting quality provided the malting conditions are closely controlled and monitored, but this can lead to labour-intensive systems. Several commercial micro-maltings have recently been released that are fully automated and thus reduce labour costs, although they are relatively expensive. Despite the cost of the labour and the equipment, micro-malting remains the best means of determining a barley's malting quality.

Standard malt analysis

Standard malt analyses are contained in all three of the official lists of methods (Anon., 1982, 1987, 1992a). Important analyses include those that measure the amount of extract that a malt will deliver into solution under standard conditions, the amount of solubilized protein in that extract and the amounts of enzymes in the malt. A number of methods have also been approved for determining the degree of modification of a malt sample.

Malt extract indicates the amount of brewhouse yield to be expected from a particular malt. The laboratory method for extracting malt simulates the mashing step of the brewing process, in that ground malt is mixed with water, the temperature of the mash is raised according to a standard temperature programme and the malt extract is then separated from the solids by a filtration step. The standard method is often criticized for failing to match commercial conditions because of differences in grind coarseness, malt-to-water ratios and the temperature programme. However, the Congress extract (Anon., 1987), as the method is often called, has become a standard method of analysis for malt quality in most parts of the world. The percentage of extract is calculated from the density of the Congress extract. A normal malt should be 76–84% (dry matter basis) soluble. The extract figure will depend on the barley variety and growth location, as well as the year of growth. Barley characteristics, such as protein content, hull thickness and kernel plumpness, will also affect extract (Narziß, 1976).

Enzymes of most interest in malt are those that degrade starch. α-Amylase and diastatic power are the most common measurements for this activity. α-Amylase attacks internal linkages of starch molecules and is important for the rapid breakdown of starch into soluble dextrins. Diastatic power is a family of enzymes that degrade starch molecules into smaller soluble sugars. α-Amylase and β-amylase are the main constituents of diastatic power, with β-amylase attacking the products of α-amylase hydrolysis to produce maltose. Limit-dextrinase and α-glucosidase are other important enzymes that form part of diastatic power. Diastatic power is especially important for hydrolysing adjunct starch in mashes.

Enzyme activities in malt samples are determined by first extracting the enzymes from the malt using specialized conditions. There are a number of official methods available that can then be used to quantify either α-amylase or diastatic power. These are nearly all colorimetric methods based on the breakdown of starch by the enzymes. Desired levels of activity for starch-degrading enzymes are dependent on the type of malt desired. Malt that is to be used with an adjunct should contain higher levels of activity than a malt that will be used in an all-malt brew. Similar to extract, the amount of starch-degrading activity in a malt sample will depend on barley variety and growth year, as well as growth location. The amount of protein in the original barley will affect enzyme activities, as will the malting process, especially kilning conditions.

The amount of soluble protein in a malt is an important quality consideration. It can influence beer processing as well as the quality of the final beer (see Section 4.5.5). Soluble protein in a Congress extract is measured using either the Kjeldahl or Dumas methods. The amount of soluble protein desired in a malt will depend on the amount of total protein in the malt. In a good malt, 40% of the protein should be soluble. This value will depend on the amount of protein in the original barley, the amount of proteolytic activity (proteases) produced during malting and on the degree of modification of the malt. The more modified a malt is, the greater percentage of soluble protein it will contain.

The fine/coarse difference test is the most common test for estimating malt modification. In this method, a malt sample is divided into two portions; one is ground coarsely and the other ground finely. The ground malts are then mashed under Congress extract conditions and their percentage extracts determined from the densities of the two extracts. The difference between these two extract values is an indication of the breakdown of the endosperm structure and thus a measurement of modification. Two newer methods, the Calcofluor-Carlsberg method (Aastrup and Erdal, 1980) and friability (Chapon *et al.*, 1980), are also recognized as official methods for estimating malt modification. In the Carlsberg method, the fluorochrome, Calcofluor, is used to stain the β-glucan in longitudinal sections of malt kernels. Modification is based on the somewhat subjective quantification of the fluorescent light given off by the stained β-glucan, with greater fluorescence indicating poorer modification of malt. A more objective method, which uses image analysis for quantifying the fluorescence, has recently been reported (Wackerbauer *et al.*, 1993). The principle of the friability estimation of modification is based on the friable nature of well-modified malt. The Friabilimeter grinds the friable portion of a kernel to flour and then separates it from the non-friable portion.

Malt homogeneity

The measurement of malt homogeneity is important but difficult. The two newer methods for measuring modification, the Calcofluor-Carlsberg method and friability, can also be used to measure homogeneity. Both methods have received official status (Anon., 1987).

4.10 FOOD BARLEY

A wide variety of food types have been made from barley over the years. Archaeological finds indicate that when barley was first cultivated, it was already being ground and used in flat breads. Whole barley was roasted, ground and then eaten with oil or other condiments. It was also cooked and used as a porridge or in soups. As time passed, the use of barley in bread decreased as wheat bread became the bread of choice. However, ground barley continued to be used when wheat was not available. Milled barley is used today in North Africa where it is pearled, coarsely ground into semolina and used for cous-cous; more coarsely ground for soups; or finely ground for use in breads (Salem and Williams, 1987). In Southeast Asia, some ground barley is incorporated into noodles. Significant quantities of pearled barley are also consumed in the form of rice extenders in Korea and Japan. In Europe and North America, the major use of barley for food is as an addition to soups in the form of pearled barley, although limited amounts of barley flour are also used as food thickeners and are processed into starch (Bhatty, 1993).

Barley is presently being promoted in many parts of the world for the ability to contribute to a healthy life. It is reputed to reduce heart disease, colon cancer and postprandial glucose concentrations (important in diabetic people; Jenkins *et al.*, 1978).

Barley has been observed to reduce blood cholesterol (hypocholesterolaemia) and thus reduce heart disease. Soluble β-glucan is believed to be the barley component most responsible for this activity, although there is some controversy to this claim (see Bhatty, 1993 for review). Oats have been a major source of β-glucan in the human diet through the 1980s and into the 1990s, but the higher level of β-glucan in barley has enticed workers to promote the consumption of barley β-glucan. Future uses of barley could include partial replacement of wheat flour, especially in doughs such as muffins and cakes where the gluten from wheat flour is not required. Increased use of pearled barley in salads as well as in soups is also being promoted. Promotions in Japan include more traditional uses of barley such as rice extenders.

Increased uses of barley for food are already being seen. Australia has had a new breakfast cereal introduced which is based on waxy barley (McIntosh *et al.*, 1993). There have also been increases in demand for rice extenders in Japan and barley flour in North America, but significant increases in demand for food barley have yet to materialize. It may take some time before appropriate barley foods are available and people are convinced of their potential benefits; however, it is unlikely that barley will ever regain the importance it once had as a food.

4.11 QUALITY OF BARLEY FOR FOOD

The quality of barley required for food is not well documented in the scientific literature. The limited use of barley for food over the past 50 years has restricted the need for scientific studies on food barley quality, although individual food companies that process barley are well aware of the quality required in their raw barley. A better understanding of barley quality is now required as new food uses are developed for barley and as traditional food barley users begin to rely more on imported barley to meet their needs. As a result, barley exporters and barley processors now need to have a better understanding of the quality needs of their food barley customers.

4.11.1 Barley pearling quality

A large proportion of the barley used for food in the world is pearled before further processing. Pearling is the mechanical removal of successive outer layers of the barley kernels. Barley is pearled to remove excess fibre, which is unsuitable for human consumption, and to produce a more uniform starting material for further processing. It removes variation in hull content by removing any adhering hulls, reduces inconsistencies in the colour and shape of the kernels, and removes barley bran. There are two reasons for removing the bran. First, it can give an undesirable colour to finished products; the colour problem exists regardless of the bran colour, so all brans should be removed by pearling (Kent, 1983). Second, the bran can cause problems with milling (see Section 4.11.2) if it is not removed prior to this process.

The quality of the pearled product is dependent on the initial barley. For example, food barley should have shallow ventral creases because traces of crease in the pearled product result in distinct black lines that can be mistaken for contamination. Pearling times can be adjusted to remove the majority of the creases, but this also increases pearling losses which is uneconomical. Barley with a blue aleurone has sometimes been implicated as a cause of poor colour in final pearled products (Kent, 1983). When blue aleurone barley is used for food processing, the colour may have to be removed with bleach or by more extensive pearling in order to remove the last traces of the aleurone.

The percentage extraction, or yield, after pearling is an important economic concern. The degree of extraction depends on the end-use of the pearled product, but the quality of the initial barley is also important. As noted above, the depth of crease and blue aleurones can affect the degree of pearling required and thus the yield of final product. Extraction rate is also affected by the plumpness of the kernels. The plumper the kernel, the more kernel that remains after pearling. The shape of the kernel can also have an effect on yield. A short, round kernel has less hull than a long, flat kernel (Narziß, 1976) and, therefore, will require less pearling.

A key requirement for pearled barley is a white end-product. Rice extenders must be white to resemble rice after cooking and flour ground from pearled barley must be white if acceptable baked products are to be produced. The colour of the pearled product is dependent on the colour of the endosperm, with a white endosperm being desired (Pomeranz, 1974). Barley with a yellow endosperm gives an undesirable product. The colour of pearled barley can also be affected by the texture of the endosperm. White, floury kernels produce a purer white product, compared to that obtained from steely (vitreous) kernels, which have a glassy, grey look to their endosperms. The colour of pearled, steely kernels is acceptable. However, a problem arises when a sample contains a mixture of floury and steely kernels, as the final product has both pure white and dull white kernels which is unacceptable.

Endosperm texture can also affect the extraction rate. This is especially true if a barley sample contains a mixture of floury and steely kernels because these two textures have different degrees of hardness. Harder steely kernels require extra pearling time to achieve the desired end-product. Therefore, when a mixture of steely and floury kernels is pearled, the steely ones will be larger and longer, and the floury kernels will be rounder and smaller (Figure 4.3). The desired product can only

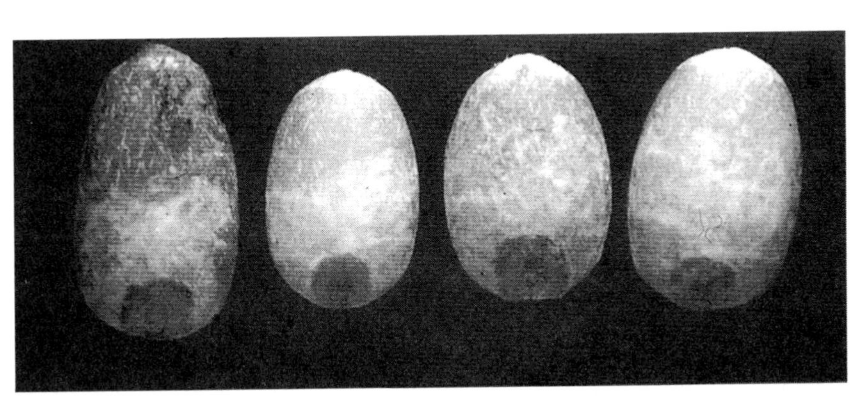

Figure 4.3 Picture showing a pearled, steely kernel (left kernel) of hull-less barley versus pearled floury kernels (three right kernels) of hull-less barley.

be achieved by a compromise. Therefore, in a mixed sample, floury kernels are over-pearled and steely kernels are under-pearled, resulting in a mixture of kernel sizes and some unwanted flour. Yield of the finished product is also reduced. These problems, which are unacceptable in the food industry, can only be overcome by pearling homogeneous, all-floury or all-steely barley.

Pearled barley for soup requires no further processing. The important concern is uniformity of finished product (Pomeranz, 1974). Consistent colour is important, but uniform size of the pearled product is more significant as it can affect cooking time as well as appearance. Uniform size is best achieved by starting with an all-steely or all-floury barley. Therefore, the uniformity of barley texture is more important than the actual steeliness or flouriness of a sample.

4.11.2 Barley milling quality

Knowledge of milling quality of barley has been extrapolated from the wheat industry. The actual milling of barley and wheat is quite similar, but there are some important differences. Covered barley contains a hull that has to be removed, usually by pearling, before any milling can be performed. Hull-less barley contains no hull and is thus more suitable for milling. It has, in fact, been milled since the early days of Mesopotamia. The lack of hull can result in a 10% increase in the milling yield as compared to covered barley (Kent, 1983).

However, the bran of both covered and hull-less barley must be removed before milling. Barley bran is brittle and shatters during milling, in contrast to wheat bran which comes off the kernel in flakes. The shattered pieces are much more difficult to separate from the flour (Bhatty, 1993). The problem can be prevented by removing the bran with pearling. Therefore, all milling barley, regardless of hull type, tends to be pearled or decorticated prior to milling.

Pearled barley needs to have certain characteristics if it is to be successfully processed into a final product. The barley required for milling in northern Africa should have steely kernels, indicative of hard kernels, which are required when milling a grain into semolina (Salem and Williams, 1987). Barley requirements for other milling uses are not so specific, although colour is again an important consideration. Reports (Melland *et al.*, 1984) indicate that final products made from barley can have objectionable colours that may require extra processing, although pearling prior to this processing may reduce the problem. It has also been indicated that the exterior of the barley should be free of weathering and microbial contamination if discoloured flour is to be avoided (Pomeranz, 1974).

4.12 MEASUREMENT OF FOOD BARLEY QUALITY

There are no official methods for analysing food quality *per se*. However, standard methods for analysing other types of cereal grains, including malting and feed barley, can be used to determine quality aspects of food barley such as protein content, moisture content and plump kernels. Standard methods used for analysing soft wheats have been used successfully to measure the milling quality of barley (McGuire, 1984). Standard methods exist for measuring various fibre fractions in other cereals, for example, β-glucan, soluble dietary fibre and total dietary fibre (Anon., 1976). These methods could be adapted for barley if health-based markets develop for food barley and ratification for food labelling is required.

Official methods for analysing the pearling quality of barley are unknown despite the importance of this food barley process. A method closely aligned with commercial pearling is used by the Japanese barley processing industry and is outlined in Figure 4.4. Using this method, a sample is sized and cleaned prior to processing. The barley is then pearled to extraction percentages similar to those commonly used in the industry (55% for covered barley and 60% for hull-less barley). Extraction times can be calculated by a preliminary run or by measuring extraction

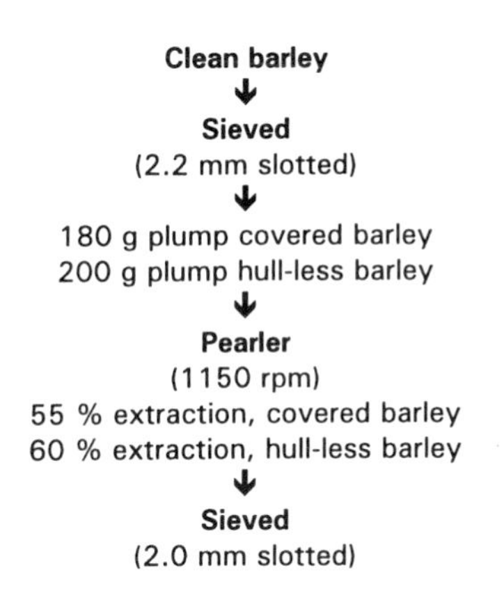

Figure 4.4 Outline of a method that can be used to investigate the pearling quality of a barley sample.

rate at an intermediate time. Results of interest include the time required to achieve the desired extraction rate, the percentages of steely kernels, as determined by an experienced eye, broken kernels and the amount of material lost as fines during the pearling process.

A sample of results for several different barley types is presented in Table 4.5. Results are presented for covered barley, hull-less barley and waxy barley. The waxy barley shows an advantage in terms of the quality parameters investigated. However, only a few waxy barley lines were tested and a wider range of samples must be studied to produce more conclusive results. McIntosh *et al.* (1993) indicate that waxy barley possesses unique food processing attributes. Such attributes could be related to unique pasting properties of waxy barley (Goering *et al.*, 1973).

Table 4.5 Ranges of pearling quality seen in three types of barley

Barley	*Pearling time** *(minutes)*	*Steely kernels (%)*	*Broken kernels (%)*
Covered (n = 15)	3.6–8.3	12–80	8–60
Hull-less (n = 6)	2.5–4.5	10–45	10–60
Waxy (n = 4)†	2.2–4.4	<–10	<5–30

*Time to reach 55% extraction (covered) or 60% extraction (hull-less).
†Waxy were all hull-less type.

There were few differences in the results obtained for the hull-less versus the covered barley samples. The hull-less samples required a shorter pearling time, but the degree of extraction was greater (Figure 4.4). However, a shorter pearling time and a higher extraction rate are benefits of using hull-less barley. The covered and hull-less samples showed a wide range in quality with respect to the percentage of steely and broken kernels. A more thorough investigation of these traits would help to determine the cause of such a range, whether it is inherent or environmental.

4.13 FEED BARLEY

More barley is used as feed than for all the other uses of barley combined. On average, over the past 20 years, 75% of the barley grown in the world (Table 4.2) has been used as feed (Gudmunds and Webb, 1994). Barley is fed to ruminants, such as cattle, sheep, goats and camels, as well as monogastrics, predominantly pigs but also poultry. Extensive use of barley as feed is due to the low price at which it can supply energy in the form of starch.

4.14 QUALITY OF FEED BARLEY

4.14.1 Plumpness

Barley plumpness is an important quality consideration for aesthetic as well as practical reasons. Variable plumpness in a sample can affect the processing of barley for ruminants. Most of the barley fed to cattle is rolled or flaked before feeding. This is necessary if maximum feeding value is to be achieved. The rolling and flaking processes can be adjusted to accommodate either thin or plump kernels, but not a mixture of the two. Rollers that are set for plump kernels allow thins to pass through in whole form, while rollers set for thin kernels crush the plump kernels. This results in the production of fines, which can cause digestive problems when fed to cattle or they can be lost during processing. Barley with a uniform size is therefore desired.

Barley plumpness is not a concern when barley is fed to monogastrics. The processing of barley for monogastrics involves hammer milling, which is not affected by kernel plumpness. There have been indications that the nutrient value of plump barley for monogastrics is greater than that of thin barley, but this has not been substantiated (Bhatty *et al.*, 1974).

4.14.2 Energy content

The digestible energy content of barley for ruminants is similar to that of other feed grains. The presence of a hull on barley is of limited concern as it is relatively digestible by ruminants. However, the indiscriminate feeding of barley to ruminants can lead to an upset in the microbial population of the rumen, a result of the quick digestion of barley starch (McAllister *et al.*, 1990), causing acidosis or bloat. This is true when cattle are started too quickly on barley or are fed too much at one time. In small ruminants, for example, goats and sheep, the problem is avoided by feeding whole barley (Barnes and Orskov, 1982). In large ruminants, problems can be avoided by starting animals slowly on feed barley and by feeding barley several times in a day versus one large feeding.

Barley causes less bloat and acidosis than when wheat is fed to ruminants. The barley hull helps to dilute the effect of the rapid starch digestion. Therefore, there would appear to be no advantage in feeding hull-less barley to ruminants. However, ruminants fed high levels of roughage (for example, dairy cows) perform well when fed hull-less barley, showing no signs of bloat or acidosis.

The barley hull is a disadvantage when barley is fed to monogastrics. It is indigestible, and so reduces the energy content of covered barley.

Monogastrics, therefore, have to consume larger quantities of barley-based diets in order to take in adequate levels of energy. Hull-less barley shows a significantly higher level of energy for monogastrics (see Table 4.6). The energy value of hull-less barley for swine approaches that of wheat, and this has led to large amounts of hull-less barley being fed to swine in Western Canada.

The energy level of barley for poultry is also affected by the polysaccharide, β-glucan. Barley β-glucan reduces the availability of energy and protein in poultry fed diets based on barley. It also causes poultry to produce sticky faeces which can lead to management problems. These problems can be overcome by supplementing poultry barley diets with enzymes that contain β-glucanase activity (Campbell and Bedford, 1992). The enzymes hydrolyse the β-glucan and thus prevent the problems of low energy and protein availability, as well as sticky faeces. The result is an increase in metabolizable energy of barley for poultry as indicated in Table 4.6. Hull-less barley also requires enzymes when included in poultry diets. When enzymes are included, the metabolizable energy of hull-less barley for chickens surpasses that of feed wheat (Table 4.6).

β-Glucan appears to have little effect on swine nutrition, although the evidence is sometimes contradictory. It would appear that the high moisture, low pH and long length of stay in a pig's stomach all combine to hydrolyse β-glucan and so eliminate the deleterious effect. The resulting β-glucan oligosaccharides have little or no effect on digestion in the small intestine (Classen and Bedford, 1991). There is a possibility of β-glucan problems in swine if barley β-glucan levels are excessively high, as is possible in waxy barley.

4.14.3 Protein and amino acid content

The protein found in feed barley is an added bonus to this energy-rich feed ingredient. Essential amino acids that are supplied by protein are required by animals for proper growth. Barley protein levels are significantly greater than those of corn and often lower than those of rye, wheat and oats (see Table 1.6). The levels of essential amino acids in barley (gram per dry matter basis) are similar to those for rye, wheat and oats as exemplified by lysine, methionine and threonine, which are the first three limiting amino acids in these cereal grains. The availability of amino acids in covered barley is somewhat lower than that of other cereal grains (Anon., 1992b).

Hull-less barley has higher levels of essential amino acids (gram per dry matter basis) than covered barley (Edney *et al.* 1992). In contrast to high-lysine lines, in which increased lysine is a result of genetics, the increased levels in hull-less barley are a result of the missing hull, which

Table 4.6 A comparison of the nutritive value of covered barley, hull-less barley and wheat for ruminant and monogastric animals

	*Energy**			*Protein*		*Lysine*	
Grain	*TDN*[†] *(%)*	*DE*[†] *(kcal/kg)*	*AME*[†] *(kcal/kg)*	*Amount*[‡] *(% dry matter)*	*Swine availability*[†§] *(%)*	*Amount*[‡] *(% dry matter)*	*Swine availability*[†§] *(%)*
Barley							
Covered	74	3140	2720 (2820)[¶]	11.5	69	0.49	67
Hulless	76	3350	2800 (3200)	15.0	78	0.53	89
Wheat	78	3430	3000 (3000)	15.8	82	0.43	72

*TDN, total digestive nutrients (ruminants); DE, digestible energy (swine); AME, apparent metabolizable energy (poultry).
[†]'As is' basis.
[‡]Edney *et al.*, 1992.
[§]Anon., 1992b.
[¶]Without β-glucanase supplement (with β-glucanase supplement).

serves as a diluent to the amino acids when it is present. Levels of the essential amino acids in hull-less barley are generally greater than in feed wheat.

Essential amino acids in hull-less barley are also more available to swine (Thacker *et al.*, 1988; Helm *et al.*, 1989). Higher availability, along with the increased levels of amino acids, reduces the need for expensive protein supplements in hull-less barley-based feeds thereby reducing the cost of these diets. The increased availability of amino acids also reduces the nitrogen content of the manure from swine fed hull-less barley. This is a very important consideration, as water pollution due to nitrogen run-off from manure spread in fields is becoming a concern in an environmentally conscious world.

4.14.4 Colour

Appearance of a barley sample is a consideration when buying barley for feed. Colour of barley has limited effect on the feeding value of the barley, although staining may indicate fungal contamination and thus the possible presence of mycotoxin. If there is evidence of fungal contamination, further testing for mycotoxins is advisable.

4.15 MEASUREMENT OF FEED BARLEY QUALITY

Nutrient composition of feed ingredients, as required for diet formulation, continues to be widely based on proximate analysis. Proximate analyses include determination of the amounts of crude fibre, crude protein, ether extract and ash. Official methods exist for all of these components (Anon., 1990). These methods are often tedious and time-consuming, and feed mills have had to revert to book values for determining the nutrient composition of feed ingredients. However, economics are demanding that feed mills obtain maximum nutrient value from their feed ingredients and, therefore, more accurate knowledge of the nutrient composition of feed ingredients is necessary. As a result, quick, inexpensive methods for determining nutrient composition are becoming more standard across the industry. Some of these methods allow feed companies to determine the nutrient quality of their ingredients as the ingredients are being unloaded at the mill. This allows more efficient use of all the nutrients in the ingredients.

4.15.1 Standard methods

Test weight (bulk density) continues to be used as a quick indication of the feed value of a barley sample. However, research indicates that the feeding value of barley is not well correlated with test weight, especially for heavier barley (Coates *et al.*, 1977; Mathison *et al.*, 1988). More meaningful methods for measuring fibre are becoming the methods used by industry to determine feeding value. These methods include acid-detergent fibre (ADF) which gives an indication of the digestibility of a barley sample by ruminants. Another method, neutral-detergent fibre (NDF), indicates the potential intake of barley by ruminants or the digestibility of barley by monogastrics (Van Soest, 1973; Anon., 1990). The ratio of NDF/ADF also gives a measurement of the feeding value of barley for ruminants. The lower the ratio (i.e., the closer to one), the better the nutritional quality, as a certain amount of ADF is important if acidosis is to be avoided.

4.15.2 Prediction equations

Equations are available for predicting the energy and amino acid content of a feed barley sample. These are based on simple standard proximate analysis such as fibre and protein for energy (Christison and Bell, 1975), and protein for amino acids (Dupchak and Hickling, 1990). Results are not as accurate as actual measurement of the parameters, but are quicker and more practical than feeding trials or amino acid analysis. Prediction equations are also of more value than relying on book values.

4.15.3 Near-infrared technology

Near-infrared (NIR) technology can rapidly determine the feed value of barley; it is already in use for predicting moisture and protein contents. Calibrations for fibre and energy in barley have also been developed (Gerstenkorn, 1985; Edney *et al.*, 1994), although the price of the scanning equipment which is required for predicting energy, and the expense of feeding trials which are required for developing the calibrations, have restricted the widespread use of such calibrations. However, the use of NIR for measuring nutritional quality of barley is sure to increase as calibrations become more available and as cost of NIR equipment drops.

4.15.4 Amino acid availability

Availability of amino acids in feed ingredients for monogastrics is becoming a greater concern. In the past, diets have been formulated on the basis of protein levels in feed ingredients. However, with the advent of inexpensive synthetic amino acids and high prices of protein supplements such as canola and soybean meals, feed companies have been forced to make better use of the amino acids in their feed ingredients. This has been achieved in two ways. First, amino acid requirements are being determined, in contrast to protein requirements, for monogastrics. Second, these amino acid requirements are then being supplied on the basis of available amino acids in feed ingredients. Several methods have been developed for measuring availability of amino acids including precision feeding (Sibbald, 1979), cannulated pigs (Sauer and Ozimek, 1986) and *in vitro* protein digestibility (Babinszky *et al.*, 1990). These methods are time-consuming and expensive, although there is the possibility of developing prediction equations for availability based on barley protein. There is also the possibility that NIR can be used to predict the availability of amino acids. However, considerable resources will need to be applied to this possibility before a practical method is available (Van Leeuwen *et al.*, 1991).

REFERENCES

Aastrup, S. and Erdal, L. (1980) Quantitative determination of endosperm modification and its relationship to the content of 1,3-1,4-ß-glucans during malting of barley. *Carlsberg Research Communication*, **45**, 369–79.

Allison, M.J., Cowe, I. and McHale, R. (1976) A rapid test for the prediction of malting quality of barley. *Journal of the Institute of Brewing*, **82**, 166–7.

Anon. (1976) *Approved Methods of the American Association of Cereal Chemists (revised Nov. 1983)*, AACC, St. Paul, MN.

Anon. (1982) *Recommended Methods of Analysis*, The Institute of Brewing, London.

Anon. (1987) *Analytica – European Brewery Convention*, Brauerei- und Getränke-Rundschau, Zurich.

Anon. (1990) *Official Methods of Analysis of the Association of Official Analytical Chemists*, AOAC, Arlington, VA.

Anon. (1992a) *Methods of Analysis of the American Society of Brewing Chemists*, ASBC, St. Paul, MN.

Anon. (1992b) *Apparent Ileal Digestibility of Crude Protein and Essential Amino Acids in Feedstuffs for Swine*, Heartland Lysine, Chicago.

Anon. (1994) *FAO Agrostat.PC (PC software)*, Food and Agriculture Organization of the United Nations, Rome.

Babinszky, L., van der Meer, J.M., Boer, H., and den Hartog, L.A. (1990) An in-vitro method for prediction of the digestible crude protein content in pig feeds. *Journal of the Science of Food and Agriculture*, **50**, 173–178.

Bamforth, C.W. (1982) Barley ß-glucans. Their role in malting and brewing. *Brewers Digest*, **57**, 22–7; 35.

Bamforth, C.W. (1994) ß-Glucan and ß-glucanase in malting and brewing: Practical aspects. *Brewers Digest*, **69**, 12–16; 21.

Bamforth C.W. and Barclay, A.H.P. (1993) Malting technology and the uses of malt, in *Barley Chemistry and Technology* (eds A.W. MacGregor and R.S. Bhatty), American Association of Cereal Chemists, St. Paul, MN, pp. 297–354.

Bang-Olsen, K., Stilling, B. and Munck, L. (1991) The feasibility of high-lysine barley breeding, in *Proceedings of the 6th International Barley Genetics Symposium* (ed L. Munck), Munksgaard International Publishers, Copenhagen, pp. 433–8.

Bansal, H.C., Srivastava, K.N., Eggum, B.O. and Mehta, S.L. (1977) Nutritional evaluation of high protein genotypes of barley. *Journal of the Science of Food and Agriculture*, **28**, 157–60.

Barnes, B.J. and Orskov, E.V. (1982) Grain for ruminants – simple processing and preserving techniques. *World Animal Review*, **42**, 38–44.

Bathgate, G.N. and Cook, R. (1989) Malting of barley for Scotch whiskies, in *The Science and Technology of Whiskies* (eds J.R. Piggott, R. Sharp and R.E.B. Duncan), Longman Scientific & Technical, Harlow, UK, pp. 19–63.

Bhatty, R.S., Christison, G.I., Sosulski, F.W., Harvey, B.L., Hughes, G.R. and Berdahl, J.D. (1974) Relationships of various physical and chemical characters to digestible energy in wheat and barley cultivars. *Canadian Journal of Animal Science*, **54**, 419–27.

Bhatty, R.S. (1993) Nonmalting uses of barley, in *Barley Chemistry and Technology* (eds A.W. MacGregor and R.S. Bhatty). American Association of Cereal Chemists, St. Paul, MN, pp. 355–417.

Briggs, D.E., Hough, J.S., Stevens, R. and Young, T.W. (1981) *Malting and Brewing Science, Vol. 1, Malt and Sweet Wort*, Chapman & Hall, London.

Buckee, G.K. (1994) Determination of total nitrogen in barley, malt and beer by Kjeldahl procedures and the Dumas combustion method – collaborative trial. *Journal of the Institute of Brewing*, **100**, 57–64.

Burger, W.C. and LaBerge, D.E. (1985) Malting and brewing quality, in *Barley* (ed D.C. Rasmusson), American Society of Agronomy, Madison, WI, pp. 367–401.

Byrne, H. and Letters, R. (1992) Technical factors in selecting a barley adjunct, in *Proceedings 22nd Institute of Brewing, Australian and New Zealand Section*, pp. 125–8.

Campbell, G.L. and Bedford, M.R. (1992) Enzyme applications for monogastric feeds: A review. *Canadian Journal of Animal Science*, **72** 449–66.

Chapon, L., Gromus, J., Erber, H.L. and Kretschemer, H. (1980) Physical methods of the determination of malt modification and their relation to wort properties, in *BEBC Monograph VI, Helsinki*, pp. 45–70.

Christison, G.I. and Bell, J.M. (1975) An assessment of bulk weight and other simple criteria for predicting the digestible energy values of feed grains. *Canadian Journal of Plant Science*, **55**, 515–28.

Classen, H.L. and Bedford, M.R. (1991) The use of enzymes to improve the nutritive value of poultry feeds, in *Recent Advances in Animal Nutrition* (eds W. Haresign and D.J.A. Cole), Butterworth-Heinemann, pp. 95–116.

Coates, B.J., Slinger, S.J., Summers, J.D. and Bayley, H.S. (1977) Metabolizable energy values and chemical and physical characteristics of wheat and barley. *Canadian Journal of Animal Science*, **57**, 195–207.

Dupchak, K.M. and Hickling, D. (1990) Regression estimates for lysine and other amino acids in Manitoba barley. *Canadian Journal of Animal Science*, **70**, 333–6.

Edney, M.J., Tkachuk, T. and MacGregor, A.W. (1992) Nutrient composition of the hull-less barley cultivar, Condor. *Journal of the Science of Food and Agriculture*, **60**, 451–6.

Edney, M.J., Morgan, J.E., Williams, P.C. and Campbell, L.D. (1994) Analysis of feed barley by near-infrared reflectance technology. *Journal of Near-Infrared Technology*, **2**, 33–41.

Enari, T.-M., Mikola, J. and Linko, M. (1964) Restriction of proteolysis in mashing by using a mixture of barley and malt. *Journal of the Institute of Brewing*, **70**, 405–10.

Erdal, K. and Gjertsen, P. (1967) β-Glucans in malting and brewing. II The fate of β-glucans during mashing, in *EBC Proceedings, Congress 1967*, pp. 295–302.

Foster, A. (1992) Total nitrogen in brewing grains by combustion method. *American Society of Brewing Chemists*, **50**, 147–8.
Foster, A. (1993) Nitrogen in wort and beer by combustion method. *American Society of Brewing Chemists*, **51**, 183–5.
Gerstenkom, P. (1985) NIR spectroscopy as a practical method for measurement of crude fibre in barley. *Getreide Mehl und Brot*, **3**, 167–70.
Goering, K.J., Eslick, R. and DeHaas, B.W. (1973) Barley starch. V. A comparison of the properties of waxy Compana barley starch with the starches of its parents. *Cereal Chemistry*, **50**, 322–8.
Gudmunds, K. and Webb, A. (1994) *PS&D View (PC software)*, United States Department of Agriculture: Production, Supply and Distribution of Data, Washington, DC.
Harlan, J.R. (1968) On the origin of barley, in *Barley: Origin, Botany, Culture, Winterhardiness, Genetics, Utilization, Pests.* Agriculture Handbook 338, United States Department of Agriculture, Washington, DC, pp. 9–31.
Helm, J.H., Spicer, H., and Aherne, F.X. (1989) Nutritional value of Condor hull-less barley for swine. *Barley Newsletter*, **36**, 150–3 (cited with permission).
Henry, R.J. (1985) Use of a scanning near-infrared reflectance spectrophotometer for assessment of the malting potential of barley. *Journal of the Science of Food and Agriculture*, **36**, 249–54.
Ingversen, J., Koie, B. and Doll,. H. (1973) Induced seed protein mutant of barley. *Experientia*, **29**, 47–56.
Jenkins, D.J.A., Wolever, T.M.S., Leeds, A.R., Gassull, M.A., Haisman, P., Dilawari, J., Goff, D.V., Metz, G.L. and Alberti, K.G.M.M. (1978) Dietary fibres, fibre analogues, and glucose tolerance: importance of viscosity. *British Medical Journal*, **1**, 1392–4.
Katz, P.C. (1979) National patterns of consumption and production of beer, in *Fermented Food Beverages in Nutrition* (ed. C.F. Gastinea), Academic Press, New York, pp. 143–55.
Kent, N.L. (1983) *Technology of Cereals*, Pergamon Press, Oxford, England.
Ko, H.L., Henry, R.J., Graham, G.C., Fox, G.P., Chadbone, D.A. and Haak, I.C. (1994) Identification of cereals using the polymerase chain reaction. *Journal of Cereal Science*, **19**, 101–6.
MacGregor, A.W. and Fincher, G.B. (1993) Carbohydrates of the barley grain, in *Barley Chemistry and Technology* (eds A.W. MacGregor and R.S. Bhatty), American Association of Cereal Chemists, St. Paul, MN, pp. 73–130.
MacLeod, A.M. (1979) The physiology of malting, in *Brewing Science, Volume I* (ed. J.R.A. Pollock), Academic Press, London, pp. 145–232.
Marchylo, B.A. and Kruger, J.E. (1984) Identification of Canadian barley cultivars by reversed-phase high-performance liquid chromatography. *Cereal Chemistry*, **61**, 295–301.
Mathison, G.W., Kerrigan, B., Hironaka, R., Milligan, L.P. and Weisenburger, R.D. (1988) Ruminant feed evaluation unit: effects of barley bulk density on energy values. *University of Alberta Agriculture & Forestry Bulletin*, **67**, 29–30.
McAllister, T.A., Rode, L.M., Cheng, K.-J., Schaefer, D.M. and Costerton, J.W. (1990) Morphological study of the digestion of barley and maize grain by rumen micro-organisms. *Animal Feed Science and Technology*, **30**, 91–105.
McGuire, C.F. (1984) Barley flour quality as estimated by soft wheat testing procedure. *Cereal Research Communications*, **12**, 53–8.
McIntosh, G.H., Le Leu, R.K., Kerry, A. and Goldring, M. (1993) Barley grain for human food use. *Food Australia*, **45**, 392–4.
Melland, R., Newman, R.K., McGuire, C.F. and Eslick, R.F. (1984) The effects of

bleach treatment on pasta made from a series of barley genotypes. *Cereal Research Communications*, **12**, 201–7.
Munck, L., Karlsson, K.E., Hagberg, A. and Eggum, B.O. (1970) Gene for improved nutritional value in barley seed protein. *Science*, **168**, 985–7.
Narziß, L. (1976) *Die Technologie der Malzbereitung*, Ferdinand Enke Verlag, Stuttgart, Germany.
Narziß, L. (1985) *Die Technologie der Würzebereitung*, Ferdinand Enke Verlag, Stuttgart, Germany.
Narziß, L., Reicheneder, E. and Edney, M.J. (1989a) Studying beer filtration with an accurate beta-glucan assay. *Monatsschrift für Brauwissenschaft*, **42**, 277–85.
Narziß, L., Reicheneder, E. and Edney, M.J. (1989b) Importance of beta-glucan size and concentration in malting. *Monatsschrift für Brauwissenschaft*, **42**, 430–7.
Narziß, L., Reicheneder, E. and Edney, M.J. (1990) The control of beta-glucan in the brewery. *Monatsschrift für Brauwissenschaft*, **43**, 66–76.
Palmer, G.H. (1975a) Influence of endosperm structure on extract development. *American Society of Brewing Chemists*, **33**, 174–80.
Palmer, G.H. (1975b) A rapid guide to endosperm malting potential of barleys using a sedimentation procedure. *Journal of the Institute of Brewing*, **81**, 71–3.
Pomeranz, Y. (1974) Food uses of barley. *Critical Reviews in Food Technology*, **4**, 377–94.
Rossnagel, B.G., Bhatty, R.S. and Harvey, B.L. (1981) Developing high-energy hull-less feed barley for Western Canada, in *Proceedings of the 4th International Barley Genetics Symposium*, Edinburgh University Press, Edinburgh, pp. 293–98.
Salem, M.B. and Williams, P.C. (1987) Barley as human food. *Rachis*, **6**, 46–7.
Sauer, W.C. and Ozimek, L. (1986) Digestibility of amino acids in swine: Results and their practical applications. A review. *Livestock Production Science*, **15**, 367–88.
Sibbald, I.R. (1979) A bioassay for available amino acids and true metabolizable energy in feeding stuffs. *Poultry Science*, **58**, 668–75.
Thacker, P.A., Bell, J.M., Classen, H.L., Campbell, G.L. and Rossnagel, B.G. (1988) The nutritive value of hull-less barley for swine. *Animal Feed Science and Technology*, **19**, 191–6.
Van Leeuwen, P. , Verstegen, M.W.A., van Lonkhuijsen, H.J. and van Kempen, G.J.M. (1991) Near infrared reflectance (NIR) spectroscopy to estimate the apparent ileal digestibility of protein in feedstuffs, in *Proceedings of the Vth International Symposium on Digestive Physiology in Pigs* (eds M.W.A. Verstegen, J. Huisman, L.A. den Hartog), Pudoc, Wageningen, Netherlands, pp. 260–5.
Van Soest, P.J. (1973) Collaborative study of acid-detergent fiber and lignin. *Journal of the Association of Official Analytical Chemists*, **56**, 781–4.
Von Wettstein, D., Nilan, R.A., Ahrenst-Larsen, B., Erdal, K., Ingversen, J., Jende-Strid, B., Kristiansen, K.N., Larsen, J., Outtrup, H. and Ullrich, S.E. (1985) Proanthocyanidin-free barley for brewing: progress in breeding for yield and research tool in polyphenol chemistry. *MBAA Technical Quarterly*, **22**, 41–52.
Wackerbauer, K., Carnielo, M. and Hardt, R. (1993) Video/image analysis system for the Calcofluor malt modification method. *EBC Proceedings, Congress 1993*, pp. 479–86.
Wendorf, F., Schild, R., Hadidi, N.E., Close, A.E., Kobusiewicz, M., Wieckowska, H., Issawi, B. and Haas, H. (1979) Use of barley in the Egyptian late Paleolithic. *Science*, **205**, 1341–7.

5

Sorghum and millets

L.W. Rooney

5.1 INTRODUCTION

Sorghum and millets constitute a major source of calories and protein for millions of people in Africa and Asia. These cereals are mainly considered as subsistence crops because of their unique tolerance to drought and adaptation to dry tropical and subtropical ecosystems throughout the world. They rank fifth and sixth in terms of total production.

Sorghum bicolor L. Moench is produced mainly for food in Africa and Asia while it is used for feed in the West. Major production centres are India, China, USA, Argentina, Mexico, West Africa, East Africa, Australia and Southern Africa.

The grasses known collectively as millets are a set of highly variable small seeded plants indigenous to many areas of the world (Table 5.1). Millets are of special value in semi-arid and tropical regions because of their hardiness, short growing season and productivity under heat and drought conditions. Pearl millet is the most widely grown on an estimated 27 million hectares followed by finger millet. World production of millets has been stable during the last decade because it is mainly cultivated as a subsistence crop. In 1991, 37.1 million hectares were sown with an average yield of 781 kg/ha totalling 28.9 million metric tons (Anon., 1992). Thirty, 56 and 13% of the production was harvested in Africa, Asia and the former USSR, respectively. Major producers were India (31.1%), China (15.8%) and the former USSR (13%).

5.1.1 Utilization of sorghum and millets

Sorghum and millets are used for human food, livestock feeds and industrial processing into a wide variety of products. The stalks and leaves are used for forages, fuel, building materials and even alcohol in

Table 5.1 Probable cytogenetic origin and scientific and common names of sorghum and major types of millets (adapted from Dendy, 1995)

Scientific name	*Common names*	*Cytogenetic origin*	*Location grown*
Sorghum bicolor	Sorghum, milo, Jowar, kafir corn, Guinea corn, cholam	Equatorial Africa	Mexico, Africa, India, China, United States
Pennisetum glaucum *P. americanum* *P. typhoides*	Pearl, bajra, cattail, bulrush, candlestick, sanyo, munga, seno, souna	West Africa	Africa, India
Eleusine coracana	Finger, ragi, Africa, bird's foot, rapoko, Hansa	Originated in Africa and domesticated in India	East and Central Africa, India, China
Setaria italica	Foxtail, Italian, kangni, navane, German, Siberian, Hungarian	Eastern Asia	Asia (China, India, Japan), North Africa, Southeast Europe, Near East
Panicum miliaceum	Proso, common, Hershey, panivarigu, broomcorn, hog, samai, Russia	China	Eastern Asia, China, Mongolia, Middle East
Echinochloa frumentacea *E. crus-galli* *E. utilis*	Japanese, barnyard, sanwa, kweichou, kudiraivali, sawan, Korean	Java/Malaysia	East Asia, India, Egypt
Paspalum scrobiculatum *P. commersoni*	Kodo, varagu, bastard, ditch, naraka	Africa or India	India
Eragrostis tef	Teff	Ethiopia	East Africa (Ethiopia)
Digitaria exilis *D. iburua*	Fonio, fundi, hungry rice, acha, crabgrass, raishan	Domesticated in Nigeria	West Africa (savanna)

some areas. Sorghum is dry milled into grits and meals for brewing, baking, snack foods and other uses. It is malted and used to produce opaque and lager beer, breakfast foods and other products. Sorghum starch has properties equal to corn starch and is produced in the Sudan through wet milling.

The major portion of sorghum production in the world is used for livestock feed. It is considered to have excellent feeding value although care must be taken to properly process the grain. It is nearly all used for feed in the Western Hemisphere. Food use has decreased during the past 25 years to approximately 35% of total production. The traditional foods made from sorghum and millet (Rooney *et al.*, 1986; Rooney and McDonough, 1987; Murty and Kumar, 1995) have been classified into several categories including fermented and non-fermented breads, thick and thin porridges, steamed products, alcoholic and non-alcoholic beverages and various snacks.

5.1.2 General references

Significant reviews on the chemistry, quality, nutritional value and technology of sorghum and millets have been published: Hulse *et al.* (1980), Rooney *et al.* (1980, 1982, 1986), Hoseney *et al.* (1987), Rooney and Serna-Saldivar (1991), Serna-Saldivar *et al.* (1991), Dendy (1995).

5.2 CLASSIFICATION

Sorghum is a cereal of remarkable genetic variability. More than 30 000 selections are present in the World Sorghum Collection in India. The plant originated in equatorial Africa and is distributed throughout the tropical, semi-tropical and arid regions of the world.

Sorghum is classed as grain sorghum, forage sorghum, grass or Sudan grass sorghum and broomcorn. The latter is grown for its long, fibrous panicle branches that are manufactured into brooms. Grain sorghum refers to short, combine-harvested cultivars grown in developed countries. However, in most countries, sorghum that produces tall stalks and grain are preferred because the stems and leaves are used for fodder, fuel and building materials. The grassy types are used primarily for grazing and hay.

5.2.1 Grain standards

The current US Grain Standards puts sorghum into four market classes: sorghum (formerly yellow sorghum), tannin sorghum (formerly brown sorghum) , white sorghum and mixed sorghum. The sorghum class consists of sorghums with red, white or yellow pericarp colour but cannot contain more than 3% of sorghum kernels with a pigmented testa or undercoat. Tannin sorghums are sorghums that have a pigmented testa in the kernel. They may appear brown, white, reddish brown or other colours. The white class contains 98% kernels with a white pericarp without a pigmented testa. Mixed sorghum contains blends of sorghum with and without pigmented testa. Appearance alone is not a reliable method of classifying sorghums. In market sorghums, it is difficult to distinguish tannin sorghums from those with a red, brownish red or even white pericarp. The Federal Grain Inspection Service uses a Chlorox bleach test to distinguish sorghums with a pigmented testa from those without (Anon., 1994). The Chlorox bleach method removes the pericarp, and the pigmented testa appears as a black layer. Appropriate standards should be used in the Chlorox test to ensure that it is reliable.

US No. 2 sorghum is usually the class of grain sold in the grain trade. Test weight, damaged kernels, broken kernels and foreign material affect the numerical grade assigned. In South Africa, the standards recognize a malting sorghum class and discriminate against brown sorghums which are undesirable for malt. Argentina produces a high percentage of brown (tannin) sorghum which affects quality significantly.

5.2.2 Appearance and genetics of sorghum grain

Several interacting factors affect the colour and overall appearance of sorghum caryopses. Appearance is mainly affected by pericarp colour and thickness, presence, thickness and colour of pigmented testa and endosperm colour (Rooney and Miller, 1982). Plant and glume colour also influence appearance of the kernel. This is especially true for weathered sorghums. Thus, wet weather (or high humidity), insect and mould damage play a major role in the condition and appearance of the grain.

Pericarp colour is genetically controlled by the *R* and *Y* genes. The combination of these genes can produce white (*R_yy or rryy*), lemon yellow (*rrY_*) or red (*R_Y_*) colour. The presence of the testa is controlled by the $B_1\_B_2\_$ genes while a Z gene controls pericarp thickness. The seed coat is absent when these genes are homozygous recessive ($b_1b_1b_2b_2$). The *tp* genes control testa colour which can be brown or purple. Sorghums with a pigmented testa ($B_1\_B_2\_$) that contain a dominant spreader (*S_*) gene have the most bird resistance and highest tannin content (Hahn and Rooney, 1986). These sorghums contain condensed tannins and are classified as type III (also called brown, bird-resistant) sorghums. Sorghums with homozygous recessive (*zz*) genes possess a thick mesocarp which usually masks the colour of the testa and endosperm. Kernels look chalky due to the presence of small starch granules (Earp and Rooney, 1982; Rooney and Miller, 1982).

Endosperm colour is mostly affected by the maternal parent. Yellow endosperm cultivars contain high levels of carotenoid pigments. Endosperm colour affects grain appearance, especially in kernels with a thin pericarp without a pigmented testa. When the pericarp is thin and colourless and the testa is absent, the kernel appears yellow. When the kernel has a thick, colourless pericarp, the kernel appears white. Bronze sorghums contain a thin, red pericarp with yellow endosperm colour. Heteroyellow endosperm sorghums are hybrids from crossing a yellow endosperm with a non-yellow endosperm parent (Rooney and Miller, 1982).

5.2.3 Endosperm mutants

Yellow, sugary, high lysine and waxy endosperm sorghum mutants exist. Waxy endosperm cultivars contain three recessive (*wx*) genes in the endosperm. Heterowaxy genotypes contain one or two of these genes in the dominant form, whereas non-waxy endosperm cultivars contain all three *Wx* genes in the dominant form. Waxy cultivars contain nearly 100% amylopectin and an endosperm that appears waxy to the naked eye (Rooney and Miller, 1982). The yellow endosperm characteristic has been used extensively in modern sorghum hybrids. Selected sugary sorghums are used for roasting in the hard dough stage in India and Ethiopia (*hl*). A black sorghum found in western Sudan has very high levels of polyphenols and condensed tannins. Special scented sorghums are used in India (Murty and Kumar, 1995).

5.3 GRAIN PROPERTIES AND MORPHOLOGY

The sorghum kernel is considered a naked caryopsis although some African types retain their glumes after threshing. Kernels differ widely in size, shape, kernel weight (3–80 mg), density (1.15–1.42 g/cm^3), and hardness (completely soft to completely hard). Commercial US sorghums are generally 4 mm long, 2 mm wide and 2.5 mm thick with a 25–35 mg kernel weight, and 1.28–1.36 g/cm^3 density.

The caryopsis consists of three distinctive anatomical components: pericarp (outer layer), germ (embryo) and endosperm (storage tissue). The relative proportion of these components varies depending upon variety (Rooney and Serna-Saldivar, 1991; Serna-Saldivar and Rooney, 1995). Sorghum is the only cereal to have significant amounts of starch in the mesocarp when the z gene is homozygous recessive.

The seed coat (or testa) is derived from the ovule integuments and is pigmented only in sorghums with dominant B_1 and B_2 genes. When the sorghum kernel contains a dominant spreader gene (S_) in the presence of *B1_B2_* genes, the pericarp is brown and contains condensed tannins. The thickness and colour of the testa varies among kernels and within a given kernel (Blakely *et al.*, 1979; Earp and Rooney, 1982).

Factors affecting the structure of pearl millet are only partially understood. In general, its structure is similar to sorghum but the kernel is one tenth the size and has a much larger germ (Rooney and McDonough, 1987; McDonough and Rooney, 1989). The endosperm of pearl millet contains pigments that affect product colour depending upon pH. Some yellow and white endosperms exist, but most have a slate grey appearance.

The sorghum kernel has a significant amount of peripheral or sub-aleurone area that affects processing quality significantly. The strong interaction of sorghum proteins and starch in the endosperm affects digestibility and must be disrupted during processing.

5.4 CHEMICAL COMPOSITION

Sorghum and pearl millet proximate composition varies significantly due to genetics and environment (Serna-Saldivar and Rooney, 1995). The protein and starch contents are usually the most variable. Sorghum is similar in composition to maize (*Zea mays*). In general, sorghum contains slightly more protein and tryptophan than maize. Sorghum is slightly lower than maize in oil content and contains more waxes which gives it a slightly lower energy level. Pearl millet has higher protein and oil content than sorghum or maize, and has significantly higher energy levels. All three grains are similar in fibre and ash content.

5.4.1 Carbohydrates

Sorghum starch has properties and uses similar to maize starch for production of derived foods or industrial products (Watson, 1984). The apparent amylose content of normal sorghum and millet starches range from 21–29% of the starch. The gelatinization temperature of sorghum starch is a few degrees higher than that for corn starch which affects the cooking time required to process sorghum for brewing and other applications. Sorghum is the only cereal with starch in the pericarp. Waxy and heterowaxy starches have reduced levels of amylose.

5.4.2 Proteins

Protein content and composition varies due to agronomic conditions (water availability, soil fertility, temperatures and environmental conditions during grain development) and genotype. Sorghum and millet grains vary more than other grain because of their harsh, variable growing conditions.

Prolamines comprise the major protein fraction in sorghum and pearl millet followed by glutelins. These fractions are located within the protein bodies and protein matrix of the starchy endosperm, respectively. Nitrogen fertilization significantly increases the accumulation of prolamines during maturation of sorghum and millet (Warsi and Wright, 1973).

Sorghum and millet grain proteins are deficient in the essential amino acids, lysine, threonine and tryptophan. The pearl millet proteins have higher lysine than sorghum and maize proteins probably because of the increased proportion of germ in the millet caryopsis. High-lysine sorghum cultivars with improved protein quality contain higher amounts of albumins, globulins and glutelins with corresponding lower proportions of prolamins. The high-lysine sorghum cultivars developed so far have significantly reduced grain yields and softer kernels; work continues to develop harder endosperm kernels with increased levels of lysine (Axtell, 1994). A source of higher lysine has not been found in pearl millet.

Approximately 80, 16 and 3% of the sorghum protein is located in the endosperm, germ and pericarp, respectively (Taylor and Schussler, 1986). The germ is rich in albumins and globulins, whereas the endosperm contains the kafirins and glutelins. Kafirins are mainly located within the protein bodies which increase with increased levels of protein in the kernel (Warsi and Wright, 1973).

The sorghum prolamine fraction is divided into three subfractions: α, β and τ-kafirins. These fractions are similar to the zein subfractions of maize (Shull *et al.*, 1991). α-Kafirins (extracted with 40–80% alcohol) are rich in glutamic acid. They have molecular weights ranging from 22–25 kDa. β-Kafirins have lower molecular weight (16–20 kDa), are richer in sulfur-containing amino acids than α-kafirins and are soluble in 10–60% alcohol solutions. The τ-kafirins are extracted with alcohol and reducing agents (mercaptoethanol) because they are highly cross-bonded with disulfide bonds. They are the highest molecular weight (28–29 kDa) prolamins and are rich in proline and sulfur-containing amino acids. The τ-kafirins are distributed within the dark staining areas in the core and periphery of the protein bodies. They have many disulfide bonds and resist digestion by proteases. Importantly, τ-kafirins become less digestible after cooking due to additional intramolecular cross-bonding. *In vitro* protein digestibility (pepsin assay) was greatly enhanced when reducing agents (i.e. metabisulfite, mercaptoethanol) that break disulfide bonds were used (Watterson *et al.*, 1990).

5.4.3 Lipids

The lipids of sorghum, maize and pearl millet are very similar in fatty acid composition; they contain high levels of oleic acid with some linoleic acid (Rooney, 1978). The sorghum kernel contains up to 0.2% carnauba waxes which is significantly higher than millet or maize. Sorghum, maize and pearl millet have approximately 3.5, 4.5 and 5.5% ether extract, respectively. The large germ of sorghum and pearl millet is difficult to remove during processing so these cereals are processed almost daily to avoid rancidity. Genetic variation in lipid content exists in sorghum and pearl millet.

5.4.4 Tannins

Sorghum is unique among major cereals because some cultivars produce polymeric polyphenols known as tannins (Butler, 1990). All sorghums contain phenols and most contain flavonoids. However, only cultivars with a pigmented testa ($B_1\_\ B_2\_$ genes) produce condensed tannins. Most cultivated sorghums do not contain any condensed tannins, even though the literature often indicates they do. These erroneous conclusions occur when total phenols are assayed and reported as tannins. Tannic acids (hydrolysable tannins) are not found in sorghum although they are reported in the literature by some authors.

Phenolic compounds have been divided into three major categories: phenolic acids, flavonoids and tannins. Phenolic acids are derivatives of benzoic or cinnamic acid. Flavonoids consist of two units: a C6–C3 fragment from cinnamic and a C6 fragment from malonyl-CoA. The major groups of flavonoids in sorghum are the flavans. Flavan 3-en 3-ols with double bond between C3 and C4, hydroxylated at the C3, are anthocyanidins. Tannins are polymers of –7 flavan-3-ol units (catechin) linked through acid labile carbon–carbon bonds (Hahn *et al.*, 1984; Mehansho *et al.*, 1987; Butler, 1989; 1990). Sorghum tannins occur only in the pericarp and testa layers or seed coat (Hahn and Rooney, 1986).

Tannins protect the grain against insects, birds and weathering (Waniska *et al.*, 1992). Rate of preharvest germination or early sprouting is significantly lower in most high tannin cultivars. These beneficial effects ensure that brown sorghums will continue to be produced in certain pest-ridden areas of the world (Butler, 1990). The agronomic advantages are accompanied by nutritional disadvantages and reduced food qualities.

5.4.5 Phenols and product colour

In general, white sorghums produce food products with the most acceptable colour. Food colour is the result of factors such as grain colour and type (pericarp colour, pigmented testa, endosperm colour), degree of milling and pH of the food system. Pigmentation in the pericarp and testa is primarily due to phenolic compounds. The colour intensity greatly depends on pH. Anthocyanins are very unstable in acid medium and are readily converted to the corresponding anthocyanidin under slight acidic conditions. Pericarp colour of sorghum appears to be due to a combination of anthocyanin and anthocyanidin pigments as well as other flavonoid compounds. Kambal and Bate-Smith (1976) found no flavonoids in the pericarp of a white sorghum whereas Nip and Burns (1971) found in four major anthocyanins in the pericarp of six white sorghums. Lutoforol was identified as the major pigment in red pericarp sorghums (Nip and Burns, 1969). Cooking in alkali (i.e. alkaline tô, tortillas) usually promotes off-colour especially in coloured grains. The glume or plant colour also affects the colour of the grain and food products. Grains with straw coloured glumes consistently produce better products than similar grains with coloured glumes. This is especially true when fabricating lime-cooked products such as tortillas and chips (Serna-Saldivar *et al.*, 1987).

5.4.6 Analysis of tannins

Several techniques have been used to characterize phenolic compounds in sorghum grain (Earp *et al.*, 1981; Hahn *et al.*, 1984; Hagerman and Butler, 1989). Most tannin assays measure total phenols, which may or may not be condensed tannins. In general, these assays correlate well with feeding trials. However, they can be misleading because the use of general phenol assays undesirably over-estimates tannin content. Different assays are likely to yield different tannin values because they respond to different chemical parts of the tannin molecule (Hagerman and Butler, 1989).

The vanillin–HCl assay measures condensed tannins directly and is the best assay for condensed tannins. The vanillin test is standardized against commercial catechins. Blanks to remove indigenous absorbance should be included in all tests. The absolute amount of tannin present in the sorghum kernel is virtually impossible to determine because a significant proportion cannot be extracted and assayed (Hahn *et al.*, 1984; Butler, 1990) and a suitable standard for sorghum tannins is unavailable. The highly polymerized or condensed tannin molecules are the most difficult to extract which may account for the decreased levels of tannins in grain as it matures.

The best way to analyse for tannin content in sorghum is to use a chlorox bleach test to determine the percentage of kernels that have a pigmented testa. If brown kernels are not present, there is no need to analyse for tannin content. If brown kernels are present, the vanillin–HCl test should be used to quantitatively measure tannins and a blank should be used to correct for non-tannin compounds.

Tannins have not been reported in pearl millet grain; certain cultivars of ragi or finger millet have condensed tannins.

5.5 INDUSTRIAL PROCESSING AND UTILIZATION

Sorghum and millets are utilized for a wide array of industrial and food products depending upon their relative price and availability. Major industrial processes include dry milling and utilization of these products.

5.5.1 Decortication

Decortication by mortar and pestle or with abrasive dehullers removes the outer layers; therefore, decorticated or 'pearled' kernels have reduced levels of fibre, ash and fat contents. Sorghum and millets are usually decorticated to remove from 12–30% of the grain. Increased decortication causes increased losses of fibre, ash and fat. Decortication slightly reduces protein and lysine content due to partial degermination. Amino acid analysis of decorticated grains showed a progressive decrease in the content of lysine, histidine and arginine. Eggum *et al.* (1983) reported that the removal of 20–25% of the sorghum weight by traditional manual decortication decreased lysine levels by 40%. Serna-Saldivar *et al.* (1987, 1988) reported losses of 14% lysine in sorghum decorticated to remove 10% of its initial grain weight.

5.5.2 Malting

Malting is a common practice in sorghum and millet producing areas (Daiber and Taylor, 1995). The effect of sprouting on the chemical composition of sorghum and millets was recently reviewed by Chavan and Kadam (1989). Grains are malted for production of opaque beers, weaning foods and other traditional dishes. Malting causes 8–30% dry matter loss (Chavan and Kadam, 1989), decreased levels of prolamins, fat, tannins and starch and increased levels of free amino acids, albumins, lysine, reducing sugars and most vitamins including synthesis of vitamins B_{12} and C (Almeida-Dominguez *et al.*, 1993; Malleshi and Desikachar, 1986; Taylor *et al.*, 1985). The activation of intrinsic amylases, proteases, phytases and fibre-degrading enzymes disrupts protein bodies, protein matrix, starch granules, cell walls to a limited extent, phytin and other polymers; thereby, nutrient accessibility is increased. The kafirins and glutelin fractions decrease during malting while soluble nitrogen and albumins/globulins increase.

Sorghum malt is produced on an industrial basis in Southern Africa and in Nigeria (Daiber and Taylor, 1995; Palmer, 1989). Sorghum malt cannot totally replace barley malt for lager beer because it has low β-amylase activity. Sorghum malting conditions are quite different from barley malting. Sorghum requires a high temperature (25°C) during steeping and germination. Sorghum loses 20% or more dry matter during malting; the endosperm cell walls are not completely modified. The yield of extract from sorghum malt is very low because the starch gelatinization is significantly higher than barley starch. Thus to achieve higher yields of sorghum extract the malt must be mashed at higher temperatures using special procedures followed by addition of extra malt to achieve hydrolysis. Often sorghum malt may be supplemented with commercial amylases to improve conversion. Malts for opaque beer have adequate enzyme activities in general. Malts for production of lager or clear beer are inadequate.

The sorghum quality critically affects malt quality. Weathered sorghum generally has low levels of germination and moulds grow rapidly and adversely affect malt quality. Brown bird-resistant sorghums are not desirable for industrial malting because they affect enzymes in microorganisms that sour the opaque beer. In traditional home brewing and malting, brown sorghums are often preferred for malting and brewing because they are not as useful for food products. In general, a red sorghum with intermediate to soft texture is preferred.

A sorghum variety for malting should have upon germination a good diastatic activity, low polyphenols, and low dry matter losses. Between 18 and 50% of the amylolytic activity of sorghum malts is attributed to the action of β-amylase. In contrast, barley malt has a good balance of α- and β-amylases. Sorghum does not respond to gibberellins during malting.

The sorghum diastatic activity developed during malting depends on temperature, moisture content, duration of malting, % germination and type of grain used. Tribal sorghum malting involves the following steps: (i) steeping (1–3 days); (ii) germination (2–6 days); (iii) sun drying; (iv) grinding (mortar and pestle). In contrast, the industrial process steeps the sorghum in special aerated tanks under controlled temperatures. Germination is conducted in controlled, aerated germinators or the grain is placed on concrete floors where it is watered and turned regularly until it is modified sufficiently. Temperature, moisture and degree of modification are difficult to control with the floor procedures. The best malt with highest diastatic activity is produced using the controlled aeration systems. Sorghum malt drying (kilning) is generally done with warm air; roasting is undesirable because too many enzymes are destroyed.

Malted sorghum (non-diastatic) is marketed as a breakfast cereal in

South Africa, Zimbabwe and Malawi. Malt extracts are used in a wide array of non-alcoholic beverages, infant foods and other products.

The lysine content of germinated normal and high lysine sorghums increased from 2.2 to 3.2 and 3.0 to 7.8 g/100 g protein. This is due to a large increase in albumins and a large decrease in kafirins and cross-linked kafirins (Wu and Wall, 1980). Germinated sorghum grain has better nutritional value due to the chemical changes mentioned above. Sorghum malt has higher available lysine, tryptophan and methionine, relative nutritional value and protein efficiency ratios compared to the ungerminated grain.

Sorghum breeders have recently begun to develop sorghum varieties with good malting characteristics.

5.5.3 Brewing

Sorghum is used to produce traditional opaque sweet and sour beers, European beer as an adjunct with barley malt, and clear beer (lager) composed of sorghum or maize–sorghum blends.

Lager beer – grits

Sorghum is often used as an adjunct in brewing barley beers (Hallgren, 1995). The adjunct is an inexpensive source of fermentable carbohydrates. Sorghum kernels are generally decorticated and milled into grits. The most desirable grit has light colour and bland flavour. Breweries in Mexico and Africa are currently using sorghum grits. Sorghum utilization can be enhanced if white sorghums with tan plant colour (uncoloured glumes) and improved milling properties are utilized through identity preserved grain acquisition. These sorghums can be decorticated to a lesser extent and the grits have a lighter colour. Brewing properties of waxy sorghum may be an advantage; but it has not been used commercially.

Clear sorghum beer

Some African governments (i.e. Nigeria) have restricted the importation of barley so new technologies have been developed to produce European beers out of sorghum malt, a combination of barley and sorghum malt and by using commercial enzymes to convert sorghum and maize blends into fermentable worts (Hallgren, 1995). With the sorghum varieties available today, it is almost always necessary to add saccharifying enzymes. Three main steps are followed to produce clear sorghum beer: (i) preparation of sorghum malt; (ii) preparation of the medium to be fermented (wort); and (iii) fermentation that leads to the final beer. The sorghum and maize grits or ground meal are solubilized by a mashing procedure (cooking-enzymatic activity of malt). The wort is obtained after filtration of the mash and fermented with yeast (*Saccharomyces cerevisiae*) to yield beer. The alcohol content (3.9%), pH (4.6), colour and foam stability of sorghum clear beer are similar to barley lager beer.

Opaque sorghum beer

Opaque beer is made out of malted sorghum and several other starchy materials (adjuncts) and is consumed while undergoing fermentation. The beer is a soured, viscous, alcoholic, effervescent, opaque beverage with a pale buff to pink colour. Sorghum beer is served warm and has a very short shelf life. Sorghums with bright red pericarp without a testa and with an intermediate endosperm are preferred for industrial production. Brown, bird-resistant sorghums are undesirable; they are treated with formaldehyde if they must be used.

In contrast to European brewing, in African opaque beer brewing there are two mashing and two fermentation steps. Sorghum brewing starts with a lactic acid fermentation (souring) of sorghum malt and water. In the second step, the adjunct (usually maize grits) is boiled in water to gelatinize the starch, cooled to 60°C and more sorghum malt is added to hydrolyse the adjunct. Control of the mash pH is critically important; if acidity is too high, the viscosity will be too high; if acidity is too low, the viscosity will be too low and sugar and alcohol production too high. Then the wort is processed to remove the coarse particles, roots, shoots and pericarp prior to inoculation with yeast to initiate alcoholic fermentation. In contrast with European brewing, the wort is not pasteurized; no hops are utilized.

Sorghum opaque beer contains 2–4% alcohol, 0.3–0.6% lactic acid, 4–10% solids with a pH of 3.3–3.5. The beer is a good source of vitamins, minerals, proteins and carbohydrates that are solubilized during malting and brewing (Diaber and Taylor, 1995). Beer spoilage (after souring) occurs quickly due to residual enzyme activity which promotes growth of microorganisms that are normally destroyed during pasteurization.

The opaque sorghum beer industry is of major significance in the utilization of sorghum in Southern and some East African countries. Most of the sorghum grown in South Africa is malted for opaque beer brewing. Home or tribal brewing is still very popular in Africa. Sorghum malt is packaged in small to large bags for home brewing. Different types of opaque beer are produced (i.e. Reef, Juba, Kimberley). Several companies produce sorghum beer powder for production of opaque beer at home. The consumer adds water to the powder and has beer after 12–24 h.

Alcohol production

The potential to utilize either sorghum grain or sweet sorghum biomass for ethanol production has been reviewed (Creelman *et al.*, 1982). One ton of maize grain produces 387 l of 182-proof alcohol whereas the same amount of sorghum grain produced 372 l. A bushel of sorghum grain (56 lbs) produces about 2.5 gallons of alcohol. Distillers grain with a protein content of 30% is a major byproduct.

Sweet sorghum crops have the potential to yield 22–45 tons of biomass per hectare in 110 days with 11.2% fermentable solids (2.5–5 tonnes/ha). These solids are about 80% soluble sugars and 20% starch. Therefore, to optimize ethanol production both liquefying and saccharifying enzymes are required. One ton of sorghum stalks has the potential to yield 74 l of alcohol (200 proof). The technology to use sweet sorghums for alcohol has been developed in Brazil.

5.5.4 Dry milling

Most of the sorghum used by the feed and food industry is dry milled before processing. The most highly refined products are produced by abrasive removal of the pericarp, followed by degermination and physical separation and/or classification of the dry milled fractions. Most of the traditional African and Indian foods are produced from decorticated-milled grain. Industrial sorghum decortication involves the use of mills with abrasive discs. Whole or decorticated sorghum flour can be produced by the use of stone, hammer, pin or roller mills. Indian villages still use stone mills to produce a coarse high extract (95%) flour that is further processed into Roti and Sankati. Hammer, pin and roller mills are used to produce flours with a finer particle size.

Most of the sorghum milled into brewers adjuncts is decorticated, degerminated by impaction, separated from germ by gravity tables, roller milled and sieved into appropriate size particles.

Milling quality

The milling quality of sorghum and millet is determined mainly by the kernel size and shape, density, hardness, structure, colour, presence of a pigmented testa, pericarp thickness and colour. Kernels with a high proportion of hard endosperm, white pericarp, no pigmented testa and a thick pericarp give outstanding dehulling properties. Soft floury kernels disintegrate during dehulling and cannot be milled efficiently. For hand dehulling, a thick starchy mesocarp (zz) reduces labour 50% or more. Long slender pearl millet kernels have very poor dehulling properties. White kernels give the highest yields of light colour flour which is preferred.

White, tan plant sorghum hybrids grown in the USA produce significantly improved yields of grits and other products with a light colour. The tan plant colour gives the lightest, brightest products.

Milling quality evaluation

In general, abrasive milling techniques are effective. The barley pearler, Kett Mill, TADD and Satake Rice Pearler have been used to determine milling properties of sorghum and millets (Reichert *et al.*, 1988; Munck, 1995) from the breeding programme. In effect these techniques are all similar with varying degrees of force applied to the grain to abrasively remove the pericarp. Good milling cultivars retain their integrity and allow the pericarp to be removed to produce high yields of white decorticated kernels. Hardness and density are strongly positively related to good milling properties.

5.5.5 Wet milling

Sorghum has been successfully processed into starch, glucose, and other products using wet milling in Mexico, the USA and the Sudan. In 1975, wet milling was discontinued in Texas because sorghum prices increased to a level similar to maize.

The procedure for sorghum wet milling is similar to the one used for maize (Watson, 1984; Rooney and Serna-Saldivar, 1991). The pigments of the sorghum pericarp sometimes give sorghum starch a light pink colour. Bleaching with $NaClO_2$ produces acceptable colour but the cooked starch gels have an undesirable appearance.

According to Watson (1984), the major difference between maize and sorghum wet milling is the way in which the starch and gluten separates. Sorghum pericarp is more fragile than the pericarp of maize, so it impedes the separation of these major components. The protein is more difficult to remove from sorghum starch and recovery is generally lower than for maize starch. The properties of sorghum starch and oil are similar to those of corn. However, sorghum starch must be bleached to remove polyphenols, takes more energy to cook and it is more difficult to recover in high yields. Other significant economic differences include a lower sorghum oil yield and the lack of carotenoid pigments in the gluten, which are highly desired by the broiler industry. Due to its smaller kernel size, sorghum requires a different design in the plates of the degerminating mills (more grinding teeth and with a triangular end configuration).

Sorghum hybrids with improved wet milling properties exist. Soft, yellow endosperm kernels with a clear or white pericarp without an undercoat gave the best wet milling starch recoveries and produced good quality starches (Norris, 1971). Sorghum starches without pigments can be produced easily.

5.5.6 Sweet sorghum – molasses and syrups

Sweet sorghum varieties which produce extra quantities of sugar in the juice are used to produce sorghum syrup (molasses). The tall plants are generally harvested when the grain is in the dough stage (Coleman, 1970). Earlier harvesting usually causes problems during syrup clarification due to excess chlorophyll pigments. Therefore, the colour, clarity and viscosity of the syrup improves with plant maturity. The stalks stripped of leaves are immediately crushed after harvest with rollers to extract juice which is evaporated to produce high quality syrups. Various methods are used to clarify the juice to improve syrup clarity. Starch causes clarification difficulties: it may be removed by enzymes. Breeders select tall cultivars with high sugar and juice yields and low levels of lignin and starch. A good quality sorghum syrup should be mild and sweet with a light colour and unique flavour.

5.6 TRADITIONAL FOOD USE OF SORGHUM AND MILLET

Progress has been made in relating the general physical and structural properties of sorghum to the major categories of traditional food products (Rooney *et al.*, 1986; Murty and Kumar, 1995). Shape, size, proportion and nature of the endosperm, germ and pericarp; presence or absence of subcoat; and colour of the pericarp are all genetically controlled (Rooney and Miller, 1982). Endosperm texture, i.e., the relative proportion of hard to soft endosperm, plays a major role in determining the quality of traditional sorghum foods (Cagampang and Kirleis, 1984; Rooney *et al.*, 1986). In sorghums with a higher percentage of corneous endosperm, the pericarp (bran) is more readily separated from the endosperm. The particle size, amount of starch damage, and other important factors relating to food quality of sorghum flours is due in part to the intact cells from the peripheral and corneous endosperm.

Three classes of sorghum based on endosperm texture have been proposed: (i) hard – suitable for thick porridges and couscous, (ii) intermediate – unfermented breads, boiled rice-like products, malting and brewing and (iii) soft-fermented breads. Thus, plant breeders selecting for food quality within a specific type of food category can select visually for certain kernel characteristics and texture. In general, within each hardness group or class, the preferred sorghum should have a white pericarp and tan plant colour and should not have a pigmented subcoat. Exceptions do exist. For example, the preferred pericarp colour of sorghum for beer is red, with a dominant intensifier gene that gives a very bright, clear red.

Plant breeders use grain hardness, density and ease of pericarp removal for early generation selection for sorghum and millet quality. Then, laboratory milling and cooking tests can be conducted in advanced generations followed by large-scale processing and cooking trials for advanced breeding materials. The assays that should be applied for each food category have been tested (Rooney *et al.*, 1986) and will evolve as more experience is gained in selecting for sorghum food quality and in developing physicochemical tests to predict it. Testing is necessary because, in several instances, 'improved' varieties and hybrids were released that had unacceptable quality for traditional foods. Consequently these improved yielding types were not grown by farmers. Brown bird-resistant sorghums with high tannin content are unsuitable for industrial malting or brewing. In South Africa, a formal programme to evaluate malting quality of sorghum is pursued. The brown sorghums are lower priced than other sorghums, and premiums are paid for red sorghums without a pigmented testa that have good malting quality.

Generally, the brown bird-resistant sorghums are undesirable for the traditional food systems. However, they are often grown where birds are a major problem. Special processes have been used to convert the brown sorghums into foods. In some areas, the brown sorghums are steeped in wood ashes, germinated, and used to produce thick porridges. Sometimes special porridges made from brown sorghum are given to new mothers or are consumed by farmers doing strenuous work. Brown sorghum porridge is said to 'stay with' the farmer longer, possibly because the condensed tannins affect digestibility.

5.6.1 Traditional milling properties

Sorghum and millet are decorticated by hand in a mortar with a pestle and then reduced into flour, meal or grits. Dehulling properties are critically important. A spherical kernel, a thick pericarp and a hard endosperm are critically important kernel attributes for decortication. Abrasive principles are used to remove the pericarp so long slender kernels are not desired nor should the kernels break during milling. In general, Souna pearl millets have very long kernels that give very poor milling yields (Rooney and McDonough, 1987). For traditional hand pounding in mortar and pestles, the women prefer sorghum with a thick starchy mesocarp because it has a 50% reduction in decortication time. The lightest and brightest products are made from white, hard tan plant sorghums with straw colour glumes. Special varieties of sorghum are grown to produce rice-like products for popping or expanding. Pearl, foxtail, finger and proso millets have white and yellow endosperm cultivars that process into attractive, light colour products.

The decorticated grain is reduced into flour by hand pounding or attrition milling usually with stones or steel plate mills. Hammer-milled flour has sharp, angular particles that give an unacceptable sandy texture in most traditional products unless special care is taken to produce the correct particle size and shapes. Sorghum and millet flours have short shelf stability because the germ is incompletely removed. Pearl millet develops an off-aroma caused by enzyme attack on precursors that develop 2-pyrrolidine (Seitz *et al.*, 1993). Heating the grain eliminates or reduces the flavour.

5.7 FOOD UTILIZATION

Sorghum and millet can be processed into a wide variety of very acceptable commercial food products provided the proper sorghum and millet cultivars are used. These grains can be extruded to produce a great array of snacks, ready to eat breakfast foods, instant porridges and other products. The flakes of a waxy sorghum obtained by micronizing can be used to produce granola products with excellent texture and taste. Tortillas and tortilla chips have been produced from sorghum alone or maize blends. The sorghum products have a bland flavour while pearl millet products have a distinctive flavour.

Neither sorghum nor millet have gluten proteins. Therefore to produce yeast leavened breads they are usually substituted for part of the wheat flour in the formula. The level of substitution varies depending upon the quality of the wheat flour, the baking procedure, the quality of the sorghum or millet flour and the type of product desired. In biscuits, up to 100% of the flour can be sorghum or millet flour. The non-wheat flour tends to give a drier more sandy texture so modifications to the formula must be made.

The major constraints to the use of sorghum and millet flours is the unavailability of stable good quality flours on a consistent basis. In addition, the large supplies of wheat, maize and rice often relegates sorghum to use as feeds. However, the new white, tan plant, food sorghum hybrids that have been developed recently are providing consistent supplies of high quality grain for processing (Young *et al.*, 1990). Pearl millet flour remains difficult to obtain. The major problem that limits sorghum food processing in many sorghum consuming countries is the lack of a reliable supply of good quality grain for processing. In those areas, sorghum is produced on small farms where it is processed traditionally. The grain sold in the market varies greatly in properties, colour and soundness so it is nearly impossible to produce consistently good quality processed products.

5.8 FEED PROCESSING AND UTILIZATION

Sorghum is used in a wide array of livestock feeds as part or all of the cereal grain in the ration. Sorghum has at least 95% of the feeding value of yellow dent maize. However it must be properly processed to obtain improved feed efficiency. For poultry and swine, sorghum is ground using hammermills to reduce the particle size to improve digestibility. It may be pelleted for poultry and pet foods. For broilers and laying hens, sorghum diets are too low in carotenoid pigments. Thus, dehydrated alfalfa, marigold meal and yellow corn gluten feed is added to the ration. In some areas where non-pigmented broilers and eggs are desired, sorghum is the preferred feed grain.

Sorghum for high concentrate feedlot cattle rations must be processed thoroughly to improve its feed efficiency. Steam flaking, micronizing, exploding, popping, reconstitution and grinding are methods of processing used to prepare sorghum for cattle and sheep. The most common method is steam flaking which consists of tempering and steaming clean grain until it reaches more than 20% moisture and flaking the kernels through pressurized corrugated rolls. The flakes are mixed with the other components of the ration and fed. Properly steam flaked sorghum has 95% or more the value of maize.

5.8.1 Quality for feed

Brown sorghums have significantly lower feed efficiency than sorghums without a pigmented testa for all livestock species. The major anti-nutritional effects of tannins are decreased protein digestibility and feed efficiency in rats, hamsters, swine, poultry and ruminants (Maxson *et al.*, 1973; Cousins *et al.*, 1981; Muindi and Thomke, 1981; Mehansho *et al.*, 1987; Knabe, 1990). Animals consume the bird-resistant sorghum rations readily. In fact they consume more feed to produce about the same gain. Thus, feed to gain ratio increases. The amount depends on the species, the feeding and processing methods and other factors. There are no differences in feeding value among sorghum grains with different colour pericarps that do not have pigmented testa.

Waxy sorghums have been shown to have improved feed efficiency for ruminants (Rooney and Pflugfelder, 1986) although the effect is decreased for swine and is not observed for poultry. The waxy sorghums steam flake more readily than non-waxy. They produce larger, more durable flakes with excellent appearance and greater dry matter digestibilities. Harder and softer sorghums do not flake properly. Sorghum that is weathered forms friable, dusty flakes that are highly undesirable.

5.8.2 Quality control

For steam flaking, the bulk density is used as an index of flake quality. A bulk density of 24–26 lb/bu is desired to obtain proper performance. Flake durability can be evaluated by using a pellet hardness tester to break the flakes; fines are measured by sieving. Sorghum flakes must be thin to achieve optimum feed efficiency which requires more energy and attention to quality control than flaking of maize. The optimum size screens are required to properly grind sorghum for feed. Coarse grinding is unacceptable. Screens used for maize do not work for sorghum.

5.9 HUMAN FOOD NUTRITIONAL VALUE

Sorghum has lower protein digestibilities than other cereal grains which is why sorghum must be properly processed prior to consumption (Rooney *et al.*, 1986). Sorghum proteins are thought to be less digestible because the protein bodies have a relatively enzyme-resistant outer shell which consists of a disulfide bound protein called gamma-kaffirin (Hamaker, 1994). He has found several sorghum lines with higher *in vitro* protein digestibility which could lead to sorghums with improved nutritional value. However, the endosperm of these grains appears to be soft which means they will be very susceptible to attack by moulds. In most climates, the sorghum kernel must have a hard endosperm to be agronomically acceptable. In the long term new hard endosperm sorghums with improved digestibility may be developed, but processing will be the major way of improving sorghum digestibility.

Many African and Indian foods are processed from decorticated sorghum. Decortication reduces the amount of fibre, minerals, proteins, and lysine (Eggum *et al.*, 1983; Serna-Saldivar *et al.*, 1987, 1988, 1994). Losses are 40% and 14% of lysine in sorghum decorticated by hand (decorticated sorghum yield 75–80%) and by a mechanical dehuller (decorticated sorghum yield 90%), respectively. Nutrient digestibilities of decorticated sorghum are slightly better than their respective whole grains but nitrogen retention and protein efficiency ratios are considerably lower because of the partial loss of germ. The development of white hybrids with thick pericarp has led to the production of light attractive milled products that have more nutritional value because the milling yields are higher.

Decortication of brown sorghums significantly reduces the amount of condensed tannins (Mwasaru *et al.*, 1988) and reduces the adverse effect of tannins on the nutritional value of sorghum grain. However, brown sorghums are soft and cannot be decorticated because of low yields of milled products.

Malting, germination, extrusion, fine grinding, and fermentation alone or in combination significantly enhance the overall nutritional value of sorghum and millet foods. Sprouting and treatment with wood ashes are used in Uganda to improve the nutritional value of brown sorghums. The treatment of high tannin sorghum grain with CaO, K_2CO_3, NH_4OH, $NaHCO_3$ lowers the amount of assayable tannins and improves the nutritional value of the grain to a level that approaches that of sorghum without tannins (Price *et al.*, 1979).

5.10 EFFECT OF MOULDS, INSECTS AND WEATHERING ON GRAIN QUALITY

The environmental conditions during grain maturation greatly affect the appearance of the grain because the sorghum head is exposed to insects, moulds and moisture. The kernels are often attacked by grain feeding bugs which provides an ideal place for attack by fungi. An environment that is hot and humid during and post maturation negatively affects grain quality. Moulds discolour the grain, break down the endosperm and significantly affect processing qualities. Mould damaged or weathered grain cannot be decorticated and the flour or grits are badly discoloured and cannot be used for food. Mouldy sorghums are impossible to malt for use in beer. Major fungi are *Fusarium, Alternaria and Curvularia* species. Sorghum mould and weather damage are the most important limitations to sorghum improvement worldwide.

Aflatoxin does not readily develop in sorghum in the field. This is in contrast to maize where aflatoxin is a very serious problem in the field prior to harvest. Sorghum like other grains can develop high levels of aflatoxins during improper storage. Sorghums with open panicles, with thin pericarp, condensed tannins, corneous endosperm and large-tight glumes are generally considered more resistant to moulds and weathering (Waniska *et al.*, 1992). Antimicrobial proteins have been found in sorghum that may lead to the production of white sorghums with more tolerance to moulds.

Efforts to improve sorghum cultivars in West Africa have been seriously hampered by head bugs that attack the grain post-anthesis which gives total grain loss or produces soft, shrunken, discoloured kernels that disintegrate during dehulling. The local sorghum landraces are photosensitive and are not badly damaged by head bugs. Tolerance to head bugs has been found in some cultivars; efforts to breed resistant cultivars are underway. The best short-term solution is to breed improved local photosensitive sorghums with tan plant colour and straw glumes which mature after the rains.

Weathering and moulding of pearl millet does occur but it is not as significant a problem.

REFERENCES

Almeida-Dominguez, H.D., Serna-Saldivar, S.O., Gomez-Machado, M.H. and Rooney, L.W. (1993) Production and nutritional value of weaning foods from mixtures of pearl millet and cowpeas. *Cereal Chemistry*, **70**(1), 14–18.

Anon. (1992) *Food and Agriculture Organization Production Yearbook*, FAO Statistics **45**(145), Rome, Italy.

Anon. (1994) *The Official United States Standards for Grain*, Federal Grain Inspection Service, Inspection Division, Washington DC.

Axtell, J.D. (1994) Breeding sorghum for increased nutritional value, in *Sorghum and Millet Collaborative Research Program Annual Report* (ed. J. Yohe), INTSORMIL, University of Nebraska, Lincoln, NE, pp. 93–9.

Blakely, M.E., Rooney, L.W., Sullins, R.D. and Miller, F.R. (1979) Microscopy of the pericarp and the testa of different genotypes of sorghum. *Crop Science*, **19**, 837–42.

Butler, L.G. (1989) Effects of condensed tannins on animal nutrition, in *Chemistry and Significance of Condensed Tannins* (eds R.W. Hemingway and J.J. Karchesy), Plenum Press, New York, pp. 391–402.

Butler, L.G. (1990) The nature and amelioration of the antinutritional effects of tannins in sorghum grain. *Proceedings of the International Conference on Sorghum Nutritional Quality*, February 26–March 1, 1990, Purdue University, West Lafayette, IN, pp. 191–205.

Cagampang, G.B. and Kirleis, A.W. (1984) Relationship of sorghum grain hardness to selected physical and chemical measurements of grain quality. *Cereal Chemistry*, **61**(2), 100–5.

Chavan, J.K. and Kadam, S.S. (1989) Nutritional improvement of cereals by sprouting, in *Critical Reviews in Food Science and Nutrition*, **28**(5), 401–37.

Cousins, B.W., Tanksley, T.O., Knabe, D.A. and Zebrowska, T. (1981) Nutrient digestibility and performance of pigs fed sorghums varying in tannin concentration. *Journal Animal Science*, **53**, 1524–37.

Creelman, R.A., Rooney, L.W. and Miller, F.R. (1982) Sorghum, in *Cereals: A Renewable Resource, Theory and Practice* (eds by Y. Pomeranz and L. Munck), American Association of Cereal Chemists, Minneapolis, MN, pp. 395–426.

Coleman, O.H. (1970) Syrup and sugar from sweet sorghum, in *Sorghum Production and Utilization* (eds J.S. Wall and W.M. Ross), AVI Publishing, Westport, CT.

Daiber, K.H. and Taylor, J.R.N. (1995) Opaque Beer, in *Sorghum and Millets Chemistry and Technology*, (ed. D.A.V. Dendy), American Association of Cereal Chemists, St. Paul, MN. pp. 299–323.

Dendy, D.A.V. (1995) *Sorghum and Millets Chemistry and Technology*, American Association of Cereal Chemists, St. Paul, MN.

Earp, C.F. and Rooney, L.W. (1982) Scanning electron microscopy of the pericarp and testa of several sorghum varieties. *Food Microstructure*, **1**, 125–34.

Earp, C.F., Akingbala, J.O., Ring, S.H. and Rooney, L.W. (1981) Evaluation of several methods to determine tannins in sorghums with varying kernel characteristics. *Cereal Chemistry*, **58**, 234–8.

Eggum, B.O., Monowar, L., Bach Knudsen, K.E. Bet al. (1983) Nutritional quality of sorghum and sorghum foods from Sudan. *Journal of Cereal Science*, **1**, 127–37.

Hagerman, A.E. and Butler, L.G. (1989) Choosing appropriate methods and standards for assaying tannin. *Journal of Chemical Ecology*, **15**, 1795–1810.

Hahn, D.H., Rooney, L.H. and Earp, C.F. (1984) Tannins and phenols of sorghum. *Cereal Foods World*, **29**, 776–9.

Hahn, D.H. and Rooney, L.W. (1986) Effect of genotype on tannins and phenols of sorghum. *Cereal Chemistry*, **63**, 4–8.

Hallgren, L. (1995) Lager beers from sorghum, in *Sorghum and Millets Chemistry and Technology* (ed. D.A.V. Dendy), American Association of Cereal Chemists, St. Paul, MN, pp. 283–97.

Hamaker, B.R. (1994) Food and nutritional quality of sorghum, in *Sorghum and Millet Collaborative Research Support Program Annual Report* (ed. J. Yohe), INTSORMIL, University of Nebraska, Lincoln, NE, pp. 138–42.

Hoseney, R.C. Andrews, D.U. and Clark, H. (1987) Sorghum and pearl millet, in *Nutritional Quality of Cereal Grains: Genetic and Agronomic Improvement* (eds R.A. Olson and K. Frey), American Society of Agronomy, Inc., Crop and Soil Science Societies of America, Madison, WI.

Hulse, J.H., Laing, E.M. and Pearson, O.E. (1980) *Sorghum and the Millets: Their Composition and Nutritive Value*, Academic Press, San Francisco, CA.

Kambal, A.E. and Bate-Smith, E.C. (1976) A genetic and biochemical study on pericarp pigments in a cross between two cultivars of grain sorghum, *Sorghum bicolour*, *Heredity*, **37**, 413–16.

Knabe, D.A. (1990) Sorghum as swine feed, in *Proceedings of International Conference on Sorghum Nutritional Quality*, February 26–March 1, 1990, Purdue University, West Lafayette, IN.

Malleshi, N.G. and Desikachar, H.S.R. (1986) Nutritive value of malted millet flours. *Quality Plant Foods in Human Nutrition*, **36**, 191.

Maxson, W.E., Shirley, R.L., Bertrand, J.E. and Palmer, A.Z. (1973) Energy values of corn, bird resistant and non-bird resistant sorghum grain in rations fed to steers. *Journal of Animal Science*, **37**, 1451–7.

McDonough, C.M. and Rooney, L.W. (1989) Structural characteristics of *Pennisetum americanum* using scanning electron and fluorescence microscopies. *Food Microstructure*, **8**, 137–49.

Mehansho, H., Butler, L.G. and Carlson, D.M. (1987) Dietary tannins and salivary proline-rich proteins: interactions, induction, and defense mechanisms. *Annual Review of Nutrition*, **7**, 423–40.

Muindi, P.J. and Thomke, S. (1981) Metabolic studies with laying hens on Tanzanian sorghum grains of different tannin contents. *Swedish Journal of Agricultural Research*, **11**, 17–21.

Munck, L. (1995) Milling technologies and processes, in *Sorghum and Millets Chemistry and Technology* (ed. D.A.V. Dendy), American Association of Cereal Chemists, St. Paul, MN., pp. 223–81.

Murty, D.S. and Kumar, K.A. (1995) Traditional uses of sorghum and millets, in *Sorghum and Millets Chemistry and Technology* (ed. D.A.V. Dendy), American Association of Cereal Chemists, St. Paul, MN, pp. 185–221.

Mwasaru, M.A., Reichert, R.D. and Mukuru, S.Z. (1988) Factors affecting the abrasive dehulling efficiency of high tannin sorghum. *Cereal Chemistry*, **65**, 171–4.

Nip, W.K. and Burns, E.E. (1969) Pigment characterization in grain sorghum. I. Red varieties. *Cereal Chemistry*, **46**, 490–5.

Nip, W.K. and Burns, E.E. (1971) Pigment characterization in grain sorghum. II. White varieties. *Cereal Chemistry*, **48**, 74–80.

Norris, J.R. (1971) Chemical and Physical and Histological Characteristics of Sorghum Grain Related to Wet Milling Properties, PhD Dissertation, Texas A&M University, College Station, TX.

Palmer, G.H. (1989) Cereals in malting and brewing, in *Cereal Science and Technology* (ed. G.H. Palmer), Aberdeen University Press, Great Britain, pp. 216–25.

Price, M.L., Butler, L.G., Rogler, J.C. and Featherston, W.R. (1979) Overcoming the nutritionally harmful effects of tannin in sorghum grain by treatment with inexpensive chemicals. *Journal of Agricultural and Food Chemistry*, **27**, 441–5.

Reichert, R.D., Mwasaru, M.A. and Mukuru, S.Z. (1988) Characterization of coloured grain sorghum lines and identification of high tannin lines with good dehulling characteristics. *Cereal Chemistry*, **65**, 165–70.

Rooney, L.W. (1978) Sorghum and pearl millet lipids. *Cereal Chemistry*, **55**, 584–90.

Rooney, L.W., Earp, C.F. and Khan, M.N. (1982) Sorghums and millets, in *CRC Handbook of Processing and Utilization in Agriculture* (ed. I.A. Wolf), CRC Press, Boca Raton, FL., 2, 123.

Rooney, L.W., Khan, M.N. and Earp, C.F. (1980) The technology of sorghum products, in *Cereals for Food and Beverages: Recent Progress in Cereal Chemistry* (eds E. Inglett and L. Munck), Academic Press, New York, NY.

Rooney, L.W., Kirleis, A.W. and Murty, D.S. (1986) Traditional foods from sorghum: their production, evaluation and nutritional value, in *Advances in Cereal Science and Technology*, Vol. VIII (ed. Y. Pomeranz), American Association of Cereal Chemistry, St. Paul, MN.

Rooney. L.W. and McDonough, C.M. (1987) Food quality and consumer acceptance of pearl millet, in *Proceeding International Pearl Millet Workshop* (eds J.R. Witcombe and S.R. Beckerman), ICRISAT, Patancheru, India.

Rooney, L.W. and Miller, F. (1982) Variation in the structure and kernel characteristics of sorghum, in *International Symposium on Sorghum Grain Quality* (eds L.W. Rooney and D.S. Murty), ICRISAT, Patancheru. A.P., India, p. 143.

Rooney, L.W. and Pflugfelder, R.L. (1986) Factors affecting starch digestibility with special emphasis on sorghum and corn. *Journal of Animal Science*, **63**, 1607–23.

Rooney, L.W. and Serna-Saldivar, S.O. (1991) Sorghum, in *Handbook of Cereal Science and Technology* (eds K.J. Lorenz and K. Kulp), Marcel Dekker, New York, NY, pp. 233–70.

Seitz, L.M., Wright, R.L., Waniska, R.D. and Rooney, L.W. (1993) Contribution of 2-acetyl-1-pyrroline to odors from wetted ground pearl millet. *Journal of Agriculture and Food Chemistry*, **41**, 955–8.

Serna-Saldivar, S.O. and Rooney, L.W. (1995) Structure and chemistry of sorghum and millets, in *Sorghum and Millets: Chemistry and Technology* (ed. D.A.V. Dendy), American Association of Cereal Chemists, St. Paul, MN, pp. 69–124.

Serna-Saldivar, S.O., Clegg, C. and Rooney, L.W. (1994) Effects of parboiling and decortication on the nutritional value of sorghum (*Sorghum bicolour* L. Moench) and pearl millet (*Pennisetum glaucum* L.). *Journal of Cereal Science*, **19**, 83–9.

Serna-Saldivar, S.O., Knabe, D.A., Rooney, L.W. and Tanksley, T.D. (1987) Effects of lime cooking on energy and protein digestibilities of maize and sorghum. *Cereal Chemistry*, **64**, 247–52.

Serna-Saldivar, S.O., Knabe, D.A., Rooney, L.W. *et al.* (1988) Nutritional value of sorghum and maize tortillas. *Journal of Cereal Science*, **7**, 83–94.

Serna-Saldivar, S.O., McDonough, C.M. and Rooney, L.W. (1991) The millets, in *Handbook of Cereal Science & Technology* (eds Lorenz, K.J. and K. Kulp), Marcel Dekker, pp. 271–300.

Serna-Saldivar, S.O., Tellez-Giron, A. and Rooney, L.W. (1987) Production of tortilla chips from sorghum and maize. *Journal of Cereal Science*, **8**, 275–84.

Shull, J.M., Watterson, J.J. and Kirleis, A.W. (1991). Proposed nomenclature for the alcohol soluble proteins (kafirins) of *Sorghum bicolour* L. Moench based on molecular weight, solubility and structure. *Journal of Agricultural and Food Chemistry*, **39**, 83–7.

Taylor, J.R.N., Novellie, L. and Liebenberg, N.W. (1985) Protein body degradation in the starchy endosperm of germinating sorghum. *Journal of Experimental Biology*, **36**, 1287–95.

Taylor, J.R.N. and Schussler, L. (1986) The protein composition of the different anatomical parts of sorghum grain. *Journal of Cereal Science*, **4**, 361–3.

Waniska, R.D., Forbes, G.A., Bandyopadhyay, B. *et al.* (1992) Cereal chemistry and grain mold resistance, in *Sorghum and Millet Diseases, a Second World Review* (eds W.J. deMilliano, *et al.*), International Crops Research Center for the Semi-Arid Tropics, Patancheru, India, pp. 265–72.

Warsi, A.S. and Wright, B.C. (1973) Effects of rates and methods of nitrogen application on the quality of sorghum grain. *Indian Journal of Agricultural Science*, **43**, 722–6.

Watson, S.A. (1984) Corn and Sorghum Starches: Production. *Starch Chemistry and Technology* (eds R.L. Whistler, J.N. Bemiller and E.F. Paschall), 2nd edn, Academic Press, Orlando, FL.

Watterson, J., Shull, J.M., Mohamed, A.A. *et al.* (1990) Isolation of a high cysteine kafirin protein and its cross reactivity with gamma zein antiserum. *Journal of Cereal Science*, **12**, 137–44.

Wu, Y.V. and Wall, J.S. (1980) Lysine content of protein increased by germination of normal and high lysine sorghum. *Journal of Agricultural Food Chemistry*, **28**, 455–8.

Young, R., Haidara, M., Rooney, L.W. and Waniska, R.D. (1990) Parboiled sorghum: development of a novel decorticated product. *Journal of Cereal Science*, **11**, 277–89.

6

Oats

F.H. Webster

6.1 INTRODUCTION

The first records of cultivated oats can be traced to about 2000 BC or the Early Bronze Period. Botanists generally maintain that the origin of a species is most likely to be found in the area of the greatest progenitor diversity. Murphy and Hoffman (1992) characterized the region of greatest oat diversity as an area situated between 25 and 45°N lat. and 20°W and 90°E long., extending from the Canary Islands, throughout the Mediterranean basin, the Middle East, to the Himalayan mountains. The centres of origin of the various *Avena* spp. are unknown but are thought to be located within this geographic area. Unfortunately, precise information on the time, place and processes associated with oat domestication are lost to antiquity (Coffman, 1961).

Cultivated oats are members of the genus *Avena* L., comprising a polyploid series of wild, weedy and cultivated species, which are found worldwide in almost all agricultural environments. According to the historical classification, cultivated oats principally belong to members of three hexaploid species: *Avena sativa, Avena byzantina* and *Avena nuda*. Additionally, a small acreage is devoted to the diploid *Avena strigosa* (Coffman, 1961). *Avena sativa* is commonly referred to as white oats, or in some cases as yellow oats, and *Avena byzantina* as red oats. On the basis of this classification, more than 75% of the total cultivated oat acreage is sown to *Avena sativa* species while most of the remainder is sown to *Avena byzantina*. However, there are considerable differences of opinion on what constitutes a separate species. Further discussion is beyond the scope of this text; however, I would like to point out that in the most recent attempt at classification (Baum, 1977), the species designation *Avena byzantina* was dropped and the members of that group placed into the *Avena sativa* taxa. Similarly, the taxa *Avena nuda* was reclassified as *Avena sativa* var. *nuda*. This was brought about in part because the three species are fully cross-fertile. As a result, the differences among the species has blurred as breeders intercross lines to combine the most elite genetic materials into high yielding agronomically superior types.

The first reported references to utilization described the medicinal value of oats (Findlay, 1956), from Hippocrates (*ca.* 460–360 BC) and Dieusches (*ca.* 400 BC) and then Dioscorides (1st century) and Galen (130–200 AD). Dioscorides described oats as a healing agent, a desiccant when applied to the skin, a cough reliever, and a natural food for horses which was used by humans in times of scarcity. The first reference to human consumption was by Pliny in *Natural History*. He reported that the Germanic tribes of the first century knew oats well and 'made their porridge of nothing else'. Despite evidence of early consumption, the grain found long-term acceptance only in Scotland and Ireland, where it was used for a variety of porridges. Scottish settlers brought oats to North America although they were not widely accepted as part of the diet until the 19th century. The most famous quote with regards to oats was documented by Sir Walter Scott who quoted Samuel Johnson as saying 'Oats, a grain which in England is generally given to horses, but in Scotland supports the people.' To which Lord Elibank is reputed to reply 'True, but where can you find such horses, where such men?'

As these statements imply, oats traditionally served as animal feed and a staple for the impoverished or those in ill health. Three developments in the 19th century helped establish oats as a staple breakfast item: an enhanced oat milling and stabilization technology, movement of oatmeal from the druggist's shelf to the grocery store, and the evolution of packaging, branded products and sales promotions.

6.2 WORLD PRODUCTION

Oats rank sixth in world cereal production behind wheat, maize, rice, barley and sorghum. Table 6.1 illustrates oat production by world geographic area from 1960 to date, with projections to the year 1999. Production was level at almost 50 million metric tonnes from 1960 to 1975, but it has declined steadily since 1975 and is predicted to reach 34.5 million metric tonnes by 1999 (P.F. Sisson, The Quaker Oats Company, personal communication). The decline in production is closely linked to the poor monetary returns per hectare relative to other competitive crops. Most countries produce oats for domestic consumption and only export the grain when production exceeds domestic use. The availability of other feed grains is also a factor. Relatively speaking, oats are more expensive to ship due to their lower test weights and bulky nature relative to other grains. Additionally, the traditionally grown oat crop

Table 6.1 World oat production* (million metric tonnes)

	1960–64	*1965–69*	*1970–74*	*1975–79*	*1980–84*	*1985–89*	*1990–94*	*1995–99*
North America	20.6	18.5	15.9	12.9	10.4	8.5	6.6	4.3
South America	0.2	0.2	0.2	0.2	0.3	0.4	0.5	0.6
Latin America	1.0	0.8	0.7	0.8	0.9	1.0	1.0	1.3
Western Europe	13.2	13.0	13.1	11.6	11.5	9.5	7.3	5.4
Eastern Europe	4.4	4.3	4.6	3.6	3.5	3.4	2.8	2.8
Russia + FSU-15†	6.3	9.0	13.2	14.4	14.4	16.3	14.7	17.5
Africa	0.2	0.2	0.2	0.2	0.2	0.2	0.2	0.1
Asia	1.0	1.0	0.8	0.8	0.8	0.7	0.6	0.5
Oceania	1.3	1.4	1.2	1.3	1.5	1.7	1.8	1.9
World	48.2	48.3	49.5	45.9	43.5	41.6	35.5	34.5

*Source: USDA, Foreign Agricultural Service. Commodity production, supply, and disposition database. Unpublished monthly computer print-outs.
†15 countries of the 'former Soviet Union'.

has a much lower caloric density due to the presence of the protective hull. Finally, the rapid expansion of soybean production and utilization has created an excellent source of protein which can be utilized in livestock rations in the United States, Europe and South America. These factors, coupled with advances in nutritional research, have allowed commercial feed companies to formulate high efficiency low cost feeds with blends of alternative commodities.

The leading producing countries are Russia, Canada, United States, Germany, Australia, Poland, Sweden, Ukraine, Finland and Byelarus. Production is increasing in the former Soviet Union countries, Oceania and South America, while it is declining in the rest of the world. Collectively, the 15 countries comprising the former Soviet Union are the world leaders in oat production. By the end of the century this group is expected to produce about 50% of the world oat supply. In contrast, by 1999 North American production will have declined to 20% of its 1960 level. This change primarily reflects a major reduction in the United States oat acreage while Canadian production has exhibited a more modest decline.

Oat yields vary substantially between the various geographic regions. Western Europe, whose 1993 average yield was 3.5 metric tonnes per hectare, has by far the highest yields. The yield is indicative of the intensive agronomic practices used in those areas. Large countries, such as the former Soviet Union, Canada and the United States, tend to sacrifice yield by using fewer inputs and relegating the crop to poorer land. Average 1993 yields were 2.2 and 1.4 metric tonnes per hectare for North America and the former Soviet Union countries, respectively. These practices impact upon the quality and quantity of oats available for purchase.

Oats play a minor role in world trade with about 2.7 million metric tonnes being exchanged in the international grain markets in 1993. However, this does represent an upward trend for world commerce. Between 1960 and 1985 world oat trade averaged 1.45 million metric tonnes, but it is expected to reach 3.0 million metric tonnes by 1999. The leading exporters in order of volume are Canada, Finland, Sweden, Australia, Ukraine and Argentina, who together account for 90% of the total world trade in oats.

6.3 OAT CHARACTERISTICS

6.3.1 Oat composition and structure

Oat quality and characteristics are closely linked to its unique composition and seed structure. Like other cereals, the oat kernel is a complex structure whose primary function is to generate a complete vegetative plant. The oat kernel is organized in a similar fashion to the other members of the Gramineae (the grass family). However, oats have many distinctive chemical and structural features, such as high levels of soluble dietary fibre (β-glucan), which is the primary component of the kernel cell walls, and high levels of good quality protein. Figure 6.1 illustrates the structural features of oat enclosed in the protective hull. Fulcher (1986) pointed out that oats have three primary structures: bran, germ and endosperm. Each fraction consists of several different tissues which, in most cereals, tend to fragment through natural cleavage regions into the three classifications upon mechanical disruption. Owing to the high lipid content and the soft texture of the kernel, oats typically do not separate into clean fractions during the milling process. For example, oat bran typically has significant quantities of starchy endosperm adhering to the bran. However, it is possible to obtain fractions, which are enriched for each structure, using traditional dry milling techniques. Relatively pure fractions can be obtained by using either hand dissection techniques or chemical separations. Please note that the following discussion is relevant to both hulled and hull-less oats – the compositional and structural relationships are the same.

Oat bran has three distinct layers. The outermost layer consists of the pericarp, testa and the nucellus, which are remnants of the maternal tissue. Those structures are primarily complex carbohydrate and very fibrous probably with a significant content of lignin or related phenolic structures. The second region is a single layer of cells called the aleurone. The cell wall of this layer contains both β-glucan and an abundance of phenolic compounds. The interior is rich in protein bodies containing deposits of phytin, niacin, phenolics and carbohydrates. The protein bodies are surrounded by lipid droplets (Fulcher, 1986). The sub-aleurone region is usually much higher in protein and fat than the endosperm and represents a transition zone between the outer regions and the starchy interior cells. The cell walls in this region are typically thicker than those found in the interior and contain proportionally larger quantities of β-glucan.

The starchy endosperm is by far the largest of the three regions. Estimates of its size range from 55–70% (Youngs, 1972) to 80% (Fulcher, 1986) of the total kernel mass. This tissue contains major reserves of starch, protein and lipids. Additionally, the cell walls make a significant contribution to the total β-glucan content. The cells in the starchy endosperm are larger and, in general, have much thinner cell walls than the outer bran region. Two factors tend to distinguish the oat endosperm region from other cereals. The first is the high β-glucan content of the cell walls (with the exception of barley, other cereal endosperm cell walls tend to be nearly devoid of the mixed linkage glucans). The second factor is the relatively high fat content. Wheat

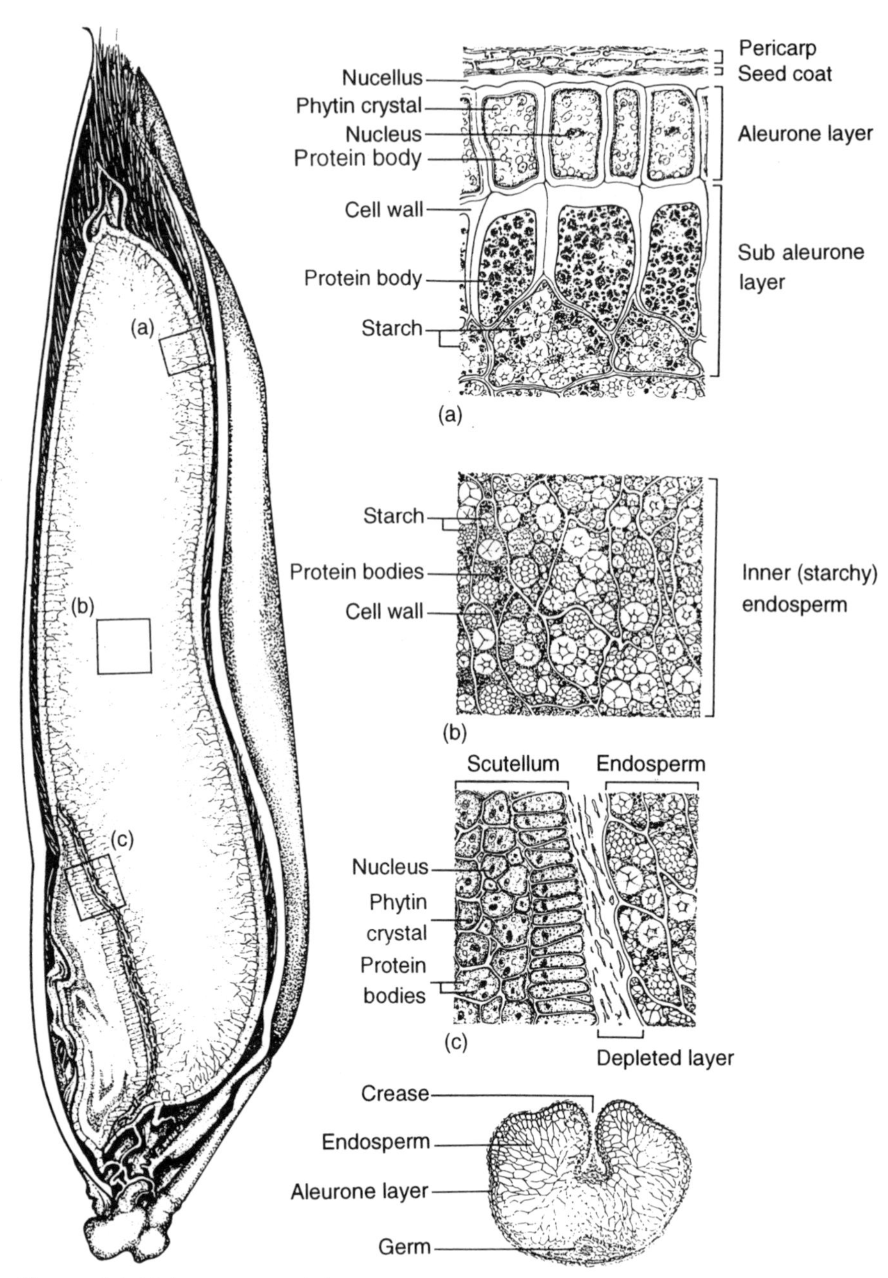

Figure 6.1 Major structural features of the oat kernel. On the left is a kernel (with hulls on) that has been split longitudinally to reveal the approximate size and location of the major tissues. At the lower right is a view of a cross-section of the groat. (a), (b), and (c), Higher magnifications of portions of the bran, starchy endosperm, and germ. (Reprinted, with permission, from R. G. Fulcher, 1986.)

endosperm, for example, contains only 0.8–1.0% lipid with the largest portion being the polar fraction (Hargin *et al.*, 1980). In contrast, oat endosperm can contain up to 6–8% lipid, which represents approximately 50% of the total oat lipids, with a large portion being the neutral glyceride fraction (Youngs, 1972).

The tissues which comprise the germ fraction account for about 2.8–3.7% of the oat kernel and contain less than 10% of the total oat lipid (Youngs, 1972). Protein makes up about 30–40% of the germ region depending upon the cultivar. In contrast to wheat and corn milling, and because of their soft texture and the nature of their milling process, oats do not provide a commercially viable quantity of the germ fraction.

6.3.2 Nutritional quality

The nutritional quality of oat products can exhibit significant variation due to differences in the starting raw materials. Table 6.2 illustrates the observed variation in oat composition. Oats are typically purchased as a bulk commodity (a blend of varieties) rather than on an identity-preserved basis; thus, the processor will observe the extremes in composition only on rare occasions. It has been our experience that the largest variations occur when oats are obtained from different geographic areas; for example, oats produced in Australia, Canada, Europe and South America tend to average 1–3% lower in protein and 1–2% higher in fat than the United States production.

The key nutritional components, fat, protein and β-glucan, of commercial cultivars exhibit about a 1.5 to 2X variation in content.

Table 6.2 Composition of US oat groats

Component	*Average value %*	*Typical range%*
Protein*	15.2	11–20
Fat†	7.6	5–9
Starch‡	51.1	44–61
β-Glucan‡	4.2	2.2–6.6
Ash*	1.9	1.3–2.3
Total dietary fibre*	8.9	7.0–11.0
Free sugars§	1.1	0.9–1.3
Moisture	10.0	9–14

*The Quaker Oats Company
†Youngs (1986)
‡D. Peterson (1992)
§MacArthur-Grant (1986)

However, as with all biological systems, these compositional traits are under genetic control and oat breeders could select for higher or lower levels of any trait. Historically, due to limited resources, the major focus of oat breeders has been on yield and agronomic traits.

The methods as described by Lane (1990) are recommended for determining the proximate composition of oats. However, some of the newer rapid methods, such as near infrared reflectance, are very desirable for the rapid and simultaneous analysis of components, such as fat, protein and moisture. Analysis of β-glucan is a laborious process and requires painstaking attention to detail. Two procedures were identified in a collaborative trial for β-glucan determination (Munck *et al.* (1989). The McCleary and Glennie-Holmes (1985) enzymatic method utilizes lichenase to break down the polymer and β-glucosidase to convert the oligosaccharides to glucose for final quantitation. This method is used widely in β-glucan analysis. The second method is the more rapid flow injection procedure (FIA) system of Jorgensen (1988). For a more detailed discussion of the analytical issues see Wood (1993).

6.3.3 Functional attributes

Functional or processing attributes are another important parameter for grain. In the case of oats, there is a paucity of information available on processing quality. Historically, oats have never been purchased on the basis of functional attributes. Consumer acceptance factors such as flavour, colour and texture are important grain quality parameters. However, colour is the only characteristic of this group that has played a significant role in the commercial acceptance of oat varieties. Properties such as water absorption capacity are important in some oat products, but no useful information is available which relates oat composition to this or other functional attributes.

6.4 OAT UTILIZATION

6.4.1 Usage

The primary use of oats is as a feed grain. Murphy and Hoffman (1992) stated that on a world wide basis 75% of oats are used for animal feed, 22% for food and seed and the remainder is exported (presumably for the same end-uses). The portion of the oat crop used in animal feed has declined from about 90% in 1950 to the current level, while the quantity used for food and seed applications has increased by about 20%. The increase in food utilization has been accelerated by the public's interest in health and nutrition. Two observations formed the basis for this interest. The first was the report by deGroot *et al.* (1963) that oats had hypocholesterolaemic properties. The second was the fact that incorporation of soluble fibre into the diet slowed the rise in blood sugar following a meal. This can be very important in the management of the insulin response in normal and diabetic individuals (Jenkins *et al.*, 1981; Heaton *et al.*, 1988). For a more in depth review of the effects of oat soluble fibre on health and nutrition, refer to one of the following articles: Anderson and Chen (1986), Anderson and Bridges (1993) or Shinnick and Marlett (1993).

Historically, oats, like other cereals, have been recognized as an important component of the diet. In addition to providing calories to meet our energy needs, oats are recognized as having the highest protein content and the best quality protein of all cereal grains (consult Peterson, 1992; or Peterson and Brinegar, 1986 for a detailed review of oat protein characteristics). Oats also contain several essential vitamins, minerals and fatty acids. Due to these attributes, oats offer several unique benefits relative to other cereals when consumed as part of a balanced diet. They are an excellent source of soluble fibre, a good source of protein and they contain a powerful antioxidant complex.

6.4.2 Food usage

Oats are unique in their food uses and attributes in comparison to most other cereal grains. First, the primary use of oats is as a whole-grain flour or flake; in contrast, the germ and significant portions of the bran are removed from most other cereals prior to their introduction into food systems. Second, oats are heat-processed to develop the characteristic toasted-oaty sensory notes and to inactivate the complex enzyme systems present in oats. Untreated oat flours rapidly develop soapy and bitter flavours because of the action of lipase, lipoxygenase and peroxidase (Hutchinson *et al.*, 1951; Kazi and Cahill, 1969; Biermann and Grosch, 1979). This treatment also has a significant impact upon the potential functionality of oat products; for example, protein solubility is dramatically reduced as the result of this treatment.

Typical commercial oat products are rolled oats, steel-cut groats, quick oats, baby oat flakes, oat bran, and oat flour. Many intermediate flakes and flours are produced by oat processors to meet individual customer specifications. Descriptions of typical products follow:

1. *Rolled oats*, which are produced by flaking whole groats, are the thickest of the standard oat-flake products; flake thickness varies from 0.020–0.030 inch (0.508–0.762 mm) depending upon the intended end-use. The thicker flakes require longer cooking periods and maintain flake integrity for extended time periods.

2. *Steel-cut groats* are produced by the sectioning of groats into several pieces; they are used in the preparation of flakes and flour and as a specialty ingredient.
3. *Quick oats* are flakes produced from steel-cut groats. In this process, oat groats are typically cut into three to four pieces before the final steaming and flaking process. Quick oats, which are usually 0.014–0.018 inch (0.356–0.457 mm) thick, require less cooking time than the whole-oat flakes.
4. *Baby oats* are also produced from steel-cut groats, but the flakes are thinner and have a finer granulation than quick oats. These smaller, thinner flakes cook more rapidly than quick oats and have a smoother texture.
5. *Instant oat flakes* are produced from 'instantized' steel-cut groats. Before cutting, the groats have been subjected to a special proprietary commercial process that results in rapid-cooking flakes. The flakes are typically 0.011–0.013 inch (0.279–0.330 mm) thick.
6. *Oat flour* is produced by grinding flakes or groats into flour for use as an ingredient in a wide variety of food products. The flour granulation will vary depending upon the end-use.
7. *Oat bran* is a bran-rich fraction produced by sieving coarsely ground oat flour.

Those interested in a detailed discussion of the oat milling process should refer to one of the following references: Salisbury and Wichser (1971), Deane and Commers (1986) or Burnette *et al.* (1992). Hull-less oats can be used in place of the traditional hulled type. However, special attention must be paid to storage and handling procedures in order to assure grain quality in the absence of the protective hull.

All of the commercial oat products described above have received a heat treatment that was sufficient to stabilize the products against the development of hydrolytic rancidity. As previously indicated, oats with elevated free fatty acid contents should not be used in the production of food for human consumption. Oat product specifications typically call for the materials to be either tyrosinase or peroxidase negative. Tyrosinase is more heat stable than lipase and is much easier to measure; thus, it has been used as a reliable indicator of residual lipase activity. Peroxidase is the most heat stable of the oxidative enzymes and it can be monitored directly to determine the degree of inactivation. The method of Mulvaney (1990), which is used for peroxidases in vegetables, can be adapted to monitor cereal peroxidase activity. Kowalewski and Jankowski (1979) also described an adaptation of a peroxide analysis for evaluation of oat peroxidase.

6.4.3 Food product applications

Hot cereals

The principal use of oats is as a hot cereal. Oatmeal is consumed on a worldwide basis, although it is most popular in North America, the United Kingdom and northern Europe. Products vary from whole flake oat products, which require several minutes to cook on the stovetop, to prepare-in-the-bowl (instant) products. The larger whole flake products maintain their flake identity and exhibit the textural characteristics associated with the presence of β-glucan. The instant products are much smoother in texture and lack flake identity. Additionally, the ease of preparation and the availability of a wide variety of flavours make instant oatmeal a very versatile product. Instant oatmeal now comprises over half the hot cereal market based on the volume sold. The instant products are very convenient, sold in a variety of flavours tailored to ethnic preferences and offer the warmth and nourishment of hot cereals. In the Scandinavian countries, cold oatmeal, which has been cooked in fruit juice, is a popular way to consume oats. This product, muesli, is served with cold milk and preserves or fruit juice.

Cold cereals

Oat flour and flakes are used in a number of 'Ready-To-Eat' (RTEs) cereals. These products are sold in extruded, flaked and other expanded forms. The high fat and fibre content of oats restricts its utilization in highly expanded products. Production of oat-based cereals requires the processor to exercise control over process temperature and final product moisture in order to protect against the initiation of oxidative rancidity (Martin, 1958). Although oats have a potent antioxidant system, it can be destroyed by over-processing. Thermal breakdown is one factor which may be responsible for loss of antioxidant capacity (Collins, 1986).

Oat bran

Oat bran became a popular product and ingredient in the mid-1980s. This was driven by public recognition of the health benefits derived from consuming foods containing significant quantities of soluble dietary fibre. Oat bran is consumed as a hot cereal and has been formulated into a variety of RTEs. Oat bran is also used in a variety of bread, cookie and granola products.

Bread products

Oat ingredients are used in a wide variety of breads, muffins and cookies. Oats can provide unique flavour, moisture retention and nutritional benefits. Oats have been shown to provide excellent moisture retaining properties which keep breads fresher longer (McKechnie, 1983). Additionally, Dodok *et al.* (1982) demonstrated that oat flour was able to stabilize the fat component in breads. Since oats contain only minor amounts of gluten, they must be mixed with wheat flour to produce bread products. The texture of oat-containing bread products can be varied by using different granulations of oat ingredients. Larger flakes produce an uneven texture, smaller flakes make a smoother product while steel cut groats provide a nutty flavour and texture (Webster, 1986). Similarly, oats provide unique attributes to cookies. Water absorption is crucial in cookie doughs (Smith, 1973); thus, choice of the ingredient particle size will impact upon the characteristics of the dough and the texture of the final product.

Infant foods

Oatmeal or flour is a major component of many infant foods. Drum-dried oatmeal, in many parts of the world, is the first solid food consumed by a baby. Important considerations in selecting oats for this use include allergenicity, flavour compatibility, nutritive value, shelf-life and stability, economics and availability (Shukla, 1975). Oats are often used in conjunction with other ingredients to provide specific nutritional needs, especially to combat malnutrition. In Mexico, a group of researchers developed a soy oat formula which is nutritionally equivalent to milk-based products and is suitable for feeding to lactose intolerant infants (Mermelstein, 1983).

Other food applications of whole grain oats

Oatmeal is a major constituent in granola cereals and bars. Rolled oats are used as thickening agents in soups, sauces and gravies, as meat extenders in meat loaf and meat patties, as an ingredient in pancake mixes and proposed as a main meal side dish (Webster, 1983). Oat flour is the base for a popular South American beverage, Frescavena. Additionally, an oat-based soup is widely consumed in South America (Webster, 1986).

6.4.4 Oat antioxidants

Oats contain a potent antioxidant complex (Daniels and Martin, 1961; Collins 1986). Oats usage as an antioxidant was first proposed by Peters and Musher (1937). A special fine ground oat flour was effective in stabilizing milk powder, butter, bacon, ice cream, cereals, frozen fish and other products sensitive to oxidative deterioration. The initial investigations identified derivatives of hydroxycinnamic acid as the primary components of the antioxidant complex. Structurally, the phenolic components of these esters, caffeic and ferulic acids, are very similar to the commercial antioxidants, butylated hydroxytoluene and butylated hydroxyanisole and exhibit similar activities.

The oat antioxidant complex consists of both hydrophilic and lipophilic compounds. Oat antioxidants include tocopherols (Bauernfeind, 1980; Burton and Ingold, 1981), various hydroxycinnamic esters of long chain alcohols, ω-hydroxy fatty acids glycerol (Daniels and Martin, 1967; Duve and White, 1991), 5-avensterol (White and Armstrong, 1986) and avenathramides (Collins, 1989; Dimberg *et al.*, 1993).

The current interest in natural antioxidants could revive the interest in utilization of oat fractions for stabilizing oxidation-sensitive products. Additionally, there is a great deal of interest in the health benefits of natural antioxidant complexes. Antioxidants reportedly have a positive effect on such problems as ageing, cholesterol oxidation and cancer incidence. As the results of research on antioxidants and health become more definitive, there may be heightened interest in consuming foods which are high in complex natural antioxidants.

6.4.5 Novel processes and usages

Several unique applications have been proposed for oat fractions. Oat starch has been proposed for adhesives applications and in cosmetics. In fact, the process of Burrows *et al.* (1984) is being investigated as a means of producing cosmetic grade oat starch. Oat flour or oatmeal has been used in facial masks and in cleansing creams. Reportedly, oats have hypoallergenic properties. These attributes have led to a series of commercial products. The benefits of oat flour in cosmetics was reviewed by Miller (1977).

Several processes, both aqueous and non-aqueous, have been developed to fractionate oats into fractions rich in β-glucan, protein, oil, bran and starch (Hohner and Hyldon, 1977; Oughton, 1980a, 1980b; Burrows *et al.*, 1984). Efforts continue to commercialize some of these processes. The primary issue is process economics; suitable applications/markets are needed for all of the fractions in order to develop a viable business proposition.

One new process provides oat fractions, which have some unique non-food applications in addition to the more traditional uses. Gupta (1991, 1992) reported on the development of porous proteinaceous particles from oats, which could be used in cosmetics, environmental remediation, and agrichemicals, as well as food. Product applications have been demonstrated in time-release formulations of agrichemicals and other systems, as an absorbant for oil spill cleanups and as food antioxidants.

Oat hulls are utilized for the production of furfurals and furans (Dunlop, 1973). Oat hulls are a rich source of pentosans and thus are an economical starting material for production of these specialty chemicals. Finally, oat hulls reportedly contain an anticariogenic factor (Madsen, 1981; Lorenz, 1984).

Inglett (1991) has recently developed a process for making a fat substitute from oats. The process utilizes α-amylase treatment to create a water soluble dietary fibre preparation which can be separated from the insoluble materials. The final product has found application as a fat replacer in baked goods, salad dressings and numerous other products. Preliminary nutritional research indicates that it maintains the cholesterol lowering properties of oat products. This may represent a very beneficial and lucrative market for oat derived products.

6.4.6 Feed usage

Historically, the primary usage of oats has been as a feed grain. Until the 1960s, their excellent nutritional profile and broad adaptability made them the feed grain of choice. As farming practices began to change and production of livestock began to shift from the traditional farm to dedicated production units, oats popularity in feed rations began to decline. Several factors played prominent roles in this change. First was the abundance of large quantities of low cost alternative feed ingredients, such as corn, soymeal, cottonseed meal and more recently, canola meal. These materials in many instances can be formulated into highly efficient and more economical feed rations. For example, in the United States, a blend of alternative feed ingredients such as corn and soymeal plus a vitamin and mineral supplement has proven to be more cost-effective in many situations. Secondly, in spite of their excellent nutrition profile, oats are low in caloric density due to their high hull content (25–30% average). Also, oat dehulling adds additional costs and is not available to many feed formulators. Finally, the high levels of soluble dietary fibre present in oats seem to reduce feed performance in very young chicks or pre-weaning pigs if oats are included at too high a level in the diet. However, this is not the case for older animals.

Oats provide several unique benefits when they make up a significant percentage of animal diets. Moran (1986) reported that oats increased milk and milk fat production in dairy cows relative to diets containing wheat or barley. Martin and Thomas (1987) and Kankare and Antila (1984) reported that oats increased the level of unsaturated fat in cows milk. Inclusion of high levels of oats in the diets of laying hens resulted in an increase in egg size and the amount of unsaturated fatty acids

(Herstad, 1980; Karunajeewa and Tham, 1987; Cave *et al.*, 1989). Morris and Burrows (1986) reported that swine fed diets based on hull-less oats were similar in carcass quality and feed efficiency to those fed control rations. The diets containing the highest levels of oats produced the best carcass yields. Interestingly, Friend *et al.* (1989) and Friend *et al.* (1988) reported that sensory panel results showed that swine fed hull-less oats produced meat which was more tender, juicier and had higher levels of flavour than controls.

Oats are the preferred or standard grain for horses because they are highly palatable, easily digestible and of excellent nutritional quality. Morrison (1956) reported that the hull played a positive role in horse nutrition because it aided digestion. However, some race horse enthusiasts feel that hull-less oats enhance performance due to their higher caloric density. Nonetheless, oats in one form or the other are the preferred feed for horses. The oats used for feeding horses typically are of extremely high quality and are purchased according to standards which rival or exceed those for milling oats. The terms race horse or pony oats do not refer to special cultivars but rather to oat grain that has been thoroughly cleaned and graded to meet buyer standards for test weight normally 54 to 63.5 kg/hl, (Schrickel *et al.*, 1992). These oats are clean and bright, free of dust, weed seeds and wild oats, possess a clean aroma free of mustiness and are in many parts of the world white in colour. The colour preference is not based on any performance factor but is reputedly a criterion which enables the buyers to detect poor quality grain or grain which has been mishandled, for example, weathered or overdried grain. In poor crop years, the willingness of horse owners to pay a premium for high quality oats may reduce the availability of good milling quality oats.

Hull-less or naked oats may offer the best opportunity for oats to regain their position as the premier feed grain. This is especially true in areas of the world where corn and traditional oilseed crops are not well adapted. Hull-less oats with their high level of good quality protein and high fat levels can satisfy most feed grain requirements. This will serve to benefit local agriculture and reduce import requirements.

6.5 OAT QUALITY

6.5.1 Oat grading standards

There are no international standards for oat quality assessment. Although they are based on similar criteria, all oats producing countries have varying oat grade classifications, measurement techniques and standards which make them unique to the country of origin. Generally speaking, only the top grades are worth considering when purchasing oats for human consumption. As testing techniques, calibrations and regulatory requirements change, these standards must also change if they are to be effective in the market channels. The increased importation of oats into the United States during the early 1990s from Sweden, Finland and Canada have heightened the interest in understanding the differences in the grade classification techniques, as most of these oats have been cleaned and sized prior to shipment. This has resulted in improved mill yields as well as other benefits, such as savings in transportation costs, less byproducts and more uniformity in finished products. Similarly, oats sold as feed for racehorses have seen improvements resulting from more uniformity and less contamination by foreign materials and other grains.

The amount and type of contaminants are extremely important to the end-user. For example, barley can be extremely difficult to separate from oats. If the barley level in oats exceeds 1–2%, it is nearly impossible to extract efficiently because the size of the barley kernel is very similar to that of oat groats. This is a serious problem because barley's tightly adhering hull is not removed during the dehulling process and thus ends up in the finished product. Hulls of any kind in food products are very objectionable to consumers. Also, barley hulls are considered to be a potential problem in race horse feed by some experts.

Although each country has its own unique set of grading standards, they are nearly all based on similar physical parameters.* While these standards are utilized for the purposes of marketing of the grain, they do not reflect all of the quality factors of importance to the processor and/or end-user. Specifically, grade classifications do not include a number of factors, such as hull percentage, which are important to the process of milling and food manufacturing. It is recommended that if a producer or marketer of oats is interested in selling oats to a food processor or for race horse oats, they should first contact the end-user to learn of any and all special grade requirements, testing procedures and quality specifications.

Below are listed two sets of oats grading standards: Table 6.3 is the current grading standards for the US; Table 6.4 is for western Canada. Note that not only do the quality factors vary, but also the levels within the factors vary considerably.

One final comment relative to oat grading standards: as with all grains there are specific tolerances for biological and chemical contaminants. Aflatoxin, vomitoxin and other fungal metabolites present major quality

*Test weights are expressed in pounds per bushel (lb/bu) in the US grading system, while Canadian grading standards are expressed in kilograms per hectolitre (kg/hl). The conversions are as follows: one bushel is equal to 35.238 litres. Conversely, one hectolitre equals 2.838 bushels. Kg/hl × 0.77 = lb/bu; or lb/bu × 1.287 = kg/hl.

Table 6.3 US oat grade and grade requirements (Anon., 1990)

Grade	Minimum limits		Maximum limits		
	Test weight per bushel (%)	*Sound oats (%)*	*Heat damaged kernels (%)*	*Foreign material (%)*	*Wild oats (%)*
US No. 1	36.0	97.0	0.1	2.0	2.0
US No. 2	33.0	94.0	0.3	3.0	3.0
US No. 3*	30.0	90.0	0.1	4.0	5.0
US No. 4†	27.0	80.0	3.0	5.0	10.0
US Sample grade	US Sample grade shall be oats which – (a) Do not meet the requirements for the grades US No. 1, 2, 3, or 4. (b) Contain eight or more stones which have an aggregate weight in excess of 0.2% of the sample weight, two or more pieces of glass, three or more crotalaria seeds (*Crotalaria* spp.), two or more castor beans (*Ricinus communis* L.), four or more particles of an unknown foreign substance(s) or a commonly recognized harmful or toxic foreign substance(s), eight or more cocklebur (*Xanthium* spp.) or similar seeds singly or in combination, ten or more rodent pellets, bird droppings, or an equivalent quantity of other animal filth per $1\frac{1}{8}$ to $1\frac{1}{4}$ quarts of oats; or (c) Have a musty, sour, or commercially objectionable foreign odour (except smut or garlic odour), or, (d) Are heating or otherwise of distinctly low quality.				

*Oats that are slightly weathered shall be graded not higher than US No. 3.
†Oats that are badly stained or materially weathered shall be graded not higher than US No. 4.

issues in many grains and must be closely monitored in order to avoid contaminated grain from entering commercial channels. In the case of oats, there are grading specifications for ergot, Sclerotinium and smut. Mycotoxins and other forms of fungal contamination are always a concern; however, relative to other grains, problems resulting from fungal infections have been minimal in oats. Pesticide residues are also considered when developing grading standards for oats. Grading standards for pesticides are somewhat dependent upon the crop management and handling practices for the specific production areas. In North America, for example, fungicides are not typically used on oats in production situations. However, this is not necessarily the case in parts of Western Europe where intensive crop management practices are used

Table 6.4 Canada Western-Primary Oat Grade Determinants (Anon. 1990)

	No. 1 Canada Western	*No. 2 Canada Western*	*No. 3 Canada Western*	*No. 4 Canada Western*
Standard of quality				
Minimum test weight (kg/hl)	53.0	51.0	47.0	43.0
Variety	Any variety of oats equal to acceptable reference varieties	Any variety of oats equal to acceptable reference varieties	Any variety of oats equal to acceptable reference varieties	Any variety of oats
Degree of soundness	Well matured, of good natural colour, must be 97% sound	Reasonably well matured, of reasonably good natural colour, must be 96% sound	Fairly well matured fair colour, must be 94% sound	Excluded from higher grades on account of damage but must be 86% sound
Maximum limits				
Heated kernels	Nil	0.1%	1.0%	3.0%
Dehulled and hull-less	4.0%	8.0%	12.0%	No. limit
Fireburnt	Nil	Nil	Nil	0.25%
Total damage	0.1%	2.0%	4.0%	6.0%
Wild oats	1.0%	2.0%	3.0%	8.0%
Barley	1.0%	2.0%	6.0%	14.0%
Wheat	1.0%	2.0%	6.0%	14.0%
Cereal grains other than wheat and barley	3.0%	4.0%	6.0%	14.0%
Ergot	Nil	0.05%	0.05%	0.10%
Sclerotia	Nil	0.05%	0.05%	0.10%
Stones*	1K	2K	5K	5K
Total damage and foreign material	3.0%	4.0%	6.0%	14.0%

*K, kernels or kernel-size pieces in 500 g.

on a routine basis. Additionally, heavy metal uptake from the soil is an issue for all crops in certain production areas. In these instances, standards have been established for the maximum allowable level of chemical contaminants, such as cadium, mercury and lead. The purchasers should familiarize themselves with the grade standards and production issues of the country marketing the oats to assure obtaining the best quality grain.

6.5.2 Milling quality

Mill yield is defined as the units of grain required to produce 100 units of finished product. The primary determinants of mill yield are the physical characteristics of the grain and the amount of extraneous material present. This is an extremely important economic factor for oat processors and should be considered carefully when purchasing oats. Modern day estimates for mill yield range from 157 to 180 (Root, 1979; Salisbury and Wichser, 1971). Actual mill yields may be higher or lower depending upon the growing season and the amount of pre-cleaning used to enhance quality. Grading standards provide a basis for establishing quality specifications for milling oats. Since milling quality has major economic implications, it is worthwhile to consider this area in more detail.

Factors which affect milling quality can be divided into three categories: genetic, environmental and storage and handling. Each oat variety has a specific genetic potential for hull percentage, test weight, ratio of primary:secondary:tertiary kernels, aspect ratio (width:length), kernel weight and presence of awns. All of these characteristics in some way affect the ease of milling and/or mill yield (Symons and Fulcher, 1988). Since these traits are under genetic control, the breeding process provides an ideal mechanism to improve the milling quality of the oat crop. Unfortunately, with the exception of hull%, very little information exists regarding the relationship between kernel characteristics and their impact on the milling process. Hull% or groat content is a reliable and widely used indicator of oat quality. Typical commercial cultivars have groat proportions in the range of 68–72%, while the genetic potential exists to produce lines which could obtain values of 78–80% in ideal environments (Forsberg and Reeves, 1992). Unfortunately, selection for this trait usually occurs very late in the variety development cycle rather than in the early rounds. This is quite understandable because there is no accurate, rapid and non-destructive method for evaluating hull% on large numbers of whole grain samples. Thus, the tendency is to conduct this evaluation only in the latter stages of development (F5 or later) or even just prior to making decisions about whether to release a line as a new variety. This situation severely limits the breeder's ability to enhance milling quality. Similarly, hull% is not normally included as part of grading or purchasing specifications.

Test weight is a trait which has been widely used as factor for determining oat quality, but in reality its value as a predictor of milling quality is somewhat limited. Test weight is a measure of how tightly the hull is wrapped around the groat, the packing characteristics of the sample and how well the groat developed in response to the environment and the amount of foreign material. For example, if you evaluated a single variety of oats from a number of environments, test weight would be expected to reflect sample quality. However, if you evaluated the milling characteristics of a number of different varieties which had the same test weight, you could expect to observe a fairly wide range of mill yields due to differences in hull%, kernel size distribution and the way the hull surrounds the groat. Additionally, oats, which are being distributed through commercial channels, typically are a blend of several different varieties. Thus, the value of test weight as a means of predicting quality has significant limitations.

The grading system described earlier provides a way of assessing milling quality. However, as indicated above, it does have limitations. In North America, oat millers, in an attempt to more accurately address milling quality issues, have instituted a specification limiting the amount of 'thins' (the amount of oats which pass through a 5/64th × 3/4th inch slotted screen (1.98 × 19.05 mm). The thins or small grains are largely lost to byproduct streams in the milling process. A limit of 10% thins maximum has been established by this group.

Another important parameter is the free fatty acid (FFA) content. Oats contain significant quantities of both fat and lipase and related oxidative enzymes. The level of free fatty acids is an indicator of kernel damage and the care which has been used in the storage and distribution system. Oats which exceed 8% FFA as percentage of the total fat should not be used to produce products for human consumption (Schrickel *et al.*, 1992). Free fatty acids content can be determined by the method of Firestone (1990).

Thus, Table 6.5 would represent a typical US specification for milling quality oats. Oats which fail to meet any of these criteria would be discounted in price according to the level of the defect or rejected entirely. Deviations from these standards are very important from both an economic and a product quality point of view and must be closely monitored.

Annual variations in environmental conditions can have major effects upon grain quality. Thus, it is extremely important to be aware of the climatic conditions in key production areas. In a normal growing environment, adapted grain varieties can usually reach their genetic potential for grain quality. However, if the grain is subjected to severe

Table 6.5 Typical US specification for milling quality oats

Minimum test weight	49 kg/hl
SCO-min.	97.0%
Maximum moisture	13.0%
FM-Max.	1.0% over 1% deducted as dockage
Maximum barley	1.0%
Maximum wheat	1.0%
Maximum heat damaged	0.2%
Maximum free fatty acids	8.0%
Colour	Light white, yellow or tan/not dark

SCO, sound cultivated oat; FM, foreign material

temperature, moisture or disease stress, especially during the grain filling period, the quality of grain in that area will decline rapidly. For example, test weights will decline while hull content and percentage thins increase, resulting in inferior milling quality grains. These factors need to be taken into consideration when sourcing grain.

Hull-less oats offer an interesting alternative to the traditional hulled type for milling. There are obvious benefits with regard to transportation costs, mill yield and byproduct disposal. However, I am unaware of any technical publications covering this subject. Further, there is currently no established market for hull-less oats, at least not in North America. If an appropriate distribution and identity preservation system plus procedures for commodity management are developed, hull-less oats could find a place in the market as a specialty crop.

6.6 CONCLUSIONS

Oat production has declined steadily since 1975 due to its decline in popularity as a feed grain. However, food and other applications are increasing at a modest rate. Historically, oats have been treated like a commodity. As a result, agricultural research has sought to maximize production of a multipurpose oat which could be used interchangeably for food and feed purposes. Although it has been successful from an agronomic point of view, this approach may have hastened the decline in production by failing to address the specific quality needs of the various end-users. Recently, many breeders have started to question the heavy focus on field yield and have begun to place additional emphasis on specific quality traits. If oats are ever to see a resurgence as a feed grain, the breeders must make a concerted effort to improve its competitiveness with other feed alternatives.

Development of an oat specifically for feed usage would have significant implications for food processors. Most certainly an oat optimized for feed purposes would not be the most desirable raw material for food production. For example, a feed oat would probably have enhanced caloric density in the form of elevated fat content which would be undesirable for food oats. Thus, successful development of feed oats would create some very interesting issues for food processors. A logical end result would be two unique classes of oats; one for food and the other for feed. In the long run, this approach could benefit both the producer and the end-users.

REFERENCES

Anderson, J.W. and Bridges, S.R. (1993) Hypocholesterolemic effects of oat bran in humans, in *Oat Bran* (ed. P.J. Wood), American Association of Cereal Chemists, St. Paul, MN, pp. 139–58.

Anderson, J.W. and Chen, W.-J.L. (1986) Cholesterol-lowering properties of oat products, in *Oats: Chemistry and Technology* (ed. F.H. Webster), American Association of Cereal Chemists, St. Paul, MN, pp. 309–33.

Anon. (1990) Oats in *Grain Inspection Handbook, Book II*, US Department of Agriculture Federal Grain Inspection Service, Washington, pp. 7.1–7.22.

Anon. (1993) Oats in *Grain Grading Handbook for Western Canada*, Canadian Grain Commission, Winnipeg, pp. 88–95.

Bauernfeind, J.C. (1980) Tocopherols in foods, in *Vitamin E. A Comprehensive Treatise* (ed. L.J. Machlin), Marcel Dekker, New York, pp. 99–167.

Baum, B.R. (1977) Oats: Wild and cultivated. A monograph of the genus Avena L (Poaceae) Monograph 14, Canada Department of Agriculture, Ottawa, Ontario.

Biermann, U. and Grosch, W. (1979) Bitter tasting monoglycerides from stored oat flour. *Z. Fur Lebensmittel Untersuchung und Forschung*, 1969, 22–6.

Burnette, D., Lenz, M., Sisson, P., Sutherland, S. and Weaver, S.H. (1992) Marketing, processing, and uses of oat for food, in *Oat Science and Technology* (eds H.G. Marshall and M.E. Sorrells), American Society of Agronomy, and Crop Science Society of America, Madison, WI, pp. 247–63.

Burrows, V.D., Fulcher, R.G. and Paton, D. (1984) *Processing aqueous treated cereals*, US Patent 4,435,429.

Burton, G.W. and Ingold, K.U. (1981) Autoxidation of biological molecules. 1. The antioxidant activity of vitamin E and related chainbreaking phenolic antioxidants in vitro. *Journal of the American Chemical Society*, **103**, 6472–7.

Cave, N.A., Hamilton, R.M.G. and Burrows, V.D. (1989) Naked oats (cv. Tibor) as a feedstuff for laying hens. *Canadian Journal of Animal Science*, **69**, 789–99.

Coffman, F.A. (1961) Origin and history, in *Oats and Oat Improvement* (ed. F.A. Coffman) Agronomy Monograph 8 ASA, Madison, WI, pp. 15–40.

Collins, F.W. (1986) Oat phenolics: Structure, occurrence, and function, in *Oats: Chemistry and Technology* (ed. F.H. Webster), American Association of Cereal Chemists, St. Paul, MN, pp. 227–95.

Collins, F.W. (1989) Oat phenolics: Avenanthramides, novel substituted N-cinnamoylanthranilate alkaloids from oat groats and hulls. *Journal of Agricultural and Food Chemistry*, **37**, 60–6.

Daniels, D.G.H. and Martin, H.F. (1961) Isolation of a new antioxidant from oats. *Nature*, London, **191**, 1302.

Daniels, D.G.H. and Martin, H.F. (1967) Antioxidants in oats: Monoesters of caffeic and ferulic acids. *Journal of the Science of Food and Agriculture*, **18**, 589–95.

Deane, D. and Commers, E. (1986) Oat cleaning and processing, in *Oats: Chemistry and Technology* (ed. F.H. Webster), American Association of Cereal Chemists, St. Paul, MN, pp. 371–412.

deGroot, A.P., Luyken, R. and Pikaar, N.A. (1963) Cholesterol-lowering effect of rolled oats. *Lancet*, **II**, 303–4.

Dimberg, L.A., Theander, O. and Lingert, H. (1993) Avenanthramides – A group of phenolic antioxidants in oats. *Cereal Chemistry*, **70**, 637–41.

Dodok, L., Morova, E. and Gallova-Adaszova, M. (1982). Influence of inactivated oat flour on gluten, dough and biscuit quality. *Bulletin Potravinarskeho Vyskumu*, **21**, 45–8.

Dunlop, A.P. (1973) The furfural industry. *Industrial Uses of Cereals* (ed. Y. Pomeranz), American Association of Cereal Chemists, St. Paul, MN, pp. 229–36.

Duve, K.J. and White, P.J. (1991) Extraction and identification of antioxidants in oats. *Journal of the American Oil Chemists' Society*, **68**, 365–70.

Findlay, W. (1956) *Oats*, Oliver and Boyd, Edinburgh, Scotland.

Firestone, D. (1990) Oils and fat, in *Official Methods of Analysis*, 15th edn (ed. K. Helrich), Association of Official Analytical Chemists, Arlington, Virginia, pp. 951–86.

Forsberg, R.A. and Reeves, D.L. (1992) Breeding oat cultivars for improved grain quality, in *Oat Science and Technology* (eds. H.G. Marshall and M.E. Sorrells), American Society of Agronomy and Crop Science Society of America, Madison, WI, pp. 751–76.

Friend, D.W., Fortin, A., Poste, L.M., Butler, G., Kramer, J.K.G. and Burrows, V.D. (1988) Feeding and metabolism trials and assessment of carcass and meat quality for growing finishing pigs fed naked oats (Avena nuda). *Canadian Journal of Animal Science*, **68**, 511–21.

Friend, D.W., Fortin, A., Butler, G., Poste, L.M., Kramer, J.K.G. and Burrows, V.D. (1989) The feeding value of naked oats (Avena nuda) for boars and barrows: Growth, carcass, meat quality and energy and nitrogen metabolism. *Canadian Journal of Animal Science*, **69**, 765–78.

Fulcher, R.G. (1986) Morphological and chemical organization of the oat kernel in *Oats: Chemistry and Technology* (ed. F.H. Webster), American Association of Cereal Chemists, St. Paul, MN, pp. 47–74.

Gupta, K.L. (1991) *Time Release Protein*, US Patent 5,012,080.

Gupta, K.L. (1992) *Time Release Protein*, US Patent 5,079,005.

Hargin, K.D., Morrison, W.R. and Fulcher, R.G. (1980) Triglyceride deposits in the starchy endosperm of wheat. *Cereal Chemistry*, **57**, 320–5.

Heaton, K.W., Marcus, S.N., Emmette, P.M. and Bolton, C.H. (1988) Particle size of wheat, maize and oat test meals: Effects on plasma glucose and insulin responses and on the rate of starch digestion in vitro. *American Journal of Clinical Nutrition*, **47**, 675–82.

Herstad, O. (1980) Kombineret foring av verpehoener med kraftforblanding og havrepellets. *Meldinger far Norges Landbrukshoegskole*, **59**, 24.

Hohner, G.A. and Hyldon, R.G. (1977) *Oat groat fractionation process*, US Patent 4,028,468.

Hutchinson, J.B., Martin, H.F. and Moran, T. (1951) Location and destruction of lipase in oats. *Nature*, London, **167**, 758–9.

Inglett, G.F. (1991) *Method for making a soluble dietary fiber composition from oats*, US Patent 4,996,063.

Jenkins, D.J.A., Wolever, T.M.S., Taylor, R.H., Barker, H., Fielden, H.,

Baldwin, J.M., Bowling, A.C., Newman, H.C., Jenkins, A.L. and Goff, D.V. (1981) Glycemic index of foods: A physiological basis for carbohydrate exchange. *American Journal of Clinical Nutrition*, **34**, 362–6.

Jorgensen, K.G. (1988) Quantification of high molecular weight (1–3) (1–4)-ß-D-glucan using Calcofluor complex formation and flow injection analysis. I. Analytical principle and its standardization. *Carlsberg Research Communications*, **53**, 277–85.

Kankare, V. and Antila, V. (1984) The effect of feed grains on the fatty acid composition of milk fat. *Journal of Agricultural Science*, Finland, **56**, 33–8.

Karunajeewa, H. and Tham, S.H. (1987) The influence of oat groats and dietary level of lysine on the laying performance of crossbred hens. *Animal Feed Science and Technology*, **17**, 271–83.

Kazi, T. and Cahill, T.J. (1969) A rapid method for the detection of residual lipase. *Analyst*, London, **94**, 417.

Kowalewski, W. and Jankowski, S. (1979) Peroxidase activity of oat kernel after steaming and drying. *Annals of Peznan Agricultural University*, CVII, pp. 53–60.

Lane, R.H. (1990) Cereal Foods, in *Official Methods of Analysis*, 15th edn (ed. K. Helrich), Association of Official Analytical Chemists, Arlington, Virginia, pp. 777–801.

Lorenz, K. (1984) Cereal and dental caries, in *Advances in Cereal Science and Technology*, Vol. VI (ed. Y. Pomeranz), American Association of Cereal Chemists, St. Paul, MN, pp. 83–137.

MacArthur-Grant, L.A. (1986) Sugars and nonstarchy polysaccharides in oats, in *Oat Chemistry and Technology* (ed. F.H. Webster), American Association of Cereal Chemists, St. Paul, MN, pp. 75–92.

McCleary, B.V. and Glennie-Holmes, M. (1985) Enzymic quantification of (1–3) (1–4)-ß-D glucan in barley and malt. *Journal of the Institute of Brewing*, **91**, 285–95.

McKechnie, R. (1983) Oat products in bakery foods. *Cereal Foods World*, **28**, 635–7.

Madsen, K.O. (1981) The anticaries potential of seeds. *Cereal Foods World*, **26**, 19–25.

Martin, H.F. (1958) Factors in the development of oxidative rancidity in ready-to-eat crisp oat flakes. *Journal of the Science of Food and Agriculture*, **9**, 817–24.

Martin, P.A. and Thomas, P.C. (1987) Reduction in the saturated fatty acid content of cow's milk through diet formulation. *Proceedings of the Nutrition Society*, **46**, 114A.

Mermelstein, N.H. (1983) Soy – oats infant formula helps fight malnutrition in Mexico. *Food Technology*, **37**, 64–72.

Miller, A. (1977) Cosmetic ingredients. *Soaps, Detergents and Toiletries Review*, **7**, 21–5.

Moran, J.B. (1986) Cereal grains in complete diets for dairy cows: A comparison of rolled barley, wheat and oats and of three methods of processing oats. *Animal Products*, **43**, 27–36.

Morris, J.R. and Burrows, V.D. (1986) Naked oats in grower-finisher pig diets, *Canadian Journal of Animal Science*, **66**, 833–6.

Morrison, F.B. (1956) *Feeds and Feeding*, 22nd edition, Morrison Publishing Co., Ithaca, New York.

Mulvaney, T.R. (1990) Vegetable products, processed, in *Official Methods of Analysis* (ed. K. Helrich), 15th edn, Association of Official Analytical Chemists, Arlington, Virginia, pp. 987–98.

Munck, L., Jorgensen, F.G., Ruud-Hansen, J. and Hansen, K.T. (1989) The EBC methods for determination of high molecular weight β-glucan in barley, malt, wort and beer. *Journal of the Institute of Brewing*, **95**, 79–82.

Murphy, J.P. and Hoffman, L.A. (1992) The origin, history and production of oats, in *Oat Science and Technology* (eds H.G. Marshall and M.E. Sorrells), American Society of Agronomy and Crop Science Society of America, Madison, WI, pp. 1–28.

Oughton, R.W. (1980a) *Process for the treatment of comminuted oats*, US Patent 4,211,695.

Oughton, R.W. (1980b) *Process for the treatment of comminuted oats*, US Patent 4,211,801.

Peters, F.N., Jr. and Musher, S. (1937) Oat flour as an antioxidant. *Industrial and Engineering Chemistry*, **29**, 146–51.

Peterson, D.M. (1992) Composition and nutritional characteristics of oat grain and products, in *Oat Science and Technology* (eds H.G. Marshall and M.E. Sorrells), American Society of Agronomy and Crop Science Society of America, Madison, WI, pp. 265–92.

Peterson, D.M. and Brinegar, A.C. (1986) Oat storage proteins, in *Oats: Chemistry and Technology* (ed. F.H. Webster), American Association of Cereal Chemists, St. Paul, MN, pp. 153–203.

Root, W.R. (1979) The influence of oat (Avena sativa L.) kernel and caryopsis morphological traits on grain quality characteristics, Ph.D. dissertation, University of WI, Madison, WI, (Dissertation Abstracts International 40: 4070–B-4071 [No. 8004738]).

Salisbury, D.K. and Wichser, W.R. (1971) Oat Milling – Systems and Products. *Bulletin from the Association of Operating Millers*, May issue, 3242–7.

Schrickel, D.A., Burrows, V.D. and Ingemansen, J.A. (1992) Harvesting, storing and feeding of oat, in *Oat Science and Technology* (eds H.G. Marshall and M.E. Sorrells), American Society of Agronomy and Crop Science Society of America, Madison, WI, pp. 223–45.

Shinnick, F.L. and Marlett, J.A. (1993) Physiological responses to dietary oats in animal models, in *Oat Bran* (ed. P.J. Wood), American Association of Cereal Chemists, St. Paul, MN, pp. 113–38.

Shukla, T.P. (1975) Chemistry of oats: Protein foods and other industrial products. *Critical Review of Food Science Nutrition*, **6**, 383–431.

Smith, W.H. (1973) Oats can really 'beef up' cookies I. *Snack Food*, **62**, 33–5.

Symons, S.J. and Fulcher, R.G. (1988) Determination of variation in oat kernel morphology by digital image analysis. *Journal of Cereal Science*, **7**, 219–28.

Webster, F.H. (1983) *Method for preparing whole grain oat product*, US Patent 4,413,018.

Webster, F.H. (1986) Oat utilization: past, present and future, in *Oats: Chemistry and Technology* (ed. F.H. Webster), American Association of Cereal Chemists, St. Paul, MN, pp. 413–26.

White, P.J. and Armstrong, L.S. (1986) Effect of selected oat sterols on the deterioration of heated soybean oil. *Journal of the American Oil Chemists' Society*, **63**, 525–9.

Wood, P.J. (1993) Physiochemical characteristics and physiological properties of oat (1–3), (1–4)-β-D-glucan, in *Oat Bran* (ed. P.J. Wood), American Association of Cereal Chemists, St. Paul, MN, pp. 83–112.

Youngs, V.L. (1972) Protein distribution in the oat kernel. *Cereal Chemistry*, **49**, 407–11.

Youngs, V.L. (1986) Oat lipids and lipid-related enzymes, in *Oats: Chemistry and Technology* (ed. F.H. Webster), American Association of Cereal Chemists, St. Paul, MN, pp. 205–26.

7

Rye and triticale

D. Weipert

7.1 INTRODUCTION

In comparison with the worldwide production of all cereals, in particular with that of wheat, rye is only of minor importance. And yet, rye is indispensable in farming and eating habits in some regions, particularly in Northern Europe and in the former Soviet Union countries. In 1994 about 22 588 000 Mt rye were produced on 11 012 000 ha. This production represents 1.2% of world cereal production and about 4% of the world wheat production, yet the world trade volume (imports and exports) of rye is estimated close to 430 million US$ (Anon., 1995). In spite of this somewhat limited amount of finance, trade with rye must not be neglected and calls for quality requirements and regulations in trade and handling as wheat does.

The primary centre of origin of rye is accepted to be southwestern Asia, from where it followed two routes: northwards to Russia and from there westwards to Poland and Germany, and secondly to Europe via the Balkan Peninsula (Bushuk, 1976; Seibel, 1988a, 1988b). From these regions rye was brought by European settlers first to North and South America and then later to Australia and Africa. There rye did not spread significantly, and the first named regions of North Europe and Russia have remained the exclusive producers of rye.

The history of rye is much shorter than that of wheat. Due to its disease resistance and hardiness, rye became an important crop, being used as food and feed over centuries. Rye will keep its place in field rotations and will maintain its importance in sustainable farming. Due to its winter hardiness and resistance to drought and to diseases and pests, rye can be grown on poor soils and under adverse environments with fewer protective measures, yielding higher than wheat under the same conditions. For these reasons rye acts as a preserver and recoverer of the soil fertility and is beneficial in the field rotation.

Triticale, a man-made cereal, is a product of an enforced crossing between wheat and rye, combining the properties of both parental cereals either intermediary or as hybrid effects. The first fertile forms were reported from the beginning of this century, but significant improvement and spread of triticale were only achieved in the last 15 to 20 years. Unlike the main cereals, there is no official report on acreage and production of triticale worldwide. Such data can be only collected from different sources. In 1986 it has been estimated that triticale was grown on about 500 000 ha in Europe, mainly in France (Suijs, 1986). Further information on growing triticale has shown that this promising cereal spreads slowly, but constantly worldwide (Anon., 1989). The main efforts in breeding new triticale varieties have been made in Canada, Mexico (CIMMYT), Hungary and particularly in Poland, where the most adaptable varieties have been created and subsequently spread to all continents. Therefore, in 1990 the acreage of triticale in Poland has exceeded 1 Mha (Haber, 1994). In other countries, such as Germany, the triticale production has reached by now nearly 50% of that of rye. The production of triticale is still small and this cereal has not gained an international market. Although triticale was meant to be a new bread cereal for use in all fields of food processing, its main usage at present is as a feed cereal. Quality requirements of this new cereal were taken from those on wheat and rye in accordance with their potential use. In many aspects triticale behaves somewhere intermediate between wheat and rye, but has more similarity to rye. Regarding its susceptibility to sprouting and to intrinsic enzymic activity, the processing of triticale is more difficult than that of rye. However, triticale is well-suited to growing on poor soils including acid soils and soils with adverse contents of iron and aluminium ions.

International trade considers rye as a feed cereal and in only a few countries, where rye bread represents an important food base, does rye have the same importance and price as wheat. The standards and grades of mercantile rye are based mostly on the physical properties of its kernels, such as test weight, moisture, dockage and besatz content (impurities), and do not consider its baking performance. Minimum requirements of rye quality for bread-making purposes in the EU were defined rather late, although the relevant characteristics and the methods for their assessment have been known and followed since about 1930. The standards and the quality regulations have gained in importance since trade in the EU was opened and no rules and additional taxes disturb the development of a new European market. The German Board of Varieties issues a national catalogue of varieties of all cereals, including rye and triticale, yearly describing their farming and technical properties.

The main uses for rye are human food (different types of breads and other bakery items) and animal feed. Besides this, due to the functional properties of its constituents, rye is used for non-food and other industrial purposes.

7.2 SPECIAL CHARACTERISTICS OF RYE AND TRITICALE AND THE FUNCTIONAL PROPERTIES OF ITS KERNEL CONSTITUENTS

Rye differs from the other cereals, particularly from wheat, in a number of characteristics. Having virtually no dormancy, rye is highly susceptible to pre-harvest sprouting. This, as well as several other disadvantages (such as tendency to lodging), aggravates its farming and therefore rye requires early harvesting, additional care and the drying of the crop prior to storage. The structure of rye proteins and their inability to form gluten on the one hand, and the high water-binding capacity of rye pentosans on the other, are responsible for different types of bakery items prepared from rye, and different techniques in bread processing to wheat. The starch and α-amylase present in the kernels and in the rye milling products are also important factors influencing the crumb properties of breads. The quality requirements of rye and the methods of their assessment are therefore not the same as for wheat.

The functional properties of relevant rye constituents – starch, pentosans, protein – as discussed in Weipert and Brümmer (1988) and Weipert (1993) are shown in an overview in Table 7.1. Herein the pentosans, having a high water-binding capacity, are responsible for the viscosity and the rheological properties of rye and of its products such as: the 'gumminess' of the kernels in milling, the consistency, yield, stickiness, and behaviour of the doughs in proofing, and the resilience and 'chewability' of the crumb of the end-product bread. The pentosans are carbohydrates, but behave significantly differently in the bread-making process to starch and to proteins — they do not undergo disintegration and gelatinization like starch, and do not release water like proteins when heated, providing the crumb with 'softness' and a longer shelf life. Rye starch, being able to gelatinize earlier, i.e. at a lower temperature than wheat starch, is more exposed to the activity of the α-amylase. As a result of the high activity of the α-amylase present in the kernel and milling products, the degraded starch is not able to form an elastic and resilient crumb, which is considered to be a serious fault in bread . In contrast to this, a low and weak amylase activity produces a dense and firm crumb, which is considered a severe fault in bread, too. So far, the sprouting susceptibility and the activity of α-amylase are regarded as the most important and the most significant factors in the processing of rye and thus in its quality assessment.

Table 7.1 Functional properties of rye and triticale grain constituents in comparison to wheat**

	Wheat	*Triticale*	*Rye*
Swelling substances			
Gluten	Highly effective	Effective*	Less effective
Pentosans	Less effective	Effective	Highly effective
Water binding capacity	High (2 times its own weight)	High	Very high (up to 8 times its own weight
Physical (rheological) properties	Stretchable	Less stretchable	Not stretchable
	Viscoelastic substance forming networks and films	Viscoelastic substance Intermediate*	High viscous substance of low elasticity
Adhesivity/stickiness on surface	Low	Intermediate*	High
Gas retention	High	Intermediate*	Low
Starch			
Granules	Small ($\bar{x}$ 19 μm)	Intermediate*	Larger ($\bar{x}$ 24 μm)
Mechanical damage	High (hard structure)	Intermediate	Low (soft structure)
Gelatinization	Late, at higher temperature	Rather low	Early, at low temperature
Presence of enzymes	Low	High–very high	High
Susceptibility to enzymic degradation	High	High–very high	Very high

*Depending on variety, significant divergences possible following the dominance of one of the parents
**From Weipert, 1993a

Triticale, being a cross product of rye and wheat, takes an intermediate position between the parents regarding the functional properties of its kernel constituents. Triticale proteins are to a greater extent water soluble and form a soft and weak gluten. The gelatinization properties of triticale starch and the generally high activity of α-amylase result in a weak starch gel. The disadvantageous properties and behaviours of both triticale constituents do not encourage their use for bread-making using the same baking process as for wheat. Not many attempts at introducing triticale as a bread cereal have been successful, although it is, in principle, possible to produce edible breads from triticale.

7.3 BREAD-MAKING

The traditional use of rye has been for baking purposes. The surplus was used as a farm-grown feed in weaning and fattening of pigs and for industrial purposes. Due to the eating habits of the population in rye growing regions, and to the price development in the world cereal trade, rye for feed and industrial purposes has increased in importance.

7.3.1 Milling

Milling products

The environmental circumstances (climate, soil), which have made rye farming important in part of the world, have influenced the eating habits of the population in various regions differently. This has eventually resulted in a greater number of bakery items produced from various milling products of rye using a variety of formulae and recipes.

Prior to baking, rye has to be prepared by cleaning, grinding, and producing the milling products with desired properties. The rye milling products are flours, meals, and whole kernel meals and flours, differing in their mineral content, colour, and coarseness. The main products are flours of types 997 and 1150, having a content of minerals of about 1.0 to 1.2%, and the meals of preferably medium and coarse granularity. The number of flour types in eastern European countries including the countries of the former Soviet Union is smaller – mostly flours with a mineral content of 0.75, 1.45 and 1.75% are used for bread-making purposes.

Due to the viscous properties of rye kernels, rye flours are generally finer than wheat flours. The tough (vitreous) rye kernels produce less coarse flours of higher pentosan content than mealy kernels. The remedies for improving the baking quality of a sprout damaged rye are limited. Usually the flour yield (extraction) would be decreased, as a result of which the enzyme rich parts of the rye kernel will be sieved to the brans. More coarse milling products would be produced, as to slow down and to lessen the enzymic action on flour particles.

Triticale behaves in the milling process like wheat rather than like rye. However, the nature of its kernel structure does not allow for a profitable extraction of flours, because of shrivelling, disordered kernel structure and high mineral content. For this reason, and considering the baking quality of most triticale varieties, the processing of triticale to whole kernel meals and flours is recommended (Weipert *et al.*, 1986). Most of the more recently released new varieties are free of such structure anomalies and the milling into flours of defined grade with reasonable yields is conceivable. Nevertheless, the flour grade striven for in triticale processing should be based on a content of minerals of 0.80%, giving an extraction yield of about 75–80% of flour.

Quality requirements of grain

Rye for milling and baking purposes has to satisfy some quality requirements, based both on commercial and hygienic demands. In the course of growing and ripening under unfavourable conditions rye kernels can undergo changes that affect its milling quality and its processing value. These also affect the forming and filling of the kernels with starch and proteins, and their enzymic state, mirroring the degree of sprouting damage. Besides this, rye can be infected by toxic ergot sclerotia (*Claviceps purpurea*) and other bacterial and fungal diseases. The infection with ergot sclerotia can be considerable when the flowering of the cross pollinating rye and triticale has been retarded by cold and wet weather. Hybrid varieties seem to be more affected by the ergot infection than the population varieties. For the reasons of toxicity, the ergot polluted rye and triticale have to be cleaned intensively.

For reasons of financial gain, a lot of clean and large rye kernels, with a high yield of flour, is required. As a measure of the external, outward physical properties like density and size, the test weight (hectolitre or bushel weight), 1000-kernel weight and sieving are applied. The 1000-kernel weight of rye grown in moderate maritime climates (like western Europe) is higher than 25 g and often reaches values of about 32 g or more. The severe continental climates of the USA, Canada and northern parts of the former Soviet Union, as well as the dry conditions of Australia and the Argentine produce small kernel rye with a 1000-kernel weight of 20–25 g and sometimes lower than 20 g. The 1000-kernel weight is, to some extent, a genetically linked property, whereby particularly the tetraploid rye varieties are characterized by high 1000-kernel weights. The bushel weight of Canadian rye grades is 54–58 lb, the hectolitre weight in Europe is expected to be 70–72 kg/hl; the lower or higher values are subject to price deductions and bonuses.

The cleanness of a rye lot is expressed with dockage or besatz fractions, assessed in a subjectively performed test. Besatz and dockage can be explained as components of a grain lot which differ from the normal and sound grain. Because of the toxicity (ergot, weed seeds, mycotoxin burdened kernels), of store security (broken and damaged kernels), of hygiene (impurities, foreign materials, chaff, stones etc.), of flour yield (undeveloped, shrivelled, and shrunken kernels), and of baking quality (heat and sprout damaged kernels) a rye lot should be as far as possible free of these besatz or dockage fractions. As the unsound kernels and impurities have to be cleaned out and removed prior to milling, the grading systems of mercantile rye are built on the presence and limitations in the besatz and dockage fractions, and consequently determine the potential use and price of rye lots. They vary in different countries (USA, Canada, former Soviet Union), but are probably most strict in the EU (Table 7.2).

The besatz (ICC Standard No. 103/1) and dockage assessments are carried out subjectively using visual methods, although attempts have been made to perform a mechanical analysis using a combination of sieves, trieurs (indented cylinders), and aspiration. Due to the changes in colour and other characteristics which can be noticed only visually, the mechanical analysis is of limited value in some specific cases. The content of shrunken kernels is assessed by the use of slit sieves with 1.7 or 1.8 mm openings.

Rye also has to have a low moisture content (14.0% or less), and be of typically sound odour and taste. A musty or stuffy odour indicates the

Table 7.2 External quality criteria of mercantile rye

	*Standard quality**	*Intervention*[†]
Moisture content (% objectionable grains)	14.0	14.5
Besatz fractions (%)	5.0	12.0
Particularly broken grains	2.0	5.0
grain impurities	1.5	5.0
particularly: heat damaged grains	–	3.0
sprouted grains	–	6.0
foreign impurities	0.5	3.0
particulary: toxic weeds	–	0.10
ergot	–	0.05
Test weight (kg/hl)	71.0	68.0

*Regulation order (EU) EEC 2731/75.
[†]Regulation order (EU) EEC 689/92.

deterioration of proteins and lipids and can be detected in the baked goods, making them inedible. Such rye is unsuitable for milling and baking purposes.

Except with the content of sprouted kernels, the baking behaviour of rye is not related to the external or physical kernel properties. The assessment of baking quality has recently been introduced through the EU rye standards, based on the German quality requirements for bread rye. Applying a simple and reliable method using the amylograph, rye suitable for bread-making is defined with an amylogramme with peak viscosity not lower than 200 BU (Brabender Units), and peak temperature not lower than 63.0°C (Seibel and Drews, 1973). The amylogramme has to be prepared from whole kernel meal (90 g) after appropriate grinding. Since the amylograph method is time-consuming, the suitability for baking is usually assessed by falling number. The minimum value for the acceptance or rejection of a lot of rye is subject to change. The relationship of falling number to amylogramme data on which the minimum values are based is strongly dependent on climatic conditions during the growing and ripening of rye and therefore changeable from crop to crop. The minimum falling number values acceptable for baking purposes may vary from 80 s in a dry, to 120 s in a rainy crop year with the same amylogramme values (Weipert and Bolling, 1979). Recently it was agreed to set the minimum falling number to 90 s as a replacement for the subjective assessment of the besatz fraction sprouted grains (Table 7.2). For special purposes, such as for production of crisp breads, rye of lower enzymic activity is required, with minimum values of falling numbers higher than 140 s.

With regard to the health situation of all agricultural crops and products, the recommendation has been given that wheat and rye should not have more than 0.40 mg/kg Pb, 0.10 mg/kg Cd, and 0.03 mg/kg Hg (Anon., 1992b). These values are not exceeded in general.

Since triticale is not a declared bread grain, there are no grades regulating its milling and baking quality. However, following the examples of rye and wheat, a lot of clean and large triticale kernels is preferred. The test weight has to be considered, as it assesses the shrivelling and the structure of triticale kernels, which may be detrimental to the flour yield. Triticale in general is susceptible to ergot infection and to pre-harvest sprouting, which must not be omitted in quality requirements.

Considering the intermediate position of triticale in baking behaviour, the minimum falling number for triticale is between those of wheat (250 s) and of rye (100 s) and should be 120–150 s (Stephan *et al.*, 1986). These data reflect the enzymic situation in triticale kernels, but do not describe the baking performance, which is more influenced by the quality and quantity of gluten.

7.3.2 Baking

Bakery products

Rye is generally appreciated for giving a specific taste to bakery items, improving their culinary acceptance and popularity. The appreciation of the rye taste is visible in the fact that patents have been applied for describing the production of rye flavouring (Spiel, 1985). The content of rye and the type of flours and of other ground products from rye are taken as the characteristics for classifying bakery items from rye. According to the German classification, rye bread must consist of at least 90% of rye milling products (Table 7.3). The milling products for breadmaking are flours of different types, meals and whole kernel products, as noted above (Milling products). They can be used alone or blended with other rye and wheat milling products to obtain the desirable specific effects in form, colour, and flavour of the bakery items. Eventually, the baked products are classified as rye breads, mixed breads (with predominant allotment of rye or wheat respectively), small bakery items (rolls etc.), special breads (steinmetz, multigrain breads, pumpernickel), dietary products (with reduced nutritive value or salt content), and pastry (breakfast breads, ginger breads). In other important rye consuming countries in Europe and the former Soviet Union, the same types of breads are produced. The breads from whole kernel products and dark flours of high extraction are the preferred typical products from rye in general.

Further to the classification and differentiation of the bakery items regarding the bread baking process, the rye bakery items can be differentiated as leavened by yeast and/or sourdough, leavened chemically, by whipped-in-air (cold type crisp bread proofing), and by extrusion cooking. According to the pH reached in dough and bread, besides the not-acidulated leavening by yeast, the dough can be acidulated directly (by lactic and other organic acids), indirectly (applying the sour dough formula), or by a combination of direct and indirect. The meals, depending on granularity, can be either scalded or cold soaked. Consequently, the large number of different baked goods need different quality characteristics of rye and its milling products.

Due to the functional properties of triticale kernel constituents which are intermediary between those of rye and wheat, milling products of triticale can be used in both types of bread baking, acidulated (rye bread formulas) and not acidulated (wheat bread formulas), pure or added to rye and wheat milling products, depending on the enzymic state of the flours as well as the desired type of bread. Very often improvers and additives such as gluten and emulsifiers are included to upgrade the baking performance of triticale milling products to correspond to that of the typical wheat breads.

Table 7.3 Overall baking score of rye flour breads in relation to amylogramme and falling number data (from Seibel and Stephan, 1973)

Peak viscosity (BU)	*Peak temperature (°C)*	*Falling number (s)*	*Quality (score)*	*Processing value*	*Bake volume, freshness (score)*	*Demand for acidulation*	*Flavour*
<250	<65	<100	Unsatisfactory	Unsatisfactory	Very good	High	Lightly sour
250–400	65–69	100–180	Satisfactory–good	Good	Good	Medium	Normal, pleasant
400–600	70–73	180–250	Good	Still good	Still good	Medium	Normal, pleasant
>600	>73	>250	Very good (excellent)	Satisfactory	Satisfactory	Low	Unsavoury, tasteless, stale

Quality requirements of flours

The processing value and thus the baking quality of rye, rye flours, and of other milling products is primarily determined by the gelatinization or pasting behaviour of starch in a slurry. Since the water-binding capacity is one of the most relevant functional properties of starch in the course of the baking process, the quality assessment of rye flours is carried out with viscometers such as Amylograph, Falling Number, Rapid Visco Analyser and similar methods. The use of the named instruments is standardized (ICC, AOAC, ISO, AFNOR, RACI etc.) and their repeatability and reproducibility are published. Such objective and physical or rheological methods are better, simpler, faster, and what is most important of all, more reliable than the visual estimation of sprouted kernels.

As the gelatinization properties of a starch in a flour-water slurry are both temperature and time dependent, the choice between time-consuming and rapid methods has to be made in agreement with the material (flour, meal) and the potential purpose of use. Whenever possible, the time-consuming amylograph method is recommended offering the advantage of ample information about the state of the starch and of other kernel constituents and about the enzymic activity (by initial, peak, and end viscosities and by peak temperature), compared with a single value (falling number, stirring number). The result of a rapid method can be considered useful only after the relationship of the results has been calibrated with the results of the time-consuming amylograph method (Weipert and Bolling, 1979; Weipert, 1993b).

The quality requirements of rye flour and the resulting processing value and baking performance are given in Table 7.3. Although the lowest baking quality has been described by the amylogramme data of a whole kernel meal suspension such as peak viscosity and peak temperature not less than 200 BU and 63°C respectively, a good quality rye flour is expected to be of higher amylogramme values. The best baking performance is expected in the range of 250–600 BU peak viscosity and

Table 7.4 Feed value of rye and milling by products (from Kirchgessner, 1970)

	Total protein	*Digestible protein*	*Total nutrient*	*Crude fibre*
		(g/100 g)		
Rye grain	9.0	7.6	75.1	2.7
Fine offals	14.5	11.6	74.4	2.6
Bran	14.3	9.4	58.4	7.5

65–69°C peak temperature and 100–180 s falling number (Seibel and Stephan, 1973). The differences in the quality requirements between the rye grain data (obtained in the whole kernel meal) and the flour (80% extraction, 1.0% mineral contents) are due to the changes in starch content and enzymic activity of the listed milling products. The table indicates further that the increased quality data do not denote an equal improvement of the processing value and baking performance of the rye flour. Whilst the low quality data indicate an enhanced enzymic activity that requires remedies like a deeper acidulation with sour dough, the production of breads from flours of high quality is easy, but the performance – loaf volume, freshness and taste – of such breads is unsatisfactory. To optimize the quality of the end-product bakery items, remedies like the blending of flours, application of enzymes, etc. are suggested (Weipert, 1993a).

Further, the composition, constitution, and enzymic activity present in different flour types and meals must be considered when estimating the quality state of a milling product assigned for various purposes. The higher the flour extraction, the higher the quality demand of the milling products to obtain a bread of good quality.

Since the enzymic state of rye grains has been considered the main quality factor, less attention has been given to the amount and state of the non starch-swelling substances, proteins and pentosans. Besides this, the relatively few rye varieties have not differed much in these characteristics and only recently by introducing new (particularly, hybrid) varieties has the genetically linked pentosan content shown some technologically significant differences. Although the influence of pentosan on baking performance, particularly on loaf volume, crumb resilience and freshness, has been known for years, no method of measuring the content and state of pentosans with a view to baking behaviour of rye has been introduced. Subsequently it has been established that the total pentosan does not lead to a better baking performance content in spite of the high water binding capacity (Meuser and Suckow, 1986). The water soluble component of the pentosans is most closely related to baking performance (Weipert, 1993c). The use of the NIR technique for estimation of both the total and water soluble pentosans will help to screen the rye for its baking quality regarding the enzyme activity and pentosan composition.

The most reliable and the best method for determination of quality behaviour of a grain is the baking trial itself. While cereal science has offered quite a number of baking trials for wheat adapted to practical use, the estimation of rye baking quality is somehow restricted to a single baking trial using the sour dough formula. There are some additional possibilities for performing a baking trial based on a non-acidulated method such as using yeast as the leavening agent, and a directly acidulated method based on the addition of lactic acid. The sour dough formula has, however, proved to be the most reliable test, since the majority of the baking items are prepared by sour dough and the baking trial is developed by the additional use of the farinograph for water absorption, i.e. optimal dough yield (Figure 7.1). From that, the overall score, integrating the dough yield, loaf volume, and crumb elasticity as the main characteristics of a baking result, is calculated and presented in one figure (Seibel and Stephan, 1980). Between the crumb elasticity (resilience) and the falling number and amylogramme data a strong correlation has been found, which makes the time-consuming and expensive baking trials replaceable by indirect methods.

Since triticale has not been used for industrial bread baking, the quality requirements of triticale and of its ground products have not been established. Several cooking and baking books, meant for houshold use, offer different types of triticale biscuits, but do not consider its quality state. A twin baking trial, based on non-acidulated (following the wheat bread type), and acidulated by lactic acid (following the rye bread type), and considering the triticale characteristics, has been developed and offered (Stephan *et al.*, 1986). An example of triticale breads from this type of baking trial is given in Figure 7.2 from a field trial with triticale varieties. Few new varieties seem to be of fairly good baking quality in respect to wheat baking formulae (Haber, 1994)

7.4 FEED

7.4.1 Specific properties of rye for feeding

In the light of human nutrition rye is considered a healthy source of energy. The high content of minerals and dietary fibres due to high

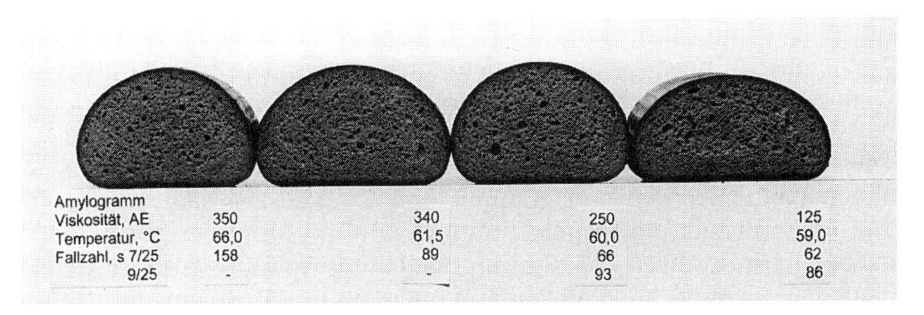

Figure 7.1 Loaves of rye bread from standard baking trial using the sour dough formula. The description denotes peak viscosity, peak temperature and falling number 7/25 (ICC-Standard) and 9/25 (9 g flour to 25 ml water).

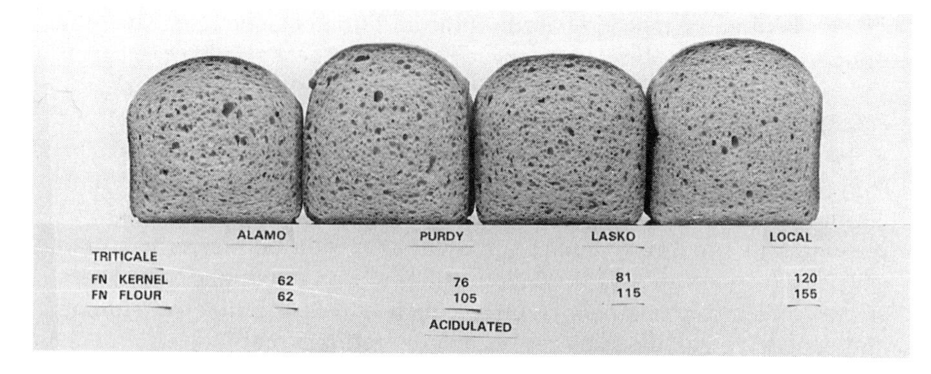

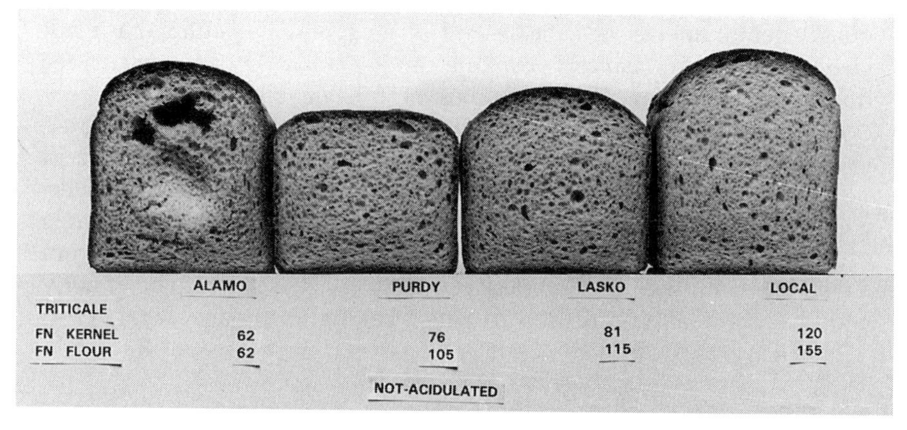

Figure 7.2 Loaves of triticale bread from standard baking trials following the acidulated (lactic acid) and non-acidulated formula.

extraction milling products and whole kernel meals, the easily digestible proteins of a favourable amino acid content (expressed by their high water solubility) and the content of B vitamins are a useful contribution to the otherwise fat and energy rich nutrition of modern mankind. This is not necessarily the case from the point of view of animal feeding and fattening. High production and a surplus of cereals in general have influenced the prices on national and international markets, have exceeded the requirements for human food, and have strongly increased the use of wheat and rye for the feeding of pigs and poultry. In this respect, rye has been considered a feed cereal in many countries and regions, but in those areas where rye was an explicit bread cereal, the use of rye as feed has increased as well.

Regarding the nutritive value of rye, the amino acid composition and high content of essential amino acids, particularly lysine, has been stressed (Seibel, 1975; Weipert, 1985). The somewhat lower protein content of rye in comparison to wheat is not of any importance, as cereals in general have to be supplemented with other sources of protein in order to provide a complete standard feed for animals. In previous times rye was the main feed source for animals, particularly pigs. Not only the farm grown grains were used for this purpose, but the offals and bran as milling byproducts were used because of their low prices and good feed value (Table 7.4). Rye, and to some extent wheat, are the cheapest farm-produced high energy feeds and are preferred in the fattening of pigs and poultry at the farm where they were harvested. In comparison to wheat and barley, rye is considered to be less effective in the fattening of pigs, having a less favourable feed conversion ratio. The feed value of rye based on contents of protein and other nutrients is given in Table 7.4. To increase the weight of pigs by 100 kg in fattening, about 350–400 kg of wheat and barley, but more than 440 kg of rye have been considered necessary (Becker and Nehring, 1965). Besides this, pigs, cattle and poultry prefer wheat to rye. More recent sources consider rye as good as the other cereals, particularly when blended with wheat or barley. The feeding of young animals with rye seems to speed their sexual maturity and food intake, and to produce a more compact firm meat, both of which contribute to the feeding result.

The content of pentosans and the other cell wall building materials such as cellulose and lignin are reported to be limiting factors in rye feeding. Due to the water binding capacity of the pentosans and to the viscosity of water solubles, the absorption of nutrients and the feeding efficiency are decreased. From these, the amount of rye in feed blends or fodder of 50% is recommended (Nieß, 1991), but should not exceed the upper limit of 75%. A number of other factors limiting the use of rye in feeding have been reported in the last 20 years. After Wieringa (1967) cautioned against 5 alkyl resorcinol (5AR) as an anti-nutritive factor in rye, this compound was found in other cereals as well, but in lower concentrations. As the 5AR is concentrated in the hyaline layer of cereal kernels, a limited use of brans in feed mixtures and breeding of 5AR free or poor varieties are desirable. In the meantime the problems in connection with 5AR are graded as of minor importance.

Trypsin as an anti-nutritive factor of equally minor importance is not a specific problem linked to rye, but common to other cereals. With regard to the availability of minerals, phytic acid is an anti-nutritional factor capturing phosphorus and calcium at the same time. It is reported that 1 g phytin-phosphorus can immobilize 0.65 g calcium and rye cultivars have been screened for intrinsic phytase activity (Fretzdorff and Weipert, 1986; Fretzdorff, 1993).

For feeding purposes the maximal ergot content of 0.1% is allowed. Higher ergot contents (0.7% or 1.4%) have caused strong reductions in feed conversion and in milk production of sows, and have revealed symptoms of toxicity with pigs (Richter *et al.*, 1992).

A new possibility for cereal usage has opened since the pet food market has gained in importance. In the case of both dry and moist pet food products cereals contribute to the mineral, dietary fibre and vitamin contents of the food and improve their texture by high water-binding capacity. In the latter case, rye can contribute not only to the texture of pet food, but also to human food (meat sausages, canned meat) more than other cereals.

Owing to the amino acid composition and often a high protein content, triticale is a sought-after, preferred and higher-priced cereal feed than most other cereals (Anon., 1989). The advantage of higher content of essential amino acids is true only in comparison to wheat, but not to rye (Table 7.5). The data for amino acid and vitamin contents for wheat and rye are consistent with those given in Souci *et al.* (1989). Due to the agricultural properties of this cereal, the availability of triticale is growing.

Table 7.5 Amino acid and vitamin contents of triticale in comparison to wheat and rye (from Anon., 1989)

Amino acid	*Wheat*	*Triticale (g/100 g of protein)*	*Rye*
Lysine	2.83	3.44	4.02
Threonine	2.98	3.55	4.06
Methionine	1.42	1.28	1.35
Isoleucine	2.68	3.45	3.70
Leucine	7.22	7.20	7.75
Phenylalanine	3.77	4.94	4.75
Valine	3.73	4.48	5.10
Tryptophan	1.10	1.02	5.10
Vitamin		*(μg/g)*	
Thiamine	9.9	9.0–9.8	7.7
Riboflavin	3.1	2.5	2.9
Niacin	48.3	16.0–17.9	15.3
Biotin	0.06	0.06–0.07	0.05
Folic acid	0.56	0.56–0.77	0.49
Pantothenic acid	9.1	8.3–9.1	6.3
Vitamin B6	4.7	4.7–4.9	3.4

7.4.2 Quality requirements of grain

Up to now, no official quality standards for rye for feed purposes have been established. A high protein content and a rye free of impurities as estimated with a besatz test are required for feeding purposes. The presence of ergot sclerotia, mould and fusarium damaged and sprouted kernels are detrimental to the feeding value of rye, being toxic and of lower nutritive value. The use of such rye is limited mostly to older animals and requires treatments such as heat or additional cleaning. For reasons of better absorption and digestion, rye is ground to meal and soaked prior to being given to the animals. Treatment with the enzymes pentosanase and β-glucanase to increase the intake of nutrients and the feeding value are being tested. This way of improving the nutritive value of cereals, particularly of barley, has proved successful with poultry and has been extended to other monogastric animals (Pettersson *et al.*, 1991). The contamination of rye by high lead, cadmium and copper has not been observed yet and rye can be considered a useful, effective and healthy feed, as long as the recommendations for its use are followed (Anon., 1992b).

7.5 INDUSTRIAL AND OTHER USES

7.5.1 Products of industrial use

In comparison to the major uses for bread-making and as animal feed, the use of rye for industrial and other purposes is small. Since rye is an indispensable cereal in crop rotation in several regions, but the human requirements and the need for feed are limited, additional uses for rye have been sought. Since rye is a cereal not differing too much in its content of relevant kernel constituents from other cereals, the possible industrial uses are the same. Due to specific functional properties of the rye constituents, particularly pentosans, the industrial processing and the application of end-products processed from rye are specific. These facts can explain why the industrial production of starch from rye has not yet been introduced, although rye starch has a possible use for the same purposes as wheat and maize starch.

The production of whisky, particularly of Canadian rye whiskey, has one of the oldest and richest traditions in the industrial use of rye. There is ample proof in patents that rye and rye milling products are being used in the production mostly of ethyl alcohol, but also of aceton-butyl-alcohol and for controlling the fermentation of fruit juices. The presence of pentosans has probably impeded the use of rye for brewing purposes, although it is not impossible to produce rye beer.

By far the greatest number of publications and patents is dedicated to the production of glues and adhesives from rye (Pomeranz, 1973). The adhesives are suitable for use in composite wood products and in veneering as well as in the finishing of paper and hardening of resins. The production of furfural from rye pentosans, which again is a base for further production of different alcohols for technical purposes, is an old and widely used industrial process. A report has been published on the special use of rye germ for production of an anti-fusarium substance and of vitamin E preparations and of ethanol extracts from rye as an insecticide. A comprehensive overview of particular contributions with respect to non-food use of cereals is given in Pomeranz (1973).

Rye has been used to improve the properties of building materials. It has been used as a lightener for bricks making them porous after burning in brick ovens, as well as an additive to gypsum-based plasters and to wall-tile materials. Using the techniques of high pressure high temperature cooking with extruders, products from rye have been obtained, that can be used for insulation and as packing material, which can be easily used as feed and thus be recycled and made harmless after use. Research work on a wider use of rye components – starch, proteins, and pentosans – is being done. The results with respect to the production of easily-degradable plastics as packing material are promising.

There are no signs in the literature that triticale has been used for industrial purposes, except for distillation of alcohol and in a few tests for beer brewing. For use in these areas, the previous comments for rye are valid for triticale too. Being a preferred feed, however, an undamaged triticale lot is too expensive for industrial uses.

7.5.2 Quality requirements

As the industrial use of rye is based on chemical and thermal modifications, there are not many minimum values in the quality and properties of rye. In many cases damaged rye is preferred for its low purchase price. Those who hold patents on the processing of specific products from rye do not reveal the quality demands and limitations based on content and state of the grain contents.

REFERENCES

Anon. (1989) *Triticale: A Promising Addition to the World's Cereal Grains.* National Academy Press, Washington DC.

Anon. (1992b) Standard limiting values for toxic residues in food. *Lebensmittelkontroleur*, 7(3), 119–120 (in German).

Anon. (1995) *FAO Production and Trade Yearbook Statistics.*

Becker, M. and Nehring, K. (eds) (1965) *Feed Handbook*, P. Parey, Hamburg and Berlin (in German).

Bushuk, W. (1976) History, World Distribution, Production and Marketing, in *Rye – Production, Chemistry and Technology* (ed. W. Bushuk,), AACC, St. Paul, Minnesota, USA.
Fretzdorff, B. (1993) Phytic acid reduction in cereal processing. Bioavailability '93 Proceedings Part 2. *Berichte der Bundesanstalt für Ernährung*, 104–08 (in German).
Fretzdorff, B. and Weipert, D. (1986) Phytic acid in cereals. Part. 1 Phytic acid and phytase in rye and rye products. *Zeitschrift für Lebensmittel – Untersuchung und Forschung*, **182**(4), 287–93 (in German).
Haber, T. (1994) Characteristics and potential use of polish triticale varieties. *Getreide Mehl und Brot*, **49**(2), 9–14 (in German).
Kirchgessner, M. (1970) *Animal Feeding*, DLG-Verlag, Frankfurt (in German).
Meuser, F. and Suckow P. (1986) Influence of rye pentosans on baking properties. *Getreide Mehl und Brot*, **40**(11), 332–6 (in German).
Nieß, E. (1991) Feeding trials for estimation of the maximal amounts of rye and oats in diet for pig fattening. *Züchtungskunde*, **65**(5), 385–98 (in German).
Pettersson, D., Graham H. and Aman P. (1991) The nutritive value for broiler chicken of pelleting and enzyme supplementation of a diet containing barley, wheat and rye. *Animal Feed Science and Technology*, **33**, 1–14.
Pomeranz, Y. (1973) Industrial utilization of cereals – annotated bibliography 1948–1972 in *Industrial uses of cereals* (ed. Y. Pomeranz), Symposium proceedings held in conjunction with AACC Annual meeting 1973 St. Louis, MS, USA.
Richter, W., Röhrmoser, G., Rindle, C. and Wolff, J. (1992) Efficiency losses caused by ergot in feed for pigs. *Schule und Beratung*, **8**, p. IV5–10 (in German).
Seibel, W. (1975) Nutritive and physiological significance and technological valuation of rye. *Allgemeiner Mühlenmarkt*, **76**, 51–4 (in German).
Seibel, W. (1988a) Significance in agriculture, in *Rye – Production, Processing, Trade* (eds Seibel, W. and W. Steller), Behrs Verlag Hamburg, p. 15–23 (in German).
Seibel, W. (1988b) Food from Rye, in *Rye – Production, Processing, Trade* (eds Seibel, W. and W. Steller), Behrs Verlag Hamburg (in German).
Seibel, W. and Drews E. (1973) Characteristics of the quality class 'bread rye'. *Mühle*, **110**(31), 483–4 (in German).
Seibel, W. and Stephan, H. (1973) Valuation of flour quality by the use of rheological methods. *Getreide Mehl und Brot*, **27**(6), 206–13 (in German).
Seibel, W. and Stephan, H. (1980) Processing value of rye milling products – Valuation of the baking trials. *Getreide Mehl und Brot*, **34**(8), 203–06 (in German).
Souci, S.W., Fachmann, W. and Kraut, H. (1989) *Food Composition and Nutrition Tables 1989/90*, Wiss. Verlagsgesellschaft, Stuttgart.
Spiel, A. (1985) *Rye flavoring*. US patent 4,560,573.
Stephan, H. (1970) Prediction of processing value of flours and meals upon the laboratory tests. *Brot und Gebäck*, **24**, 231–9 (in German).
Stephan, H., Bremmer, J.M. and Weipert, D. (1986) Use of triticale for production of bread and rolls. *Getreide Mehl und Brot*, **40**, 208–12 (in German).
Suijs, L.W. (1986) Recognition of 6 × Winter Triticale in some European Countries, in *Proceedings of International Triticale Symposium* (ed. L.N. Darvey), Australian Institute of Agricultural Science, Sydney, p. 573.
Weipert, D. (1985) Triticale – properties and possible usage. *Getreide Mehl und Brot*, **39**(10), 291–8 (in German).
Weipert, D. (1993a) Cereals and its products, in *Food Rheology* (eds Weipert, D., H.D. Tscheuschner and E. Windhab), Behrs Verlag Hamburg (in German).

Weipert, D. (1993b) Assessment of sprouting resistance in rye varieties. *Getreide Mehl und Brot*, **47**(4), 1–9; **47**(5), 63 (in German).
Weipert, D. (1993c) *Baking performance of rye as influenced by pentosans.* Paper presented at the 1993 AACC Annual Meeting, Miami Beach, USA (in press).
Weipert, D. and Bolling, H. (1979) A notice to the relationship Falling Number – Amylogramme Data with Rye. *Die Mühle und Mischfuttertechnik*, **116**(36), 485–7 (in German).
Weipert, D. and Brümmer, J.M. (1988) Quality assessment of rye and its milling products, in *Rye – Production, Processing, Trade* (eds Seibel, W. and W. Stiller), Behrs Verlag Hamburg (in German).
Weipert, D., Zwingelberg, H. and Stephan, H. (1986) Milling aspects of triticale processing. *Getreide Mehl und Brot*, **40**(4), 107–16 (in German).
Wieringa, G.W. (1967) *On the occurrence of growth inhibiting substances in rye.* Publ. No. 156, Institute for storage and processing of agricultural produce, Wageningen, The Netherlands.

Part Two

Chemistry and Biochemistry of Cereal Quality

8

Cereal grain proteins

P.R. Shewry

8.1 HISTORY AND CLASSIFICATION

Cereal seed proteins were among the earliest of all proteins to be studied, with Beccari (1745) reporting the isolation of wheat gluten. Subsequently Einhof (1805, 1806) reported the first isolation of barley and rye storage proteins, but it was the pioneering work of T.B. Osborne that placed the study of cereal proteins, and other proteins of plant origin, on a systematic basis. In a series of detailed studies reported between about 1880 and 1920 (see Osborne, 1924) he classified proteins into groups on the basis of their extraction and solubility in a series of solvents: water (albumins), dilute aqueous salts (globulins), alcohol/water mixtures (prolamins) and dilute acids or alkalis (glutelins). This has provided a framework for the modern study of cereal seed proteins and, indeed, for the classification of proteins in general.

It was apparent even before Osborne's studies that prolamins were a unique group of proteins present only in cereal grains, where they usually account for a substantial proportion of the total grain proteins. It has subsequently been shown that they correspond to the major storage proteins in most cereals, and also form the major component of the gluten fraction which plays a key role in determining the functionality of wheat doughs. They have consequently been the focus of most of the research on cereal proteins carried out over the last century.

Although it is inevitable that the prolamins will comprise a major part of this chapter, other types of protein (both storage and non-storage) which are significant components of the mature grain will also be discussed. The focus will be on their biological properties, chemical characteristics and impact on grain functionality and quality.

8.2 STARCHY ENDOSPERM STORAGE PROTEINS

The starchy endosperm is the major storage tissue of the cereal grain, containing most of the starch and storage proteins. The latter account for about half of the total grain nitrogen and fall into two groups: prolamins and globulins/glutelins (Figure 8.1). The prolamins are the major storage proteins in all cultivated cereals except oats and rice, although smaller amounts of globulins may also be present. The converse is the case in oats and rice, where globulin-type proteins are the major components with smaller amounts of prolamins. Both of these types of protein can be regarded as true storage proteins in that they have no other known function.

8.2.1 Prolamins

The prolamin storage proteins are classically defined by their insolubility in water or dilute solutions of salts but solubility in alcohol/water mixtures, and by their high contents of proline and glutamine which together account for between 30 and 70 mol% of the total amino acid residues in individual prolamin proteins. It was the high contents of proline and amide nitrogen (derived from glutamine) which led Osborne to coin the name prolamin (see Osborne, 1924). However, it is now known that some prolamins are insoluble in aqueous alcohols without prior reduction of interchain disulfide bonds, while others may be soluble in water or dilute salt solutions under such conditions. It is therefore more accurate to define prolamins as cereal seed storage proteins that are insoluble in water or aqueous salts in the native state, and are soluble in alcohol/water mixtures either in the native state *or* after reduction of interchain disulfide bonds.

The prolamins are given trivial names which are usually based on the latin generic name, such as hordein (barley), secalin (rye) and zein (maize). An exception is wheat in which the gluten proteins are classified into two groups. The gliadins are monomeric proteins and are readily soluble in aqueous alcohols, while the glutenins consist of polymers stabilized by interchain disulfide bonds. Although these polymers are usually insoluble in aqeous alcohols, their component subunits are soluble after reduction of the interchain disulfide bonds. The gliadins and glutenin subunits are now known to be structurally related and both groups are considered to be prolamins (Shewry *et al.*, 1986).

Cereal prolamins exhibit wide variation in their structures but most have two features in common. The first is that their sequences can be divided into two or more separate regions, or domains, which differ in their amino acid compositions, secondary structures, and evolutionary origins. The second is the presence in at least one of these domains of repeated sequences based on one or more short peptide motifs and/or stretches consisting predominantly or wholly of one amino acid residue. These domains are largely responsible for the unusual amino acid compositions and solubility properties of the whole proteins.

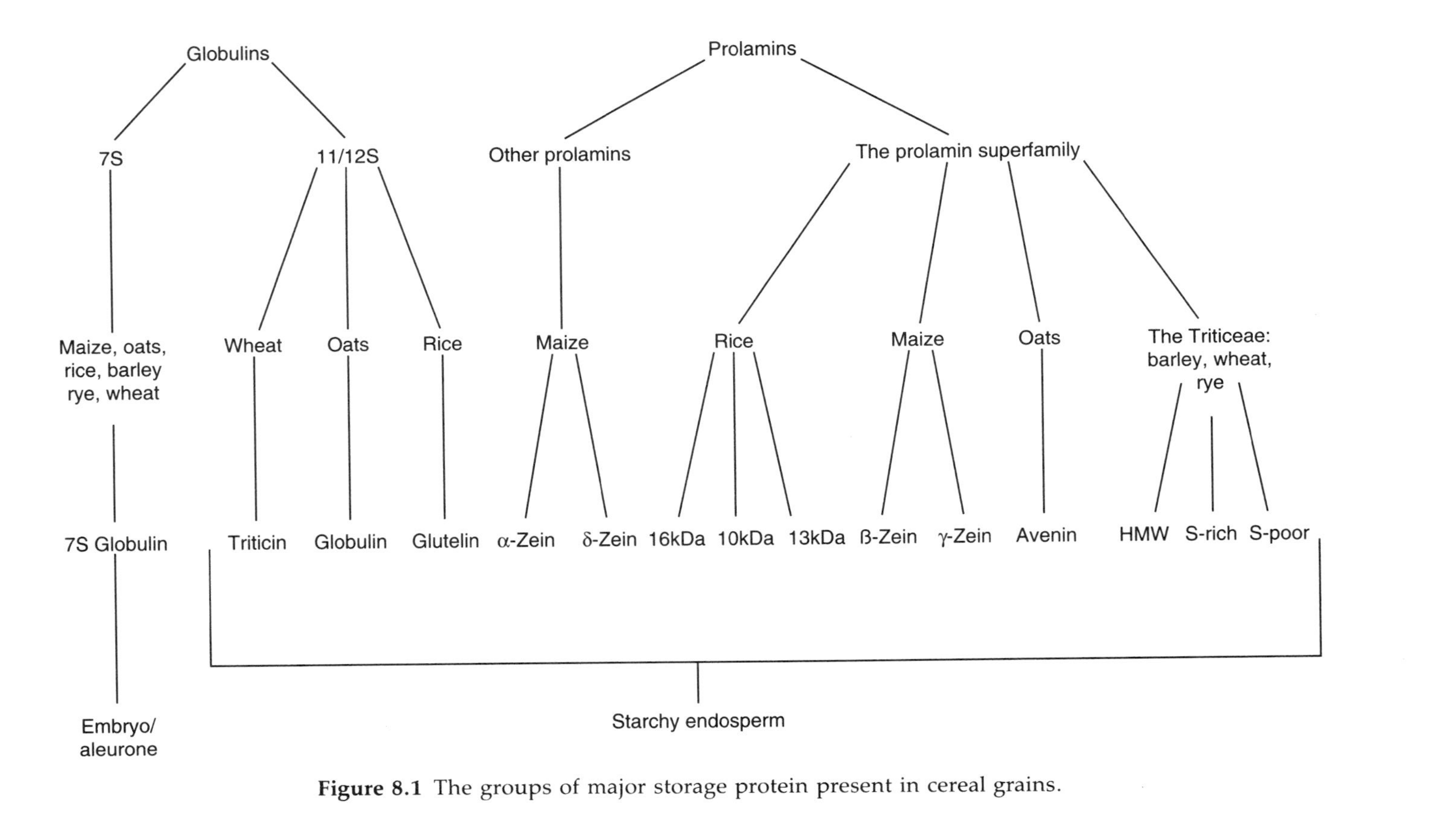

Figure 8.1 The groups of major storage protein present in cereal grains.

Despite wide variation in structure all of the prolamins of the tribe Triticeae (barley, wheat, rye and related species) are structurally related, forming a **prolamin superfamily**. These will be discussed first, followed by discussions of the prolamins present in the other major cereals: oats, rice and maize. Sorghum and most species of millet (including pearl millet and Job's tears) are classified with maize in the grass subfamily Panicoideae. These appear to have essentially similar groups of prolamins to those present in maize and will not be discussed further.

8.2.2 The prolamin superfamily of the Triticeae

The prolamins of the Triticeae are complex polymorphic mixtures, with about 20–30 components being separated by two-dimensional isoelectric focusing/SDS-PAGE in barley and rye, and over 50 components in hexaploid bread wheat (Figure 8.2). These vary in their relative molecular mass (M_r) by SDS-PAGE from about 30 000 to over 100 000, but comparison of their amino acid sequences shows that they fall into three groups: these are called the high molecular weight (HMW), sulfur-rich (S-rich) and S-poor prolamins (Shewry *et al.*, 1986; Shewry and Tatham, 1990) (Figure 8.2 and Table 8.1). The S-rich prolamins are the quantitatively major group and can be further subdivided on the basis of their amino acid sequences and their presence as monomers or polymers (Figure 8.3).

Schematic structures of typical members of the three groups are compared in Figure 8.3 All contain extensive repeated sequences. In the S-rich prolamins these are located in the N-terminal part of the protein, and account for between a third and a half of the whole sequence. These repeats are rich in proline and glutamine and are based on short

Table 8.1 Summary of the groups of prolamins present in the Triticeae: barley, wheat and rye

	Barley	*Rye*	*Wheat*
S-rich			
Ancestral γ-type (monomeric)	γ-type hordeins	40K γ-secalins	γ-type gliadins
Derived γ-type (polymeric)	absent	75K γ-secalins	absent
Polymeric type	B hordein	absent	LMW subunits
α-Type (monomeric)	absent	absent	α-type gliadins
S-poor (monomeric)	C hordein	ω-secalins	ω-gliadins
HMW (polymeric)	D hordein	HMW secalins	HMW subunits

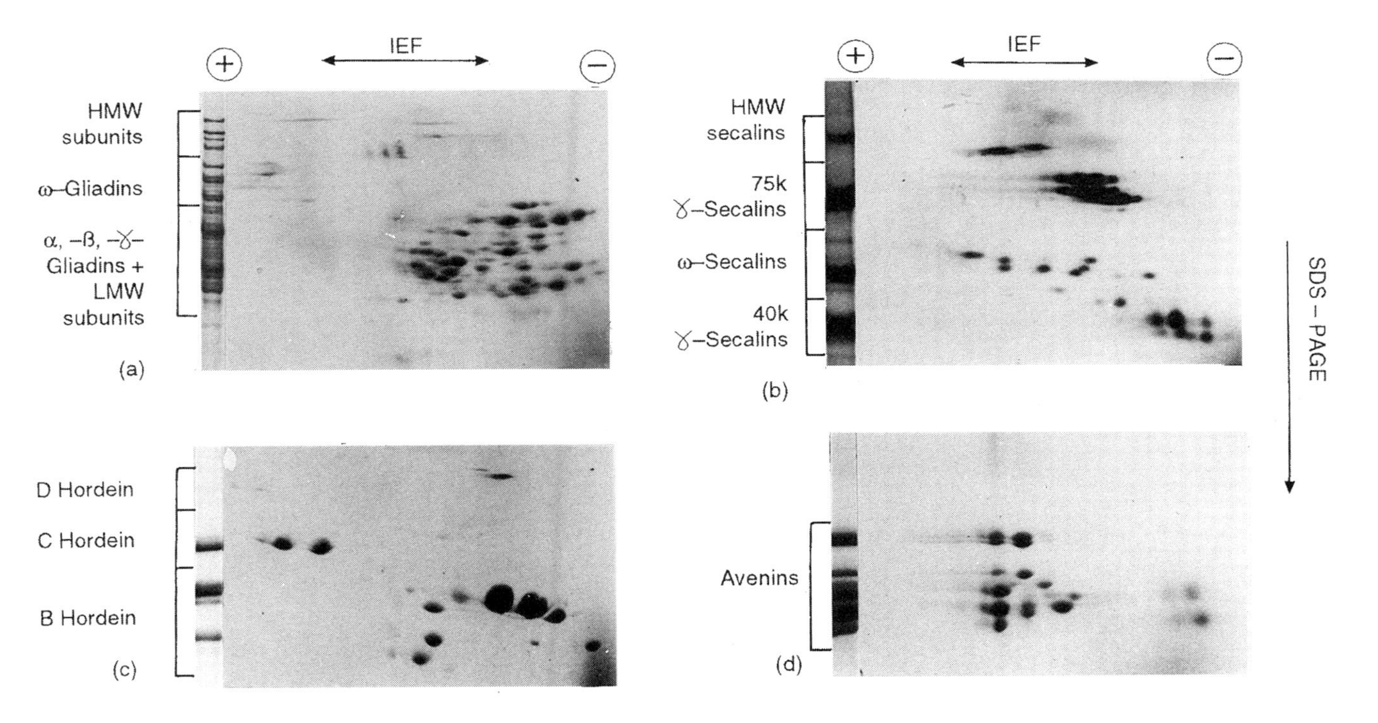

Figure 8.2 Two-dimensional analysis (IEF/SDS-PAGE) of total prolamin fractions from (a) wheat cv. Chinese Spring; (b) rye inbred line MPI 109; (c) barley line P12/3 and (d) oats cv. Goodland. Taken from Shewry *et al.*, 1988.)

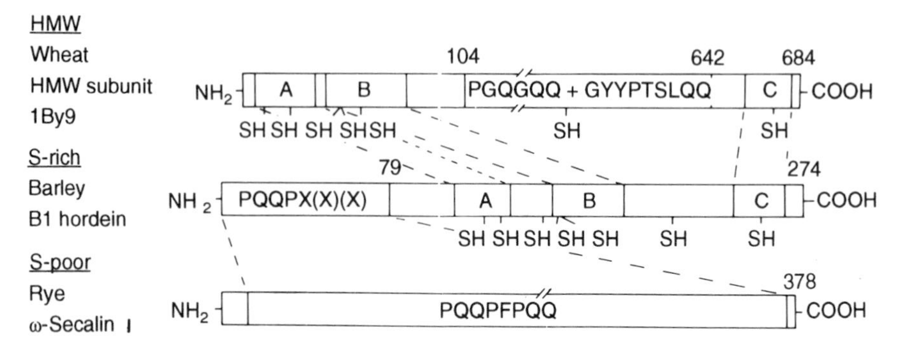

Figure 8.3 Schematic structures of typical S-rich, S-poor and HMW prolamins. Homologous domains, including three conserved regions called A, B and C, are indicated by the dashed lines.

repeated peptides. Peptide motifs based on the conserved tetrapeptide ProGlnGlnPro are always present, but in some S-rich prolamins these are interspersed with second motifs which are also rich in proline and glutamine. The C-terminal domains of the S-rich prolamins are non-repetitive, poor in glutamine and proline, and contain most or all of the cysteines and charged residues.

Repeats which are rich in glutamine and proline are also present in the S-poor prolamins, where they account for almost the entire sequence. The consensus repeat motif in C hordeins, ω-secalins and probably also most ω-gliadins is the octapeptide ProGlnGlnProPheProGlnGln. These are clearly related to the ProGlnGlnPro-based repeats present in the S-rich prolamins, implying an evolutionary relationship. The S-poor prolamins contain no cysteine residues.

The HMW subunits of bread wheat have been studied in detail because of their role in determining bread-making quality (see p. 21) (Shewry *et al.*, 1989, 1992). They consist of 627–827 amino acid residues (M_r 67 500 to 87 700), with most of the variation in size arising from differences in the extent of the central repetitive domain (481–681 residues). This domain is flanked by non-repetitive domains of 81–104 residues and 42 residues at the N- and C-termini, respectively. The repeats in the HMW subunits are unrelated to those present in the S-rich and S-poor prolamins, being based on hexapeptides (consensus ProGlyGlnGlyGlnGln), nonapeptides (GlyTyrTyrProThrSerPro or LeuGlnGln) and, in some subunits only, tripeptides (GlyGlnGln). However, three conserved regions present in the N- and C-terminal domains (called A, B and C, see Figure 8.3) are clearly related to sequences in the C-terminal domains of the S-rich prolamins, indicating that the two groups of prolamin arose from related proteins by the insertion of blocks of unrelated repeats. As in the S-rich prolamins, the cysteine residues are largely restricted to the non-repetitive domains.

The three groups of prolamins differ in their ability to form disulfide-bonded polymers. The S-poor prolamins are monomeric proteins, with no cysteine residues and hence no disulfide bonds, while the HMW prolamins are always present in high M_r polymers stabilized by interchain disulfide bonds. The S-rich prolamins vary in their behaviour, some being monomeric with intra-chain disulfide bonds (e.g. α-gliadins and γ-gliadins), and others present in disulfide-stabilized polymers (e.g. B hordeins of barley and low molecular weight (LMW) subunits of wheat glutenin) (Figure 8.3). The polymeric S-rich prolamins probably form intrachain as well as interchain disulfide bonds, with the LMW subunits forming interchain bonds with HMW subunits as well as with other LMW subunits. The relevance of these polymers to the functionality of wheat gluten is discussed briefly in Section 8.6.3.

8.2.3 The prolamins of oats and rice

Although oats and rice are not closely related botanically, they are similar in that prolamins only represent about 5–10% of the total grain proteins. However, the types of prolamins present in the two species differ considerably. The prolamins (avenins) of oats consist of about 160–200 residues, with M_r of 18 500 to 23 000 (Figure 8.2) (Egorev, 1988; Chesnut *et al.*, 1989). They are most closely related in structure to the prolamin superfamily of the Triticeae, which is consistent with the classification of the tribe Aveneae in the same subfamily (Festucoideae) of the grasses (Gramineae). In particular, conserved non-repetitive sequences related to regions A, B and C are present, and repetitive sequences rich in proline and glutamine. However, the repetitive sequences are present in two blocks close to the N- and C-termini, respectively, rather than in one block, and the precise repeat motifs differ from those in the Triticeae.

Rice is not closely related to any of the other major cereals, and has distinct groups of prolamins. Three groups can be recognized, with M_r by SDS-PAGE of about 10 000, 13 000 and 16 000 (Kim and Okita, 1988a, 1988b; Masumura *et al.*, 1989, 1990; Barbier and Ishihama, 1990; Horikoshi *et al.*, 1991; Shyur *et al.*, 1992). None contain repeated sequences while glutamine and proline together account for 25% or less of the total residues. They have no close homologues in other cereals, but appear to be distantly related to each other and to members of the prolamin superfamily of the Triticeae.

8.2.4 The prolamins (zeins) of maize

SDS-PAGE of zeins shows bands with M_r of about 10 000, 14 000, 16 000, 19 000, 22 000 and 28 000, although more components are resolved by isoelectric focusing (IEF) or 2-D IEF/SDS-PAGE (Wilson *et al.*, 1981) (Figure 8.4). Comparisons of sequences show that the component proteins fall into four discrete groups called α, *β*, γ and δ zeins (Pederson *et al.*, 1982, 1986; Marks and Larkins, 1982; Marks *et al.*, 1985; Kirihara *et al.*, 1988; Prat *et al.*, 1987).

The α-zeins are the major components, and comprise the M_r 19 000 and M_r 22 000 bands. Sequencing of corresponding cDNA and genomic clones shows that the M_r 19 000 α-zeins consist of about 210–220 residues with true molecular masses of ≈23–24 000, and the M_r 22 000 α-zeins of 240–245 residues with true molecular masses of ≈ 26 500–27 000. The M_r 19 000 and M_r 22 000 components have similar structures, consisting of repetitive domains flanked by short unique domains at the N-terminus (36–37 residues) and C-terminus (10 residues) (Figure 8.5). The repeats consist of blocks of about 20 residues and are highly degenerate in sequence. Nine blocks are present in most M_r 19 000 zeins and 10 blocks in most M_r 22 000 zeins, accounting for the differences in molecular weight. The α-zeins do not appear to be related to any other proteins, except the major prolamins of other panicoid cereals (i.e. sorghum and millets).

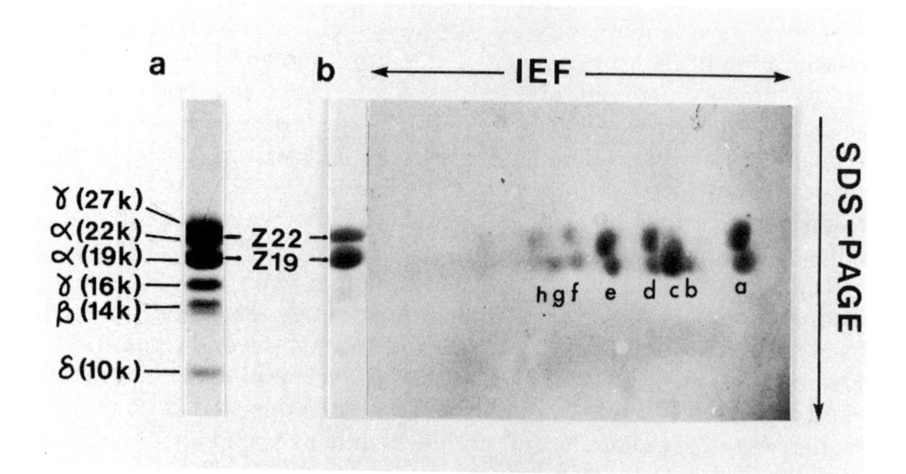

Figure 8.4 One-dimensional SDS-PAGE of total zeins of maize (part a) and 2-D IEF/SDS-PAGE of the alcohol-soluble Zein I fraction (part b), showing the major groups. The letters a–h indicate the α-zein (Z19 + Z22) bands resolved by 1-D IEF. (Part b is taken from Wilson *et al.*, 1981.)

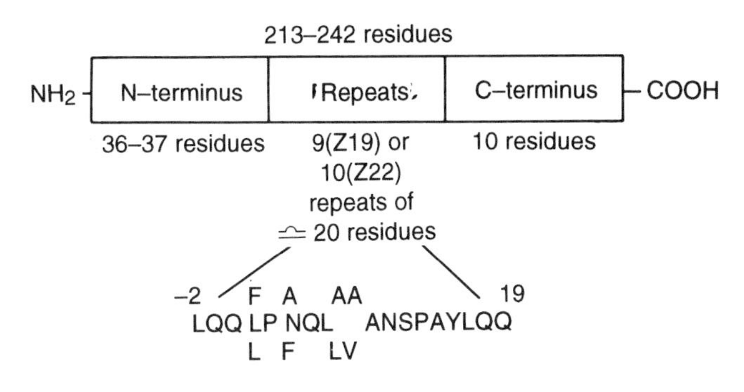

Figure 8.5 Schematic structures of the M_r 19 000 (Z19) and M_r 22 000 (Z22) α-zeins of maize (Redrawn from Shewry and Tatham, 1900.)

In contrast the *β*-zeins and γ-zeins are clearly related to each other and to the prolamin superfamily of the Triticeae. The *β*-zeins correspond to the M_r 14 000 and M_r 16 000 bands separated by SDS-PAGE and the major components consist of about 160 residues with true molecular masses of about 17 500. They contain no repeated sequences, but do have a methionine-rich region towards the C-terminus. The γ-zeins correspond mainly to the M_r 28 000 band and consist of about 200 residues with true molecular masses of about 22 000. They contain blocks of conserved hexapeptides (ProProProValHisLeu) and regions rich in proline and cysteine. Both also contain sequences which are related to the three conserved regions (A, B and C) present in the prolamins of the Triticeae.

The final zein type, δ-zein, is a minor component, and consists of 129 residues with a true molecular mass of 14 400. It does not contain repeated sequences, but may also be distantly related to the prolamin superfamily of the Triticeae. δ-Zein contains 17 methionine residues, most of which are present as doublets (i.e. MetMet) in a central region.

The zeins resemble the prolamin superfamily of the Triticeae in that the component groups vary in their aggregation behaviour and in their contents of sulfur-containing amino acids. The α-zeins contain only one or two cysteine residues per molecule, and are present either as monomers or oligomers. In addition they also have low contents of methionine (0 to about 2 mol%). In contrast the *β*-, γ- and δ-zeins are sulfur-rich, containing high levels of cysteine and/or methionine, and all form alcohol-insoluble polymers which are only extracted under reducing conditions. In fact, the γ-zeins are readily soluble in water under such conditions which is probably related to the presence of positively-charged repeated sequences. The reduced *β*- and δ-zeins are water-insoluble, which is probably related to their high contents of methionine.

8.2.5 Globulins and glutelins

The legumin-type 11–12S globulins are major storage proteins in many non-graminaceous species including legumes, crucifers, cucurbits and composites. Related proteins also form the major storage protein fraction in oats and rice, where they are classified as globulins and glutelins respectively. In addition, less closely related but clearly homologous proteins are also present in wheat.

The oat globulins have the classical legumin structure of six pairs of disulfide bonded subunits (see Casey *et al.*, 1986). Each subunit pair consists of an α-subunit ($M_r \approx 21\,700$) and a β-subunit ($M_r \approx 31\,000$), giving a holoprotein of M_r about 322 000 with an $S_{20.w}$ value of 12.1 (Peterson, 1978).

Whereas the oat globulins are extractable in salt (1 M NaCl at pH 8), the glutelins of rice can only be extracted under denaturing conditions. However, they also consist of pairs of disulfide-bonded subunits (of M_r $\approx$ 37–39 000 and 22–23 000), which have about 70% sequence identity with those of the oat globulin. Both fractions are polymorphic mixtures of components which are encoded by small families of genes.

Legumin-like proteins are also present in wheat, where they are called triticins (Singh and Shepherd, 1985; Singh *et al.*, 1988, 1993). They consist of large and small subunits of M_r about 40 000 and 22–23 000 respectively, although the large subunits exhibit M_r of 52 000 and 58 000 by SDS-PAGE. These anomalously high M_r could result from glycosylation, as consensus glycosylation sites are present in the amino acid sequences. These subunit pairs appear to be further assembled into dimeric structures with M_r by SDS-PAGE of 150–160 000. The subunits have significant sequence similarity to the oat globulin, but the large subunits also contain four copies of a novel repeated decapeptide containing two lysine residues. The present of related proteins in barley and rye has not so far been demonstrated.

8.3 EMBRYO AND ALEURONE STORAGE PROTEINS

In botanical terms the aleurone layer is part of the endosperm, and is therefore triploid (in contrast to the diploid embryo). However, in biological terms it is closer in function to the scutellum of the embryo, remaining alive in the mature grain (unlike the starchy endosperm) and synthesizing hydrolytic enzymes during germination. It also resembles the embryo in its storage compounds, lacking starch but containing oil bodies and different types of protein body. The latter appear to contain

7S globulin-type storage proteins which are unrelated to the 11S globulins/glutelins present in the starchy endosperm, but may be related to 7S globulins present in a range of other plants.

7S Globulins with sequence similarity to the 'vicilin-type' globulins of legumes and cottonseed have been identified in embryos of maize (where they account for 10–20% of the mature protein) (Belanger and Kriz, 1989; Wallace and Kriz, 1991), the embryos of oats and rye (Burgess *et al.*, 1983; Burgess and Shewry, 1986) and the aleurones and embryos of wheat and barley (Burgess and Shewry, 1986; Quatrano *et al.*, 1986; Yupsanis *et al.*, 1990). These proteins consist mainly of subunits with M_r between 45 000 and 65 000, but may undergo post-translational processing to give smaller subunits. A number of embryo-located γ-globulins have also been reported in rice (Morita and Yoshida, 1968), but their relationships to those present in other cereals are not known.

8.4 THE SYNTHESIS AND DEPOSITION OF STORAGE PROTEINS

The prolamin and globulin storage proteins discussed above are located within the developing grain in discrete deposits called protein bodies. This facilitates their storage in a highly concentrated form, and avoids interference with other cellular processes. The transport of proteins and their deposition into protein bodies involves complex processes of sorting and targeting which are still incompletely understood. However, it appears that two mechanisms of protein body formation occur in cereals, which may operate at different times or for different protein types.

In both cases the first stage is the synthesis of the protein on the rough endoplasmic reticulum (ER) and transport of the nascent polypeptide into the ER lumen with the cleavage of an N-terminal signal sequence. This is followed by protein folding and disulfide bond formation, which may be assisted by molecular chaperones (such as homologues of the immunoglobulin binding protein BiP) and the enzyme protein disulfide isomerase, respectively. The proteins may then aggregate to form deposits directly within the lumen of the ER, or may be transported via the Golgi apparatus and Golgi vesicles to form protein bodies of vacuolar origin. These two pathways are summarized in Figure 8.6

The storage globulins are almost certainly always transported via the Golgi route, similar to the homologous 7S and 11/12S globulins of legumes and other species (Craig, 1988). In contrast the prolamins may take either route. In maize (Larkins and Hurkman, 1978) and rice all of the prolamins appear to be retained within the ER, which results in rice in two discrete populations of protein bodies: ER-derived protein bodies containing prolamins and vacuolar protein bodies containing glutelins (Krishnan *et al.*, 1986). Only one population of protein bodies is present

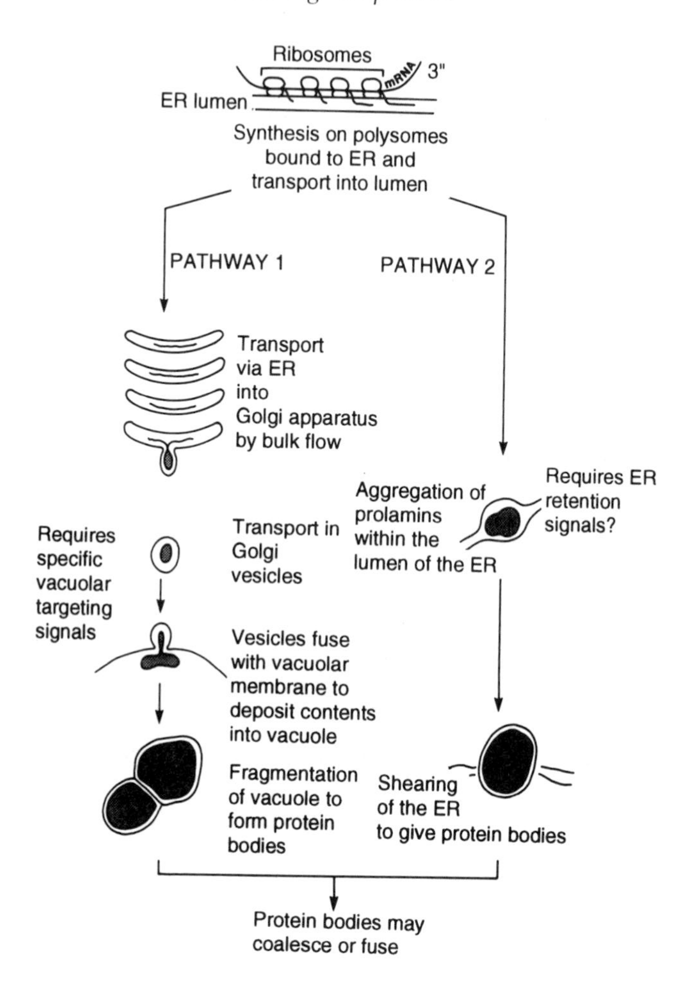

Figure 8.6 Two pathways of storage proteins deposition in developing cereal grains. Pathway 1 may operate for globulin storage proteins in all cereals and for some prolamins in barley and wheat. Pathway 2 may operate for all prolamins in maize and rice and some prolamins in barley and wheat. (Modified from Shewry, 1993.)

in maize starchy endosperm, which does not apparently contain 11/12S globulins or glutelins. In barley and wheat some prolamins appear to be transported via the Golgi to the vacuole, which in wheat results in the formation of biphasic protein bodies containing inclusions of triticins embedded in a matrix of prolamins (Bechtel *et al.*, 1991). However, a second population of prolamin-containing protein bodies also forms by direct accumulation within the ER. In addition these may fuse with protein bodies of vacuolar origin, either due to compression as the endosperm cells become distended with starch, or possibly by a novel route which bypasses the Golgi as proposed by Levanony *et al.* (1992). There is also evidence that the HMW glutenin subunits are preferentially deposited via the ER route, and the gliadins via the Golgi route (Rubin *et al.*, 1992). Biphasic protein bodies, but consisting of inclusions of prolamin in a globulin matrix, are also present in oats, but their origin is still unclear (Lending *et al.*, 1989).

Studies of other systems have shown that specific sequence information is required to ensure that proteins are either transported to the vacuole or retained within the ER. Such signals have not yet been identified in cereal prolamins and globulins, but vacuolar targeting sequences appear to be present in legume globulins (Chrispeels and Raikhel, 1992; Muntz *et al.*, 1993). It is anticipated that such signals are also present in the cereal prolamins and globulins that are transported via the Golgi route to the vacuoles, whereas direct retention within the ER could result solely from aggregation to form insoluble deposits which are physically impossible to transport.

The protein bodies in the starchy endosperm may become disorganized during the later stages of grain maturation and drying, resulting in a continuous proteinaceous matrix surrounding the starch granules. This may be relevant to the formation of the gluten matrix in doughs made from wheat flours.

8.5 OTHER PROTEINS

Cereal grains contain a multitude of proteins in addition to the major storage prolamins and globulins discussed above. These may play structural or metabolic roles (for example proteins present in cell walls and membranes, enzymes and components of transport systems), or be minor storage components present in protein bodies. In some cases they may contribute to resistance mechanisms, inhibiting feeding or infection by pests and pathogens. Only a small proportion of the total have been characterized in detail, and the present discussion will be restricted to components of particular interest in relation to their biological or functional properties. Their tissue specificity has not been examined in all cases, but with few exceptions the proteins discussed here are located predominantly or solely in the starchy endosperm.

8.5.1 Hydrolytic enzymes

The vast majority of the enzymes that hydrolyse the grain storage reserves (protein, lipid and carbohydrate) during germination are synthesized *de novo* in the scutellum and aleurone layer. However, two important carbohydrases appear to be synthesized during grain development and stored until germination, at least in barley where they have been studied in relation to malting quality. β-Amylase is only synthesized during barley grain development, and consists of a mixture of isoenzymes of M_r about 57 000 (Shewry *et al.*, 1988; Shewry 1993). It is synthesized in the aleurone and starchy endosperm, and consists of a mixture of forms which are readily extractable ('free') or are only extracted in the present of a reducing agent or by partial proteolysis ('bound'). The latter form is probably associated with other proteins by disulfide bonds, and proteolysis by endogenously produced proteases may result in the release of active enzyme during germination. The second enzyme, limit dextrinase (debranching enzyme), also accumulates in a bound form during grain development and is slowly released during germination (probably by partial proteolysis). However, there is also evidence of *de novo* synthesis in germinating aleurone layers (Fincher and Stone, 1993).

One barley protein which is associated with β-amylase as heterodimers is termed protein Z. This is an albumin of M_r about 40 000, and also occurs in free and bound forms. Its biological role in barley is not known, although it has sequence similarity to a family of trypsin inhibitors found in animals (Brandt *et al.*, 1990). However, its main interest to maltsters and brewers is that it is the major protein of barley origin which is present in intact form in beer, where it may contribute to foam stability or haze formation (Hejgaard, 1977). α-Amylase and protein Z are also of interest in that they are present at elevated levels in the high lysine barley line Hiproly, and contribute significantly to the increased content of lysine (Hejgaard and Boisen, 1980).

8.5.2 Inhibitors of hydrolytic enzymes

Cereal seeds are rich sources of low M_r proteins which inhibit the activities of proteases and/or carbohydrases. These have been most widely studied in barley and wheat, but are also present in other species. They can be classified into a number of groups (or families) on the basis of their sequence relationships and specificities. It is not possible to include detailed discussions of all these proteins here, but the reader is referred to several recent review articles (Garcia-Olmedo *et al.*, 1987, 1992; Richardson, 1991; Shewry, 1993). I will instead briefly discuss several components which are of particular interest or relevance to grain utilization.

The barley α-amylase subtilisin inhibitor (BASI) (Svendsen *et al.*, 1986) and its homologues in wheat (WASI) (Mundy *et al.*, 1984) and rice (RASI) (Ohtsubo and Richardson, 1992) are proteins of about M_r 20 000.

They belong to the widespread family of Kunitz protease inhibitors, and are unusual in that they inhibit endogenous cereal α-amylases as well as α-amylases from some other organisms. Zawistowska *et al.* (1988) showed that BASI could be used to improve the bread-making quality of wheat flour mixed with malted barley flour. It is therefore possible that BASI or WASI could be used to limit the effects of pre-harvest sprouting, either by addition of exogenous protein (e.g. produced by fermentation) or by engineering the plants to synthesize protein during sprouting.

The barley chymotrypsin inhibitors CI-1 and CI-2 belong to the potato inhibitor 1 family and are low M_r ($\approx$ 9000) proteins which contain no cysteine residues, but 9.5 and 11.5 g lysine/100 g protein respectively. They are present at increased levels (×6- to ×8-fold) in Hiproly and derived high lysine lines and contribute significantly to the elevated lysine contents of the whole grains (Hejgaard and Boisen, 1980). Their low M_r and lack of cysteine have also made them attractive targets for structural and protein engineering studies (Campbell, 1992).

The most abundant and complex group of inhibitors present in cereal grains are referred to as the **cereal inhibitor superfamily**. These have M_r ranging from about 12 000 to 16 000 and may be present as monomers or as dimeric or tetrameric complexes. The individual subunits may exhibit activity against trypsin, against exogenous α-amylases (notably from insect larvae), against both types of enzyme, or lack inhibitory activity. Several of the components present in barley, wheat and rye were initially defined on the basis of their solubility in chloroform/methanol mixtures, and are therefore often called CM proteins. A number of sequences have been determined by isolation of cDNA and/or genes, showing some similarity to the non-repetitive domains of the S-rich and HMW prolamins of the prolamin superfamily of the Triticeae (see above). This is particularly apparent in the presence of regions corresponding to A, B and C, containing conserved cysteine residues.

The major interest in the CM proteins of wheat and barley resides in their roles in grain utilization. In pasta wheat two CM proteins, called CM3 and CM16, are associated with the glutenin fraction, and have been called durum sulfur-rich glutenins (DSGs) (Kobrehel and Alary, 1989a,b; Gautier *et al.*, 1989). There is a strong correlation between the pasta surface quality on cooking (i.e. stickiness) and the total contents of sulfydryl groups and disulfide bonds present in these two proteins (Kobrehel and Alary, 1989a, 1989b), which may relate to their interactions with gliadins and glutenins by strong non-covalent forces.

The CM proteins also have negative impacts on grain processing in relation to the health of workers in the milling and baking industries, being the active agents in baker's asthma. The major proteins responsible for this appear to be glycosylated forms of CM proteins: CM16 in wheat and CMb and a related M_r 14 500 protein in barley (Mena *et al.*, 1992; Sanchez-Monge *et al.*, 1992). A related protein is also a major allergen in atopic food allergy to rice (Izumi *et al.*, 1992).

8.5.3 Starch granule proteins

Proteins associated with the starch granules of wheat are of interest in relation to the milling properties of grain and to the functional properties of starches prepared from wheat. These proteins account for about 0.2% of wheat starch prepared by water washing, corresponding to about 1% of the total grain protein (Schofield and Greenwell, 1987). They can be divided into two groups: proteins associated with the granule surface, and proteins present within the granule (intrinsic proteins). The latter comprise five major polypeptides with M_r ranging from 59 000 to 149 000, and almost certainly include major enzymes involved in starch biosynthesis. For example, the absence of the major M_r 59 000 intrinsic starch granule protein from mutant waxy (100% amylopectin) starches of maize, barley and rice indicates that it may correspond to the starch granule bound starch synthase which catalyses the synthesis of amylose.

The proteins associated with the starch granule surface also comprise five major components, but with lower M_r than the intrinsic proteins ($\approx$ 5000 to 30 000). None of these has yet been ascribed a function, but one is of particular interest because its presence in bread wheat is invariably associated with grain softness (Greenwell and Schofield, 1986, 1989). This is an M_r 15 000 protein which has been called friabilin. It is absent from tetraploid wheats or barley, but its presence in a range of diploid wheats and in rye is unexpected and so far unexplained (Morrison *et al.*, 1992).

Despite work in several laboratories we still know little about the properties of friabilin or the molecular basis for its association with softness. However, it is clear that friabilin is not a single protein but a mixture of proteins (Morris *et al.*, 1994), and that at least some of these are related to the cereal inhibitor superfamily and the puroindolines (see below) (Jolly *et al.*, 1993; Rahman *et al.*, 1994). In addition, immunochemical studies demonstrated that friabilin is also synthesized in kernels of hard genotypes of wheat, but is present in lower amounts and does not bind as strongly to starch granules as in soft genotypes (Jolly *et al.*, 1993).

It has been postulated that friabilin acts as a 'non-stick' agent on the granule surface, preventing adhesion to the protein matrix and facilitating separation during milling. The relationship to puroindoline, a lipid binding protein, indicates that friabilins may also bind lipids, and that this may play a role in determining softness (Rahman *et al.*, 1994).

This is also consistent with the observation that softness is associated with a higher content of 'free' polar lipids (Morrison *et al.*, 1989).

8.5.4 Antimicrobial and cysteine-rich proteins

Cereal grains, in common with many other storage tissues, contain a range of proteins that are active (at least *in vitro*) against pests and pathogens. These include the inhibitors of hydrolytic enzymes discussed above, which may inhibit digestive enzymes of insect pests (CM proteins) and bacterial subtilisin (BASI, WASI and RASI). Others include endochitinases, β-glucanases, protein synthesis inhibitors (ribosome inactivating proteins), peroxidases and lectins. These have been studied in most detail in barley, where they have been reviewed by Shewry (1993). In many cases they have been demonstrated to inhibit the growth of pests and pathogens, either alone or in combination, and *in vitro* and/or in transgenic plants. For example, the barley protein synthesis inhibitor confers resistance to soil borne fungal pathogens in transgenic tobacco plants (Logemann *et al.*, 1992). This discussion will not cover all these proteins in detail, but will focus on three types of cysteine-rich proteins which display lipid binding and antimicrobial activities.

The thionins consist of 45–47 residues including eight cysteines. They were initially purified from petroleum ether extracts and thought to be lipoproteins but it is now considered that the interaction with lipids was an artefact of the extraction procedure. At least three distinct forms are present in endosperms of wheat and barley (α, β, γ) (Garcia-Olmedo *et al.*, 1989; Colilla *et al.*, 1990; Mendez *et al.*, 1990; Shewry, 1993), while related proteins are present in leaves where their synthesis may be induced by infection with pathogens (Bohlmann *et al.*, 1988). They have been reported to have a range of activities including inhibition of insect α-amylases, toxicity to cultures of mammalian cells, yeasts and bacteria, induction of membrane leakiness and inhibition of cell free protein synthesis (Shewry, 1993), and almost certainly play a role in resistance to pests and pathogens. They may also play a role in redox systems, perhaps in combination with thioredoxin (Johnson *et al.*, 1987), and could therefore contribute to disulfide exchange during wheat dough mixing and development.

The lipid transfer protein (LTP) of barley was initially identified as the product of an aleurone-specific cDNA (Mundy and Rogers, 1986). The protein sequence showed some similarity to the α-amylase inhibitor I-2 from finger millet (Campos and Richardson, 1984), and it was therefore called probable amylase/protease inhibitor (PAPI). Subsequent studies showed homology with a well-characterized lipid transfer protein from maize (Bernhard and Somerville, 1989), and it was therefore re-named as an LTP. The LTPs of barley and wheat have been shown to transfer phosphatidylcholine between liposomes and mitochondrial membranes *in vitro* (Breu *et al.*, 1989; Dieryck *et al.*, 1992) but their *in vivo* role is still unclear. The lack of specificity in their lipid transfer activity, their synthesis with a signal peptide (indicating passage through the secretory system) and their restriction (at least in barley) to the aleurone would argue against a role in membrane biogenesis and maintenance.

The wheat and barley LTPs exhibit limited sequence homology with the CM proteins of the cereal inhibitor superfamily and with a third cysteine-rich wheat protein called puroindoline (Blochet *et al.*, 1993). This is a basic protein of 115 residues, which contains a unique amphiphilic region consisting of five tryptophan and three lysine residues.

The thionins (see above), phospholipid transfer proteins (Terras *et al.*, 1992) and puroindoline (Blochet *et al.*, 1993) may all inhibit microbial growth by binding to and destabilizing membranes. Their ability to bind lipids may also be exploited in the food industry, for example to develop novel protein emulsifiers.

8.6 THE IMPACT OF CEREAL PROTEINS ON GRAIN QUALITY

Although proteins only account, on average, for about 10% of mature cereal grain, they have major effects on quality, whether for food, feed or processing.

8.6.1 Nutritional quality

The prolamin storage proteins are low in certain essential amino acids, which limits the overall nutritional quality for monogastric animals of all those cereals in which prolamins are the major storage proteins (i.e. all except oats and rice). Thus lysine is the major limiting amino acid in barley, wheat and maize, followed by threonine in barley and wheat and tryptophan in maize. In practical terms, this may result in inefficient use of cereals feeds when used as the sole source of protein for rapidly growing livestock such as pigs and poultry, but it is debatable whether there is any relevance to human nutrition. Exploitation of spontaneously occurring high lysine mutations has led to some success in producing improved lines of maize, but genetic engineering may ultimately prove to be the best strategy for small grain cereals.

8.6.2 Malting quality

Malting quality is a highly complex character, and is influenced by a number of interacting factors including the proteins present in the mature grain. These include two enzymes which are subsequently involved in the mobilization of starch: β-amylase and limit dextrinase. In addition, the hordein storage proteins may affect malting in two respects: negatively by physically restricting starch hydrolysis and contributing to the formation of chill haze, and positively in contributing hydrophobic beer peptides responsible for foaming and the cling of the beer to the glass (Lazzeri and Shewry, 1993; Kauffman *et al.*, 1994).

8.6.3 Processing properties of wheat

White flour is derived from the starchy endosperm of the wheat grain, and therefore contains mainly starch and prolamin (i.e. gliadin and glutenin) storage proteins. When flour is wetted and kneaded to form dough the gliadins and glutenins form a network, which can be isolated as a cohesive proteinaceous mass by washing to remove most of the starch granules and other dough components. This is the gluten fraction, which exhibits a combination of two physical properties: elasticity and extensibility (viscosity). A precise combination of these properties is essential to entrap carbon dioxide during bread-making, and to provide a cohesive dough for other products such as pasta, noodles and couscous. The structure of wheat gluten is not understood in detail, but it is generally considered that the glutenins form an elastic network stabilized by interchain disulfide bonds, while the gliadins interact with this via non-covalent forces (principally hydrogen bonds and hydrophobic interactions), and contribute mainly to viscosity. In addition other dough components undoubtedly contribute to dough viscoelasticity by interacting with gluten; for example lipids almost certainly act as plasticizers. The role of gluten in wheat quality has been a major stimulus to its study, and optimization of gluten quality is a major target for plant breeders, agronomist and genetic engineers.

REFERENCES

Barbier, P. and Ishihama, A. (1990) Variation in the nucletide sequence of a prolamin gene family in wild rice. *Plant Molecular Biology*, **15**, 191–5.

Beccari, J.B. (1745) *De Frumento. De Bononiensi Scientarium et Artium.* Instituto atque Academia Commentarii 2 (Part 1), Bologna, L. Vulpe, 122–7.

Bechtel, D.B., Wilson, J.D. and Shewry P.R. (1991) Immunocytochemical localization of the wheat storage protein triticin in developing endosperm tissue. *Cereal Chemistry*, **68**, 573–7.

Belanger, F.C. and Kriz, A.L. (1989) Molecular characterization of the major maize embryo globulin encoded by the *Glb* 1 gene. *Plant Physiology*, **91**, 636–43.

Bernhard, W.R. and Somerville, C.R. (1989) Coidentity of putative amylase inhibitors from barley and finger millet with phospholipid transfer proteins inferred from amino acid sequence homology. *Archives of Biochemistry and Biophysics*, **269**, 695–7.

Blochet, J-E., Chevalier, C., Forest, E., Pebay-Peyroula, E., Gautier, M-F., Joudrier, P., Pézolet, M. and Marion, D. (1993) Complete amino acid sequence of puroindoline, a new basic and cystine-rich protein with a unique tryptophan-rich domain, isolated from wheat endosperm by Triton X-114 phase partitioning. *FEBS Letters*, **329**, 336–40.

Bohlmann, H., Clausen, S., Behnke, S., Giese, H., Hiller, C., Reimann-Philipp, U., Schrader, G., Barkholt, V. and Apel, K. (1988) Leaf-specific thionins of barley – A novel class of cell wall proteins toxic to plant-pathogenic fungi and possibly involved in the defence mechanism of plants. *EMBO Journal*, **7**, 1559–65.

Brandt, A., Svendsen, I. and Hejgaard, J. (1990) A plant serpin gene. Structure, organization and expression of the gene encoding barley protein Z. *European Journal of Biochemistry*, **194**, 499–504.

Breu, V., Guerbette, F., Kader, J.C., Kannangara, C.G., Svensson, B. and Von Wettstein-Knowles, P. (1989) A 10 kD barley basic protein transfers phosphatidylcholine from liposomes to mitochondria. *Carslberg Research Communications*, **54**, 81–4.

Burgess, S.R. and Shewry, P.R. (1986) Identification of homologous globulins from embryos of wheat, barley, rye and oats. *Journal of Experimental Botany*, **37**, 1863–71.

Burgess, S.R., Shewry, P.R., Matlashewski, G.J., Altosaar, I. and Miflin, B.J. (1983) Characteristics of oat (*Avena sativa* L.) seed globulins. *Journal of Experimental Botany*, **34**, 1320–32.

Campbell, A.F. (1992) Protein engineering of the barley chymotrypsin inhibitor 2, in *Plant Protein Engineering* (eds P.R. Shewry and S. Gutteridge), Cambridge University Press, Cambridge, pp. 257–68.

Campos, F.A.P. and Richardson, M. (1984) The complete amino acid sequence of the α-amylase inhibitor I-2 from seeds of ragi (Indian finger millet, *Eleusine coracana* Gaertn). *FEBS Letters*, **167**, 300–4.

Casey, R., Domoney, C. and Ellis, N. (1986) Legume storage proteins and their genes, in *Oxford Surveys of Plant Molecular and Cell Biology*, Vol. 3 (ed. B.J. Miflin), Oxford University Press, Oxford, pp. 1–95.

Chesnut, R.S., Shotwell, M.A., Boyer, S.K. and Larkins, B.A. (1989) Analysis of avenin proteins and the expression of their mRNAs in developing oat seeds. *The Plant Cell*, **1**, 913–24.

Chrispeels, M.J. and Raikhel, N.V. (1992) Short peptide domains target proteins to plant vacuoles. *Cell*, **68**, 613–16.

Colilla, F.J., Rocher, A. and Mendez, E. (1990) γ-Purothionins: Amino acid sequence of two polypeptides of a new family of thionins from wheat endosperm. *FEBS Letters*, **270**, 191–4.

Craig, S. (1988) Structural aspects of protein accumulation in developing legume seeds. *Biochemie und Physiologie der Pflanzen*, **183**, 159–71.

Dieryck, W., Gautier, M-F., Lullien, V. and Joudrier, P. (1992) Nucleotide sequence of a cDNA encoding a lipid transfer protein from wheat (*Triticum durum* Desf.). *Plant Molecular Biology*, **19**, 707–9.

Egorev, T.A. (1988) The amino acid sequence of the 'fast' avenin component (*Avena sativa* L.). *Journal of Cereal Science*, **8**, 289–92.

Einhof, H. (1805) Chemical analysis of rye (*Secale cereale*). *Neues Allgemeines Journal de Chemie*, **5**, 131–53.

Einhof, H. (1806) Chemical analysis of small barley (*Hordeum vulgare*). *Neues Allgemeines Journal de Chemie*, **6**, 62–98.

Fincher, G.B. and Stone, B.A. (1993) Physiology and biochemistry of germination in barley, in *Barley: Chemistry and Technology*, Chapter 6 (eds A.W.

MacGregor and R.S. Bhatty), American Association of Cereal Chemists, St Paul, MN, pp. 247–95.
Garcia-Olmedo, F., Salcedo, G., Sanchez-Monge, R., Gomez, L., Royo, J. and Carbonero, P. (1987) Plant proteinaceous inhibitors of proteinases and α-amylases, in *Oxford Surveys of Plant Molecular and Cell Biology*, Vol. 4 (ed. B.J. Miflin), Oxford University Press, Oxford, pp. 275–334.
Garcia-Olmedo, F., Rodriguez-Palenzuela, P., Hernández-Lucas, C., Ponz, F., Marañia, C., Carmona, M.J. Lopez-Fando, J., Fernandez, J.A. and Carbonero, P. (1989) The thionins: a protein family that includes purothionins, viscotoxins and crambins, in *Oxford Surveys of Plant Molecular & Cell Biology*, Vol. 6 (ed. B.J. Miflin), Oxford University Press, Oxford, pp. 31–60.
Garcia-Olmedo, F., Salcedo, G., Sanchez-Monge, R., Hernandez-Lucas, C., Carmona, M.J., Lopez-Fando, J.J., Fernandez, J.A., Gomez, L., Royo, J., Garcia-Maroto, F., Castagnaro, A. and Carbonero, P. (1992) Trypsin/α-amylase inhibitors and thionins: possible defence proteins from barley, in *Barley* (ed. P.R. Shewry), CAB International, Wallingford, pp. 335–50.
Gautier, M.-F., Alary, R., Kobrehel, K. and Joudrier, P. (1989) Chloroform/methanol-soluble proteins are the main components of *Triticum durum* sulfur-rich glutenin fractions. *Cereal Chemistry*, **66**, 535.
Greenwell, P. and Schofield, J.D. (1986) A starch-granule protein associated with endosperm softness in wheat. *Cereal Chemistry*, **63**, 379–80.
Greenwell P. and Schofield, J.D. (1989) The chemical basis of grain hardness and softness in wheat end-use properties, in *Proceedings of the 1989 International Association of Cereal Science Technology Symposium Helsinki*, Helsinki Press, pp. 59–71.
Hejgaard, J. (1977) Origin of a dominant beer protein. Immunochemical identity with a *β*-amylase-associated protein from barley. *Journal of the Institute of Brewing*, **83**, 94–6.
Hejgaard, J. and Boisen, S. (1980) High lysine proteins in Hiproly barley breeding: Identification, nutritional significance and new screening methods. *Hereditas*, **93**, 311–20.
Horikoshi, M., Kobayashi, H., Yamazoe, Y., Mikami, B. and Morita, Y. (1991) Purification and complete amino acid sequence of a major prolamin of rice endosperm. *Journal of Cereal Science*, **14**, 1–14.
Izumi, H., Adachi, T., Fujii, N., Matsuda, T., Nakamura, R., Tanaka, K., Urisu, A. and Kurosawa, Y. (1992) Nucleotide sequence of a cDNA clone encoding a major allergenic protein in rice seeds. *FEBS Letters*, **302**, 213–16.
Johnson, T.C., Wada, K., Buchanan, B.B. and Holmgren, A. (1987) Reduction of purothionin by the wheat seed thioredoxin system. *Plant Physiology*, **85**, 446–51.
Jolly, C.J., Rahman, S., Kortt, A.A. and Higgins, T.J.V. (1993) Characterisation of the wheat M_2r 15000 'grain-softness protein' and analysis of the relationship between its accumulation in the whole seed and grain softness. *Theoretical and Applied Genetics*, **86**, 589–97.
Kauffman, J.A., Mills, E.N.C., Brett, G.M., Fido, R.J., Tatham, A.S., Shewry, P.R., Onishi, A., Proudlove, M. and Morgan, M.R.A. (1994) Immunological characterisation of barley polypeptides in lager foam. *Journal of the Science of Food and Agriculture*, **66**, 345–55.
Kim, W.T. and Okita, T.W. (1988a) Nucleotide and primary sequence of a major rice prolamine. *FEBS Letters*, **231**, 308–10.
Kim, W.T. and Okita, T.W. (1988b) Structure, expression, and heterogeneity of the rice seed prolamines. *Plant Physiology*, **88**, 649–55.
Kirihara, J.A., Petri, J.B. and Messing, J. (1988) Isolation and sequence of a gene

encoding a methionine-rich 10-kDa zein protein from maize. *Gene*, **71**, 359–70.

Kobrehel, K. and Alary, R. (1989a) The role of a low molecular weight glutenin fraction in the cooking quality of pasta. *Journal of the Science of Food and Agriculture*, **47**, 487–500.

Kobrehel, K. and Alary, R. (1989b) Isolation and partial characterization of two low molecular weight durum wheat glutenins. *Journal of the Science of Food and Agriculture*, **48**, 441–52.

Krishnan, H.B., Franceschi, V.R. and Okita, T.W. (1986) Immunochemical studies on the role of the Golgi complex in protein-body formation in rice seeds. *Planta*, **169**, 471–80.

Larkins, B.A. and Hurkman, W.J. (1978) Synthesis and deposition of zein in protein bodies of maize endosperm. *Plant Physiology*, **62**, 256–63.

Lazzeri, P. and Shewry, P.R. (1993) Biotechnology of cereals, in *Biotechnology & Genetic Engineering Reviews* (ed. M. Tombs), Intercept Ltd., UK, pp. 79–146.

Lending, C.R., Chesnut, R.S., Shaw, K.L. and Larkins, B.A. (1989) Immunolocalization of avenin and globulin storage proteins in developing endosperm of *Avena sativa* L. *Planta*, **178**, 315–24.

Levanony, H., Rubin, R., Altschuler, Y. and Galili, G. (1992) Evidence for a novel route of wheat storage proteins to vacuoles. *Journal of Cell Biology*, **119**, 1117–28.

Logemann, J., Jack, G., Tommerup, H., Mundy, J. and Schell, J. (1992) Expression of a barley ribosome-inactivating protein leads to increased fungal protection in transgenic tobacco plants. *Bio/Technology*, **10**, 305–8.

Marks, M.D. and Larkins, B.A. (1982) Analysis of sequence microheterogeneity among zein messenger RNAs. *Journal of Biological Chemistry*, **257**, 9976–83.

Marks, M.D., Lindell, J.S. and Larkins, B.A. (1985) Nucleotide sequence analysis of zein mRNAs from maize endosperm. *Journal of Biological Chemistry*, **260**, 16451–59.

Masumura, T., Shibata, D., Hibino, T., Kato, T., Kawabe, K., Takeba, G., Tanaka, K. and Fujii, S. (1989) cDNA cloning of an mRNA encoding surfur-rich 10 kDa prolamin polypeptide in rice seeds. *Plant Molecular Biology*, **12**, 123–30.

Masumura, T., Hibino, T., Kidzu, K., Mitsukawa, N., Tanaka, K. and Fujii, S. (1990) Cloning and characterization of a cDNA encoding a rice 13 kDa prolamin. *Molecular and General Genetics*, **221**, 1–7.

Mena, M., Sanchez-Monge, R., Gomez, L., Salcedo, G. and Carbonero, P. (1992) A major barley allergen associated with baker's asthma disease is a glycosylated monomeric inhibitor of insect α-amylase: cDNA cloning and chromosomal location of the gene. *Plant Molecular Biology*, **20**, 451–8.

Mendez, E., Moreno, A., Colilla, F., Pelaez, F., Limas, G.G., Mendez, R., Soriano, F., Salinas, M. and de Haro, C. (1990) Primary structure and inhibition of protein synthesis in eukaryotic cell-free system of a novel thionin, γ-hordothionin, from barley endosperm. *European Journal of Biochemistry*, **194**, 533–9.

Morita, Y. and Yoshida, C. (1968) Studies on γ-globulin of rice embryo. Part I, Isolation and purification of globulin from rice embryo. *Agricultural and Biological Chemistry*, **32**, 664–70.

Morris, C.F., Greenblatt, G.A., Bettge, A.D. and Malkawi, H.I. (1994) Isolation and characterization of multiple forms of friabilin. *Journal of Cereal Science* **21**, 167–74.

Morrison, W.R., Law, C.N., Wylie, L.J., Coventry, A.M. and Seekings, J. (1989) The effect of group-5 chromosomes on the free polar lipids and breadmaking quality of wheat. *Journal of Cereal Science*, **9**, 41–51.

Morrison, W.R., Greenwell, P., Law, C.N. and Sulaiman, B.D. (1992) Occurrence of friabilin, a low molecular weight protein associated with grain softness, on starch granules isolated from some wheats and related species. *Journal of Cereal Science*, **15**, 143–9.

Mundy, J. and Rogers, J.C. (1986) Selective expression of a probable amylase–protease inhibitor in barley aleurone cells: Comparison to the barley amylase-subtilisin inhibitor. *Planta*, **169**, 51–63.

Mundy, J., Hejgaard, J. and Svendsen, I. (1984) Characterization of a bifunctional wheat inhibitor of endogenous α-amylase and subtilisin. *FEBS Letters*, **167**, 210–14.

Müntz, K., Jung, R. and Saalbach, G. (1993) Synthesis, processing, and targeting of legume seed proteins, in *Seed Storage Compounds, Biosynthesis, Interactions, and Manipulation* (eds P.R. Shewry and K. Stobart), Oxford University Press, Oxford, pp. 128–46.

Ohtsubo, K-I. and Richardson, M. (1992) The amino acid sequence of a 20 kDa bifunctional subtilisin/α-amylase inhibitor from bran of rice (*Oryza sativa* L.) seeds. *FEBS Letters*, **309**, 68–72.

Osborne, T.B. (1924) *The vegetable proteins*, Longmans, Green & Co, London.

Pedersen, K., Devereux, J., Wilson, D.R., Sheldon, E. and Larkins, B.A. (1982) Cloning and sequence analysis reveal structural variation among related zein genes in maize. *Cell*, **29**, 1015–26.

Pederson, K., Argos, P., Naravana, S.V.L. and Larkins, B.A. (1986) Sequence analysis and characterization of a maize gene encoding a high-sulfur zein protein of M_r 15 5000. *The Journal of Biological Chemistry*, **261**, 6279–84.

Peterson, D.M. (1978) Subunit structure and composition of oat seed globulin. *Plant Physiology*, **62**, 506–9.

Prat, S., Perez-Grau, L. and Puigomenech, P. (1987) Multiple variability in the sequence of a family of maize endosperm proteins. *Gene*, **52**, 41–9.

Quatrano, R.S., Litts, J., Colwell, G., Chakerian, R. and Hopkins, R. (1986) Regulation of gene expression in wheat embryos by ABA. Characterization of cDNA clones for the EM and putative globulin proteins and localization of the lectin wheat germ agglutinin, in *Molecular biology of seed storage proteins and lectins* (eds L.M. Shannon and M.J. Chrispeels), American Society of Plant Phytologists, pp. 127–36.

Rahman, S., Jolly, C.J., Skerritt, J.H. and Wallosheck, A. (1994) Cloning of a wheat 15-kDa grain softness protein (GSP). GSP is a mixture of puroindoline-like polypeptides. *European Journal of Biochemistry*, **223**, 917–25.

Richardson, M. (1991) Seed storage proteins: The enzyme inhibitors, in *Methods in Plant Biochemistry*, Vol. 5 (ed. L.J. Rogers) Academic Press, London, pp. 259–305.

Rubin, R., Levanony, H. and Galili, G. (1992) Evidence for the presence of two different types of protein bodies in wheat endosperm. *Plant Physiology*, **99**, 718–24.

Sanchez-Monge, R., Gomez, L., Barber, D., Lopez-Otin, C., Armentia, A. and Salcedo, G. (1992) Wheat and barley allergens associated with baker's asthma. *Biochemical Journal*, **281**, 401–5.

Schofield, J.D. and Greenwell, P. (1987) Wheat starch-granule proteins and their technological significance, in *Cereals in a European context*, (ed. I.D. Morton), Ellis Horwood, Chichester, pp. 407–20.

Shewry, P.R. (1993) Barley Seed Proteins, in *Barley: Chemistry and Technology* (eds J. MacGregor and R. Bhatty), American Association of Cereal Chemists, St Paul, MN, pp. 131–97.

Shewry, P.R. and Tatham, A.S. (1990) The prolamin storage proteins of cereal seeds: structure and evolution. *Biochemical Journal*, **267**, 1–12.

Shewry, P.R., Tatham, A.S., Forde, J., Kreis, M. and Miflin, B.J. (1986) The classification and nomenclature of wheat gluten proteins: a reassessment. *Journal of Cereal Science*, **4**, 97–106.

Shewry, P.R., Parmar, S., Buxton, B., Gale, M.D., Liu, C.J., Hejgaard, J. and Kreis, M. (1988) Multiple molecular forms of α-amylase in seeds and vegetative tissues of barley. *Planta*, **176**, 127–34.

Shewry, P.R., Parmar, S. and Field, J.M. (1988) Two-dimensional electrophoresis of cereal prolamins: Applications to biochemical and genetic analyses. *Electrophoresis*, **9**, 727–37.

Shewry, P.R., Halford, N.G. and Tatham, A.S. (1989) The high molecular weight subunits of wheat, barley and rye: genetics, molecular biology, chemistry and role in wheat gluten structure and functionality, in *Oxford Surveys of Plant Molecular and Cell Biology*, Vol. 6. (ed. B.J. Miflin), Oxford University Press, Oxford, pp. 163–219.

Shewry, P.R., Halford, N.G. and Tatham, A.S. (1992) The high molecular weight subunits of wheat glutenin. *Journal of Cereal Science*, **15**, 105–20.

Shyur, L-F., Wen, T-N. and Chen, C-S. (1992) cDNA cloning and gene expression of the major prolamins of rice. *Plant Molecular Biology*, **20**, 323–26.

Singh, N.K. and Shepherd, K.W. (1985) The structure and genetic control of a new class of disulphide-linked proteins in wheat endosperm. *Theoretical and Applied Genetics*, **71**, 79–92.

Singh, N.K., Shepherd, K.W., Langridge, P., Gruen, L.C., Skerritt, J.H. and Wrigley, C.W. (1988) Identification of legumin-like proteins in wheat. *Plant Molecular Biology*, **11**, 633–9.

Singh, N.K., Donovan, G.R., Carpenter, H.C., Skerritt, J.H. and Langridge, P. (1993) Isolation and characterization of wheat triticin cDNA revealing a unique lysine-rich repetitive domain. *Plant Molecular Biology*, **22**, 227–37.

Svendsen, I.B., Hejgaard, J. and Mundy, J. (1986) Complete amino acid sequence of the α-amylase/subtilisin inhibitor from barley. *Carlsberg Research Communications*, **51**, 43–50.

Terras, F.R., Goderis, I.J., Van Leuven, F., Vanderleyden, J., Cammue, B.P.A. and Broekaert, W.F. (1992) In vitro antifungal activity of a radish (*Raphanus sativus L.*) seed protein homologous to nonspecific lipid transfer proteins. *Plant Physiology*, **100**, 1055–58.

Wallace, N.H. and Kriz, A.L. (1991) Nucleotide sequence of a cDNA clone corresponding to the maize globulin-2 gene. *Plant Physiology*, **95**, 973–5.

Wilson, C.M., Shewry, P.R. and Miflin, B.J. (1981) Maize endosperm proteins compared by sodium dodecyl sulphate gel electrophoresis and isoelectric focusing. *Cereal Chemistry*, **58**, 275–81.

Yupsanis, T., Burgess, S.R., Jackson, P.J. and Shewry, P.R. (1990) Characterisation of the major protein component from aleurone cells of barley (*Hordeum vulgare* L.) *Journal of Experimental Botany*, **41**, 385–92.

Zawistowska, U., Langstaff, J. and Bushuk, W. (1988) Improving effect of a natural α-amylase inhibitor on the baking quality of wheat flour containing malted barley flour. *Journal of Cereal Science*, **8**, 207–9.

9

Cereal grain carbohydrates

B.A. Stone

9.1 INTRODUCTION

Carbohydrates are quantitatively the major components of cereal grains (Table 9.1). They comprise the high molecular weight reserve polysaccharides found in the starch granules and the structural polysaccharides in the walls of cells of various parts of the grain. Together these polysaccharides may account for ~85% by weight of the whole grain. In addition there are up to ~3% of low molecular weight carbohydrates, amongst which sucrose is the predominant compound.

The functional roles of grain carbohydrates in the physiology of the plant, their behaviour in cereal technology and in the digestive tracts of monogastric animals are intimately associated with the structure and organization of the parts of the cereal grains from which they are derived. It is appropriate, therefore, before discussing the individual carbohydrate components to give a short account of the structure of cereal grains.

9.2 STRUCTURE OF CEREAL GRAINS

The cereal grain, botanically the caryopsis, is enveloped in a hull (or husk) (Figure 9.1). The hull consists of the remains of modified leaves (pales), composed of empty cells with lignified secondary walls and is sometimes cemented, as in rice, to the caryopsis. In naked (hull-less) barleys the hull is dislodged on milling. The grain itself (Figure 9.2a) consists of a fruit coat (pericarp), seed-coat (testa) and the seed. The tissues of the parentally-derived, pericarp-seed coat are multilayered structures with walls that are often lignified or cutinized. The pericarp and seed coat are joined by a cementing layer. At grain maturity, the cells of these tissues are empty, and in some layers collapsed, so that the pericarp-seed coat is essentially a composite of cell walls with different

Table 9.1 Carbohydrate composition of cereal grains (% of dry weight)

	Monosaccharides	*Disaccharides*		*Oligosaccharides*		*Polysaccharides*							
	Glucose	*Fructose*	*Sucrose*	*Maltose*	*Raffinose*	*Glucodi-fructose*	*Fructans*	*Starch*	*Cellulose*	*Heteroxylan*	*(1→3,1→4)-β-D-Glucan*	*Total carbo-hydrate*	*References*
Barley (*Hordeum vulgare*)	0.03–0.06 0.02–0.11	0.03–0.16 0.03–0.2	0.34–2 0.74–1.9	0.006–0.14 0.1	0.14–0.83 0.16–0.56	+	0.019–0.97 0.3–0.78	51.5–72.1 63–65	1.44–5 4–5	4.4–7.8	3.64–6.11	78–83.9 73–83	Henry, 1988; MacGregor and Fincher, 1993
Wheat (*Triticum aestivum*)	0.03–0.09	0.06–0.08	0.54–1.55	0.05–0.18	0.19–0.68	0.26–0.41	0.94–1.14	63.2	2.71	7.62	1		Lineback and Rasper, 1988; Pomeranz, 1988; Beresford and Stone, 1983; Mares and Stone, 1973

Oats (*Avena sativa*)	0.05	0.09	0.64	tr	0.19	0.04	0.09	43–61	?	?	2.5–6.6	?	Macleod and Preece, 1954; MacArthur-Grant, 1986; Wood, 1986
Rye (*Secale cereale*)	0.08	0.10	1.86	tr	0.42	0.75	3.94	?	?	?	?	?	MacLeod and Preece, 1954
Maize (*Zea mays*)	0.05	0.05	0.78	tr	0.18	0	0	?	?	?	?	?	MacLeod and Preece, 1954
Sorghum (*Sorghum bicolor*)	0.09–0.19	0.09–0.21	0.85–1.50 0.61–0.99	0.02	0.11–0.12 0.15–0.42	0.05–0.12 (stachyose)	?	60.9–75.9	?	2.5–4.5	?	?	Hirata and Watson, 1967; Wall and Blesin, 1970; Karim and Rooney, 1972; Murty *et al.*, 1985

tr = trace
? = no figures available

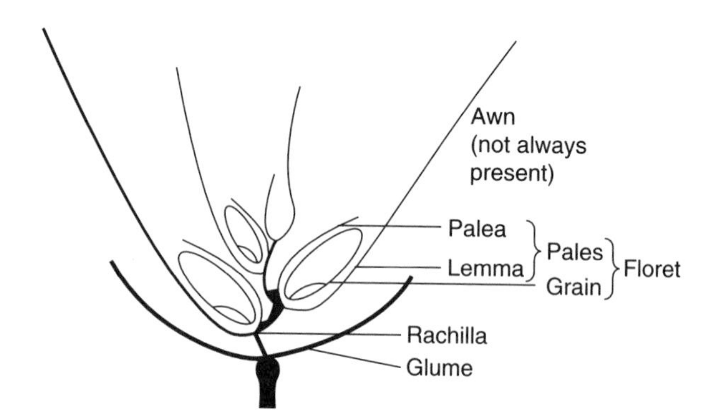

Figure 9.1 Grains of wheat on the spikelet. At maturity the interlocked palea and lemma, which are modified leaves, enclose the grain but are not adherent to it. (Reprinted, with permission, from Evers and Bechtel, 1988.)

compositions and mechanical properties. There is a close physical association between the inner seed coat surface and the outer surface of the seed but there is no chemical bonding.

The tissues of seeds are functionally concerned with the establishment and nurturing of the next generation of the plant. The embryo (Figure 9.2a) contains the primordial shoot and root tissues enveloped in the scutellum. The cells of these tissues are living and their walls are thin. The scutellar epithelium which interfaces with the endosperm is a specialized tissue that initially has a secretory and then an absorptive function during seed germination. The bulk of the seed consists of triploid endosperm tissue. The endosperm in cereals consists of two cell types, the starchy endosperm and the aleurone (Figure 9.2b). The cells of the starchy endosperm are filled with starch granules embedded in a matrix of protein. In most grains, endosperm cells have relatively thin walls but in some, e.g. oats, the walls of the sub-aleurone endosperm are thickened (Figure 9.3). The aleurone is a tissue of one or more cells thick, depending on the species, covering the surface of the starchy endosperm. The aleurone is differentiated from the starchy endosperm during development of the seed. It is characterized by its thick, bilayered, unlignified walls and by its cell contents which include a prominent nucleus, protein bodies, with phytic acid and (in some cereals) niacytin inclusions, and lipid deposits.

9.3 LOW MOLECULAR WEIGHT CARBOHYDRATES: MONO-, DI- AND OLIGOSACCHARIDES

The reducing aldohexose, D-glucose (Figure 9.4a) and the isomeric ketohexose, D-fructose (Figure 9.4b) are minor components of the low

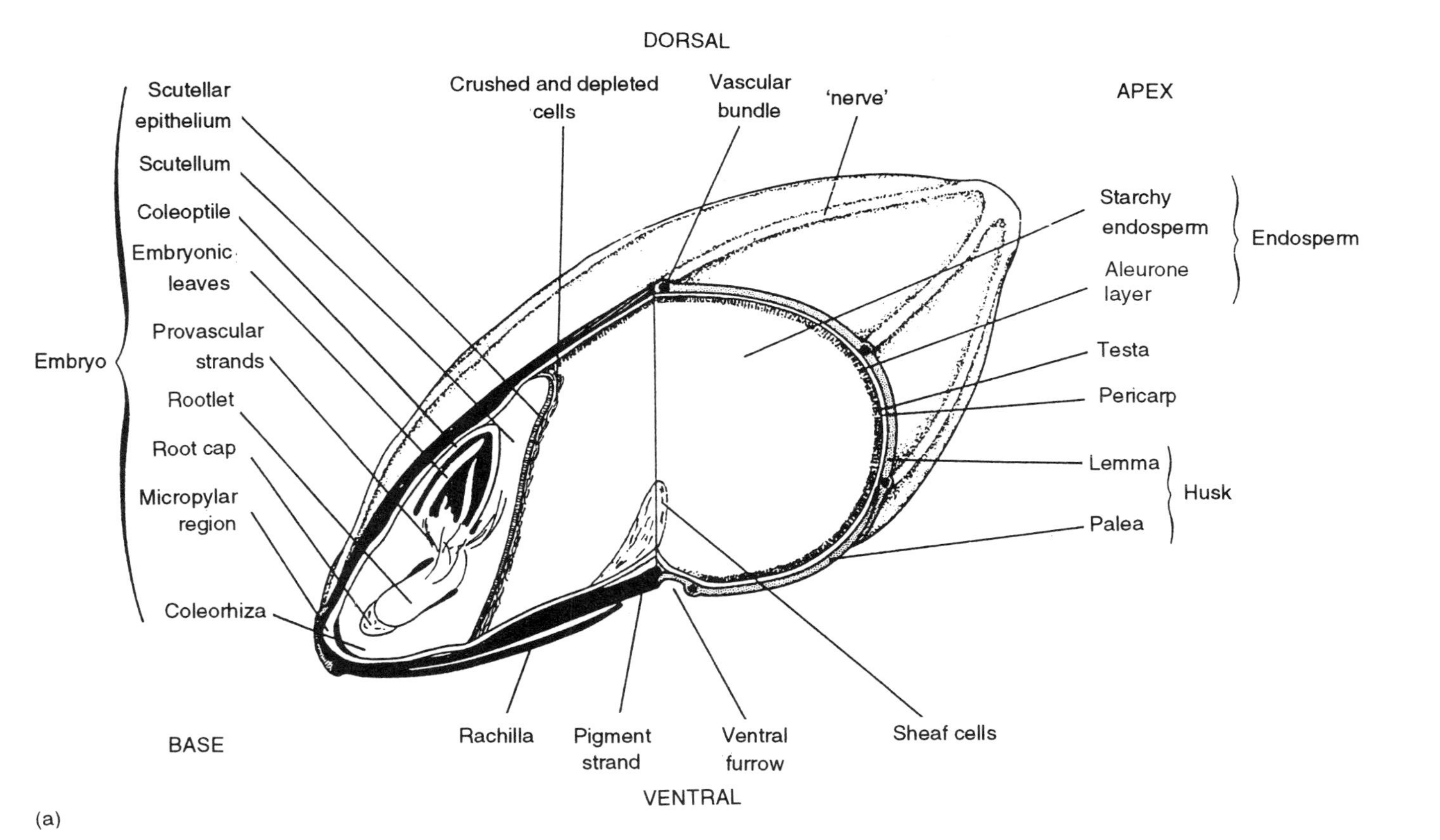

Figure 9.2 (a) Barley grain showing component tissues. (Reprinted, with permission, from Briggs, 1978.)

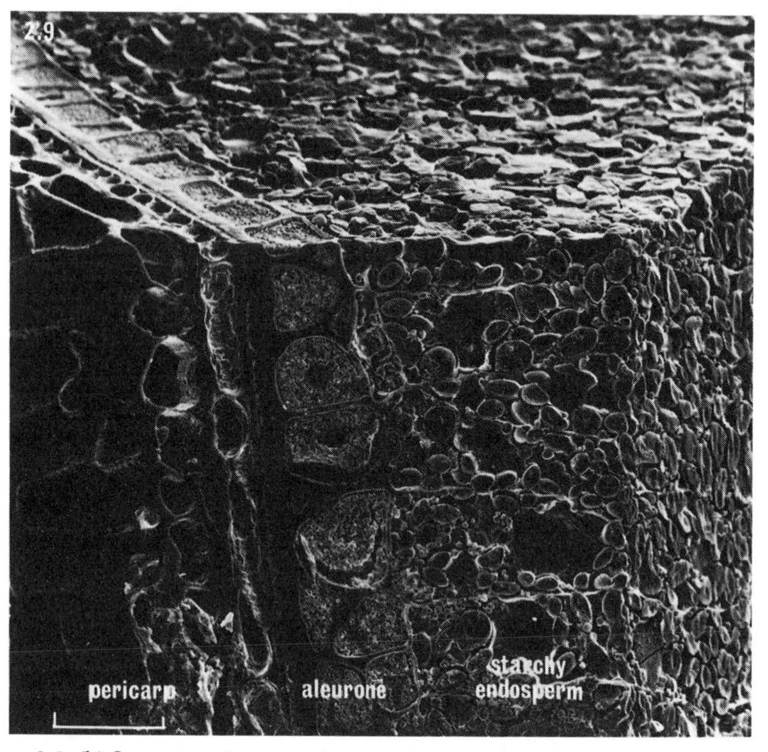

Figure 9.2 (b) Scanning electron micrograph of a portion of the outer surface of a wheat grain, showing starchy endosperm, aleurone, and overlying nucellar remnants, seed coat and inner pericarp. The outer pericarp has been lost in preparation. Preparation and electron micrograph by Susan Joyner. (Reprinted, with permission, from Fincher and Stone, 1986.)

molecular weight carbohydrate fraction. Sucrose (Figure 9.4c), a non-reducing disaccharide, composed of *β*D-glucopyranosyl and α-D-glucopyranosyl and *β*-D-fructofuranosyl residues, is the most prominent component. Its (1→6)-α-D-galactosyl derivative, raffinose (Figure 9.4d), and two *β*-D-fructosyl derivatives, kestose (Figure 9.4e) and isokestose (Figure 9.4f), are found in small amounts in some cereal grains. The reducing disaccharides maltose (Figure 9.4g) and melibiose (Figure 9.4h) may also be present in minor amounts. Together with the polymeric fructans (see Section 9.5.2) these mono- and oligo-saccharides constitute the water extractable and 80% (w/v) ethanol extractable carbohydrates of grain. The low molecular weight carbohydrates are part of the material translocated from photosynthetic or storage tissues to the developing grain. The amount of the low molecular weight carbohydrates in the grain decreases as the grain matures and at maturity reaches ~3% of its weight. Information available on the composition of low molecular weight carbohydrates in cereal grains is summarized in Table 9.1. The highest

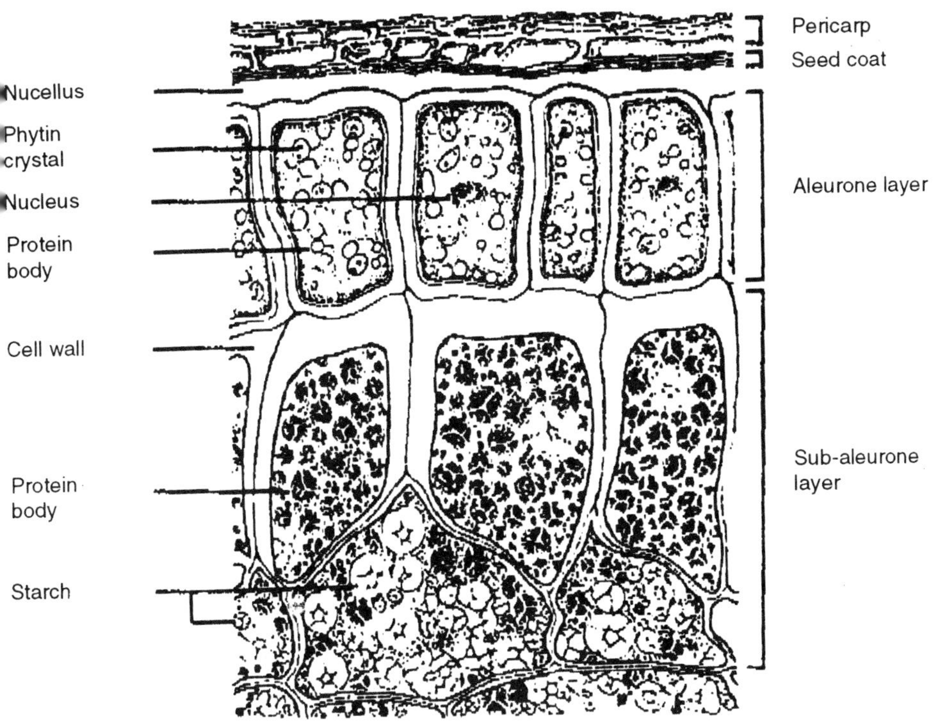

Figure 9.3 Outer layers of an oat grain showing thick-walled, sub-aleurone endosperm cells overlying cells with thinner walls. (Reprinted, with permission, from Fulcher, 1986.)

concentrations are found in the tissues of the embryo (wheat 20%; maize, 11%) but they are virtually absent from the mature pericarp-seed coat tissues.

9.4 HIGH MOLECULAR WEIGHT CARBOHYDRATES: POLYSACCHARIDES

9.4.1 Structural organization of polysaccharides

The polysaccharides of cereal grains function either as structural components of cell walls or as storage polymers. They exhibit structural patterns that can be disposed into one of four different categories. These are illustrated in Figure 9.5. The structural patterns of the cereal polysaccharides relate directly to their specific functions in the grain. Thus the essentially linear polysaccharides in the first three groups are characteristic of structural elements of cell walls whereas the branch-on-branch polymers of the fourth group are characteristic of storage polysaccharides, although exceptions are known. The organization and function of walls of cereal grains are outlined in the following sections.

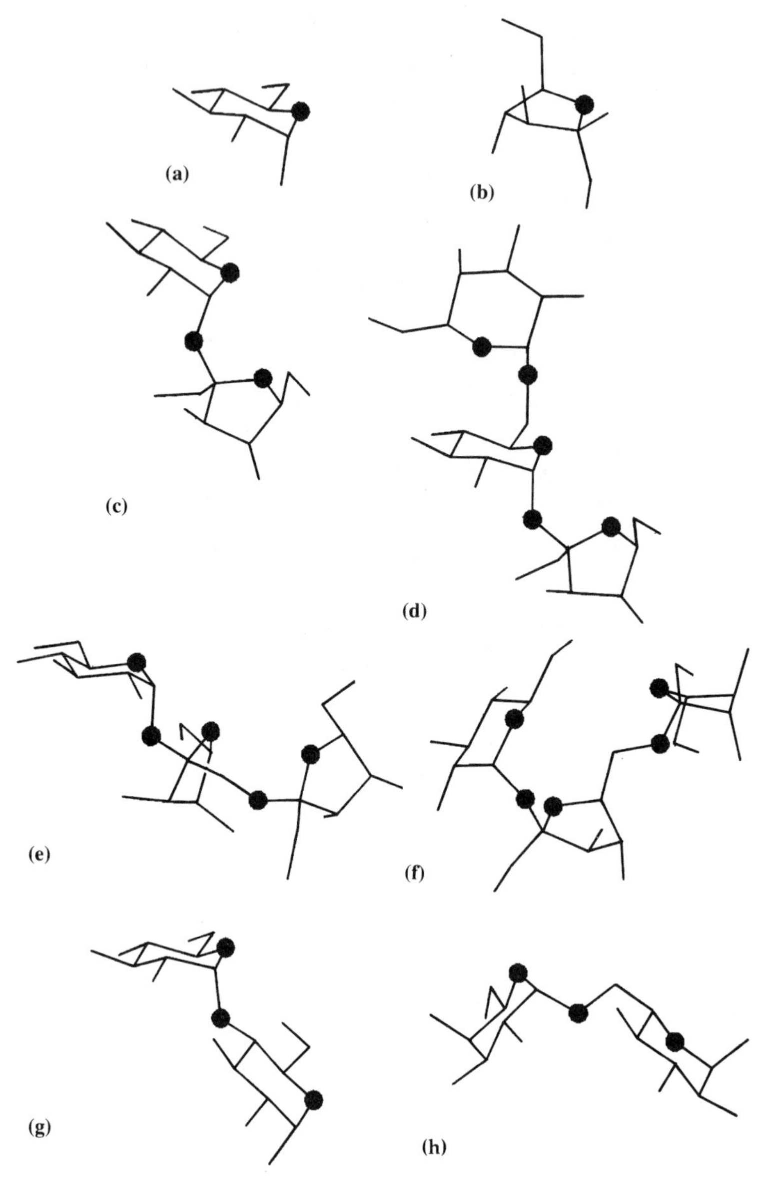

Figure 9.4 Structures of mono-, di- and oligosaccharides found in cereal grains: (a) α-D-glucopyranose, (b) *β*-D-fructofuranose, (c) sucrose (α-D-glucopyranosyl(1→2)-O-*β*-D-fructofuranoside), (d) raffinose (α-D-galactopyranosyl (1→6)-O-α-D-glucopyranosyl-O-*β*-D-fructofuranoside, (e) 1-kestose (*β*-D-fructofuranosyl(2→1)-O-*β*-D-glucopyranoside, (f) 6-kestose (*β*-D-fructofuranosyl (2→6)-O-*β*-D-fructofuranosyl(2→1)-O-α-D-glucopyranoside), (g) α-maltose (α-D-glucopyranosyl(1→4)-O-α-D-glucopyranoside), (h) α-melibiose (α-D-galactopyranosyl (1→6)-O-α-D-glucopyranoside).

Linear homopolymers (Type I)

One monosaccharide unit
One linkage position

Linear heteropolymers (Type II)

● and ◎ , Different monosaccharides.
In some examples the heterogeneity is in the linkage position

Side-chain branched heteropolymers (Type III)

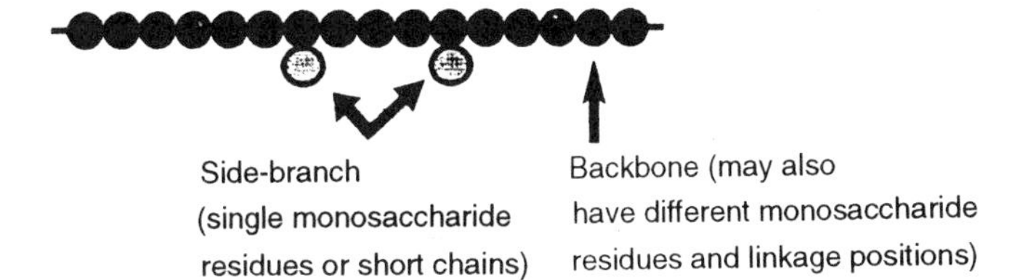

Branch-on-branch homo- or hetero-polymers (Type IV)

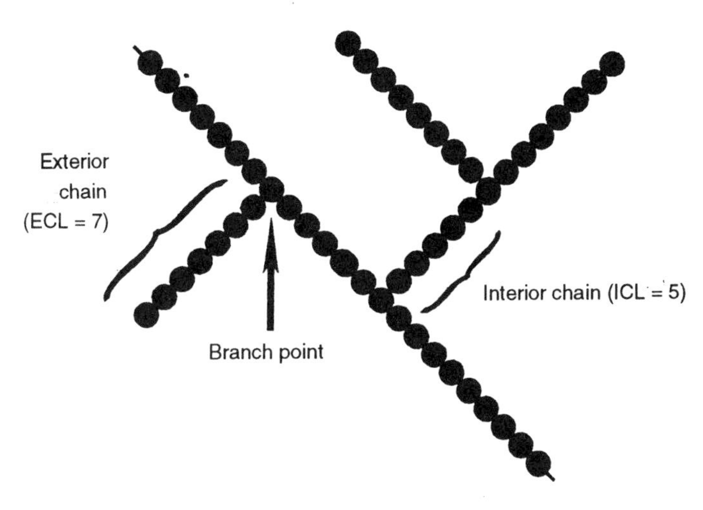

Figure 9.5 Basic organization of monosaccharide residues in polysaccharides.

9.4.2 Cell wall organization and function

Cell walls are extracellular structures overlying the plasma membrane of all plant cells. They persist even when the cell is physiologically dead. Cell walls of plants, whatever their origin, are constructed on a similar pattern. They can be considered to be reinforced gels. The gel or matrix phase, is composed of heteropolysaccharides belonging to Types II and III (Figure 9.5) that can associate with one another over some portion of their surface to form junction zones (Figure 9.6). The gel is reinforced by microfibrils which are rod-like assemblies of Type I homopolysaccharides (Figure 9.5).

A structural model of a simple primary wall incorporating these features is shown in Figure 9.7 (Bacic *et al.*, 1988; Carpita and Gibeaut, 1993; McCann and Roberts, 1991). In the wall the cellulosic microfibrils are embedded in a matrix of non-cellulosic heteropolysaccharides, proteins and glycoproteins. Some of the non-cellulosic polysaccharides (arabinoxylans, xyloglucans, glucomannans) are associated with the surfaces of the cellulosic microfibrils and may form bridges between adjacent microfibrils (Figure 9.7). The matrix phase in unlignified walls is gel-like and hydrated. This organization is typical of walls of the starchy endosperm, embryonic tissues and the scutellum. The polysaccharides in the unlignified walls of the starchy endosperm and the thick (primary) outer layer of the aleurone wall are readily accessible to

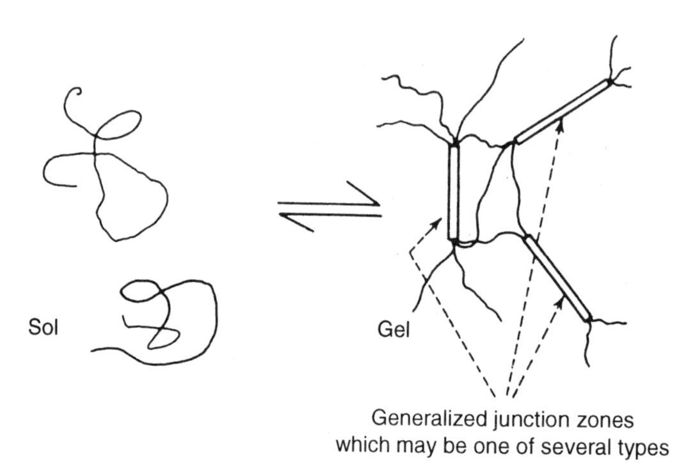

Figure 9.6 Conversion of a polysaccharide random coil (sol) to a network (gel) which is cross-linked by the conformational ordering of chain segments and their consequent association. The ordered entities (junction zones) are shown as rectangles. (Reproduced, with permission, from Rees, 1975, 1979.)

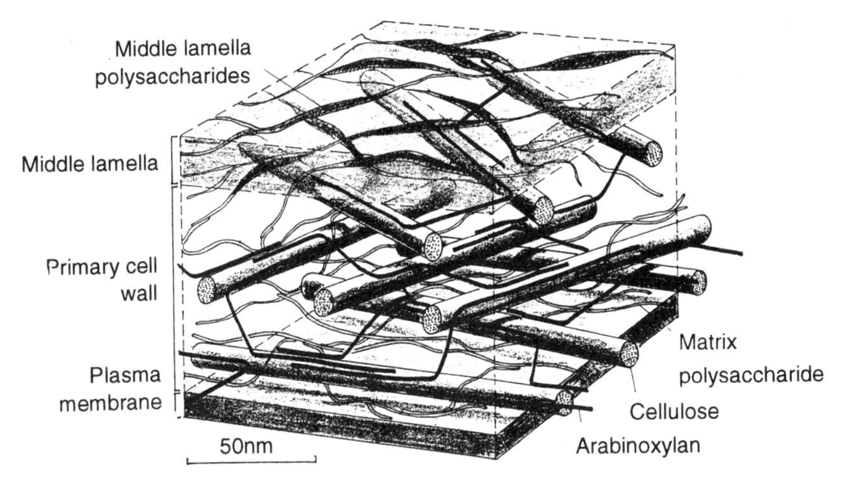

Figure 9.7 Organization of polymeric components in cell walls. A simplified, schematic representation of the spatial arrangement of polymers in a primary cell wall, for example, from a starchy endosperm cell. Note the cellulosic microfibrils are embedded in a network of non-cellullosic polysaccharides (arabinoxylans and (1→3, 1→4)-β-D-glucans) and protein. The non-cellulosic polysaccharides can form associations with surfaces of several microfibrils and provide cohesive interactions. The native primary wall will contain ~60% water but in walls of some cells (e.g. those of the pericarp-seed coat) the water is replaced by lignin, a phenylpropanoid polymer which overlies and encrusts the microfibrils and non-cellulosic polysaccharides and may be covalently bonded to the latter. (After McCann and Roberts, 1991, with permission.)

depolymerizing enzymes secreted from the aleurone during germination (see Section 9.7.2) although the thin, inner (secondary) aleurone wall layer remains intact during most of the germination period.

The walls of the tissues which comprise the pericarp-seed coat in many cases also have secondary walls in which the water in the matrix of the developing wall has been replaced by deposits of lignin or suberin which encrusts the microfibrillar and matrix polysaccharides and, in some cases, is covalently bound to the matrix polysaccharides. These walls are impermeable to water and the polysaccharides are relatively inaccessible to the approach of depolymerizing enzymes and this enhances the protective function of the pericarp-seed coat tissues. The outer surfaces of walls of epidermal cells of the palea and lemma, that make up the hull, are covered by a layer of hydrophobic cutin which is intimately associated with the underlying wall polysaccharides.

Walls have several functions in the plant depending on the cell's age and tissue location. Most importantly they serve as structural elements in the tissue or organ, imparting shape, strength and resilience to the tissue and, through their surface-to-surface interactions, contribute to the adhesion between individual cells. Walls envelop and protect the turgid protoplast and physically prevent direct mechanical damage to its surface and prevent its bursting due to the hypertonicity of the protoplast. In primary walls the water-holding capacity of the wall material prevents dehydration of the protoplasts. Walls also have a transport and communication function. The porous nature of some walls, e.g. those of the embyro and endosperm paranchyma cells, allows extracellular (apoplastic) movement of water and dissolved nutrients, phytohormones, morphogenetic and other signalling molecules between cells. The size of their pores limits the extent of molecular exchange between the protoplasts and their surroundings. Walls are also involved in metabolic events. Thus in the scutellar epithelium, the walls carry enzymes involved in pre-absorptive processing of nutrients and in post-secretory modification of wall components during wall formation. During germination the degraded components of walls of the starchy endosperm and aleurone cells serve as sources of nutrients for the developing embryo. During cell growth the walls are plastic, allowing the protoplast to expand, but they are irreversibly bonded at the end of wall growth.

The chemical and physical properties of polysaccharides found in walls of cells in cereal grain tissues are now considered in turn.

9.4.3. Polysaccharides of the cell wall

Cellulose ((1→4)-β-D-glucan)

Cellulose (Marchessault and Sundararajan, 1983) is a ubiquitous component of the walls of vascular (flowering) plants. It is a linear homopolymer (Type I, Figure 9.5) composed solely of β-D-glucopyranosyl residues, all (1→4)-linked (Figure 9.8a). The individual chains may have a degree of polymerization of up to ~15 000. The substituent hydroxyl groups on the glucosyl monomers are co-planar (equatorial) with the pyranose ring and each successive residue is rotated 180°C with respect to its neighbour. The molecular chains have a ribbon-like conformation which allows parallel packing of neighbouring chains into three-dimensional aggregates held together by intra- and inter-molecular hydrogen bonds (Figure 9.8b). These aggregates, called microfibrils, may reach diameters of ~25 nm and consist of some 100 cellulose molecules (Figure 9.8c). They are clearly visible in electron micrographs. Within the microfibrils the individual cellulose molecules may be packed into highly ordered (crystalline) arrays, which give sharp X-ray diffraction patterns, although there may be regions where the arrangement is less orderly. In particular, the surface chains may be disordered

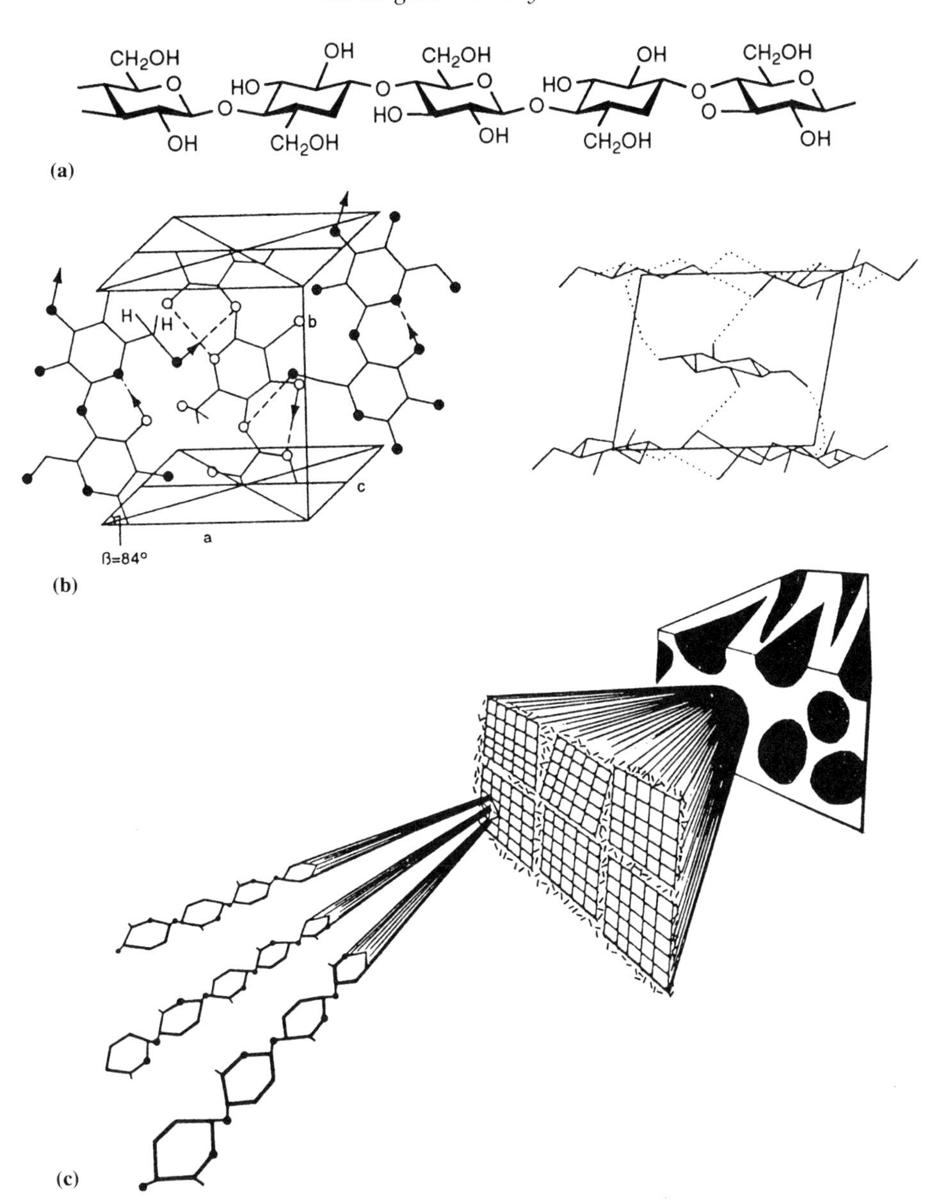

Figure 9.8 (a) Portion of a cellulose molecule. (b) Diagrams showing packing of cellulose molecules in a crystalline form. Hydrogen bonding occurs both within chains and between chains. (c) Schematic illustration of the structural continuity from cellulose molecule to microfibrils. The microfibril is a single crystal-like entity encompassing about 100 cellulose molecules. The surface chains may be disordered and interact with water and matrix polysaccharide molecules. The microfibrils are shown embedded in the matrix of polysaccharide (and lignin). (b and c, reprinted with permission from Marchessault and Sundararajan, 1983.)

and associate with water and matrix polysaccharides (see Section 9.4.2). Cellulose is quite insoluble in water but may be brought into solution by certain hydrogen-bond-breaking reagents and sodium hydroxide solutions at high concentrations.

(1→3, 1→4-β-D-Glucans (cereal β-glucans)

The (1→3,1→4)-β-D-glucans (Stone and Clarke, 1993; MacGregor and Fincher, 1993; Wood *et al.*, 1994) are components of walls of cells in both vegetative and seed tissues in grasses and cereals. They are linear polymers (Type II, Figure 9.9a) in which the β-D-glucopyranosyl monomers are joined by both (1→4)- and (1→3)-inter-unit linkages, usually in the ratio ~2:1. The glucan may be considered to be a (1→3)-β-linked copolymer of cellotriosyl residues ($G4G4G_R$), cellotetraosyl ($G4G4G4G_R$) residues and longer (1→4)-β-D-oligoglucosides (Figure 9.9b). The glucan is polydisperse with respect to molecular weight. Average molecular weights of 1.13×10^6 (rye), 2.14×10^6 (barley) and 3.00×10^6 (oats) have been recorded (Wood *et al.*, 1994). In solution the glucans have an extended, reptate (snake-like) conformation as a consequence of the interrupting effect of the (1→3)-linkages on the extended ribbon-like cellulosic stretches (Figure 9.9c). The (1→3,1→4)-β-D-glucans from different species can be characterized by the ratio of $G4G3G_R$:$G4G4G3G_R$ oligoglucosides released by (1→3,1→4)-β-D-glucan endohydrolase (EC 3.2.1.73) action (Figure 9.9b) [oats, 2.1–2.4; barley and rye, 2.8–3.3 (Wood *et al.*, 1994)]. The concentration of (1→3,1→4)-β-D-glucan in walls of cereal grains depends on the cultivar, variety, season and growth site. Barley, oat and rye grains are rich sources of (1→3,1→4)-β-D-glucan.

Part of the (1→3,1→4)-β-D-glucan can be extracted from the walls of starchy endosperm cells of barley, oats and rye by water at 40°C and further fractions are extracted at higher temperatures. Aqueous solutions of these endosperm (1→3,1→4)-β-D-glucans are quite viscous; their intrinsic viscosities are dependent on cultivar, species and solvent (barley, 18–24.7 dl g^{-1}; oats, 80 dl g^{-1} in water) (Aastrup, 1979) and for oat (1→3,1→4)-β-D-glucan in dimethylsulfoxide and 7 M urea, 17–18 dl g^{-1} (Wood *et al.*, 1978). On cooling the glucan slowly precipitates from aqueous solutions. The (1→3,1→4)-β-D-glucans in the endosperm walls of wheat and walls of wheat and barley aleurone cells are not extracted by water at 40°C but dilute alkali is effective.

(1→3,1→4)-β-D-glucans as well as other β-glucans, e.g. cellulose and callose, on complexing with the stilbene fluorochrome, Calcofluor, induce a bright blue UV-fluorescence which is useful in determining their quantitative distribution in cereal grains (Fulcher *et al.*, 1994) and in

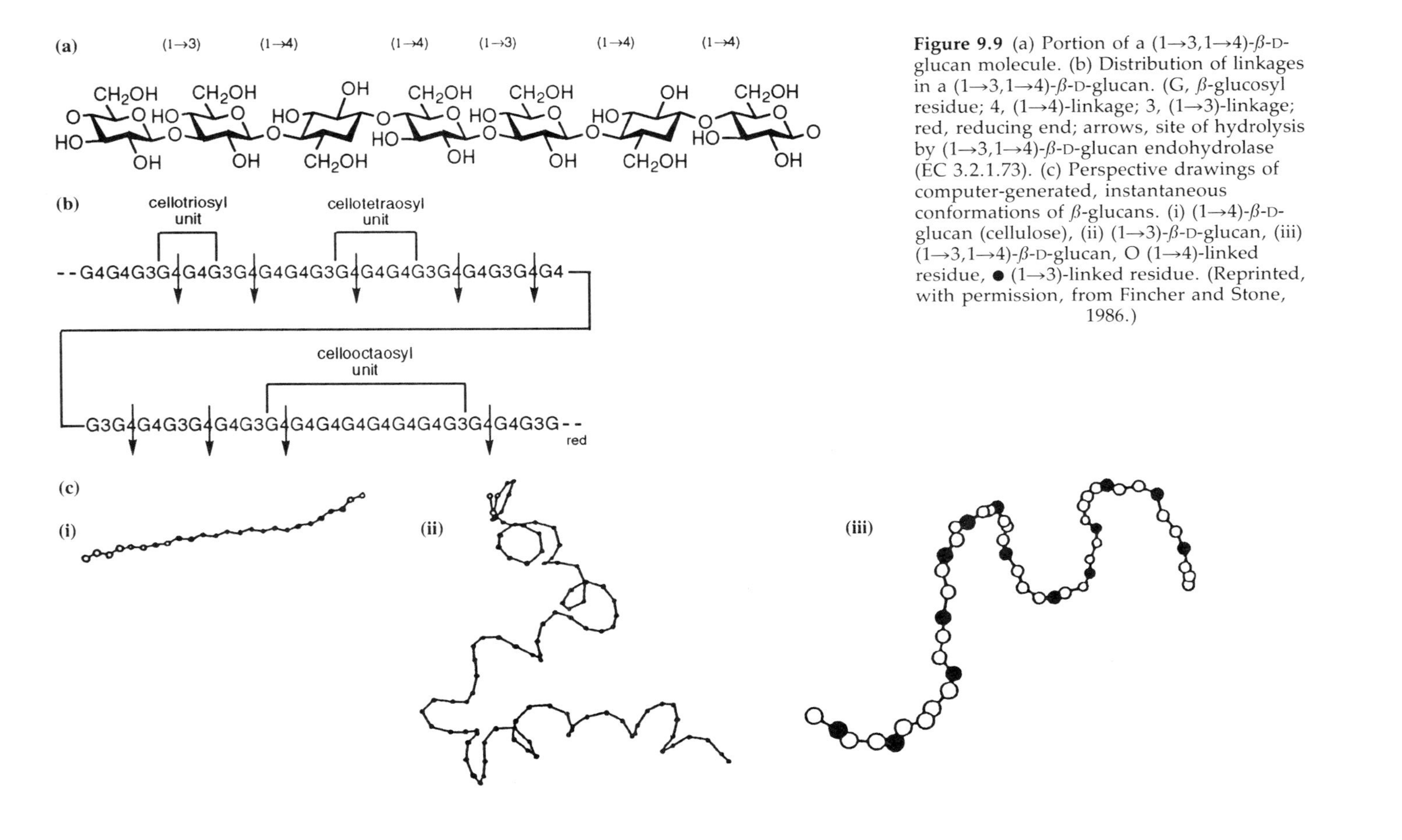

Figure 9.9 (a) Portion of a (1→3,1→4)-β-D-glucan molecule. (b) Distribution of linkages in a (1→3,1→4)-β-D-glucan. (G, β-glucosyl residue; 4, (1→4)-linkage; 3, (1→3)-linkage; red, reducing end; arrows, site of hydrolysis by (1→3,1→4)-β-D-glucan endohydrolase (EC 3.2.1.73). (c) Perspective drawings of computer-generated, instantaneous conformations of β-glucans. (i) (1→4)-β-D-glucan (cellulose), (ii) (1→3)-β-D-glucan, (iii) (1→3,1→4)-β-D-glucan, O (1→4)-linked residue, ● (1→3)-linked residue. (Reprinted, with permission, from Fincher and Stone, 1986.)

quantifying soluble (1→3,1→4)-β-D-glucan in grain extracts and products (Sendra, *et al.*, 1989).

(1→4)-β-D-Glucomannans

These linear heteropolymers (Type II, Figure 9.5) are copolymers of β-glucopyranose (~30%) and its 2-epimer β-D-mannopyranose (~70%) (Figure 9.10a) joined by (1→4)-linkages. In some examples the backbone

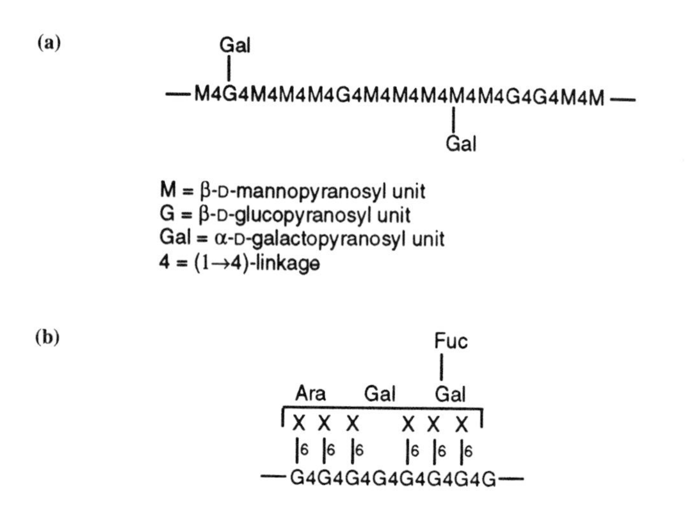

G = β-D-glucopyranosyl unit
X = α-D-xylopyranosyl unit
4 = (1→4)-linkage

Gal = β-D-galactopyranosyl unit
Ara = α-L-arabinopyranosyl unit
Fuc = α-L-fucopyranosyl unit

(c)

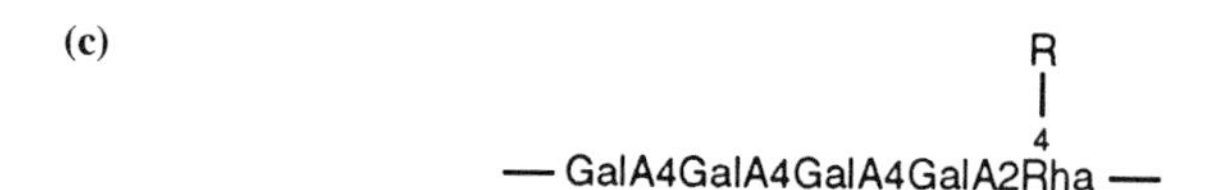

GalA = α-D-galacturopyranosyl unit
Rha = α-D-rhamnopyranosyl unit
R = oligosaccharides composed of (1→5)-L-arabinofuranosyl and/or (1→4)-D-galactopyranosyl units and other monosaccharide units
4 = (1→4)-linkage

Figure 9.10 (a) Glucomannan backbone substituted at C(O)6 with single α-D-galactopyranosyl residues ((galacto)-glucomannan). (b) Heptasaccharide units of xyloglucans consisting of cellotetraosyl residues substituted through C(O)6 by single α-D-xylopyranosyl residues. Side-chain substituents are shown. (c) Backbone of a rhamnogalacturonan showing sites of substitution.

is substituted with α-D-galactopyranose residues and esterified with acetyl residues. Their degree of polymerization range from <100 to several thousand and their conformation is similar to cellulose, thus glucomannan chains may associate strongly with surfaces of cellulosic microfibrils (see Section 9.4.2). Glucomannans are minor constituents of walls of cereal aleurone and starchy endosperm cells but in certain *indica* rice cultivars the endosperm walls contain up to 17% glucomannan (Zamorski and Shibuya, 1990).

Xylo-(1→4)-β-D-glucans (heteroglucans)

Xylo-(1→4)-β-D-glucans (heteroglucans) (Fry, 1989; Hayashi, 1989) are a family of side-chain-branched heteroglycans (Type III, Figure 9.5) in which a cellulose-like (1→4)-β-D-glucan backbone is substituted regularly by α-D-xylopyranosyl residues at C(O)6 of the glucosyl residues (Figure 9.10b). The xylosyl residues may be further substituted by other monosaccharide residues (α-L-arabinofuranosyl, β-D-galactopyranosyl and α-L-fucopyranosyl) depending on the species. Cell wall matrix xyloglucans have DPs of 600–700, are insoluble in water and readily associate with surfaces of cellulose microfibrils (see Section 9.4.2). They are of minor significance in walls of cereal grain tissues.

Arabino-(1→4)-β-D-xylans (heteroxylans)

These side-chain branched heteroglycans (Type III, Figure 9.5), have a (1→4)-β-D-xylan backbone that is homomorphous with cellulose. The

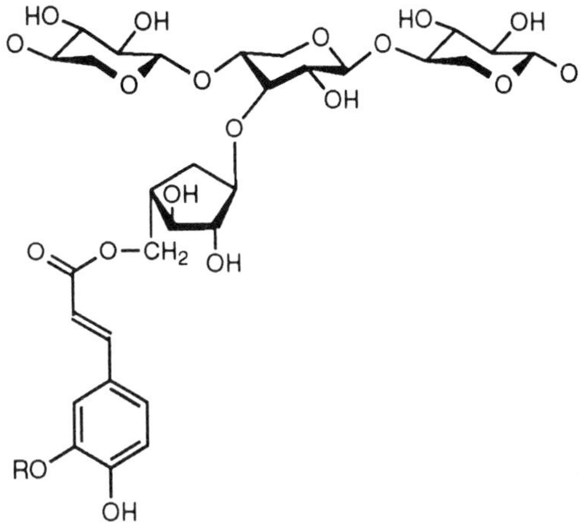

Figure 9.11 Portion of a (1→4)-β-D-xylan chain substituted at C(O)3 of one residue with 5-O-feruloyl/*p*-coumaryl-α-L-arabinofuranose (R = H, *p*-coumaric acid; R = OCH_3, ferulic acid). (Reprinted, with permission, from Fincher and Stone, 1986.)

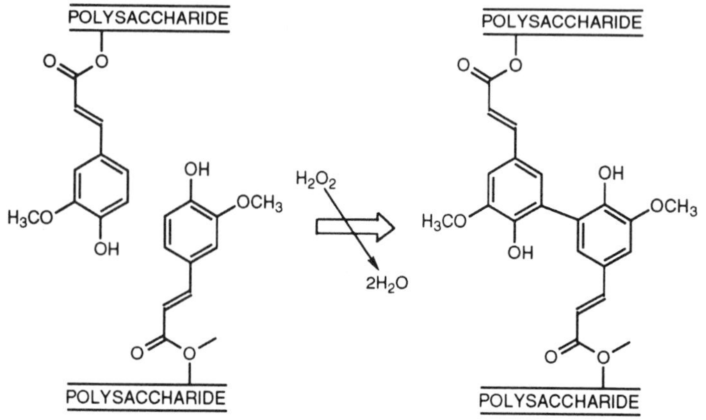

Figure 9.12 Cross-linking of arabinoxylan chains through a dehydrodiferulic acid diester. (A range of isomeric dehyrodiferulic acids has been found in walls of vegetative tissues in grasses including the 5–5′ coupled dehydrodimer shown, Ralph *et al.*, 1994.)

xylan is substituted predominantly with single L-arabinofuranosyl (Ara*f*) residues at either C(O)2 and/or C(O)3 of the xylopyranosyl (Xyl*p*) residues (Figure 9.11). The frequency of Ara*f* substitution depends on the wall type and the species (Xyl*p*:Ara*f* 0.89–13). Moreover the distribution of the Ara*f* residues along the xylan chain is not regular (Perlin, 1951; Viëtor *et al.*, 1983). Other substituents are also found on the xylan backbone; these include D-glucuronic acid and its 4-O-methyl ether (2–9%) linked to C(O)2 of Xyl*p* units, as well as various oligomeric chains. The Xyl*p* residues are often acetylated and the Ara*f* residues are commonly esterified with feruloyl (FA) and *p*-coumaroyl (*p*CA) residues (Figure 9.11) at C(O)3 (wheat aleurone walls 1.8% w/w, FA/*p*CA = 9/1; wheat endosperm walls: 0.04% w/w, mainly FA, trace *p*CA; Bacic, 1979).

Part of the arabinoxylan in walls of endosperm cells is soluble in water at 40°C and higher temperatures but alkaline extractants are required to bring all the arabinoxylan into solution. The DP of endosperm arabinoxylans ranges from 100–1200 and their aqueous solutions are somewhat viscous (intrinsic viscosity 6.12 dl/g; Andrewartha *et al.*, 1979). The removal of Ara*f* substituents, either by dilute acid or by α-L-arabinofuranosidase action, leads to aggregation and precipitation of the resulting xylan molecules (Andrewartha *et al.*, 1979). Dehydrogenative dimerization of feruloyl residues on neighbouring arabinoxylan chains, catalysed by peroxidase in the presence of H_2O_2, leads to gelation of arabinoxylan solutions (Neukom and Markwalder, 1978) (Figure 9.12). Dimerization of feruloyl substituents on arabinoxylans in walls may occur, although in walls of wheat aleurone that are rich in FA (1.8%) the amount of FA dimers is very low (0.03% w/w) (Rhodes, 1993). The covalently bound FA is responsible for the bright blue, UV-induced fluorescence of aleurone walls (O'Brien and McCully, 1981).

In walls of vegetative tissues, including hulls of grains, arabinoxylans are also crosslinked by dehydrodiferulic acid bridges (Figure 9.12) (Lam *et al.*, 1990) and in lignified walls the arabinoxylans are crosslinked to lignin, *inter alia*, through ferulic acid and dehydrodiferulic acid ester–ether bridges (Iiyama *et al.*, 1993, 1994). These covalent bridges are important determinants of the cohesiveness of the wall fabric and the resistance of lignified walls to digestion by polysaccharide hydrolases.

Pectic polysaccharides

The pectic polysaccharides (Jarvis, 1984; Bacic *et al.*, 1988) are a diverse group of side-branched heteroglycans. The backbone is usually a copolymer of (1→4)-linked α-galacturonosyl and (1→2)-linked α-L-rhamnofuranosyl residues (Figure 9.10c). This rhamnogalacturonan is variously substituted on the rhamnosyl residues by arabinan, arabino-galactan and more complex branched oligosaccharides. In addition variable amounts of (1→4)-α-linked homogalacturonan may be present. Variable proportions of the galacturonic acid carboxyl groups on these polymers are methyl esterified and some pectins are acetylated. Pectins are minor constituents of walls of vegetative tissues of grasses (Chesson *et al.*, 1995) and occur in low amounts in endosperm walls of some grains, e.g. rice (Shibuya and Nakane, 1984). In the walls they are complexed with Ca^{2+} which promotes gelation by forming Ca^{2+}-bridges between galacturonosyl residues on adjacent chains.

Callose ((1→3)-β-D-glucan)

Callose (1→3)-β-D-glucan) (Stone and Clarke, 1993) occurs as small deposits on the inner surfaces of walls of endosperm cells of grains especially in the sub-aleurone region. The content in barley is around 1% (Bacic and Stone, 1981). Callose deposits are easily recognized by the intense yellow fluorescence induced by the aniline blue fluorochrome (Stone *et al.*, 1984). The deposition of callose is a well-known consequence of stress and wounding. Thus the deposits may arise as a consequence of plasmolysis during the drying of the grain.

Arabinogalactan-peptides

These polysaccharide–peptide molecules belong to a family of arabino-galactan-proteins which are widely distributed in plants (Fincher *et al.*, 1983). The polysaccharide portion consists of a branch-on-branch, β-D-galactan joined by (1→3)- and (1→6)-linkages and substituted by single α-L-arabinofuranosyl residues. In the wheat endosperm arabinogalactan-peptide the polysaccharide is covalently linked to a peptide through hydroxyproline residues (Fincher *et al.*, 1974; McNamara and Stone, 1981; Strahm *et al.*, 1981). The wheat arabinogalactan-peptide is water-soluble and is not precipitated in saturated ammonium sulphate. Its subcellular location and function are unknown.

9.4.4 Composition of walls of cereal grains

Table 9.2 summarizes the information available concerning the polysaccharide composition of walls of aleurone and starchy endosperm cells in cereal grain tissues. The composition of walls of wheat pericarp (beeswing) and cross cells have been determined by Ring and Selvendran (1980) and DuPont and Selvendran (1987). They are composed mainly of glucuronoarabinoxylan (~60%) and cellulose (~30%) and are lignified.

9.5 STORAGE POLYSACCHARIDES

The storage components of cereal grain are the proteins and lipids of the aleurone and starchy endosperm (see Chapters 8 and 10) and the endosperm starch. Minor amounts of fructans are found in the embryo.

9.5.1 Starch

Starch (Guilbot and Mercier, 1995; Whistler *et al.*, 1984; Galliard, 1987) occurs in a granular form in the cells of the starchy endosperm of the grain but is absent from aleurone cells. At maturity some residual starch may be found in the cells of the embryonic tissues and the scutellum.

Starch granules in cereal endosperms occur in a variety of shapes and sizes, each characteristic of the species. The granules develop in a membrane-bound organelle, the plastid (amyloplast), but at maturity the plastid membrane is lost although some proteins remain associated with the granule surface (Chapter 8). Cereal starch granules show a biphasic distribution in size; in normal barleys large granules (A-type), 15–25 μm in diameter, comprise 10–20% by number (85–90% by weight) of the total, the remainder being small (B-type) <6 μm in diameter. In rice and oat endosperm the polyhedral granules (3–9 μm, rice; 3–12 μm, oats) are found in the amyloplast as clusters (20–60 μm, rice; 80–150 μm, oats).

Starch is an association of two polysaccharides, amylopectin and amylose (Table 9.3). Amylopectin (Manners, 1989), usually the most abundant of the two (~70–80% w/w of granule starch), is a branch-on-branch polymer (Type IV, Figure 9.5) composed solely of α-glucopyranosyl residues. In the inter-branch segments the glucose residues

Table 9.2 Composition of walls of aleurone and starchy endosperm cells of cereal grains (% carbohydrate)

	Cellulose	*(1→3,1→4)-β-Glucan*	*Heteroxylan*	*Glucomannan*	*Pectin*	*Phenolic acids*	*Protein*	*References*
Barley								
starchy endosperm	~2	~75	~20	~2	?	0.05	5	Fincher, 1975, 1976
aleurone	~2	~26	~67	~2	?	1.2		Bacic and Stone, 1981
Wheat								Mares and Stone, 1973
starchy endosperm	4	~20	~70	~7	?	+	+	Bacic and Stone, 1980
aleurone	~2	~29	~65	~2	?	1.8	+	Bacic and Stone, 1981
Rice								
starchy endosperm	~28	~20	~27	15	~3	+		Pascual and Juliano, 1984
	~48	+	~40	–	~10	1.2		Shibuya and Nakane, 1984
								Shibuya *et al.*, 1985

Table 9.3 Comparison of the properties of cereal amyloses and amylopectin

	Amylose	*Amylopectin*
Monomer residue	α-D-Glucopyranosyl	α-D-Glucopyranosyl
Linkage type	Interchain (1→4) (98–99%) Branch point (1→6) (1%)	Interchain (1→4) (95%) Branch point (1→6) (~5%)
Organization	Long linear chains (CL ~1000) (low degree of branching) 9–20 chains per molecule	Branch-on-branch CL = 19–24 ICL = CL-ECL-1 = 5.9 ECL = 12–16
β-Amylolysis limit	70–80% Conversion to maltose	55% Conversion to maltose
Molecular weight	10^5–10^6	$>10^8$
A/B chain ratio	–	1:1–1.5:1
Degree of polymerization	1.5–6.3 × 10^3	10^4–10^7
Crystalline pattern	–	A pattern
I_2 complex (λ_{max})	660 nm (blue)	530–550 nm (red-violet)
Solubility in H_2O	High	Low
Retrogradation tendency	High	Low

are (1→4)-linked and branching is at C(O)6 of the branch-point residue. The average distance between branch points is 20–25 glucose residues depending on the source, i.e. ~5% of the glucose residues are at branch points. Individual amylopectin molecules have DPs of 10^4–10^7. A current model of the arrangement of the branched chains in an amylopectin molecule is shown in Figure 9.13a. This 'cluster model' shows regions rich in branch points separated by regions of sparse branching. The molecule has a large number of non-reducing chain ends (the sites of action of exo-amylases, see Section 9.7.2) and a single reducing chain end. Amylopectin molecules are poorly soluble in water.

Amylose, which usually comprises ~20–30% w/w of cereal starches (Table 9.4), is also a polymer of (1→4)-linked α-glucopyranosyl residues but is sparsely branched, with long interbranch chain segments. The DP of cereal amyloses is ~4000. Amylose chains assume a helical conformation (Figure 9.13b) with a repeat of six residues per turn. The interior of the helix is large enough to accommodate 'guest' molecules such as fatty acids, fatty alcohols (*n*-butanol, *n*-pentanol) and hydrocarbon chains of monoacyl lipids and, in cereal starches (wheat, maize), the helices exist partly in such a complex with endogenous lipids. The association of amylose chains with iodine is responsible for the intense blue iodine colour characteristic of starch. The absorption maximum of the iodine complex increases towards longer wavelengths as the chain length of the (1→4)-α-glucan segment increases. Amylopectin stains red-violet due to the association of iodine with the outer chains (DP ~15) which can form a short helix.

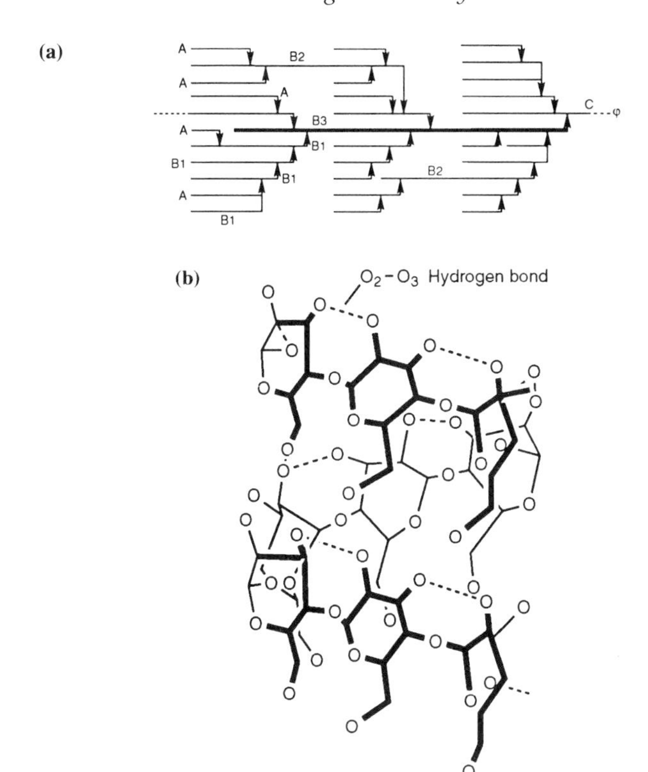

Figure 9.13 (a) Cluster model of amylopectin. The chains marked A are linked to the molecule only by a potential reducing group. The chains marked B are similarly linked but also carry one or more A chains. There is one chain C, carrying the sole reducing group. The chains marked B_1 are confined to single clusters, those marked B_2 extend to two clusters and those marked B_3 into three clusters. Other B chains (not shown) could extend into four clusters. (Reprinted, with permission, from Hizukuri, 1986.) (b) Helical structure of amylose (V-form) showing a six unit repeat. The hydrophilic hydroxyls project outwards and associate with the solvent, the C–H point inwards creating a hydrophobic core. (Reprinted, with permission, from Winterburn, 1974.)

Granule structure

Starch granules exhibit concentric ring structures, that are visible in sections through granules and in granules that have been eroded by hydrolytic enzymes. These rings represent zones of different molecular packing density. The amylopectin molecules are arranged radially in the rings. In the granule the long, outer helical chains of the amylopectin molecules align with one another and in some cases interact to form double helices. The amylose component appears to overlie the amylopectin molecules. The orderly arrangement of the starch molecules is evidenced by their birefringence; the granules exhibit a Maltese cross pattern when viewed under polarized light. They also have characteristic X-ray diffraction patterns due to the ordered packing of adjacent branches of the amylopectin component. Cereal starches all show an A-type X-ray pattern. B-type patterns are given by tuber starches, whereas a C-type pattern, intermediate between A- and B-type, is given by tapioca and legume starches. These organizational patterns can also be recognized by their solid state, ^{13}C-NMR spectra (Gidley, 1992). The crystallinity of starch depends on the environmental conditions prevailing during endosperm development.

Surface pores in starch granules of corn, sorghum and millet (Fannon *et al.*, 1992) have been shown to be openings to serpentine channels that penetrate the granule interior (Fannon *et al.*, 1993).

Amylose is more soluble than amylopectin and can be leached from the granules with hot water. Starch granules can be damaged by abrasion, for example, during milling. Damaged granules swell spontaneously in cold water giving a gel and a soluble fraction.

Gelatinization and retrogradation (Morris, 1990)

Native starch granules are insoluble in cold water but on heating a suspension, the granules absorb water and swell irreversibly to produce a paste in which enlarged granule ghosts can be seen. For cereal starches the gelatinization temperature, i.e. the range of temperature over which starch granules suspended in water start to swell irreversibly, is dependent on the species (see Table 9.4). Swelling is accompanied by loss of order (birefringence) and crystallinity and the solubilization of amylose. Even on heating to 100°C there is little evidence for the release of substantial quantities of amylopectin. Thus heating results in a fluid, composed of porous, gelatinized and swollen granules with an amylopectin skeleton, suspended in a hot amylose solution. Cooling converts the fluid into a turbid, viscoelastic paste or, at sufficiently high starch concentrations (>6%, w/w), an opaque, elastic gel. Thus the starch gel is a composite comprising a gel matrix (amylose) which interpenetrates the porous, deformable component (amylopectin). Gelation and storage of starch gels involves crystallization of the amylose component. Double helices may form between amylose chains and these helices may in turn aggregate laterally. In addition crystallization of the amylopectin component appears to be responsible for the stiffening of the swollen granules which reinforces the amylose gel matrix. The crystallization of the amylopectin is thermally reversed by heating to 90°C.

The crystallization and stiffening of starch gels on ageing is responsible for the 'staling' observed in starch-based foods. The staling behaviour is

Table 9.4 Comparative compositions of starch granules from different cereal grains

Starch source	*Amylose (%)*	*Gelatinization temperature (°C)*	*References*
Barley			
normal			
large (A)	22.1–30.0	58–63	MacGregor and Fincher, 1993
small (B)	18.2–41.3	57–64	
mutant			
high amylose	34.8–45.0	64–66	
waxy	0.4–13.6	54.1–67.5	
Wheat			
hard	24.5	62.9	Berry *et al.*, 1971
soft			
durum	27.0	56.1	Klassen and Hill, 1971
Oats	25.9–27.9	53–59	MacArthur and D'Appolonia, 1979
Rice			
non-waxy	7–37	55–79	Juliano, 1985
waxy (glutinous)	0.8–1.3		
Rye	24.0	57–70	Berry *et al.*, 1971
Triticale	23.0	55–62	Berry *et al.*, 1971
Maize			
normal	21–24	62–72	Meeuse *et al.*, 1960
mutant			
$wx^a\ wx^a\ wx^a$	2.4		
du du du	36.7		
ae ae ae	67.0		

dependent on amylopectin structure and granule structure. Amylose crystallization is an important factor in determining the resistance of starch in human diets to digestion by pancreatic α-amylase (see Section 9.8).

Mutant starches

Cereal starches with higher than normal amylose or amylopectin contents are well-known. These starches arise through mutations affecting enzymes in the starch biosynthetic pathway. High (up to 80%) amylose starches are found in mutant barley and maize varieties (as well as wrinkled peas). Maize, rice, wheat and sorghum starches consisting almost entirely of amylopectin (less than 1% w/v amylose) are referred to as 'waxy starches'. Waxy starches show the same concentric ring pattern, molecular ordering and crystallinity as normal starches supporting the view that the internal structure of the granule is due to the amylopectin component.

Phosphate esters (6-phosphoglucose phosphate) are associated with the amylopectin in waxy rice starches. It has been suggested that the 6-phosphoglucose phosphate arises by transfer of Glc-P from UDPGlc followed by hydrolysis of the glucosyl residue to leave the phosphate ester (Lomako *et al.*, 1994). In non-waxy varieties the phosphorus arises mainly from the lysophospholipids which are associated with the amylose fractions, probably as amylose inclusion complexes. There is a consistent relationship between amylose content and lipid content.

9.5.2 Fructans

Polymers of fructose (fructans) (French, 1989) are minor storage forms in grains of barley and wheat as well as the vegetative tissues of temperate cereals. They consist of short chains of β-D-fructofuranosyl residues in (2→6)-linkage to the fructosyl residue of sucrose. They are found in amounts up to 20% of barley grain during its development and are located mainly in the embryo.

9.6 CARBOHYDRATES IN ALEURONE PROTEIN BODIES

The aleurone protein bodies (Figures 9.2b and 9.3) are reserves of amino acids that are mobilized during germination and repolymerized to produce the hydrolytic and other enzymes secreted by the aleurone cells. Two types of inclusions are present in aleurone protein bodies. The first, the phytin globoids, are rich in phytic acid, the hexaphosphate of *myo*-inositol (Figure 9.14) (Cosgrove, 1980, 1989). This compound occurs as a magnesium-calcium salt and is present in grains at levels of 0.5–2%. On germination the phosphate ester groups are cleaved by the enzyme phytase (EC 3.1.3.26) arising from the aleurone cells, to provide phosphate for the developing plant. Phytic acid chelates metals such as zinc and iron quite strongly and this has nutritional implications when diets of monogastric animals include cereal fractions high in phytic acid (Lasztity and Lasztity, 1988).

The second inclusion in aleurone protein bodies are the niacytin granules seen in wheat and barley, but not in rice. Their chemical nature is not completely described but they are believed to be composed of niacin (nicotinic acid and its amide) esterified to short oligoglucosides and are associated with *o*-aminophenol (Mason *et al.*, 1973). Aleurone is thus a rich source of nicotinic acid although this vitamin is not available for absorption from the alimentary tract unless it is liberated from its

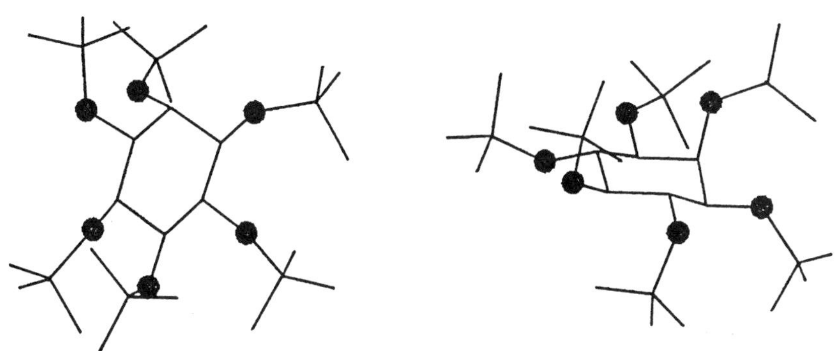

Figure 9.14 Two conformation structures of phytic acid (*myo*-inositol hexaphosphate).

esterified form, e.g. by pre-treatment with dilute alkali as practised in production of masa from maize in tortilla preparation (Serna-Saldivar *et al.*, 1990).

9.7 CARBOHYDRATES IN CEREAL TECHNOLOGY

9.7.1 Carbohydrates in bread-making

In bread manufacture (Blanshard *et al.*, 1986), a dough is formed by thorough mixing of wheat or rye flour, water, yeast and salt and sometimes sugar, shortening, emulsifiers, enzymes, oxidizing and reducing agents. The dough is allowed to develop (prove) and aerate before baking in an oven.

During dough mixing, water is absorbed by the flour to an extent that is dependent primarily on the degree of starch damage and the protein content. The highly hydratable, non-starch polysaccharides comprising water-insoluble and soluble heteroxylans and (1→3,1→4)-β-D-glucans of the endosperm walls are also important contributors to water absorption (Fincher and Stone, 1986; Rouau *et al.*, 1994). The water-extractable heteroxylans, bearing esterified ferulic acid residues, influence the rheological properties of doughs. Their viscous nature increases the resistance of the dough to extension and decreases dough extensibility. Oxidative gelation of the heteroxylans in doughs through intermolecular crosslinking between heteroxylans and proteins and heteroxylans themselves may further account for their rheological effects (Fincher and Stone, 1986). On the other hand, the water-soluble lund *et al.*, 1989) may have beneficial effects on baking quality of wheat flour.

Starch in the dough is enzymically depolymerized to provide a source of maltose and maltodextrins that are fermented to carbon dioxide by baker's yeast during the proving stage. The release of carbon dioxide baker's yeast during the proving stage. The release of carbon dioxide allows the dough to inflate during fermentation. Starch depolymerization in the dough relies on the synergystic action of endo-amylase (α-amylase) and exo-amylase (β-amylase) to produce maltose. Wheat flour contains abundant endogenous β-amylases but low amounts of α-amylases necessitating the use of supplementary α-amylase sources (e.g. germinated grain, barley malt flour or fungal enzyme preparations). It is believed that only damaged starch granules are available for amylase attack. Extensive starch degradation due to excess α-amylase, as found in flour from prematurely sprouted grain, results in loaves with sticky and weak crumb and poor water-holding capacity.

During baking the starch gelatinizes forming the matrix of the crumb in association with the gluten. In stored loaves the texture of the crumb hardens (stales) due to crystallization and stiffening (retrogradation of the starch components) as the gel ages (see Section 9.5). On retrogradation starch becomes increasingly resistant to alimentary digestion (see Section 9.8).

At baking temperature (210°C, 30 min), polysaccharides in the crust undergo chemical alteration. Thus thermal fragmentation and transglycosidation of starch in the dry conditions yields glucosaccharides with end units of (1→6)-β-anhydro-D-glucopyranose (Westerlund *et al.*, 1989).

9.7.2 Fate of grain carbohydrates during germination

At the outset of germination the access of water to the grain sets in train a series of events in the embryonic axis leading to secretion of hormones that in turn activate the dormant, but living, scutellar epithelium, and later the aleurone cells (Duffus, 1987, Fincher and Stone, 1993). This activation manifests itself in a dramatic series of morphological and biochemical changes. On activation, scutellar and aleurone cells show abundant rough endoplasmic reticulum and Golgi profiles, typical of cells synthesizing and secreting proteins. Both cell types synthesize and secrete a range of hydrolytic enzymes for proteins, nucleic acids, lipids and polysaccharides. Among the polysaccharide hydrolases synthesized in barley are (1→4)-α-D-glucan endohydrolases (α-amylases) (EC 3.2.1.1), (1→3)-β-D-glucan exo- and endo-hydrolases (EC 3.2.1.58 and EC 3.2.1.39, respectively), (1→3,1→4)-β-D-glucan endohydrolase (EC 3.2.1.73) α-L-arabinofuranosidases (EC 3.2.1.53), limit dextrinase or starch debranching enzyme (EC 3.2.1.10), β-D-glucosidases (EC 3.2.1.21) and α-D-glucosidases (EC 3.2.1.20). These enzymes are secreted into the starchy endosperm from the scutellar epithelium and aleurone cells and there digest the polysaccharides of the endosperm walls to permit access to cellular starch, protein and lipids.

During germination, endosperm wall dissolution begins in cells that face the scutellar epithelium and then proceeds over the dorsal and ventral surfaces of the endosperm. This pattern of dissolution is the result of a spatially and temporally co-ordinated secretion of hydrolytic enzymes from the scutellar epithelium and aleurone cells. The walls of the aleurone cells secreting the enzymes are degraded in a biphasic pattern. Most of the outer wall is degraded in the initial attack, but the inner wall remains and disappears more slowly. The monosaccharides released from the wall polysaccharides make a major contribution to the total energy made available to the embryo from starchy endosperm during germination, 18% of the available carbohydrate in the case of barley (Morall and Briggs, 1978).

The suite of starch-degrading enzymes secreted from the aleurone and scutellar epithelium is augmented by the (1→4)-α-D-glucan maltohydrolase (*β*-amylase) (EC 3.2.1.2) which is present in the endosperm, mostly on the outer surface of the starch granules. *β*-Amylase is activated by proteolytic enzymes from the aleurone and acts synergystically with α-amylase and limit dextrinase (debranching enzyme) to degrade the granules. Large starch granules are initially degraded at discrete sites, particularly along their equatorial axis, and eventually holes continuous with the interior of the granules can be seen. The inner region is progressively degraded leaving an outer shell that ultimately collapses and disappears. Small starch granules are degraded by a more general surface erosion.

The diffusible hydrolysis products (mono- and oligosaccharides) from starch and wall polysaccharides, together with amino acids and peptides from endosperm protein, are transported across the plasma membrane of the cells of the scutellar epithelium which at this stage of germination has an absorptive function.

Endogenous inhibitors of α-amylase and limit dextrinase, as well as proteases, are found in various cereal grains and other seeds (Silano, 1987; Richardson, 1991; MacGregor *et al.*, 1994). These are low molecular weight proteins that form complexes with the various hydrolytic enzymes and inhibit their catalytic function. It has been suggested that these inhibitors are effective against the digestive enzymes of insects, bacteria and fungi and hence are involved in defence against attack by these organisms (Whitaker, 1983). Another function, for which there is growing evidence, is in regulating the action of endogenous starch and protein depolymerizing enzymes thus preventing premature hydrolysis of endosperm reserves.

9.7.3 Grain carbohydrates in malting and brewing

Malting is essentially a controlled, mass germination of barley or less commonly sorghum or wheat, to produce malt for use in a variety of food processes including brewing, distilling and vinegar production.

The malting process commences with the steeping of the grain to raise the moisture content to a level at which germination begins. Germination, or modification, is allowed to proceed until the endosperm cell walls and the intracellular protein matrix are degraded sufficiently to expose the starch granules. Properly modified malt is friable and readily millable. The appropriately modified grain is kilned, i.e. raised to a temperature high enough to halt germination but low enough to retain the activity of hydrolases for starch, wall polysaccharides and proteins.

In brewing and distilling, the milled malt is extracted with warm water in a 'mashing' step. Mashing extends the malting process by allowing the enzymolysis of starch, following its gelatinization (65°C), to proceed yielding soluble, fermentable carbohydrate. The evenness of malting and the extent to which the wall-degrading enzymes, in particular the (1→3,1→4)-*β*-D-glucan hydrolases, have modified the endosperm walls to permit free diffusion of starch and protein hydrolases to their substrates is important for the success of the mashing step. In undermodified malt depolymerization of endosperm wall (1→3,1→4)-*β*-D-glucan is incomplete and (1→3,1→4)-*β*-D-glucans from the walls are released into solution producing high viscosity malt extracts (worts) that are difficult to filter. Furthermore, the limit dextrinase (debranching enzyme) which cleaves (1→6)-α-linkages at branch points in amylopectin is produced late in germination and, if deficient, may lead to accumulation of branched oligosaccharides which are not substrates for fermentation by brewer's yeast.

The wort is fermented in various ways to produce beer which contains low levels of polysaccharides including *β*-amylase limit dextrins and residual (1→3,1→4)-*β*-D-glucan. These components are believed to be important determinants of beer character such as viscosity (palate body) and foam stability. The (1→3,1→4)-*β*-D-glucans may have a detrimental effect on beer storage and stability because of their potential to precipitate, particularly when high alcohol beers are stored at low temperatures.

9.7.4 Cereal starch utilization

Cereal starches in physically, but not chemically, modified forms are found universally as components of foods. In the formulation of some starch-based food products it has been found to be beneficial to use starch chemically modified in various ways to improve its functional properties (Wurzburg, 1986; Doane, 1994). Thus dextrinization (partial depolymerization) by acid or heat treatment, oxidation with hypochlorite or pyrolysis as a dry powder have been used. In addition, starch molecules have been modified by cross-linking without disruption of the granule. This augments hydrogen bonding interactions and maintains the integrity of the swollen granule when the temperature is raised above that of gelatinization. In this way viscosity reduction and the development of undesirable textures in the food is minimized. Cross-linking using agents such as phosphorus oxychloride, sodium trimetaphosphate, epichlorhydrin or adipic–acetic acid mixed anhydride to give quite low degrees of modification (one cross-link in 1000–3000 glucose residues) dramatically alters the cooking behaviour of the starch.

Another problem in formulation of starch-based foods is the instability of solubilized starch due to amylose and amylopectin retrogradation (see Section 9.5.1) leading to opaque gels. Partial acetyl esterification or etherification with hydroxypropyl groups, at comparatively low degrees of substitution, prevents starch molecules from interacting. Thus the functionality of high amylopectin (waxy) starches used as thickening and texturing agents in a variety of food products is improved by cross-linking. Retrogradation of these starches may be reversed on heating but on storage at 4°C or freezing, their stability is lost. This behaviour can be overcome by increasing the level of acetylation or hydroxypropylation.

Other starch modifications involve the introduction of negatively charged groups such as monophosphate esters or half-succinate or half-maleic esters to give anionic starch. Succinic anhydride substituted with a range of hydrophobic alkenyl, alkyl, aralkyl and aralkeyl groups forms starch monoesters with both anionic and hydrophobic characteristics and properties suitable for both thickening and texturizing. For example, they are useful as stabilizers of oil–water emulsions and in encapsulating oils.

Etherification of starch with 2-diethylaminoethyl substituents yields cationic, cold-water-soluble starches that have found wide applications in paper-making since they have affinity for the negatively charged pulp surfaces.

Another application of starch is in formation of films. The amylose component, in particular, may be cast into films with good tensile strength and biodegradabilty. The film-forming property of starch is used extensively in paper sizing, corrugating adhesives and binders.

The drive to find alternatives to non-biodegradable, petrochemically-based plastics has led to the development of starch-polyvinyl alcohol films, laminated with poly(vinyl chloride) to reduce water sensitivity. When starch at 5–30% moisture is subjected to elevated pressures and temperatures above the glass transition point and melting temperature, it becomes thermoplastic. Extrusion processes based on these properties are used to prepare starch-based plastics which contain, for instance, 60–75% maize starch and up to 40% poly(vinyl alcohol) or other synthetic polymers.

Foamed starch products for use as biodegradable packaging have been produced by co-extrusion of starch ethers and small amounts of poly(vinyl alcohol). Similarly edible foamed products have been produced by co-extrusion of starch, plant fibre and food additives.

9.8 FATE OF CEREAL GRAIN CARBOHYDRATES IN THE ALIMENTARY TRACT OF MAN AND MONOGASTRIC ANIMALS

Cereal grain carbohydrates are major dietary energy and carbon sources for man and for commercially valuable monogastric animals, such as pigs and poultry (Asp *et al.*, 1993). The free monosaccharides in cereal grains, glucose and fructose, are readily absorbed by the mucosal cells of the ileum and jejunum. Sucrose is cleaved to glucose and fructose by specific hydrolases of the brush border of the intestinal mucosa. Other cereal oligosaccharides, such as raffinose, are only partly cleaved in the intestine and the residual components pass to the colon where they are fermented by the resident bacteria.

The starch components are randomly depolymerized by pancreatic endo-amylase (α-amylase) (EC 3.2.1.1) in the lumen of the small intestine and the oligosaccharide products, maltose, maltotriose and isomaltose, are further cleaved by α-glucosidases (EC 3.2.1.20) on the brush border of intestinal mucosa cells. It is noteworthy that some starches e.g. high amylose starches and staled, starch-based foods resist digestion in the upper alimentary tract and pass to the colon where they are fermented. These are notionally components of the dietary fibre complex.

None of the structural polysaccharides in cell walls of cereal grains are digested in the upper alimentary tracts of monogastric animals since the required specific hydrolytic enzymes are absent. However colonic bacteria produce the necessary hydrolases and depending on the physical state of the wall polysaccharides (water-soluble or insoluble), their association with other wall components, such as lignin and suberin, and the residence time of the digesta in the colon, the wall polysaccharides may be more or less fermented to volatile, short-chain, fatty acids (acetic, propionic and butyric acids) which are absorbed, and to some extent metabolized by the colonic mucosal cells or passed to the bloodstream for metabolism in the liver or peripheral tissues.

The wall components of cereal grains contribute to the dietary fibre complex of human diets. In particular the so-called 'soluble-fibre' components appear to have quite important effects on the glycaemic response (Jenkins *et al.*, 1981). These effects, notably due to soluble $(1\rightarrow3,1\rightarrow4)$-β-D-glucans, for example from oat and barley grain, but also contributed by soluble arabinoxylans and pectins, appear to be exerted through their ability to raise the viscosity of the intestinal contents and slow the diffusion of the mono- and oligosaccharides in the digesta to the intestinal mucosa. In addition there is now a consensus that cereal (1→3,1→4)-β-D-glucans have hypocholesterolaemic and other desirable effects when present in the diet at elevated levels.

The presence in the hind gut of fermentable carbohydrate from undigested carbohydrates from cell walls of aleurone, endosperm and pericarp-seed coat leads to high microbial activity and the microbial population accounts for increased faecal bulk due to increased water-holding capacity of the digesta (Cheng *et al.*, 1987). In chickens the presence of fermentable cell wall material from dietary grains, typically soluble (1→3,1→4)-β-D-glucan and arabinoxylan from barley, leads to decreases in the metabolizable energy of the diet. In addition the products of the resultant high caecal microbial populations may be responsible for the 'sticky dropping' phenomenon in chickens fed barley high in (1→3,1→4)-β-D-glucan. Various pre-treatments of barley diets for chickens have been employed to remove the fermentable polysaccharides and so increase productivity and avoid sticky droppings (Fincher and Stone, 1986).

REFERENCES

Aastrup, S. (1979) The relationship between the viscosity of an acid flour extract of barley and its β-glucan content. *Carlsberg Research Communications*, **44**, 289–304.

Andrewartha, K.A., Phillips, D.R. and Stone, B.A. (1979) Solution properties of wheat arabinoxylans and enzymically modified arabinoxylans. *Carbohydrate Research.*, **77**, 191–204.

Asp, N.G., Bjørck, I. and Nyman, M. (1993) Physiological effects of cereal dietary fibre. *Carbohydrate Polymers*, **21**, 183–7.

Bacic, A. (1979) Biochemical and ultrastructural studies on endosperm cell walls. Ph.D. Thesis, La Trobe University, Melbourne.

Bacic, A. and Stone, B.A. (1980) A (1→3)- and (1→4)-linked β-D-glucan in the endosperm cell walls of wheat. *Carbohydrate Research*, **82**, 372–7.

Bacic, A. and Stone, B.A. (1981) Chemistry and organisation of aleurone cell wall components from wheat and barley. *Australian Journal of Plant Physiology*, **8**, 475–95.

Bacic, A., Harris, P.J. and Stone, B.A. (1988) Structure and function of plant cell walls, in *The Biochemistry of Plants*, Vol. 14 (ed. J. Preiss), Academic Press, San Diego, pp. 297–371.

Beresford, G. and Stone, B.A. (1983) (1→3,1→4)-β-D-Glucan content of *Triticum* grains. *Journal of Cereal Science*, **1**, 111–14.

Berry, C.P., D'Applonia, B.L. and Gilles, K.A. (1971) The characterization of triticale starch and its comparison with starches of rye, durum and HRS wheat. *Cereal Chemistry*, **48**, 415–27.

Blanshard, J.M.V., Frazier, P.J. and Galliard, T. (1986) *Chemistry and Physics of Baking Materials, Processes and Products*. The Royal Society of Chemistry, London.

Briggs, D.E. (1978) Grain quality and germination, in *Barley*, Chapman and Hall, London, pp. 174–221.
Carpita, N. and Gibeaut, D.M. (1993) Structural models of primary cell walls of flowering plants: consistency of molecular structure with the physical properties of the walls during growth. *The Plant Journal*, **3**, 1–30.
Cheng, B-Q., Trimble, R.J., Illman, R.J., Stone, B.A. and Topping, D.L. (1987) Comparative effects of dietary wheat bran and its morphological components (aleurone and pericarp-seed coat) on volatile fatty acid concentrations in the rat. *British Journal of Nutrition*, **57**, 69–76.
Chesson, A., Gordon, A.H. and Scobbie, L. (1995) Pectic polysaccharides of mesophyll cell walls of perennial ryegrass leaves. *Phytochemistry*, **38**, 579–83.
Cosgrove, D.J. (1980) *Studies in organic chemistry*. Vol. 4, *Inositol Phosphates*, Elsevier, Amsterdam.
Cosgrove, D.J. (1989) *Phytates in cereals and legumes*. CRC: Press Boca Raton, Florida.
Doane, W.M. (1994) Opportunities and challenges for new industrial uses of starch. *Cereal Foods World*, **39**, 556–63.
Duffus, C.M. (1987) Physiological aspects of enzymes during grain development and germination, in *Enzymes and their Role in Cereal Technology* (eds J. E. Kruger, D. Lineback and C.E. Stauffer), American Association of Cereal Chemists Inc., St. Paul MN, USA, pp. 83–116.
DuPont, M.S. and Selvendran, R.R. (1987) Hemicellulosic polymers from the cell walls of beeswing wheat bran. Part I. Polymers solubilized by alkali at 2°. *Carbohydrate Research*, **163**, 99–113.
Evers, A.D. and Bechtel, D.B. (1988) Microscopic structure of the wheat grain, in *Wheat Chemistry and Technology* (ed. Y. Pomeranz), 3rd edn, American Association of Cereal Chemists, St Paul, Minnesota, pp. 47–95.
Fannon, J.E., Hawker, R.J. and BeMiller, J.N. (1992) Surface pores in starch granules. *Cereal Chemistry*, **69**, 284–8.
Fannon, J.E., Shull, J.M. and BeMiller, J.N. (1993) Interior channels of starch granules. *Cereal Chemistry*, **70**, 611–13.
Fincher, G.B. (1975) Morphology and chemical composition of barley endosperm cell walls. *Journal of the Institute of Brewing*, **81**, 116–22.
Fincher, G.B. (1976) Ferulic acid in barley cell walls: a fluorescence study. *Journal of the Institute of Brewing*, **82**, 347–9.
Fincher, G.B. and Stone, B.A. (1986) Cell walls and their components in cereal technology. *Advances in Cereal Science and Technology*, **8**, 207–95.
Fincher, G.B. and Stone, B.A. (1993) Physiology and biochemistry of germination in barley in *Barley – Chemistry and Technology* (eds A.W. MacGregor and R.S. Bhatty), American Association of Cereal Chemists Inc., St. Paul, MN, USA, pp. 247–95.
Fincher, G.B., Sawyer, W.H. and Stone, B.A. (1974) Chemical and physical properties of an arabinogalactan-peptide from wheat endosperm. *Biochemical Journal*, **139**, 535–45.
Fincher, G.B., Stone, B.A. and Clarke, A.E. (1983) Arabinogalactan-proteins: structure, biosynthesis and function. *Annual Review of Plant Physiology*, **34**, 47–70.
French, A.D. (1989) Chemical and physical properties of fructans. *Plant Physiology*, **134**, 125–36.
Fry, S.C. (1989) Structure and function of xyloglucans. *Journal of Experimental Botany*, **40**, 1–11.
Fulcher, R.G. (1986) Morphological and chemical organization of the oat kernel,

in *Oats: Chemistry and Technology* (ed. F.H. Webster), American Association of Cereal Chemists, St Paul, Minnesota, pp. 47–74.
Fulcher, R.G., Faubion, J.M., Ruan, R. and Miller, S.S. (1994) Quantitative microscopy in carbohydrate analysis. *Carbohydrate Polymers*, **25**, 285–93.
Galliard, T. (1987) *Starch, Properties and Potential*, John Wiley & Sons.
Gidley, M.J. (1992) Nuclear magnetic resonance analysis of cereal carbohydrates, in *Developments in Carbohydrate Chemistry* (eds R.J. Alexander and H.F. Zobel), American Association of Cereal Chemists Inc., St. Paul Minnesota, USA.
Guilbot, A. and Mercier, C. (1984) Starch, in *The Polysaccharides* (ed. G.O. Aspinall), Vol. 3, Academic Press, New York, pp. 183–247.
Hayashi, T. (1989) Xyloglucans in the primary cell wall. *Annual Review of Plant Physiology and Plant Molecular Biology*, **40**, 139–68.
Henry, R.J. (1988) The carbohydrates of barley grains – A review. *Journal of the Institute of Brewing*, London, **94**, 71–8.
Hirata, Y. and Watson, S.A. (1967) in *Starch: Chemistry and Technology*, Vol. II Industrial Aspects (eds R.L. Whistler and E.F. Paschall), Academic Press, New York.
Hizukuri, S. (1986) Polymodal distribution of chain lengths of amylopectins and its significance. *Carbohydrate Research*, **147**, 342–3.
Hoseney, R.C. (1984) Functional properties of pentosans in baked foods. *Food Technology*, **38**, 114–17.
Iiyama, K., Lam, T.B.T. and Stone, B.A. (1993) Cell wall biosynthesis and its regulation, in *Forage Cell Wall Structure and Digestibility* (eds H.G. Jung, D.R. Buxton, R.D. Hatfield and J. Ralph), American Society of Agronomy, pp. 621–83.
Iiyama, K., Lam, T.B.T. and Stone, B.A. (1994) Covalent crosslinks in the cell wall. *Plant Physiology*, **104**, 315–20.
Jarvis, M.C. (1984) Structure and properties of pectin gels in plant cell walls. *Plant, Cell and Environment*, **7**, 153–64.
Jenkins, D.J.A., Wolever, T.M.S., Taylor, R.H. *et al.* (1981) Glycaemic index of foods: a physiological basis for carbohydrate exchange. *American Journal of Clinical Nutrition*, **34**, 362–6.
Juliano, B.O. (1985) *Rice: Chemistry and Technology*, American Association of Cereal Chemists, St. Paul, MN, USA.
Karim, A. and Rooney, L.W. (1972) Pentosans in sorghum grain. *Journal of Food Science*, **37**, 365–8.
Klassen, A.J. and Hill, R.D. (1971) Comparison of starch from triticale and its parental species. *Cereal Chemistry*, **48**, 647–54.
Lam, T.B.T., Iiyama, K. and Stone, B.A. (1990) Primary and secondary walls of grasses and other forage plants: Taxonomic and structural considerations, in *Microbial and Plant Opportunities to Improve Lignocellulose Utilization by Ruminants* (eds D.E. Akin, L.G. Ljungdahl, J.R. Wilson and P.J. Harris), Elsevier, New York, pp. 43–69.
Lasztity, R. and Lasztity, L. (1988) Phytic acid in cereal technology. *Advances in Cereal Science and Technology*, **10**, 309–71.
Lineback, D.R. and Rasper, V.F. (1988) Wheat carbohydrates, in *Wheat Chemistry and Technology* (ed. Y. Pomeranz), Vol. 1, American Association of Cereal Chemists, MN, USA, pp. 277–372.
Lomako, J., Lomako, W.M., Kirkman, B.R. and Whelan, W.J. (1994) The role of phosphate in muscle glycogen. *Biofactors*, **4**, 167–71.
MacArthur, L.A. and D'Appolonia, B.L. (1979) Composition of oat and wheat carbohydrates II. Starch. *Cereal Chemistry*, **56**, 458–61.
MacArthur-Grant, L. (1986) Sugars and non-starchy polysaccharides in oats, in

Oats: Chemistry and Technology (ed. F.M. Webster), American Association of Cereal Chemists, St. Paul, USA, pp. 75–91.

McCann, M.C. and Roberts, K. (1991) Architecture of the primary cell wall, in *The Cytoskeletal Basis of Plant Growth and Form* (ed. C.W. Lloyd), Academic Press, London, pp. 109–29.

MacGregor, A.W. and Fincher, G.B. (1993) Carbohydrates of the barley grain, in *Barley – Chemistry and Technology* (eds A.W. MacGregor and R.S. Bhatty), American Association of Cereal Chemists, MN, USA, pp. 73–130.

MacGregor, A.W., Macri, L.J., Schroeder, S.W. and Bazin, S.L. (1994) Purification and characterisation of limit dextrinase inhibitors from barley. *Journal of Cereal Science*, **20**, 33–41.

MacLeod, A.M. and Preece, I.A. (1954) Studies on free sugars of the barley grain. V. Comparison of sugars and fructosans with those of other cereals. *Journal of the Institute of Brewing*, London, **60**, 46–55.

McNamara, M.A. and Stone, B.A. (1981) Isolation, characterization and chemical synthesis of a galactosyl-hydroxyproline linkage compound from wheat endosperm arabinogalactan-peptide. *Lebensmittel-wissenschaft und Technologie*, **14**, 182–7.

Manners, D.J. (1989) Recent developments in our understanding of a mylopectin structure. *Carbohydrate Polymers*, **11**, 87–112.

Marchessault, R.H. and Sudararajan, P.R. (1983) Cellulose, in *The Polysaccharides* (ed G.O. Aspinall), Academic Press, New York, Vol.2, pp. 11–95.

Mares, D. and Stone, B.A. (1973) Studies on wheat endosperm. I. Chemical composition and ultrastructure of the cell walls. *Australian Journal of Biological Sciences*, **26**, 793–812.

Mason, J.B., Gibson, N. and Kodicek, E. (1973) The chemical nature of bound nicotinic acid of wheat bran – studies of nicotinic acid-containing macromolecules. *British Journal of Nutrition*, **30**, 297–311.

Meeuse, B.J.D., Andries, M. and Wood, J.A. (1960) Floridean starch. *Journal of Experimental Botany*, **11**, 129–40.

Morall, P. and Briggs, D.E. (1978) Changes in cell wall polysaccharides of germinating barley grains. *Phytochemistry*, **17**, 1495–502.

Morris, V.J. (1990) Starch gelation and retrogradation. *Trends in Food Science and Technology*, **1**, 2–6.

Murty, D.S., Singh, V., Suryaprakash, S. and Nicodemus, K.O. (1985) Soluble sugars in five endosperm types of sorghum. *Cereal Chemistry*, **62**, 150–2.

Neukom, H. and Markwalder, H.U. (1978) Oxidative gelation of wheat flour pentosans: A new way of crosslinking polymers. *Cereal Foods World*, **23**, 374–6.

O'Brien, T.P. and McCully, M. (1981) *The Study of Plant Structure: Principles and Selected Methods*, Termarcarphi, Melbourne, Australia.

Pascual, C.G. and Juliano, B.O. (1983) Properties of cell wall preparations of milled rice. *Phytochemistry*, **22**, 151–9.

Perlin, A.S. (1951) Structure of the soluble pentosans of wheat flour. *Cereal Chemistry*, **28**, 382–93.

Pomeranz, Y. (1988) Chemical composition of kernel structures, in *Wheat: Chemistry and Technology* (ed. Y. Pomeranz), 3rd edn, Vol. I, Ch. 4, American Association of Cereal Chemists, St. Paul, MN, USA.

Ralph, J., Quideau, S., Grabber, J.H. and Hatfield, R.D. (1994) Identification and synthesis of new ferulic acid dehydrodimers present in grass cell walls. *Journal of the Chemical Society, Perkin Transactions*, **1**, 3485–3498.

Rees, D.A. (1975) Stereochemistry and binding behaviour of carbohydrate chains, in *Biochemistry of Carbohydrates* (ed. W.J. Whelan), MTP International

Reviews of Science, Biochemistry Series, Butterworths, London, Vol. 5, pp. 1–42.
Rees, D.A. (1979) Polysaccharides: conformation and properties in solution, in *Comprehensive Organic Chemistry* (ed. E. Haslam) Vol. 5, Biological Compounds, Pergamon Press, Oxford, pp. 817–30.
Rhodes, D.R. (1993) Structural analysis of wheat endosperm cell walls. PhD Thesis, La Trobe University, Melbourne.
Richardson, M. (1991) Seed storage proteins: the enzyme inhibitors, in *Methods in Plant Biochemistry*, Vol. 5 (ed. L.J. Rogers), Academic Press, London, pp. 259–305.
Ring, S.G. and Selvendran, R.R. (1980) Isolation and analysis of cell wall material from beeswing wheat bran (*Triticum aestivum*). *Phytochemistry*, **19**, 1723–30.
Rouau, X., El-Hayek, M-L. and Moreau, D. (1994) Effect of an enzyme preparation containing pentosanases on the bread-making quality of flours in relation to changes in pentosan properties. *Journal of Cereal Science*, **19**, 259–72.
Sendra, J.M., Carbonell, J.V., Gosalbes, M.J. and Todo, V. (1989) Determination of glucan in wort and beer by its binding with Calcofluor using a fluorimetric flow-injection analysis (FAI) method. *Journal of the Institute of Brewing*, **95**, 327–32.
Serna-Saldivar, S.O., Gomez, M.H. and Rooney, L.W. (1990) Technology, chemistry and nutritional value of alkaline-cooked corn products. *Advances in Cereal Science and Technology*, **10**, 243–307.
Shibuya, N. and Nakane, R. (1984) Pectic polysaccharides of rice endosperm cell walls. *Phytochemistry*, **23**, 1425–9.
Shibuya, N., Nakane, R., Yasui, A., Tanaka, K. and Iwasaki, T. (1985) Comparative studies on cell wall preparations from rice bran, germ and endosperm. *Cereal Chemistry*, **62**, 252–8.
Silano, V. (1987) α-Amylase inhibitors, in *Enzymes and their Role in Cereal Technology* (eds J.E. Kruger, D. Lineback and C.E. Stauffer) American Association of Cereal Chemists, St. Paul, USA, pp. 141–99.
Stone, B.A. and Clarke, A.E. (1993) *The Chemistry and Biology of (1→3)-β-Glucans*, La Trobe University Press, Melbourne.
Stone, B.A., Evans, N.A., Bönig, I. and Clarke, A.E. (1984) The application of Sirofluor, a chemically defined fluorochrome from aniline blue for the histochemical detection of callose. *Protoplasma*, **122**, 191–5.
Strahm, A., Amado, R. and Neukom, H. (1981) Hydroxyproline-galactoside as a protein-polysaccharide linkage in a water soluble arabinogalactan-peptide from wheat endosperm. *Phytochemistry*, **20**, 1061–3.
Victor, R.J., Angelino, S.A.G.F. and Voragen, A.G. (1993) Structural features of arabinoxylans from barley and malt cell wall material. *Journal of Cereal Science*, **15**, 213–22.
Wall, J.S. and Blesin, C.W. (1970) in *Sorghum Production and Utilization* (eds J.S. Wall and W.M. Ross) Avi Publishing Co. Inc., Westpoint, Connecticut, USA.
Westerlund, E., Theander, O. and Aman, P. (1989) Effects of baking on protein and aqueous ethanol extractable carbohydrate in white bread fractions. *Journal of Cereal Science*, **10**, 139–147.
Whistler, R.L., BeMiller, J.N. and Paschall, E.F. (eds) (1984) *Starch: Chemistry and Technology*, 2nd edn, Academic Press, New York.
Whitaker, J.R. (1983) Protease and amylase inhibitors in biological material, in *Xenobiotics in Foods and Feeds* (eds J.W. Finley and D.E. Schwass), American Chemical Society, Washington, DC, pp. 15–46.
Winterburn, P.J. (1974) Polysaccharide structure and function, in *Companion to*

Biochemistry (eds A.T. Bull, J.R. Lagnado, J.O. Thomas and K.F. Tipton), Longmans London, pp. 307–41.

Wood, P.J. (1986) Oat ß-glucan: Structure, location and properties in *Oats – Chemistry and Technology* (ed F.H. Webster), American Association of Cereal Chemists, MN, USA, pp. 121–52.

Wood, P.J., Siddiqui, I.R. and Paton, D. (1978) Extraction of high viscosity gums from oats. *Cereal Chemistry*, **55**, 1038–49.

Wood, P.J., Weisz, J. and Blackwell, B.A. (1994) Structural studies of (1→3, 1→4)-ß-D-glucans by ^{13}C-nuclear magnetic resonance spectroscopy and by rapid analysis of cellulose-like regions using high performance anion-exchange chromatography of oligosaccharides released by lichenase. *Cereal Chemistry*, **71**, 301–7.

Wurzburg, O.B. (1986) Forty years of industrial starch research. *Cereal Foods World*, **31**, 897–903.

Zamorski, R. and Shibuya, N. (1990) Genetic background of glucomannan expression in rice endosperm cell wall. *International Workshop on Frontiers in Plant and Microbial Glycans, Kyoto, Japan, Abstracts* p. 20.

10

Other grain components

Y. Fujino, J. Kuwata, Y. Mano and M. Ohnishi

This chapter reviews the lipids, inorganic components, vitamins and possible toxic substances found in cereals from a biochemical and nutritional perspective.

10.1 CEREAL LIPIDS

Lipids represent approximately 3% of whole cereal grain and decrease to almost half this level during milling (Table 10.1). Lipids can generally be classified as straight-chain or branched chain (Fujino, 1989, 1993, 1994). The main classes of lipids are shown in Figure 10.1. The triglycerides are most abundant while the phospholipids, glycolipids and terpenoids are minor components. In both the straight- and branched chain groups, lipids with no polarity in the molecule are called neutral lipids whereas the ones with polarity are refered to as polar lipids. In this section, fatty acids and triglycerides are discussed as the representative neutral lipids in cereals and phospholipids and glycolipids as characteristic of the polar lipids. Terpenoids (e.g. carotenoid, tocopherol and oryzanol) with nutritive function are described in the section on vitamins. Other minor lipids from the straight chain group such as fatty carbonyls, hydrocarbons and waxes are discussed by Morrison (1978) and Fujino (1983). Branched lipids such as those in the squalene, sterol and chlorophyll groups are described by Kuroda *et al.* (1977), Morrison (1978) and Fujino (1983).

Most classes of lipids contribute to nutritional value and functional properties of foods. Reviews of cereal lipids include those by Fujino (1978), Morrison (1978), Weber (1978), Fujino (1983) and Juliano (1983).

10.1.1 Fatty acids

Both free and combined fatty acids are found, with the combined forms overwhelmingly predominant in nature. Cereal fatty acids range from

Table 10.1 Lipid and fatty acid content of cereals (per 100 g). (Resources Council, 1982, 1990)

Cereal	*Solid (g)*	*Lipid (g)*	*Fatty acid (g)*	*Saturate :*	*Monoenoic :*	*Polyenoic :*	*Remarks*
Wheat							
Whole grain	87.0	3.1	2.4	23 :	14 :	23	Hard grain
Wheat flour	85.5	1.8	1.5	28 :	11 :	61	Hard flour
White bread	62.0	3.8	3.5	24 :	36 :	40	
Rice							
Whole grain	84.5	3.0	2.6	27 :	35 :	38	Brown (crude)
White rice, raw	84.5	1.3	1.2	36 :	26 :	38	Polished
White rice, boiled	35.0	0.5	0.5>	36 :	26 :	38	Polished
Maize							
Whole grain	85.5	5.0	–	–	–	–	
Sweet corn, raw	25.3	1.4	1.1	20 :	38 :	42	Immature
Sweet corn, boiled	25.3	0.4	0.3	20 :	38 :	42	Immature

Chain	Series	Class
Straight*	free	: fatty acid, fatty carbonyl, hydrocarbon etc.
	bound	: triglyceride, phospholipid glycolipid, wax etc.
Branched†	free	: squalene, sterol, carotenoid etc.
	bound	: chlorophyll, tocopherol, oryzanol etc.

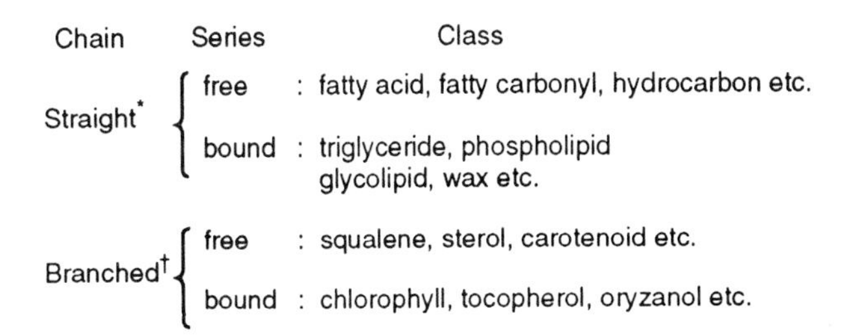

Figure 10.1 Classification of cereal lipids. *, Generally fatty acid-related (acyloid) lipids; †, generally terpene-related (terpenoid or prenoid) lipids.

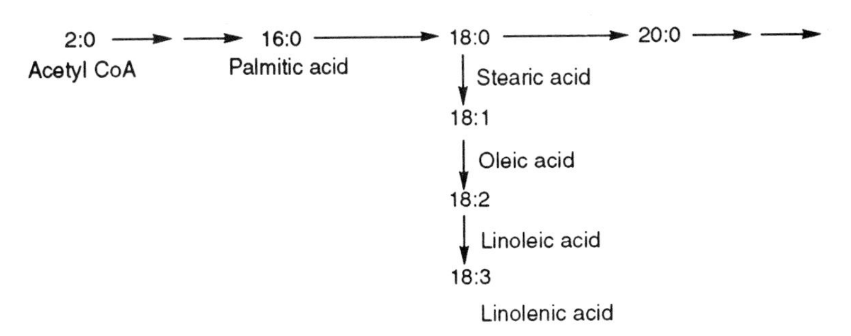

Figure 10.2 Biosynthetic relationships of fatty acids in plants.

Linoleic acid	$CH_3(CH_2)_4CH{=}CHCH_2CH{=}CH(CH_2)_7COOH$	18:2
Oleic acid	$CH_3(CH_2)_7CH{=}CH(CH_2)_7COOH$	18:1
Palmitic acid	$CH_3(CH_2)_{14}COOH$	16:0

Figure 10.3 Major fatty acids in cereals.

C_8 to C_{28} with unsaturated fatty acids being more abundant than saturated fatty acids (Table 10.1). The main saturated fatty acids are palmitic and stearic acid whereas the chief unsaturated fatty acids are oleic, linoleic and linolenic acid. The sequential biosynthesis of these fatty acids is depicted in Figure 10.2. The most abundant fatty acid in cereals is linoleic acid followed by oleic acid and palmitic acids (Figure 10.3)

Fatty acids in cereals are sources of energy in the diet and provide essential fatty acids nutritionally. Physiologically, the fatty acids contribute materials to biomembranes.

Combined fatty acids may break down to liberate fatty acids during the storage or processing of cereals. Further decomposition may produce low-molecular weight aldehydes, ketones and other compounds. These products together with those produced by the degradation of other cereal components often cause a change of flavour and quality such as that in the off-odour of rice (Yasumatsu *et al.*, 1966) and the fragrant smell in maize (Legendre *et al.*, 1978).

10.1.2 Triglycerides

The greatest part of cereal lipids is triacylglycerol. The fatty acids are generally randomly arranged with saturated acids more common at C-1 and C-3 and unsaturated acids at C-2. The principal species in rice grain are shown in Figure 10.4 (Mano *et al.*, 1989).

Small amounts of di- and mono-glyceride are also present in cereals. These glycerides are related metabolically as shown in Figure 10.5. Mixtures of triglycerides are referred to as fat or oil. Commercial cereal oils and general plant seed oils usually contain not only triglycerides but also small quantities of di- and mono-glycerides and other minor lipid components. The cereal oils could therefore more precisely be considered to be lipid mixtures that nutritionally provide a supply of energy and essential fatty acids.

(a)	(b)	(c)
H_2CO— 18:1	H_2CO— 18:1	H_2CO— 16:0
HCO— 18:2	HCO— 18:1	HCO— 18:1
H_2CO— 18:2	H_2CO— 18:2	H_2CO— 18:2

a: Oleoyl-linoleoyl-linoleoyl-glycerol
b: Oleoyl-oleoyl-linoleoyl-glycerol
c: Palmitoyl-oleoyl-linoleoyl-glycerol

Figure 10.4 Major species of triacylglycerol in rice grains. (a) Oleoyl, linoleoyl-linoleoyl-glycerol; (b) oleoyl-oleoyl-linoleoyl-glycerol; (c) palmitoyl-oleoyl-linoleoyl-glycerol.

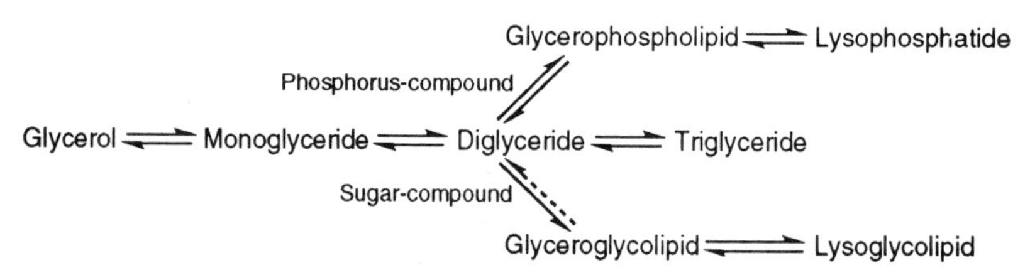

Figure 10.5 Metabolic relationships in glyceridic lipids.

Lipid	Pericarp	Aleurone layer	Embryo	Non-starchy	Starchy	Whole grain
Total lipid	42.5	220–386	269–319	237–386	138–255	916–1243
Neutral lipid	36.6	159–321	213–266	85–138	6.2–13.7	96.2–175
Phospholipid	3.0	39.4–57.3	40.0–46.5	59.9–136	130–228	289–516
Glycolipid	2.9	13.2–21.6	6.6–9.5	44.1–111	1.6–17.2	96.2–175
Neutral lipid						
Acylsterol	5.6	3.4–11.0	7.5–9.8	4.3–17.0	0.7–2.3	20.0–25.8
Triglyceride	6.6	132–290	195–243	46.1–70.0	0.5–1.2	367–568
Diglyceride	14.8	1.9–11.8	0–13.0	9.7–17.8	0–0.9	18.2–72.9
Fatty acid	9.1	5.9–14.4	2.5–5.7	7.7–20.5	3.9–8.3	3.6–110.5
Monoglyceride	0.5	1.7–2.4	2.2–4.8	6.9–26.5	0.9–1.2	13.8–30.5
Glycolipid						
Acylglycosylsterol	1.6	4.8–6.5	2.2–4.8	3.9–13.2	0–2.8	9.4–16.3
Monoglycosyldiglyceride	–	8.4–16.6		6.3–26.5	0.5–1.4	8.4–31.8
Monoglycosylmonoglyceride	–	8.4–16.6		0.9–7.4	0–2.6	4.7–14.0
Diglycosyldiglyceride	–	8.4–16.6	0–6.6	32.3–74.2	0–5.2	45.0–117
Diglycosylmonoglyceride	–	8.4–16.6		3.9–24.0	0–8.9	13.1–42.7
Phospholipid						
Acylphosphatidylethanolamine	–	0–2.1	0–0.3	8.2–49.8	0	9.0–98.2
Acyllysophosphatidylethanolamine	–	0–1.7	0–0.9	6.6–29.7	0	6.8–35.6
Diphosphatidylglycerol	–	0.3–2.3	0.6–1.7	–	0	<7.6
Phosphatidylglycerol	–	3.1–7.2	0.4–2.9	1.3–8.9	0	8.3–27.1
Phosphatidylethanolamine	–	3.1–7.2	4.3–6.9	1.3–8.9	0	8.3–27.1
Phosphatidylcholine (lecithin)	–	19.8–32.2	20.3–28.1	5.0–19.8	0–1.3	21.2–79.6
Phosphatidylinositol	–	5.5–9.6	5.5–8.5	–	0	<26.1
Lysophosphatidylglycerol	–	–	–	–	4.8–15.4	6.7–23.7
Lysophosphatidylethanolamine	–	–	1.0–1.2	0.6–5.4	14.2–28.6	19.1–31.3
Lysophosphatidylcholine (lysolecithin)	–	2.2–5.2	1.0–4.6	12.4–29.9	106–188	159–227
Others	–	0.2–1.1	–	1.1–3.5	1.2–29.4	6.0–22.1

Table 10.3 Distribution of lipids in rice grains. (Chouduhry and Juliano 1980)

Lipid	*Whole grain (brown rice)*	*Bran**	*Embryo*	*Polish*[†]	*Sub-aleurone layer*	*Inner endosperm*
	%	%	%	%	%	%
Proportion of whole grain	100	5.9–6.4	1.3–1.5	4.1–4.4	4.9–5.2	82.5–83.8
Proportion of lipid content	2.9–3.4	19.4–25.5	34.1–36.5	10.2–15.0	5.6–8.5	0.4–0.8
Proportion of lipid						
Neutral lipid	85–87	88–90	91–92	86–88	82–86	66–76
Triglyceride	69–71	75–76	77–79	70–74	58–62	30–37
Phospholipid	8–9	7–8	6–7	8–9	8–12	12–17
Glycolipid	4–6	4–5	2–3	4–5	5–7	12–18
Fatty acid	6–7	4–5	4	5–8	13–17	27–29
Acylglycosylsterol	2–3	2	1	2–3	2–4	5–6
Glycosylsterol	<1	<1	<1	1	1–2	2–3
Diglycosyldiglyceride	<1	<1	<1	<1	1	1–2
Phosphatidylcholine	4	3–4	3–4	3–4	4	3–5
Phosphatidylethanolamine	3–4	3	3–4	3	3–4	3–5
Lysophosphatidylcholine	–	–	–	–	1–2	2–4
Lysophosphatidylethanolamine	<1	<1	<1	<1	1–2	2–4
Others	8–11	8–9	8–9	9–12	11–12	15–17

*Corresponding to a mixture of coat, parts of aleurone layer and embryo.
[†]Corresponding almost to aleurone layer.

10.1.3 Phospholipids

Glycerophospholipids and sphingophospholipids are found in nature. The former are well described in cereals whereas the latter are not so well characterized. Nutritionally phospholipids supply energy, essential acids, phosphorus and vitamins and contribute to biomembranes.

Glycerophospholipids

These lipids combine phosphorus compounds with glycerides. More than 40 classes of these lipids, distinguished by their phosphorus moiety, are found in nature. About 10 of these are found in cereals (Tables 10.2, 10.3 and 10.4), phosphatidylcholine (lecithin), phosphatidylethanolamine (popularly cephalin) and phosphatidylinositol being representative (Figure 10.6) (Miyazawa *et al.*, 1977).

The distribution of fatty acids in the molecule is random at both C-1 and C-2 with unsaturated acids more common at C-2 rather than C-1.

Table 10.4 Distribution of lipids in maize grains. (Calculated from Weber, 1979)

Lipid	*Embryo μg/grain*	*Endosperm μg/grain*
Total lipid	8100	1561
Neutral lipid	2479	1008
Phospholipid		
Acylphosphatidylethanolamine	8	5
Acyllysophosphatidylethanolamine	8	4
Diphosphatidylglycerol	9	3
Phosphatidylglycerol	20	6
Phosphatidylethanolamine	51	20
Phosphatidylcholine	252	154
Phosphatidylinositol	96	14
Phosphatidylserine	4	–
Lysophosphatidylethanolamine	1	7
Lysophosphatidylcholine	2	81
Others	9	2
Glycolipid		
Acylglycosylsterol	3	76
Monoglycosyldiglyceride	18	138
Cerebroside and glycosylesterol	111	293
Monoglycosylmonoglyceride	–	142
Diglycosyldiglyceride	24	320
Sulfoquinovosyldiglyceride and diglycosylmonoglyceride	5	296

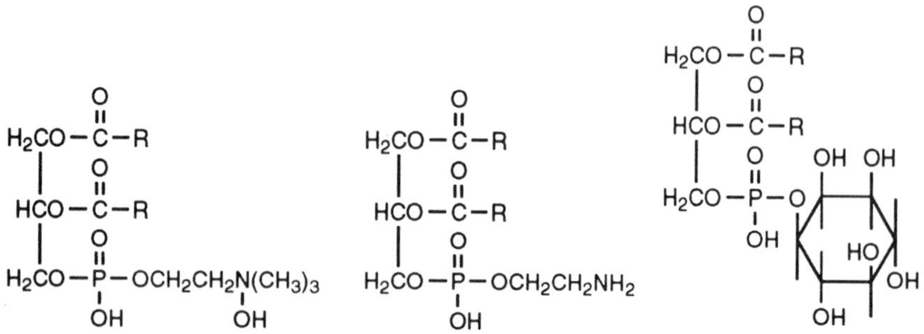

Phosphatidylcholine (lecithin)

Phosphatidylethanolamine

Phosphatidylinositol

Figure 10.6 Principal phospholipids in rice.

Lysolipids such as lysolecithin and lysocephalin, probably derived from enzymatic hydrolysis which gives more 1-acyl than 2-acyl, can usually be found in cereals. These glycerophospholipids are metabolically related to diglycerides as indicated in Figure 10.5.

Sphingophospholipid

Ordinary sphingophospholipids are not found in cereals, but the precursors and successors of these lipids have been reported (Fujino, 1983).

C4-Saturate series (sphinganine* type):

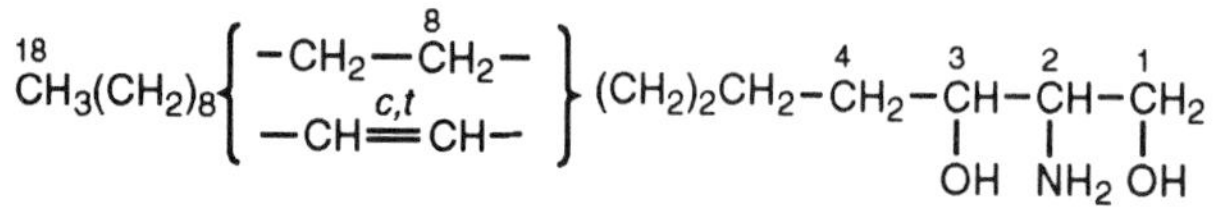

C4-Unsaturate series (sphingenine† type):

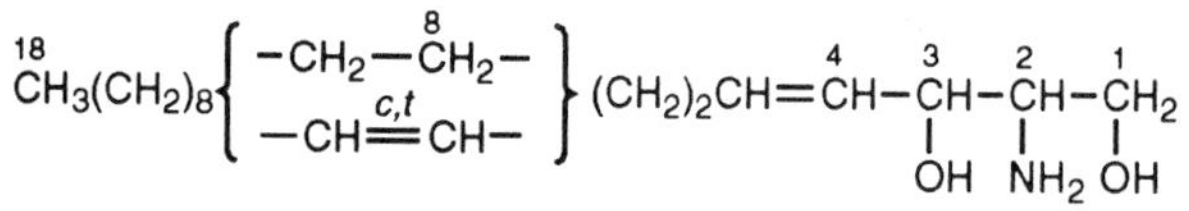

C4-Hydroxysphinganine‡ type):

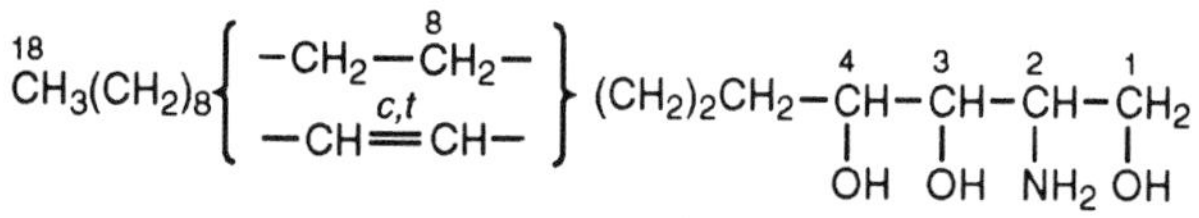

Figure 10.7 Fundamental structures of sphingoids. *, Dihydrosphingosine; †, sphingosine; ‡, phytosphingosine.

Sphingoid

There are free and bound types, the latter being overwhelmingly predominant in nature. Free sphingoid has a basic structure of a C_{12-22}, mainly C_{18}, aminoalcohol with a 1,3-dihydroxy-2-amino configuration. Among the more than 60 types of natural sphingoid, several from the C_{18} group have been found in cereals.

Sphingoids are classified according to the structure at C-4 into three series and include forms that are either saturated or unsaturated at C-8 (Figure 10.7) (Fujino, 1983; Morrison, 1983). The configuration in nature is only *trans* for the unsaturates at C-4 whereas both *cis* and *trans* unsaturates are found at C-8. Generally at C-4 both saturated and unsaturated forms are widespread in nature; 4-hydroxy and 8-unsaturated sphingoids are much more common in plants, including cereals, than in animals. Sphingoids can be formed by the condensation of fatty acids with the amino acid, serine (Figure 10.8).

Ceramide

This occurs widely in nature, especially in plants – including cereals. The molecule has an *N*-acylsphingoid where the main component acid is C_{24} with a 2-hydroxy and the sphingoid is of C_{18} with a 4-hydroxy in the usual ceramides found in cereals (Figure 10.9; Fujino, 1983).

Ceramidephosphorylcholine (sphingomyelin) is a typical sphingophospholipid in animals but is not found in plants. However, ceramidephosphate (Hoshi *et al.*, 1973) and ceramidephosphorylinositol (Smith and Lester, 1974) have been found in yeast and sphingophosphoglycolipid has been detected in cereals, so that there is a possibility that

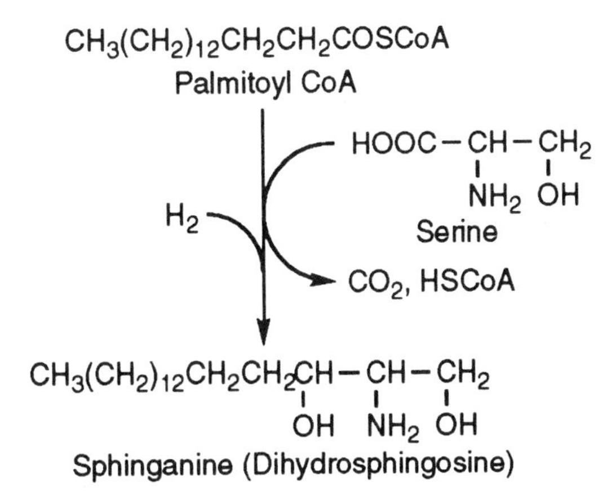

Figure 10.8 Biosynthesis of sphingosine.

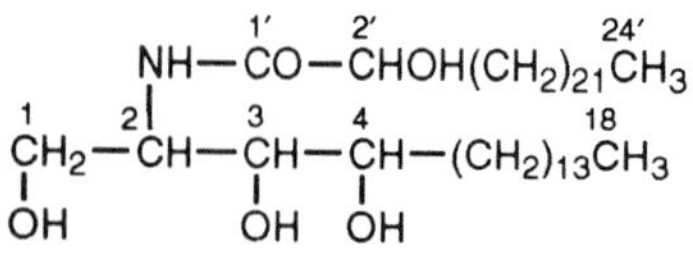

$$NH{-}\overset{1'}{C}O{-}\overset{2'}{C}HOH(CH_2)_{21}\overset{24'}{C}H_3$$
$$\overset{1}{C}H_2{-}\overset{2}{C}H{-}\overset{3}{C}H{-}\overset{4}{C}H{-}(CH_2)_{13}\overset{18}{C}H_3$$
$$OH \quad OH \quad OH$$

N-2′-Hydroxylignoceroyl-4-hydroxysphinganine

Figure 10.9 Typical ceramide species in cereals.

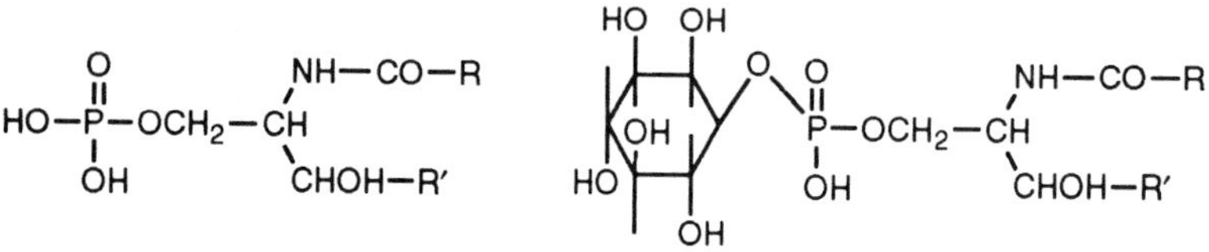

Phosphorylceramide (Ceramide phosphate)

Inositylphosphorylceramide (Ceramide phosphorylinositol)

Figure 10.10 Possible sphingophospholipids in cereals.

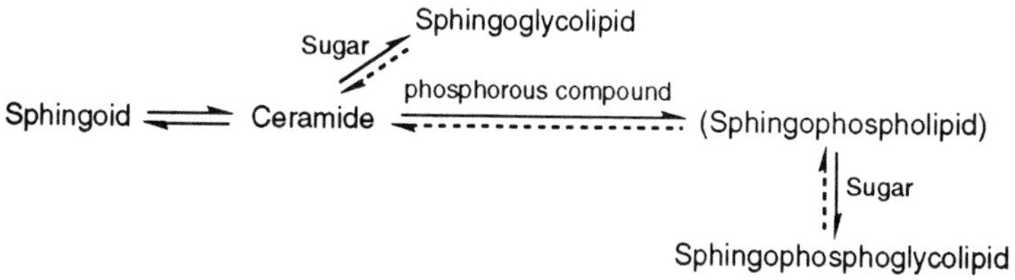

Sphingoid ⇌ Ceramide ⇌ (Sphingophospholipid) — phosphorous compound; Ceramide ⇌ Sphingoglycolipid — Sugar; (Sphingophospholipid) ⇌ Sphingophosphoglycolipid — Sugar

Figure 10.11 Metabolic relationships of sphingolipids in cereals.

ceramidephosphate and ceramidephosphorylinositol may be found in cereals as metabolic intermediates in small amounts (Figure 10.10). Generally, ceramide is known as an intermediate in the metabolism of sphingophospholipid and sphingoglycolipid and it is postulated that in plants ceramide will be a precursor to both sphingoglycolipid and sphingophosphoglycolipid (Figure 10.11).

10.1.4 Glycolipids

Naturally occurring glycolipids include glyceroglycolipids, sphingoglycolipids and sterylglycoside, all of which are found in cereals. Only the first two are discussed here, though the latter has been well studied (Fujino and Miyazawa, 1979; Fujino and Ohnishi, 1979a,b; Ohnishi and Fujino, 1980).

Glyceroglycolipid

These are glycosylglycerides where glyceride combines with neutral or acidic saccharide moieties. Although not abundant in nature, glyceroglycolipids can be a source of energy, essential fatty acids and sometimes even sulfur, nutritionally. Glyceroglycolipids are involved in photosynthesis in the plant and are metabolically derived from diglycerides (Figure 10.5).

(Neutral) glyceroglycolipid

More than 50 glyceroglycolipids occur in plants and microbes, varying in the structure of the sugar component. Several species are found in cereals including monosaccharide- and oligosaccharide-containing forms, especially diglycosyl types where galactose is a common sugar (Figure 10.12) (Kondo *et al.*, 1974; Fujino and Miyazawa, 1979; Weber, 1979). These molecules contain saturated and unsaturated fatty acids, among which the latter, particularly linoleic acid, predominates. The most abundant diglyceridic species is 18:2-18:2. Partial hydrolysis of cereal glyceroglycolipid gives chiefly 1-acyl lysolipids (Figure 10.5).

Glycerosulfoglycolipid

This is a class of sulfolipids with a sulfonyl sugar and an acidic glycolipid found in cereals (Kondo *et al.*, 1974). The sugar constituent is usually sulfoquinovose, or 6-sulfo-deoxy-glucose. The characteristics of the sulfolipids, originating in chloroplasts, differ considerably from those of the neutral glyceroglycolipids (Figure 10.12).

Sphingoglycolipid

These are glycosylceramides where ceramide combines mostly with neutral saccharides, which sometimes link with sulfur or phosphorus,

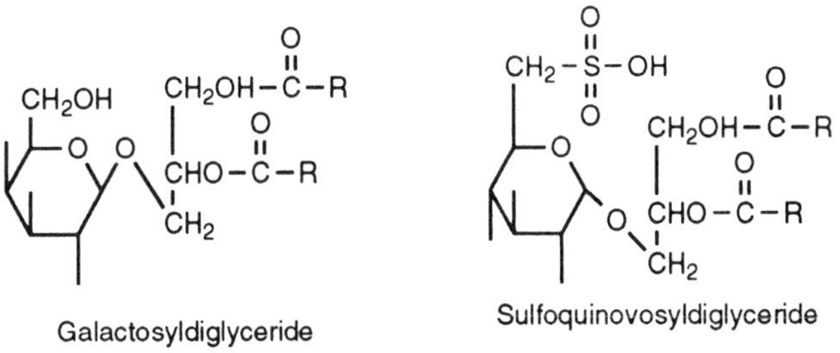

Galactosyldiglyceride

Sulfoquinovosyldiglyceride

Figure 10.12 Principal glycolipids in cereals.

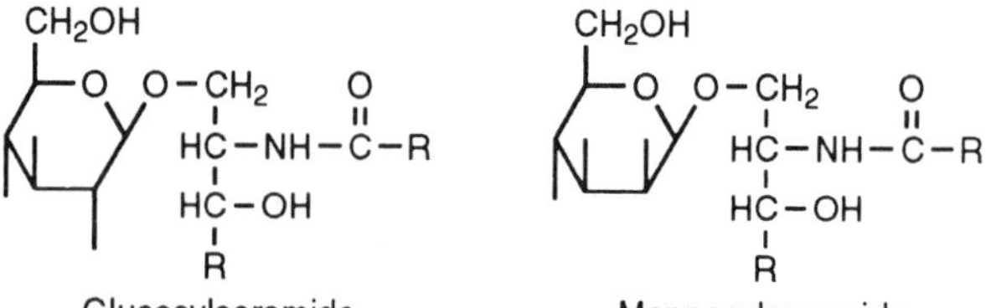

Glucosylceramide

Mannosylceramide

Figure 10.13 Typical cerebrosides in cereals.

though the former has not been detected in plants. Sphingoglycolipids are produced from ceramide as depicted in Figure 10.11.

(Neutral) sphingoglycolipid

More than 100 species exist, mainly in the animal kingdom, differing in the sugar component. Species reported in cereals include mono- and oligohexosyl types, especially the former as cerebrosides with glucose or mannose as the hexose (Figure 10.13). It has been shown in rice and wheat that the fatty acid component is principally C_{16} or C_{24}-hydroxy saturated whereas the component base is chiefly C_{18} unsaturated sphingenine (Laine and Renkonen, 1974: Fujino and Ohnishi, 1982).

Sphingophosphoglycolipid

This is an acidic glycolipid, where ceramidephosphorylinositol combines with several sugars. A compound containing a tetrasaccharide of mannose-inositol-glucuronic acid-glucosamine, called phytoglycolipid (Carter *et al.*, 1969), is distributed widely in the leaves and seeds of plants including cereals (Figure 10.14) (Ito *et al.*, 1985). Cereal phytoglycolipid has been reported to contain a 2-hydroxy C_{20} acid component and a 4-hydroxy C_{18} sphingoid base (phytosphingosine).

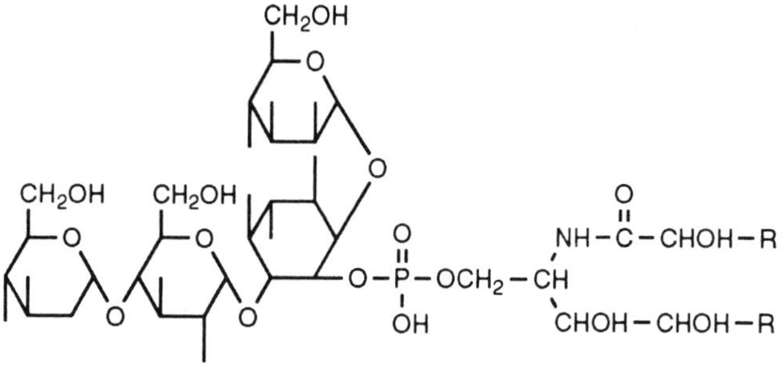

Aminoglucosyl-glucuronyl-(mannosyl-)inosityl-ceramide

Figure 10.14 Phytoglycolipids in plants.

10.1.5 Distribution of lipids in cereals

Lipids are distributed universally but unevenly throughout the grain in tissues, cells and organelles.

Tissue

The distribution of lipids between tissues such as the embryo and endosperm and endosperm fractions such as the aleurone, sub-aleurone and starchy endosperm is similar for species such as wheat and rye, rice and barley, and maize and millet.

Wheat

The distribution of lipids in the wheat grain is shown in Table 10.2 as an example. Lipids are found in all tissues with triglycerides being more abundant in the aleurone and embryo and present in low levels in the endosperm while phospholipids and glycolipids are present at higher concentrations in the endosperm.

Rice

The distribution of lipids in rice grain is shown in Table 10.3. Lipids are found in greatest quantities in the bran and embryo followed by the polish, approximately corresponding to the aleurone layer, sub-aleurone layer and inner endosperm. Neutral lipids decrease in the same order whereas phospholipids and glycolipids increase. Bran is quite rich in lipid (20–25%) and is often used to produce rice bran oil or rice oil.

Maize

Data on the distribution of lipids in maize grain is given in Table 10.4. Lipids, especially neutral lipids and phosopholipids, are abundant in the embryo whereas glycolipids are present at higher concentrations in the endosperm. Maize germ, being rich in lipid (35–50%) is frequently used for processing to yield corn oil.

Cellular location

Lipids are found as intracellular components in cereals, principally in spherosomes, protein bodies and starch granules. The coexistence of these organelles in the sub-aleurone of cereal grains is depicted in Figure 10.15.

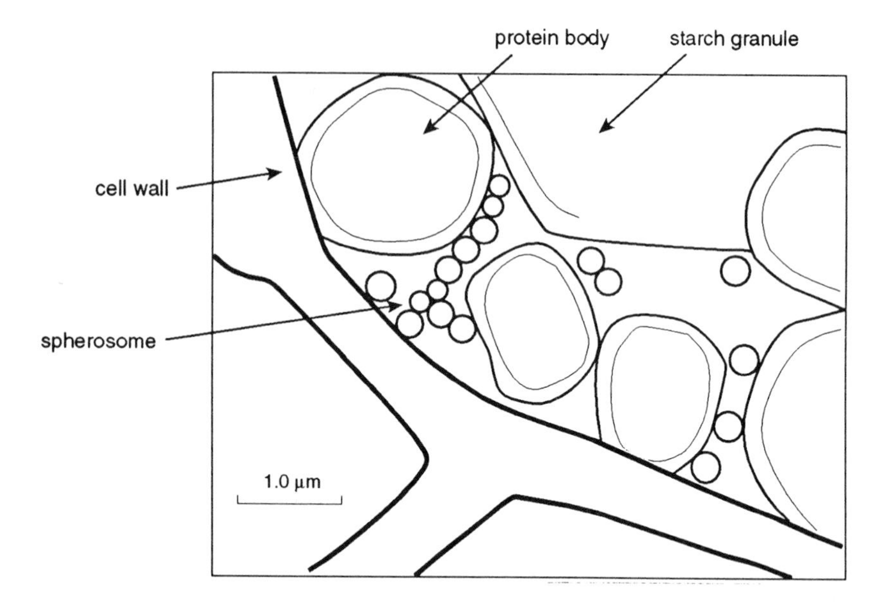

Figure 10.15 Electron microscopic pattern of sub-aleurone layer in cereal grains. (Modified from Bechtel and Pomeranz, 1978).

Spherosomes

These small lipid bodies are distributed widely but are especially abundant in the aleurone, sub-aleurone and embryo. Spherosomes contain most of the total lipid in the cereal grain with predominantly neutral lipid, or triglycerides and smaller quantities of phospholipid and glycolipid as in the whole grain (Tables 10.2. 10.3 and 10.4).

Protein bodies

Protein bodies are found in the cells of the outer and inner endosperm. These bodies are mainly protein with lipid, carbohydrate, phytin and minerals as minor components. The lipids of protein bodies are mainly neutral lipids with minor amounts of polar lipids consisting chiefly of glycosylsterol and phospholipid mainly lysolecithin and lysocephalin (Hirayama and Matsuda, 1971). The lipid composition is intermediate between that of the spherosomes and starch granules.

Starch granules

Starch granules are found in the cells of the inner endosperm and consist of starch for the greater part with lipid, protein and minerals in smaller quantities. Lipids represent approximately 1% of the starch

Table 10.5 Internal lipid in cereal starch (mg/100 g). (Drawn from Morrison, 1983)

Lipid	*Wheat*	*Rice*	*Maize*
Triglyceride, diglyceride and acylsterol	1–2	2–12	3
Fatty acid	15–54	221–355	379
Monoglyceride and acylglycosylsterol	1–2	19–31	9
Mono- and diglycosylmonoglyceride	4–12	20–40	13
Lysophosphatidylglycerol	30–41	38–48	7
Lysophosphatidylethanolamine	41–75	86–112	20
Lysophosphatidylcholine	451–734	453–513	262
Others	27–62	–	9

granule, consisting mainly of external and internal fractions (Fujino, 1978). The outer lipid is associated with the terminals of the amylopectin chains with a 'broom-structure' on the surface of the granules. The internal lipid is bound inside the amylose helix-structure within the starch granules. The external lipid being easily removed is at times called non-starchy lipid whereas the internal lipid being more difficult to remove is sometimes called true-starchy lipid. 'Fat-by-hydrolysis' of starch in the old literature almost corresponds with the internal lipid fraction of the starch granule.

Both external and internal lipids are rich in monoacyl compounds. In the internal lipid this is principally lysolecithin in wheat and lysolecithin and fatty acids in rice and maize (Table 10.5). The monoacyl lipid which is included inside the helix of amylose forms an amylose-lipid complex (Acker and Becker, 1971; Fujino, 1978; Morrison, 1978; Ito *et al.*, 1979). The monocayl group includes both saturated and unsaturated fatty acids with the latter predominating. These lipids are postulated to serve as a template in the biosynthesis of the helix structure of amylose in starch granules (Morrison, 1978).

10.2 INORGANIC SUBSTANCES

The inorganic components of the grain include water and minerals. The inorganic composition of grain is shown in Table 10.6. These components contribute to the nutritional and other value of the grain (Juliano, 1972; Obara, 1981; Kurasawa, 1982; Nagao, 1984).

10.2.1 Water

Biological systems contain free and bound water, both contributing to factors such as cereal quality, taste and storage properties.

Table 10.6 Composition of inorganic substances in cereals (per 100 g). (Resources Council, 1982)

Cereal				*Mineral*				
	Water (g)	*Solid (g)*	*Ash (g)*	*Ca (mg)*	*P (mg)*	*Fe (mg)*	*Na (mg)*	*K (mg)*
Wheat								
Whole grain	13.0	87.0	1.6	30	390	3.9	3	350
Wheat flour	14.5	85.5	0.4	20	75	1.0	2	80
White bread	38.0	62.0	1.7	36	70	1.0	520	95
Wheat germ	9.2	90.8	4.1	65	1200	6.6	12	1100
Rice								
Whole grain	15.5	84.5	1.3	10	300	1.1	2	250
White rice, raw	15.5	84.5	0.6	6	140	0.5	2	110
White rice, boiled	65.0	35.0	0.1	2	30	0.1	2	27
Rice bran	13.5	86–5	8.9	46	1500	6.0	5	1800
Maize								
Whole grain	14.5	85.5	1.3	5	290	2.3	3	290
Sweetcorn, raw	74.7	25.3	0.7	3	80	0.6	0	300
Sweetcorn, boiled	74.7	25.3	0.7	3	65	0.6	0	290

Free water

The free water is not associated with other components and can be easily removed by drying. Free water is the major component of the total water in foods. The water content given in food analysis is usually the free water value. Cereals contain of the order of 15% water in the raw state depending on variety and growth conditions but this usually increases to 40–80% in the food depending upon processing (Table 10.6). The free water content influences the storage life of grain and the taste, especially the texture of processed cereals.

Bound water

Bound or combined water is associated with other components in the cereal and cannot be easily removed by drying. The bound water is a minor part of the total water content of the food. Absolute determination of bound water is difficult but water activity (A_w: vapour pressure P of food/vapour pressure P_0 of water) can be measured. A low A_w signifies a higher proportion of the water is bound (Table 10.7; e.g. Fujino, 1977). Raw cereal is a less-watery food whereas cooked cereals are medium or more watery foods.

The A_w necessary for the growth of microbes is around 0.95 for bacteria, 0.85 for yeast and 0.75 for mould. Thus raw cereals are resistant to microorganisms but most processed cereal foods are not. Lipid oxidation, non-enzymatic browning and other reactions occur more at high A_w as found in the cereal prepared for consumption.

10.2.2 Minerals

Minerals represent about 1.5% of the grain and decrease to much lower levels on milling (Table 10.6). Phosphorus is the most abundant mineral

Table 10.7 Water activity in food

Food	*Water content (%)*	*Water activity* (A_w)
Pure water	100	1.00
More-watery food	90–60	0.99–0.97
Medium-watery food	60–20	0.97–0.75
Less-watery food	20–10	0.75–0.60
Dried food	10–2	0.60–0.10
Non-watery food	0	0.00

being followed by potassium whereas calcium, sodium and iron are present in low amounts. Most cereal mineral is probably present in an organic form and some in inorganic forms.

Inorganic forms

Alkaline metals such as potassium and magnesium are considered to exist in the form of phosphates, carbonates, chlorides and sulfates, especially as phosphates. The soluble forms may contribute to the sense of taste of some cereals.

Organic forms

Phosphorus, the major mineral in cereals, actually exists in organic forms as phytin, phospholipid, nucleotide or some other compound. The phosphorus in phytin represents almost 80% of the total phosphorus in cereal (Nelson *et al.*, 1968). Phosphorylation of inositol gives phytic acid which combines with alkaline minerals to form phytin, a major form of stored phosphorus in cereals (Figure 10.16; Kurasawa *et al.*, 1969). Phytic acid is associated with protein, carbohydrate and even lipid via the phosphoryl moiety (Thompson, 1986; Lasztity and Lasztity, 1990). Phytin and phytic acid are sources of inositol and phosphorus in the diet but may inhibit the absorption of minerals such as Ca, Mg, Cu and Fe in the small intestine.

10.2.3 Distribution of inorganic components

The water content of raw cereals is almost equivalent to that of milled cereals (Table 10.6). This indicates that water is relatively uniformly distributed in the grain.

In contrast, minerals are distributed unevenly throughout the grain. Minerals are abundant in the outer layers such as the pericarp, aleurone layer and embryo but present in low concentrations in the endosperm.

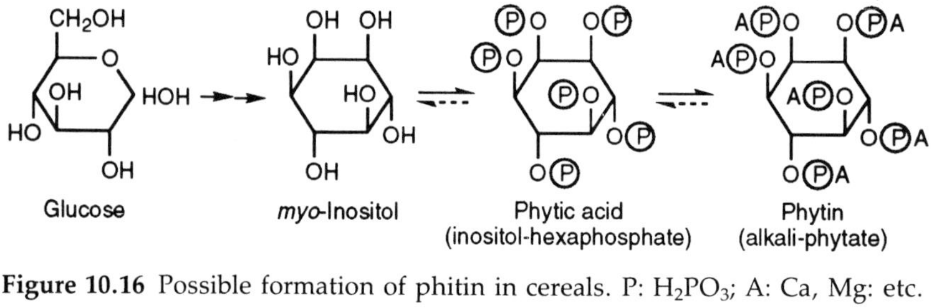

Glucose → *myo*-Inositol ⇌ Phytic acid (inositol-hexaphosphate) ⇌ Phytin (alkali-phytate)

Figure 10.16 Possible formation of phitin in cereals. P: H_2PO_3; A: Ca, Mg: etc.

Phytic acid is present in high concentrations in the embryo but is scarce in the endosperm (Lasztity and Lasztity, 1990).

10.3 CEREAL VITAMINS

Vitamins may be divided into fat-soluble and water-soluble groups. The water-soluble vitamins include B_1 (thiamine), B_2 (riboflavin), niacin, pantothenic acid, inositol and C (ascorbic acid), whereas the fat-soluble group includes A_1 (retinol), D_2.D_3 (calciferol), E (tocopherol) and K (phylloquinone). The water-soluble group are not systematically related but the fat-soluble group belong to the terpenoid series (Figure 10.1). Many of these are found in cereals (Juliano, 1972; Obara, 1981; Kurasawa, 1982; Nagao, 1984) as shown in Table 10.8.

10.3.1 Water-soluble vitamins

The main water-soluble vitamins in cereals are B_1, B_2 and niacin, especially B_1. These decrease greatly on milling (Table 10.8) and further on cooking and processing. Cereals, particularly polished rice and wheat flour, are often enriched in these vitamins because of the losses during preparation. An insoluble B_1 compound is often added to rice to prevent dissolution of the B_1 during processing (Figure 10.17). B_1 in the form of thiamine pyrophosphate is involved in the metabolism of glucose whereas B_2 in the form of nucleotides is involved in oxidation–reduction reactions.

10.3.2 Fat-soluble vitamins

The main fat-soluble vitamins in cereals are tocopherol (E group) and carotenoids (provitamin A group), especially the former. Chemically, tocopherol is a bound diterpene whereas carotenoids are in the tetraterpene series. These vitamins decrease markedly on milling and

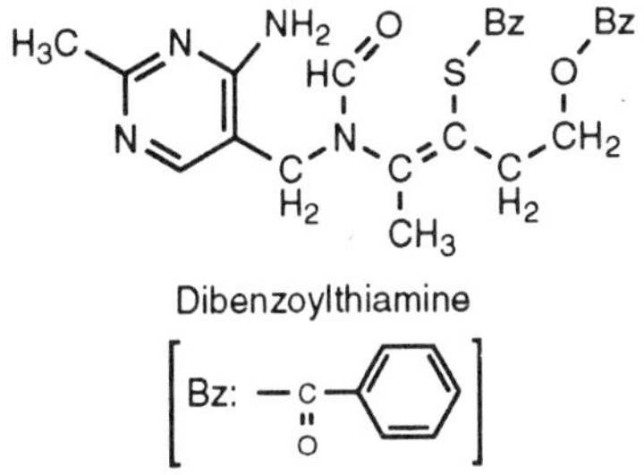

Dibenzoylthiamine

[Bz: —C(=O)—C₆H₅]

Figure 10.17 Artificial vitamin B_1 (example).

Table 10.8 Principal vitamins in cereals (per 100 g). (Resources Council, 1982: 1990)

Cereal	*Lipidic*		*Non-lipidic*			
	A^* *Potency* *(IU)*	$E^\dagger$ *Potency* *(mg)*	B_1 *(Thiamine)* *(mg)*	B_2 *(Riboflavin)* *(mg)*	*Niacin* *(mg)*	C *(Ascorbic acid)* *(mg)*
Wheat						
Whole grain	0	1.4	0.46	0.09	5.0	0
Wheat flour	0	0.4	0.10	0.05	0.9	0
White bread	0	0.5	0.07	0.07	0.7	0
Wheat germ	0	29.3	2.10	0.60	7.0	0
Rice						
Whole grain	0	1.6	0.54	0.06	4.5	0
White rice, raw	0	0.4	0.12	0.03	1.4	0
White rice, boiled	0	0.2	0.03	0.01	0.3	0
Rice bran	0	–	2.50	0.50	25.0	0
Maize						
Whole grain	100	–	0.30	0.10	2.0	0
Sweetcorn, raw	24	0.4	0.16	0.14	2.4	10
Sweetcorn, boiled	24	0.4	0.16	0.14	2.3	6

*Carotenoid.
†Total tocopherol homologues.

often decline further on cooking and processing. Wheat and rice both lack vitamin A and are frequently enriched in this vitamin.

Tocopherol

The fundamental structure is formed by condensation of a hydroquinone with a non-cyclic diterpene. Eight homologues, α-, β-, γ-, δ-tocopherol and α-, β-, γ-, δ-tocotrienol, are known. In addition, oryzanol, a substance similar to tocopherol, is found in rice. Oryzanol has a structure in which ferulic acid is esterified with a cyclic triterpenic alcohol (Shimizu and Kurokawa, 1988). Feruloylcycloartenol is typical of about 10 analogues. There are close relationships between tocopherol and oryzanol structurally and biochemically (Figure 10.18) (Fujino, 1993, 1994). Both compounds have a terminal phenolic-OH which is conducive to inhibition of lipid oxidation.

Carotenoid

These can be divided into carotene (hydrocarbon) and xanthophyll (non-hydrocarbon) groups. Both, especially lutein and zeaxanthin (giving the yellowish colour to maize) from the latter group, are found in cereals.

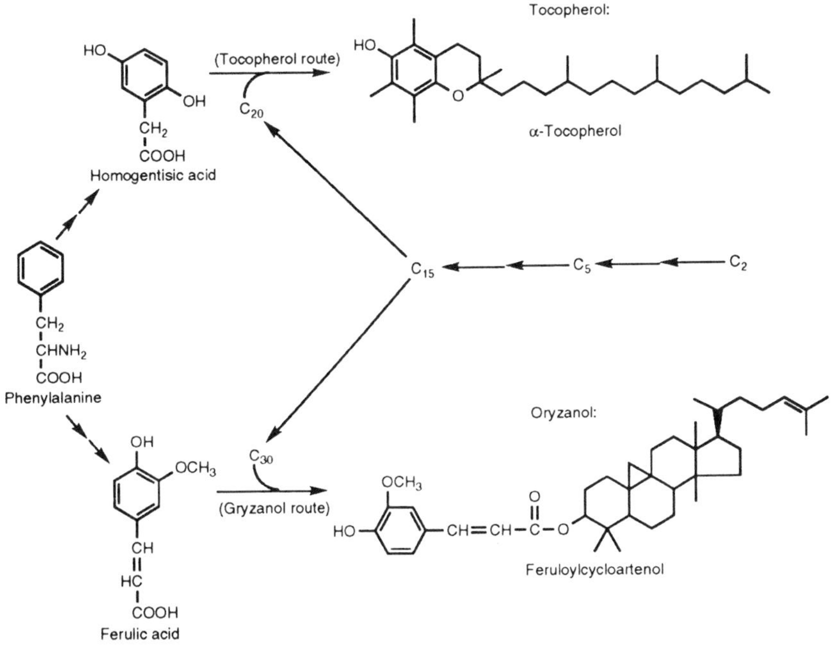

Figure 10.18 Biosynthetic relationships postulated for tocopherol and oryzanol.

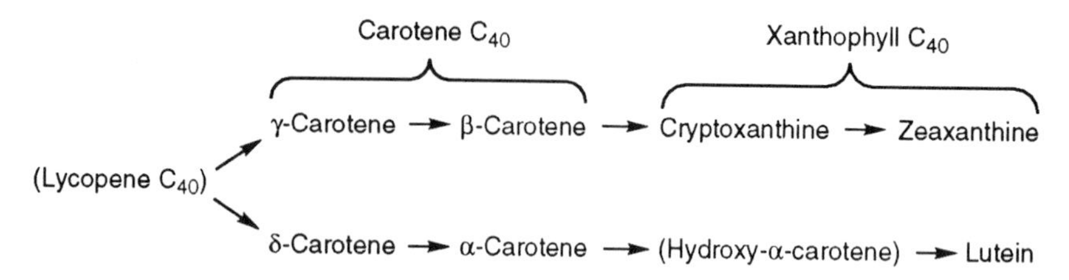

Figure 10.19 Biosynthesis of carotenoids in cereals.

These carotenoids are related biosynthetically (Figure 10.19) (Fujino, 1983). Some carotenoids are metabolized within the body to form retinol and homologues. *β*-Carotene itself has recently been reported to have antioxidative and anticarcinogenic properties.

10.3.3 Distribution of vitamins in cereals

Generally the water-soluble vitamins are found more in the outer parts of the grain. This tendency is stronger in rice than in wheat. Rice bran and germ is rich in B_1, B_2 and niacin whereas the coat and aleurone and especially the endosperm are poorer and these vitamins are almost absent from boiled white rice (Table 10.8). Almost half of the tocopherol in wheat is reported to be in the embryo (Hall and Laidman, 1968). Carotenoid, especially xanthophyll, is abundant in the inner parts of maize such as the endosperm but lacking in the bran and germ (Quackenbush *et al.*, 1961).

10.4 TOXIC OR ANTI-NUTRITIONAL SUBSTANCES IN CEREALS

Toxic or at least anti-nutritional substances are sometimes found in cereals. These can be natural (of cereal origin) or environmental (of non-cereal origin).

10.4.1 Naturally occurring components

Toxic proteins and lipids may be present in small amounts in cereals.

Proteinaceous components

Allergens and enzyme inhibitors are known in cereals. These may be similar proteins (Gomez *et al.*, 1990; Matsuda and Adachi, 1992) and may be largely denatured by heating or decomposed by digestion.

Antigens

A gliadin fraction from wheat causes coeliac disease (gluten sensitivity). Proteins from wheat, rice, maize and barley may give rise to allergic reactions such as atopic dermatitis and asthma (Nakamura, 1987). A hypoallergenic rice has recently been developed using enzymatic techniques (Watanabe *et al.*, 1990).

Enzyme inhibitors

Proteins which inhibit α-amylase and/or trypsin are found in cereals such as wheat, maize and rye. A family of α-amylase and trypsin inhibitors are present in these species (Barber *et al.*, 1986). The main allergen in wheat and barley is an α-amylase inhibitor (Gomez *et al.*, 1990).

Lipid components

Some cereal lipids such as monoglycerides, diglycerides, lysophospholipids and lysoglycolipids exhibit haemolytic activity. Lysolecithin and lysocephalin, in particular, have been implicated but can be degraded in the digestive system.

10.4.2 Contaminants

Cereals are sometimes contaminated with agricultural, industrial and biological pollutants (Kawasiro, 1977; Kurata and Uchiyama, 1979; Fujino, 1980). This type of contaminant cannot necessarily be detoxified in the usual cereal production and consumption process and so should be strictly avoided.

Agricultural chemicals

Agricultural chemicals are commonly applied in cereal production in the field and post-harvest. Thus , inorganic contaminants (e.g. As-, Br-, Pb-series etc.) and organic contaminants (BHC, DDT, parathion, etc., Figure 10.20) can remain in cereals as toxins.

Industrial chemicals

Diverse inorganic (Cd-, Hg-, Pb-series, etc.) and organic (PCB, benzopyrene, etc., Figure 10.21) industrial pollutants may contaminate cereals. Brown rice has, for example, been contaminated with cadmium or PCB.

Biological contaminants

Cereals may be exposed to small animals (mice, insects, etc.) or microorganisms (moulds, yeasts, bacteria, etc.) during cultivation, transport and storage. Damage by small animals can result in microbial contamination. Moulds are an especially common problem in cereals. Toxins produced by moulds are collectivelly called mycotoxins. Cereal mycotoxins that have been reported include ergot-toxin (lysergic acid) produced by ergot fungi on wheat and rye, aflatoxins (B_1, B_2 etc.) produced by *Aspergillus flavus* on wheat, rice, maize and other species, yellowed rice-toxins (citrinin, etc.) produced by *Penicillium* species on rice and toxins produced by *Fusarium* on wheat and other cereals (Figure 10.22).

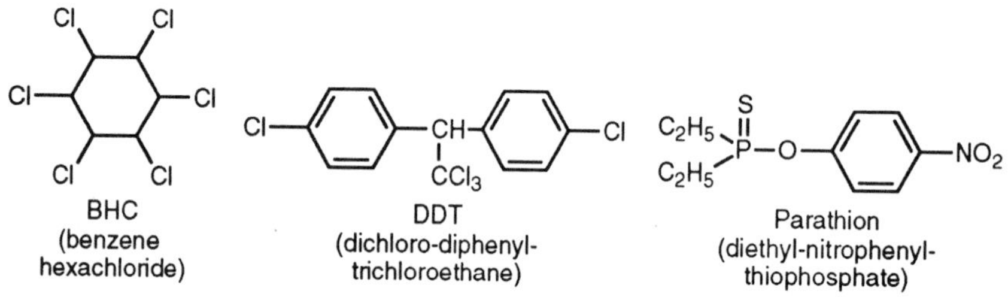

BHC (benzene hexachloride)
DDT (dichloro-diphenyl-trichloroethane)
Parathion (diethyl-nitrophenyl-thiophosphate)

Figure 10.20 Agricultural chemicals that may contaminate cereals.

PCB [$m + n$: 2–10] (polychloro-biphenyl)
Benzo [α] pyrene (3,4-benzopyrene)

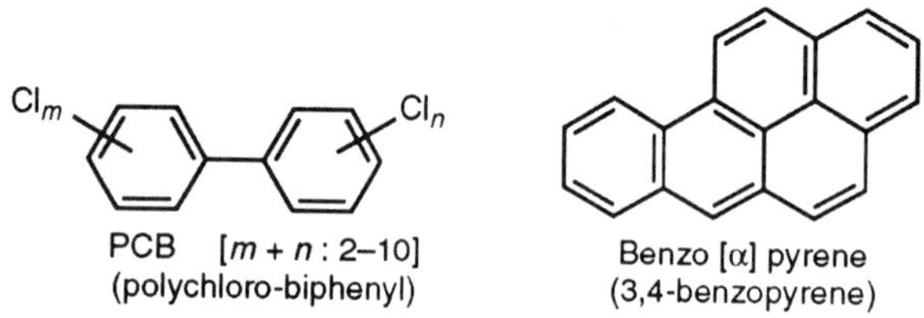

Figure 10.21 Industrial chemicals that may contaminate cereals.

Lysergic acid
Aflatoxin B_1
Citrinin

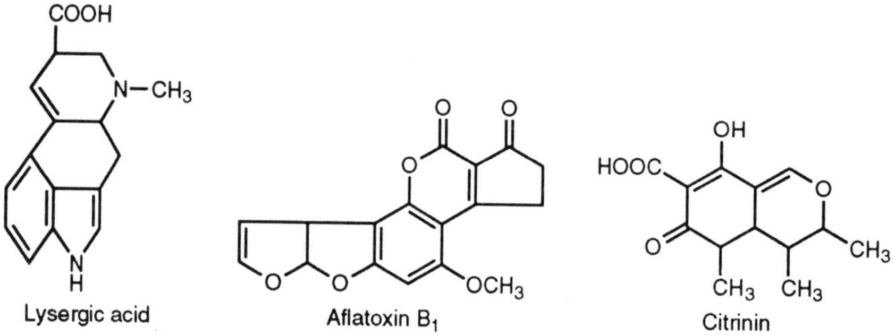

Figure 10.22 Mycotoxins found in cereals.

Table 10.9 Residue limits for agricultural chemicals (ppm). (Example from Japan (Ministry of Health & Welfare, 1994))

Agricultural chemicals	*Cereal*				*Legume*	
	Rice	*Wheat*	*Maize*	*Buckwheat*	*Soybean*	*Pea*
Inorganic						
Br	–	50	–	–	–	–
As-, Pb-	–	–	–	–	–	–
Organic (C1)						
BHC	0.2	0.2	0.2	0.2	0.2	0.2
DDT	0.2	0.2	0.2	0.2	0.2	0.2
Dorin series	ND	ND	ND	ND	–	ND
Organic (P)						
Parathion	ND	0.3	0.3	0.3	0.3	0.3
Malathion	0.1	8.0	2.0	2.0	0.5	0.5

ND, not detected.

Table 10.10 Differences in tolerance of agricultural chemical residues (ppm)

Chemical	*Japan M*	*Japan E*	*FAO/WHO*	*USA*
Organic P series				
Malathion				
Brown rice	0.1*	–	8.0	8.0
Wheat	8.0	0.5*	8.0	8.0
Wheat flour	1.2	0.5*	2.0	–
Parathion				
Brown rice	ND*	–	–	1.0
Wheat	0.3*	–	–	1.0
Chlorpyrifosmethyl				
Brown rice	(0.1)	0.01*	0.1	6.0
Wheat	(0.1)	–	10.0	6.0
Wheat flour	–	0.01	2.0	–
Organic non-P series				
Imazalil				
Brown rice	0.05	–	–	–
Wheat	0.01	–	0.01	0.05

M, Ministry of Health & Welfare; E, Environmental Agency; FAO, Food & Agricultural Organization; WHO, World Health Organization; ND, not detected

*, value having been settled up to 1993.

10.4.3 Distribution of contaminants in the grain

Natural toxins may be distributed uniformly in the grain but other environmental contaminants are more likely to be found on the grain surface. Residual limits for agricultural chemicals in cereals have been set (Table 10.9) but are sometimes confusingly different (Table 10.10). Adequate standards should be set for the security of human health and the environment.

REFERENCES

Acker, L. and Becker, G. (1971) Recent studies on the lipids of cereal starches II. Lipids of various types of starch and their binding to amylose. *Starch*, **23**, 419–24.

Barber, D., Scanchez-Monge, R., Garcia-Olmendo, F., Salcedo, G. and Mendez, E. (1986) Evolutionary implications of sequential homologies among members of the trypsin/alpha-amylase inhibitor family (CM-proteins) in wheat and barley. *Biochimica et Biophysica Acta*, **873**, 147–51.

Bechtel, D.B. and Pomeranz, Y. (1978) Implications of the rice kernal structure in storage, marketing and processing: a review. *Journal of Food Science*, **43**, 1538–52.

Carter, H.E., Strobach, H.E. and Hawthorne, J.N. (1969) Biochemistry of the sphingolipids XVIII. Complete structure of tetrasaccharide phytoglycolipid. *Biochemistry*, **8**, 383–8.

Chouduhry, N.H. and Juliano, B.O. (1980) Effect of amylose content on the lipids of mature rice grain. *Phytochemistry*, **19**, 1385–9.

Fujino, Y. (1977) *Outline of Food Chemistry (Jpn.)*, Shoka-bo, Tokyo.

Fujino, Y. (1978) Rice lipids. *Cereal Chemistry*, **55**, 559–71.

Fujino, Y. (1980) Toxic matters in Japanese rice, in *Reports of the International Association for Cereal Chemistry*, Victoria Druck, Vienna, pp. 53–8.

Fujino, Y. (1983) Lipids in cereals. *Oil Chemistry (Jpn.)*, **32**, 67–81.

Fujino, Y. (1989) Comprehensive review of glycolipids in nature. *Reports of Japanese Conference on Biochemistry of Lipids*, **31**, 1–3.

Fujino, Y. (1993) Non-food uses of rice as a whole, in *Proceedings of the ICC Symposium* (R. Lasztity and M. Karpati, eds), International Association of Cereal Science and Technology, Budapest, pp. 43–59.

Fujino, Y. (1994) Utilization of rice in Japan. *Ernährungsforschung*, **39**, 65–81.

Fujino, Y. and Miyazawa, T. (1979) Chemical structures of mono-, di-, tri- and tetraglycosylglycerides in rice bran. *Biochimica et Biophysica Acta*, **572**, 442–51.

Fujino, Y. and Ohnishi, M. (1979a) Isolation and structure of diglycosylsterols and triglycosylsterols in rice bran. *Biochimica et Biophysica Acta*, **574**, 94–102.

Fujino, Y. and Ohniski, M. (1979b) Novel sterylglycosides, cellotetraosylsitosterol and cellopentaosylsitosterol in rice grain. *Proceedings of the Japan Academy*, **55B**, 243–6.

Fujino, Y. and Ohnishi, M. (1982) Species of sphingolipids in rice grain. *Proceedings of the Japan Academy*, **58B**, 36–9.

Gomez, L., Martin, E., Hernandez, D., Sanchez-Monge, R., Barber, D., del Pozo, V., de Andres, B., Armentia, A., Lahoz, C., Salcedo, G. and Palomino, P. (1990) Members of the alpha-amylase inhibitors family from wheat endosperm are major allergens associated with baker's asthma. *FEBS Letters*, **261**, 85–8.

Hall, G.S. and Laidman, D.L. (1968) The determination of tocopherols and isoprenoid quinones in the grain and seedlings of wheat (Triticum vulgare) *Biochemical Journal*, **108**, 465–73.
Hargin, K.D. and Morrison, W.R. (1980) The distribution of acyl lipids in the germ, aleurone, starch and non-starch endosperm of four wheat varieties. *Journal of Science of Food and Agriculture*, **31**, 877–88.
Hirayama, O. and Matsuda, H. (1971) lipid components and distribution in brown rice. *Bulletin of Japanese Agricultural Chemical Society (Jpn.)*, **47**, 371–7.
Hoshi, H., Kishimoto, Y. and Hignite, C. (1973) 2,3-Erythyo-dihydroxy-hexacosanoic acid homologues: Isolation from yeast cerebrin phosphate and determination of their structures. *Journal of Lipid Research*, **14**, 406–14.
Ito, S., Sato, S. and Fujino, Y. (1979) Internal lipid in rice starch. *Starch*, **37**, 217–21.
Ito, S., Kojima, M. and Fujino, Y. (1985) Occurrence of phytoglycolipid in rice bran. *Agricultural and Biological Chemistry*, **49**, 1873–5.
Juliano, B.O. (1972) The rice caryopsis and its composition, in *Rice Chemistry and Technology* (D.F. Houston, ed.), American Association of Cereal Chemists, St. Paul, MN, pp. 16–74.
Juliano, B.O. (1983) Lipids in rice and rice processing, in *Lipids in Cereal Technology* (P.J. Barnes, ed.), Academic Press, London pp. 305–30.
Kawasiro, I. (1977) *Food Hygienics (Jpn.)*, Kosei-kan, Tokyo.
Kondo, Y., Ito, S. and Fujino, Y. (1974) Sulphoquinovosyldiglyceride and triglycosyldiglyceride in rice grains. *Agricultural and Biological Chemistry*, **38**, 2549–52.
Kurasawa, F. (1982) *Rice and its processing (Jpn.)*, Kenpaku, Tokyo.
Kurasawa, F., Hayakawa, T. and Watanabe, S. (1969) Change of inositol phosphate in rice seed during germination in dark. *Bulletin of Japanese Agricultural Chemistry Society (Jpn.)*, **43**, 55–9.
Kurata, H. and Uchiyama, M. (1979) *Food poisoning (Jpn.)*, Kosei-kan, Tokyo.
Kuroda, N., Ohinishi, M. and Fujino, Y. (1977) Sterol lipids in rice bran. *Cereal Chemistry*, **54**, 997–1006.
Lasztity, R. and Lasztity, L. (1990) Phytic acid in cereal technology, in *Advances of Cereal Science and Technology*, Vol. X (Y. Pomeranz ed.), American Association of Cereal Chemists, St. Paul, MN, pp. 309–71.
Laine, R.A. and Renkonen, O. (1974) Ceramide di- and trihexosides of wheat flour. *Biochemistry*, **12**, 2887–43.
Legendre, M.G., Dupuy, H.P., Ory, R.L. and McIlrath, W.O. (1978) Instrumental analysis of volatiles from rice and corn products. *Journal of Agricultural and Food Chemistry*, **26**, 1035–8.
Mano, Y., Ohnishi, M., Sasaki, S., Kojima, M., Ito, S. and Fujino, Y. (1989) Molecular species of triacylglycerols in rice bran and some oil-bearing seeds. *Journal of Japanese Nutrition and Food Society (Jpn.)*, **42**, 251–8.
Matsuda, K. and Adachi, T. (1992) Alpha-amylase and trypsin inhibitor family. *Chemistry and Biology (Jpn.)*, **30**, 763–5.
Miyazawa, T., Yoshino, Y. and Fujino, Y. (1977) Studies on phospholipids in non-glutinous and glutinous rice bran. *Journal of the Science of Food and Agriculture*, **28**, 889–94.
Morrison, W.R. (1978) Cereal lipids, in *Advances in Cereal Science and Technology*, Vol.II (Y. Pomeranz, ed.), American Association of Cereal Chemists, St. Paul, MN, pp. 221–348.
Morrison, W. R. (1983) Acyl lipids in cereals, in *Lipids in Cereal Technology* (P.J. Barnes, ed.), Academic Press, London, pp. 11–32.

Nagao, S. (1984) *Wheat and its Processing (Jpn.)*, Kenpaku, Tokyo.
Nakamura, R. (1987) Allergens in cereals. *Chemistry and Biology (Jpn. Rev.)*, **25**, 739–41.
Nelson, T.S., Ferrara, L.W. and Storer, N.L. (1968) Phytate phosphorus content of feed ingredients derived from plants. *Poultry Science*, **47**, 1372–4.
Obara, T. (1981) *Non-staple cereals (Jpn,)*, Jusonbo, Tokyo.
Ohnishi, M. and Fujino, Y. (1980) Structural study on new sterylglycosides in rice bran, cellotetraocylsitosterol and cellopentaosylsitosterol. *Agricultural and Biological Chemistry*, **44**, 333–8.
Quackenbush, F.W., Firch, J.G., Rabourn, W.J., McQuistan, M., Petzold, E.N. and Kargl, T.E. (1961) Composition of corn – analysis of carotenoids in corn grain. *Journal of Agricultural and Food Chemistry*, **9**, 132–4.
Resources Council (ed.) (1982) *Standard Tables of Food Composition in Japan (Jpn.)*, Government Printing Bureau, Tokyo.
Resources Council (ed.) (1990) *Tables of Lipidic Components of Foods in Japan (Jpn.)*, Government Printing Bureau, Tokyo.
Shimizu, S. and Kurokawa, J. (1988) Rice oil, in *Handbook of Fat and Fatty Foods (Jpn.)* (Y. Abe, ed.) Saiwai-shobo., Tokyo, pp. 139–57.
Smith, S.M. and Lester, R.L. (1974) Inositol phosphorylceramide, a novel substance and the chief member of a major group of yeast sphingolipids containing a single inositol phosphate. *Journal of Biological Chemistry*, **249**, 3395–405.
Thompson, L.V. (1986) Phytic acid: Factor influencing on starch digestibility and blood glucose response, in *Phytic Acid: Chemistry and Applications* (E. Graf, ed.) Pilatus Press, Minneapolis, pp. 173–94.
Watanabe, M., Miyakawa, J., Ikezawa, Z., Suzuki, Y., Hirano, T., Yoshizawa, T. and Arai, S, (1990) Production of hypoallergenic rice by enzymatic decomposition of constituents proteins. *Journal of Food Science*, **55**, 781–3.
Weber, E.J. (1978) Corn lipids. *Cereal Chemistry*, **55**, 572–84.
Weber, E.J. (1979) The lipids of corn germ and endosperm. *Journal of American Oil Chemists Society*, **56**, 637–41.
Yasumatsu K., Moritaka, S. and Wada, S. (1966) Studies on cereals IV. Volatile carbonyl compounds in cooked rice, V. Stable flavor of stored rice. *Agricultural and Biological Chemistry*, **30**, 478–82; 483–6.

Part Three

Breeding for Cereal Quality

11

Breeding cereals for quality improvement

C.W. Wrigley and C.F. Morris

11.1 PHILOSOPHY AND STRATEGY FOR QUALITY IMPROVEMENT

11.1.1 The breeder's role in crop improvement

Grain quality is increasingly becoming an important factor in grain trading, both internationally and within grain-producing countries. No longer does the term 'quality' merely refer to bulk density (plumpness of grain), the absence of extraneous material (e.g. dockage or pesticides) and general soundness. Instead, an increasingly technical market is looking to the breeder to tailor varieties to the wide range of specific processing requirements detailed in Chapters 1 to 7. Not only is the breeder required to 'build in' these quality attributes, but also to add genetically-based resistance or tolerance to pests and environmental vagaries which reduce yield and quality.

Indeed, the primary effort of most breeders is directed towards assembling resistance or tolerance to a myriad of fungi, bacteria, insects, viruses and nematodes in what is known as 'adapted' material. Adaptation, put another way, is simply the fitness of crop growth and development in a particular environment. Examples include winter hardiness of winter-sown cereals, early heading and maturation in environments where climate shifts from mild and moist to hot and dry, and short stature for high yield, intensive management systems. Although pests and environment are traditionally considered yield-limiting factors, the breeder is increasingly asked to limit their adverse impact on quality; e.g. reducing the occurrence or severity of pre-harvest sprouting by incorporating genes for seed dormancy, mitigating the effects of heat stress on grain storage protein functionality, and even to consider the introduction of genes for allergy-free grain or resistance to insect attack on the stored grain (Garcia-Casado *et al.*, 1994).

All these advantages are required 'for free', delivered integrally with the seed for sowing, preferably with improved yield and the appropriate protein content, providing the grower at harvest with a premium payment due to delivery of grain that is better suited to processing requirements than the 'run-of-the-mill' grain.

The process whereby the breeder attempts to achieve these miracles involves appropriate choice of parent lines (each having a set of desirable attributes that would be highly valuable if combined into one genotype), cross-fertilization between the parents, and selection throughout successive generations in search of those few new genotypes which embody the spectrum of desirable attributes in a homozygous background.

Very large numbers of lines are produced in the first few generations (at F2, F3, F4) of this complex process, as is shown for the example of rice breeding in Figure 11.1. This example is typical of similar breeding strategies for other self-pollinated cereals. Severe culling is obviously essential in these early stages to reduce populations to more manageable numbers at later generations when more intensive testing is warranted, and when there is the need to replicate selected lines at a range of sites. The ultimate product, shown in Figure 11.1 as a single cultivar, may take many generations (continuing to F7, F8, or more) and many years. See Simmonds (1979), Lupton and Derera (1981), Patterson and Allan (1981), Coffman *et al.* (1961), Stoskopf (1985), Pomeranz (1987) and Poehlman (1979) for general background to breeding methodologies.

The breeding process, complex as it may be, is only the first step in a long sequence (illustrated in Figure 11.2), leading to the consumer. Consideration of this chain of events provides the perspective against which breeding for appropriate grain quality should be seen. The breeder must be aware of the needs of the consumer and the processor to ensure success at the critical stage of 'Sale of Grain'. Given difficulties in selling because of quality deficiencies, the farmer will seek seed of a more appropriate quality type. Thus the sequence is not linear (as shown in Figure 11.2), but a series of feedback loops. Information generated at all stages assists the breeder in improving the selection of more appropriate genotypes for release, the ultimate grain value being a combination of yield and quality.

A further consideration in a globally-competitive marketplace is that no direct premium may be offered for higher grain quality. Rather, the end-user may demand improved grain quality at the existing price. In this way grain of lesser quality is displaced while grain of better quality secures greater market share. In the absence of artificial price supports, the grain of lesser quality will assume a lesser value due to the reduced demand. Similarly, grain of poorer quality, say, of markedly lower bulk density, may receive severe discounts in the marketplace while grain of

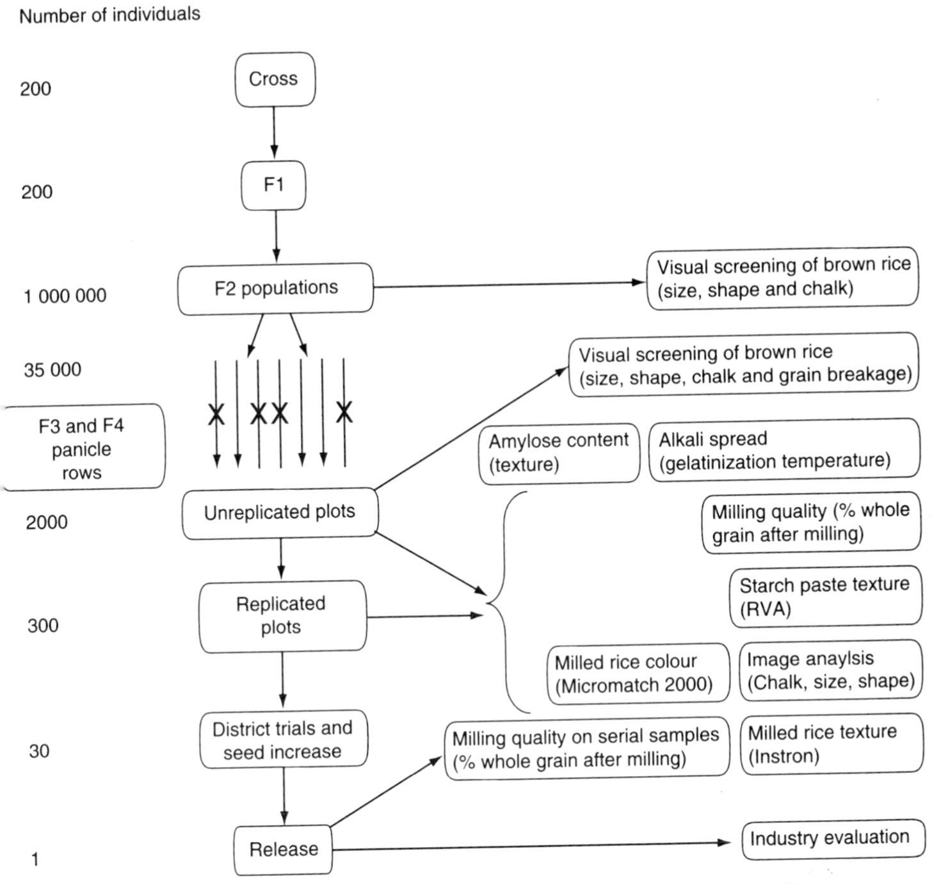

Figure 11.1 Selection for grain quality in a rice-breeding programme. (Diagram provided by A. B. Blakeney, Yanco, NSW, Australia.)

satisfactory bulk density receives no particular premium. Regardless of premium or discount, the key feature is value differentiation.

11.1.2 Yield increase *versus* quality improvement

Traditionally, the breeder's primary concern has been 'to fill bags' – to increase grain yield; attention to grain quality has been a secondary consideration. The diversification of grain markets according to quality type has increased the accent on quality, but it has not necessarily changed the need to continue breeding for yield. Some studies indicate that for wheat, at least, breeding for an improvement in many quality

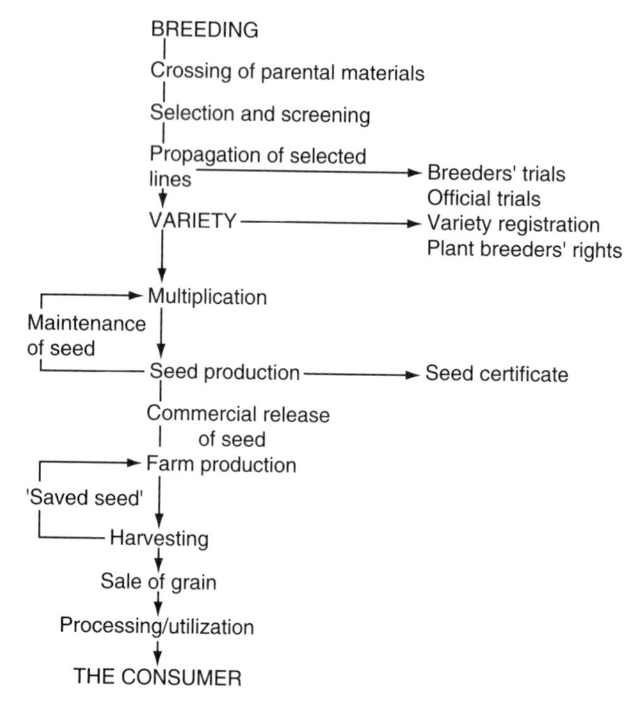

Figure 11.2 The setting of breeding in the overall sequence of events through to the consumer. (Adapted from Cooke – Chapter 1 in Wrigley, 1995.)

attributes does not necessarily impose a yield penalty (O'Brien and Ronalds, 1984, 1986). For example, altering the composition of the grain protein to give appropriate dough properties may not prejudice yield; whereas most attempts to increase the genetic potential for protein content have involved yield penalties. Even if there were no conflict between these two breeding objectives, the addition of so many extra characteristics of grain quality imposes a much greater burden on the breeder with respect to the larger number of screening tests and the more exacting requirements for parent lines.

Priority setting between yield and quality is potentially resolved by knowing the premium payments expected for grain of appropriate quality type. This consideration is also an essential part of the economics of considering the extent to which a yield penalty might be tolerated to achieve a quality objective. The practical problem for the breeder is that these market considerations must be anticipated many years ahead – both with respect to market-price differential according to quality type, and prediction of which aspects of quality will be appropriate to the markets – not just the markets of today but markets that will be operating when the results of today's cross-fertilization reach advanced-line status several years hence.

An overriding consideration in the quality–yield nexus is the likelihood that a high-yielding variety may not achieve government or industry approval if it does not measure up to minimum quality attributes. This last consideration commands the basic philosophy that quality selection in breeding should at least involve culling below-minimum lines as early as is practicable.

11.1.3 Early *versus* late-generation selection for quality

Traditionally, breeders have selected for grain yield and agronomic suitability in the earlier stages of the breeding process, waiting several generations before considering quality-based selection, even when quality has been a major objective. This approach has two distinct advantages. First, large numbers of lines with little potential of future grower acceptance are easily culled, and second, by more advanced generation the number of lines is reduced and a reasonably large sample size (several hundred grams or more) is available, facilitating medium-size testing, generally as a small-scale version of the appropriate process (e.g. baking or malting). In this way, the number of samples for quality testing is reduced, and this is a necessity since the small-scale processing tests are generally expensive and labour-intensive. However, this approach involves the likelihood that poor-quality lines have been propagated unnecessarily. Therefore, it is desirable that quality characteristics be considered at an earlier stage to improve overall breeding efficiency. O'Brien and Ronalds (1986) reported that the cost per breeding line could be reduced by selecting for quality prior to selecting for yield. Fischer *et al.* (1989) found a useful response in the seventh generation (F7) to selection in F3 for various quality characteristics. Brennan and O'Brien (1991) have further reported on the economic aspects (with respect to breeding costs) of early-generation testing for quality, concluding that the decision depended largely, as argued above, on the likely market differentials for quality versus yield, and the bias of the variety-licensing system for grain quality.

11.1.4 Moving towards early-generation selection for quality

Our ability to capitalize on the potential advantages of early-generation selection for quality depends on the practical aspects of predictive quality testing methodology and biological considerations of genetics, the influence of the environment and their interactions. The most

(a)

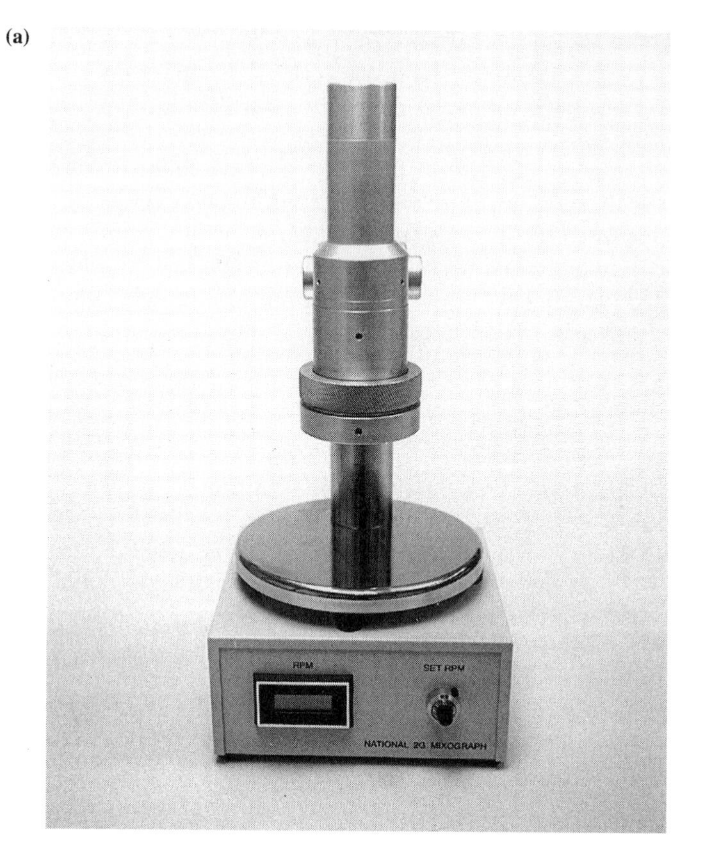
RPM
SET RPM
NATIONAL 2G MIXOGRAPH

(b)

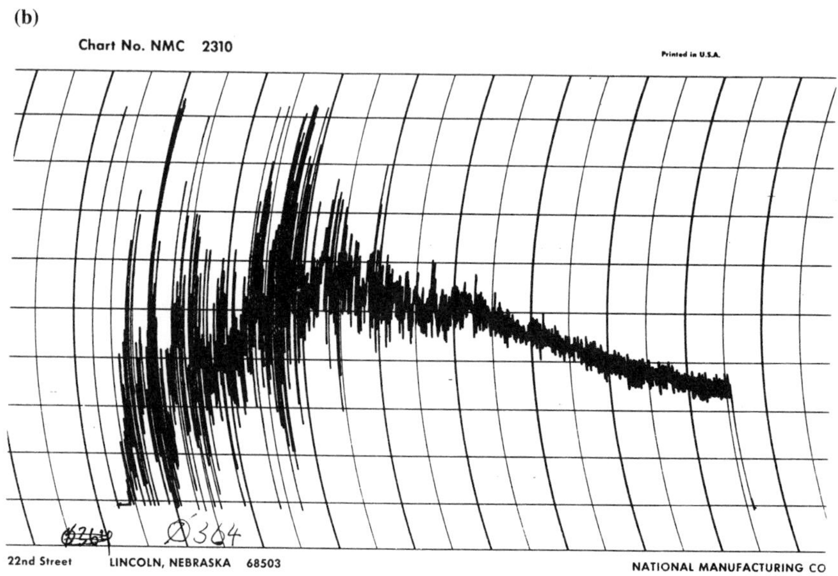
Chart No. NMC 2310
Printed in U.S.A.
Ø364
22nd Street
LINCOLN, NEBRASKA 68503
NATIONAL MANUFACTURING CO

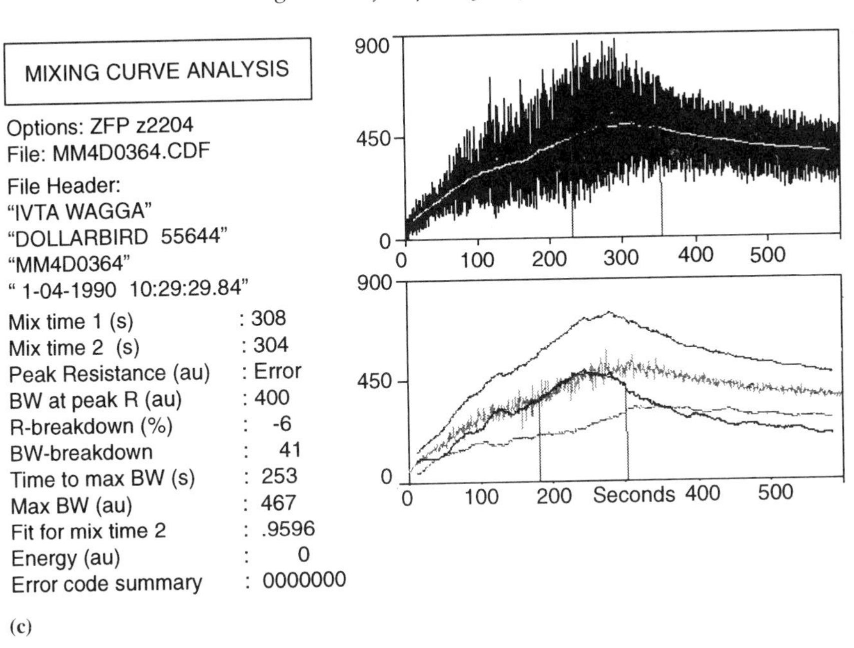

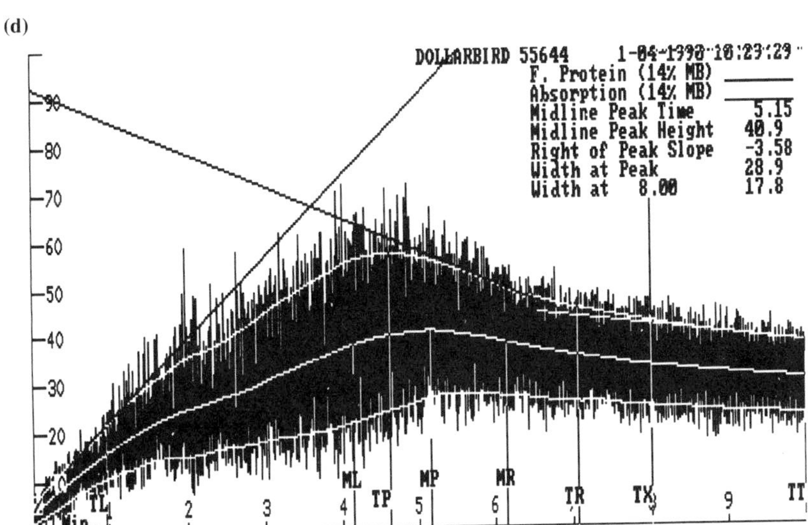

Figure 11.3 The 2-g direct-drive Mixograph (a), and one-page printouts of results for dough samples made from the same flour sample (b) using the traditional mechanical recording system (in this case for the 35-g Mixograph), and (c) and (d) the screen report from the 2-g direct-drive Mixograph, with smoothed curves and parameters derived with either (c) *Mixograf* software (Rath *et al.*, 1990) or (d) *MixSmart* software (from TMCO National, Lincoln, Nebraska, USA). (Mixing traces provided by P. W. Gras, CSIRO, Sydney, NSW, Australia.)

obvious restrictions are the enormous numbers of lines (many thousands, probably) and the small sample size (possibly the product of a single plant, some of which must serve as seed for the next generation). The biology of the system imposes additional constraints. Under conventional systems of cross-fertilization, earlier generations possess a high degree of heterozygosity. Further, quantitative traits and traits with low heritability, that is traits that are highly influenced by the environment, pose a particular challenge. Often the cereal technologist is faced with differentiating breeding lines from unreplicated trials against a background of environmental influence (Morris and Raykowski, 1995). Part of the general 'heritability' difficulty is the greater degree of heterozygosity at early generation – particularly a problem for 'quantitative' characters involving many genes (such as overall processing quality).

In response to the need to cope with small sample size, there has been progress in miniaturizing some processing tests, e.g. computerized testing of two grams of flour (or even wheatmeal) in the micro Mixograph (Figure 11.3) (Rath *et al.*, 1990). Using this equipment, Gras and O'Brien (1992) have reported that major aspects of dough quality have high heritability if tested as early as F3. Another example is the very small-scale testing of wheat-milling quality, using the USDA mill (Figure 11.4a) (Shoup *et al.*, 1957). Figure 11.4b shows the distribution of milling performance of 399 F4 lines resulting from a cross between lines with good and poor milling (Seeborg and Barmore, 1957). Two operators can process 400–600 5 g samples per day using this mill.

The introduction of doubled haploid breeding (Luckett and Darvey, 1992) provides the opportunity to take advantage of 'instant' homozygosity and cull at early generation for end-use quality. Constraints owing to testing methodology and heritability of traits remain. To date, the labour- and technology-intensive nature of doubled haploid breeding, and the limited number of anther- and microspore-responsive genotypes, has precluded widespread adoption of this approach. At issue, too, is the value of heterogeneity associated with F3-derived lines in broad environmental adaptation of cultivars. The concept of obtaining environmental stability (especially to pathogens) through genetic heterogeneity is the basis for 'multilines', cultivars that are a physical mixture of many individual genotypes. Although reasonable from an ecological viewpoint, these gene deployment strategies pose unique challenges to quality evaluation.

11.1.5 Identification of molecular markers for quality attributes

The greatest potential for early-generation quality selection lies with the use of molecular markers. Two general types of molecular markers exist.

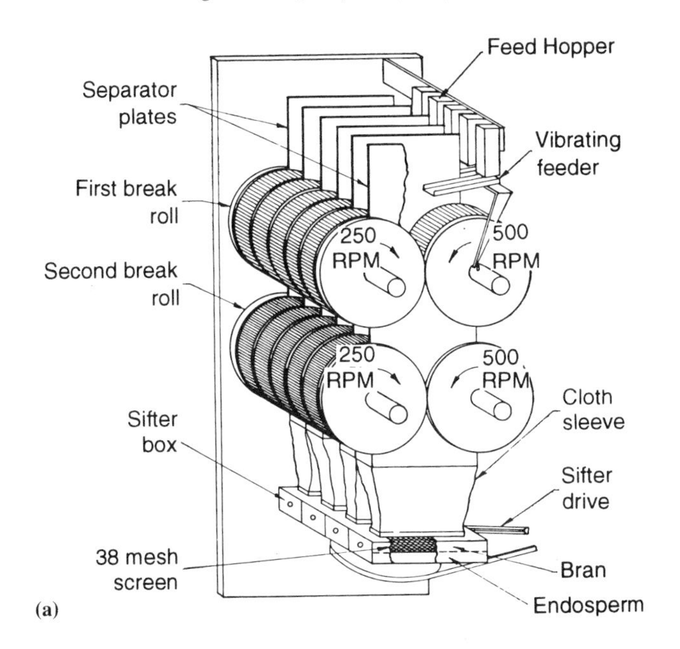

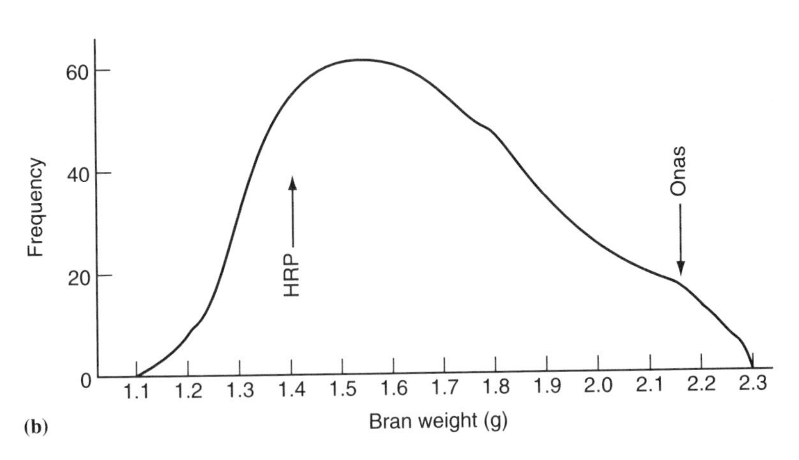

Figure 11.4 (a) Schematic of the USDA micro mill. Four 5–10-g samples are milled and sifted simultaneously, and feed rate is held constant by a vibratory feeder. Rolls, 15.2 cm diameter, are modified from a Bühler MLU-202 test mill. (Taken from Shoup *et al.*, 1957). (b) Distribution of milling performance of 399 F4 lines resulting from a cross between lines with good (HRP) and poor (Awned Onas) milling quality. Parental values are indicated by arrows. (Taken from Seeborg and Barmore, 1957).

The first involves genes or gene products directly related to the trait of interest. The second involves markers that are more or less tightly linked to the gene of interest such that in the case of very tight linkage, the marker is always inherited (and present) with the gene of interest. As genetic linkage decreases, the marker and the gene of interest are more likely to segregate and not occur in the same individual. In the case of unlinked traits, segregation is independent and gene assortment random. As long as linkage is relatively tight, either marker type is effective.

The first step towards using this approach is to break processing quality down to specific quality attributes, each of which might be expected to be tested far more simply than the overall process, and each of which might be more simply inherited. Then the next step would be, if possible, to identify an aspect of chemical composition that would indicate the status of the quality attribute. Analysis for the presence/absence of this chemical marker would be performed on a much smaller sample than would the attribute it marks, probably on a half grain leaving the germ end available for sowing.

This approach has the further potential advantage of leading to specific genes that code for the polypeptide markers, permitting screening to be performed at this level, even using DNA from a leaf fragment well before grain is produced and avoiding the need to harvest grain from lines that are rejected. Alternatively, the screening process to identify a marker might be performed at the gene level, leading to the identification of a chemical marker as a consequence. In any case, the process of marker identification is likely to proceed via statistical correlation involving many genotypes. This process may identify a relationship that is useful in breeding, though not necessarily causative. It is valuable, if possible, to carry on the identification process to the further step of determining the nature of the relationship between the gene(s), the chemical marker(s) and quality attributes.

For example, Table 11.1 also shows for wheat how the process of breaking processing quality down to a few attributes in combination can indicate overall quality suitability for many diverse products. One of these is dough strength, the object of intense research, focused on the polypeptides of glutenin, and now a prime example of successful chemical analysis at early generation (Payne, 1987; MacRitchie *et al.*, 1990). Analysis for relevant aspects of protein composition in this case generally involves gel electrophoresis (Gupta and Shepherd, 1990), but the more automatic processes of HPLC (Kruger and Bietz, 1994) and capillary electrophoresis (Werner *et al.*, 1994) have been adapted to this analysis with the potential of providing the breeder with less labour-intensive alternatives. Analysis of plant DNA for this trait based on RFLP or PCR analysis for the genes for the high-molecular-weight

Table 11.1 Quality attributes preferred in wheats for specific products. (Adapted from Wrigley, 1994)

Product	*Protein content*	*Grain hardness*	*Dough strength*
Breads			
Pan bread	>13%	Hard	Strong
Flat bread	11–13%	Hard	Medium
Steamed–Northern China	11–13%	Hard	Medium/strong
Steamed–Northern China	10–12%	Soft/medium	Medium
Noodles			
Alkaline	11–13%	Hard	Medium
White	10–12%	Medium/soft	Medium
Instant	11–12%	Medium	Medium
Biscuit/cake	8–10%	Very soft	Weak
Starch/Gluten	>13%	Hard (soft preferred)	Strong
Pasta	>13%	Very hard	Strong

(HMW) subunits of glutenin is also available as an alternative approach (Reddy and Appels, 1993; D'Ovidio and Anderson, 1994), as explained in Chapter 12.

A second example of marker-based quality prediction in early generations is the use of friabilin to assess endosperm texture. Friabilin is a family of proteins associated with water-washed starch from soft, but not hard wheat (Greenwell and Schofield, 1986). As described in Sections 1.4 and 1.5, wheat grain texture is important in wheat processing and thus in wheat trade, hard grain being required for many bread products, soft wheat being preferred for cakes, biscuits and grocery flour (Table 11.1). The use of this protein marker (via visualization after gel electrophoresis) requires the aqueous isolation of starch, but has been adapted to the analysis of individual wheat grains (Bettge *et al.*, 1995). Since the occurrence of friabilin on the surface of water-washed soft starch is probably an artifact of starch isolation (Jolly *et al.*, 1993; Greenblatt *et al.*, 1995), gene probing or antibody-based technologies on whole grain or flour would not be considered useful.

The identification of protein markers opens up the possibility of even further simplifying the screening process by generating antibodies specific to the marker proteins and applying immunological technologies developed for medical immunodiagnostics. This approach, attempted for the softness marker, friabilin, was not fully successful (though it produced a valuable test to distinguish durum from bread-wheat products) (Greenwell, 1992). The use of antibodies reactive with glutenin polypeptides has been successfully applied to increasing the efficiency of screening many wheat-grain samples for dough strength (Andrews *et al.*, 1993).

11.1.6 Characteristics desirable in tests for quality selection

Table 11.2 is a 'breeder's wish list' showing desirable characteristics for screening tests to cull for quality, especially at early generation. The vast numbers in the early stages (Figure 11.1) demand that large sample numbers can be processed. Ideally, a test should be simple to perform, with low labour requirement, and it should be safe. The latter stipulation refers partly to the need for radioactive probes in early techniques for gene probing, many now being replaced by enzyme-linked visualization methods. Simplicity and speed will be satisfied by the third request — the possibility of taking a scoopful of flour or meal (with, say ±50% accuracy) to avoid the tedium of accurate sample weighing.

Many of these requirements are met by two of the example tests listed (Table 11.2) – near infrared analysis (broadly) and antibody-based kits (such as that of Andrews *et al.*, 1993). The other example, the 2 g Mixograph, is not a mass screening test, although it can be applied to finely ground wholemeal as well as to flour. Most importantly, the Mixograph gives direct information about dough properties, as distinct from the other two tests, whose results are interpreted via calibration to larger samples that have been analysed by more conventional means. Other examples of test systems that might be examined for these criteria include gel electrophoresis, various forms of HPLC, and the SDS and

Table 11.2 Requirements of quality-test systems for breeding, with examples given for three experimental approaches, namely, the 2-g Mixograph (Mixo), near infrared spectroscopy (NIR) and immunoassay with quality-specific antibodies (Ab)

	Mixo	*NIR*	*Ab*
Large numbers: >1000/day	×	✓	✓
Flexibility: grain, meal or flour	✓	✓	✓
Independent of sample size ±50%	×	✓	✓
Simple: a few steps	✓	✓	✓
Safe to use: not radioactive	✓	✓	✓
Value of information	Direct		Correlative
Independent of environment	×	?	?

Zeleny sedimentation tests, these lying between direct dough testing and NIR/antibody screening in complexity and ability to handle large numbers, but still being largely correlative in the information given.

11.1.7 Non-destructive analysis of whole-grain samples

A method that appears to satisfy most of the criteria in Table 11.2 is the analysis of whole-grain samples by non-destructive means. An example is near infrared transmission (NIT) spectroscopy, now in routine use for composition analysis (Ronalds *et al.*, 1991, 1992), especially for grain moisture, protein content (an important quality attribute for wheat – see Table 11.1) and oil content. There have been reports that near infrared methodology has been extended to the analysis of wheat grade (Delwiche and Norris, 1993), of malting quality in barley grain, and of hardness and dough quality in wheat (Ronalds *et al.*, 1991, 1992; Delwiche and Weaver, 1994). In these cases, it may be difficult to elucidate the chemical bases of the relationships, but if they prove to be reproducible, they will be of great value in breeding since they satisfy many of the breeder's requirements. Image analysis is another non invasive method with the potential to provide grain-quality data on whole-grain samples, leaving the grain intact for further analysis or for propagation (Sapirstein, 1995).

11.1.8 Genotype versus environment

Finally, it is important for the breeder to know the extent to which the information provided is modified by the growth environment. Ideally, the breeder needs to know about the genetic potential of lines to produce grain of specific quality. In addition, however, it must also be known to what extent this genetic potential is modified by growth conditions. Environmental influences will be absent or minimal for tests involving genetic probing (such as RFLP, PCR based and RAPD methods, as is explained in Chapter 12) and for electrophoretic methods based on the presence or absence of specific proteins. In general, growth conditions (and protein content) will influence quality-testing results and they will need to be taken into account when interpreting results.

Protein content itself is a prime example of the difficulty experienced in breeding, being greatly influenced as it is by growing conditions. There has long been the general understanding that there is a nexus between protein content and grain yield. Yet, there are continuing endeavours by breeders to increase the genetic potential of new cultivars to attain target levels of grain protein content without a yield penalty (Beninati and Busch, 1992; Stein *et al.*, 1992). For a review of the subject, refer to Konzak (1977).

11.1.9 Late-generation testing

As the breeder narrows down the number of promising lines later in the breeding process, there are larger quantities of grain available (e.g. over a kilogram), permitting more extensive testing, especially lab-scale versions of commercial testing. As the selection process nears its final goal of one or a few lines with ideal quality (Figure 11.1), it is important to extend the testing process beyond the breeder's laboratory, often to obtain testing in processors' hands, and generally via participation in collaborative growing and testing in a consortium of laboratories.

A factor to consider at this stage, as well as to plan for throughout the process, is whether a particular variety will be readily distinguishable from other varieties (especially those that differ in quality type) by conventional methods of variety identification (Wrigley and Batey, 1995).

11.1.10 Computer-based collation of results

The process of applying so many tests to large numbers of samples inevitably produces enormous amounts of data, for which computer-based analysis becomes virtually essential. Several systems have been developed for this purpose. The one developed at the USDA Western Wheat Quality Lab (WAS, Morris and Raykowski, 1993) is illustrated in section 11.2.8 in relation to selection for wheat-grain quality.

11.2 WHEAT

Wheat is the only material suited to the production of leavened bread and associated products. Because of mankind's desire for this family of foods, wheat is a major cereal in world grain production, and the grain for which quality attributes are most critical. It is thus the grain that attracts most attention in breeding for grain quality.

11.2.1 Quality testing – half grain or half tonne

The wheat breeder needs a range of quality tests to suit sample sizes all the way from half a grain up to half a tonne of wheat. Throughout this range, the question to be answered is the same: How well will this sample suit processing requirements for manufacture into a certain product? More specifically: what is its genetic potential to do so, given a certain range of growing conditions?

At the earliest stage of the quality-selection process, the breeder has one precious grain . . . and a dilemma. If it is destroyed in the test process, there is nothing left to grow on into another generation. Non-destructive testing is thus attractive, but also limited in scope. Visual examination is the obvious first step, relying on the breeder's experience (Table 11.3). New techniques are being developed to objectively analyse grain samples non-destructively; image analysis can do so on individual grains; near infrared transmittance (NIT) can determine moisture and protein content and various other qualities on whole grain – potentially one at a time, but generally averaging over a bulk sample.

As a new plant forms from the germ only of the grain, it is possible to cut the grain in half, leaving the germ end for propagation and the rest of the grain for testing. The range of chemical tests available can be performed on small samples much less than half a grain.

Ultimately, however, the breeder wants an answer about quality on a scale that approximates full industrial processing. That is the 'half-tonne' end of the spectrum, fortunately the end when the selection range has been reduced from many thousands to less than a dozen candidates. At this stage, it becomes apparent how successful has been the earlier selection process of testing for component quality attributes likely to contribute to integrated processing quality.

11.2.2 Choice of parent lines

The successful wheat cultivar embodies a myriad of genetic traits. These traits may be grouped into the broad categories of grain yield, resistance to biotic and abiotic stresses (climate, insects, pathogens), agronomic performance (plant height), and end-use quality. The plant breeder must consider each when making crosses and selecting progeny, ensuring that each is at a level acceptable to the farmer or end-user. In this regard, the breeder may find the attainment of satisfactory end-use quality particularly troublesome. The problem is exacerbated if sufficient attention was not devoted to selection of parents for crossing. Why? Quality is the sum total of many complexly-inherited, quantitative traits that possess variable levels of heritability. Although one may argue that adaptation is equally complex, in the case of adaptation, progeny can be screened and selected much more easily: plants thrive, survive, or they are killed by drought, heat, frost, insects, diseases, etc.

Likewise, the yield of a plant, or plot of plants, is directly influenced by adaptation. While several hundred progeny may be planted in pathogen-infested soil to assess which have genetic resistance, this type of 'mass' selection is rarely possible for quality. Consequently, it is the successful wheat breeder who carefully selects parental lines for crossing. Often, the most successful strategies for quality involve including one parent with excellent quality, 'top-crossing' the resultant F1 with another good quality parent, or repeatedly 'back-crossing' using the excellent quality parent. In practice, crosses generally involve elite germplasm, such as lines that just barely did not make cultivar release,

Table 11.3 Quality testing of lines of wheat in the breeding process

Quality attribute	*Test material*	*Method or equipment*	*References*
At early generation (F3–F5)			
Colour (red/white)	Whole grains	Visual; Image analysis;	Bason *et al.* (1994)
		NaOH soak	Kimber (1971)
		Colour metre	Bason *et al.* (1994)
		NIT	Ronalds and Blakeney (1995)
		Image analysis	Bason *et al.* (1993); Sapirstein (1995)
Hardness	Whole grains	Visual, Perten SKCS 4100, NIT	Martin *et al.* (1993); Delwiche (1993)
	Wholemeal	NIR	Anon. (1983) Method 39–70A
	Grain	Particle size index	Symes (1961); Yamazaki and Donelson (1983)
Grain size	Grain	Visual, Thousand-kernel weight, Perten SKCS 4100	Wrigley (1995); Martin *et al.* (1993)
Protein content	Grain	NIT, Kjeldahl	Anon. (1983)
	Meal or flour	NIT, Dumas	Anon. (1983)
Milling	Grain	Quadrumat Junior	Whan (1974)
		USDA micro mill	Shoup *et al.* (1957)
Dough strength	Flour (wholemeal)	MicroMixograph	Finney and Shogren (1972)
			Rath *et al.* (1990)
	Half grain, meal	SE-HPLC (size distribution)	Batey *et al.* (1991)
	Half grain, meal	SDS gel electrophoresis (glutenin subunits)	Payne (1987); Gupta and Shepherd (1990)
	Half grain, meal	Antibody test kit	Andrews *et al.* (1993)
Baking quality	Flour	SDS sedimentation	Anon. (1983)
		Zeleny sedimentation	Anon. (1983)
		Pelshenke fermention	Anon. (1983) Method 56–60
Noodle, starch manufacture	Meal, flour, starch	RVA pasting	Anon. (1983), Konik *et al.* (1993)
Baking water absorption	Flour	Alkaline water absorption	Anon. (1983); Kitterman and Rubenthaler (1971)
Weather damage potential	Meal, flour	RVA,	Walker *et al.* (1988)
		Falling Number	Anon. (1983)
Intermediate (F6–F7) and advanced generations (beyond F8)			
Most of the above, although some should no longer be required e.g. grain colour and hardness, plus checks that the samples are sound, clean and of adequate test weight. Further testing details are provided in Chapter 1.			
Milling quality	Grain	Lab-scale mill	Anon. (1983)
Flour colour	Flour	Colour meter	
Dough-mixing properties	Flour	Mixograph	Anon. (1983)
		Farinograph	Anon. (1983)
		Extensigraph	Anon. (1983)
Pan-bread baking	Flour	Bake test (a range)	Anon. (1983)
Cookie baking	Flour	Bake test	Anon. (1983)
Noodles	Flour	Small-scale simulations	Chapter 1
	Flour, meal, starch	RVA, Amylograph	Konik *et al.* (1993)
Flat bread	Flour, meal	Small-scale simulations	Chapter 1

and/or successful cultivars. To this gene pool is introgressed new sources of yield, pathogen and insect resistance, etc., as noted above. The cereal chemist can contribute to the cultivar-development process in several important ways: first, by defining and interpreting the quality needs of end-users, by identifying superior parental lines (or the defects in parental lines), and by developing quicker, easier methods of quality assessment with greater power of accurately predicting end-use quality.

11.2.3 Early generation selection

Traditionally, a wheat breeder needed not only a good eye (to select grain by appearances), but also a good set of teeth. Before the availability of micro-scale testing equipment, it was usual for the breeder to chew the grain – one or a few grains at a time – thus to determine its hardness, possibly even its milling potential, and to form a tiny dough ball in the mouth from the crushed grain, thus to roughly evaluate dough strength by pulling it with the fingers (Buller, 1919).

Visual examination is still probably the main selection tool for most breeding programmes in early generations (F2 to F4). In fact, grain appearance must be appropriate at all stages of breeding and production. Specific attributes to look for at early generation include grain colour, hardness and size/shape. The need for doing so depends, of course, on the parent lines chosen: are progeny the result of crosses between red and white or between hard and soft parents? As grain colour and hardness are highly heritable attributes and not greatly influenced by environment, it is valid to select for them early. Definitive testing of grain colour involves soaking for about 15 minutes in 4% sodium hydroxide solution to accentuate differences in bran colour between red and white wheats (Kimber, 1971), though this test may prevent propagation of the seed tested. Illumination from below the untreated grain helps in the visual distinction between red and white grains (Bason *et al.*, 1994).

Grain hardness and milling quality

Visual examination is less satisfactory (though probably adequate) for distinguishing hard- from soft-grained genotypes, another basic quality-type characteristic like grain colour. Laboratory methods of quantitatively determining grain hardness include those applicable to 10 g samples such as the particle-size-index test (Cutler and Brinson, 1935; Symes 1961; Yamazaki and Donelson, 1983) (grind and sieve), the pearling-index test (McCluggage, 1943; Chesterfield, 1971) (abrasion of outer layers) and other tests of grinding difficulty. Recent USDA research

(Massie *et al.*, 1993; Martin *et al.*, 1993) has led to the commercial development of a single-kernel hardness tester (the 'Single Kernel Wheat Characterization System – SKWCS', see Instruction Manual, Perten Instruments Inc., Reno, Nevada, USA). This instrument weighs each kernel, measures the outer dimension perpendicular to the long axis, measures the moisture content based on electrical conductivity, and measures the hardness based on a 'crushing profile'. Although the standard test precludes the further possible advantage of propagation after testing, half seeds have provided reliable hardness measurement, thus saving the germ end for planting (Bettge and Morris, personal communication).

When there is adequate sample (say, a minimum of 50g), the most commonly used hardness test involves near infrared reflectance spectroscopy (NIR) of the ground wholemeal sample, based on calibration with samples tested by one of the above lab tests (AACC Method 39–70A of the American Association of Cereal Chemists (AACC); Brown *et al.*, 1993; Carver, 1994). Near infrared techniques are used simultaneously to determine protein content and moisture (needed for other testing). The single-kernel hardness tester mentioned above is also designed in a version to determine a range of composition attributes by near infrared technology grain-by-grain, including grain texture/hardness, thus preserving the grains intact (Delwiche, 1993).

Some breeders claim to be able to accurately predict milling quality and protein content by visual examination, based on long experience. Perhaps it will some day be possible to preserve such subjective judgement as an objective, routine procedure using digital image analysis – the statistical analysis of images of grains captured by video camera (Sapirstein, 1995), but such technologies are still in their formative stages. Image analysis has however been reported to successfully distinguish red- from white-grained wheats (Bason *et al.*, 1993).

As mentioned in Section 11.1.4, milling may be conducted on as little as 5 g of grain using the USDA Western Wheat Quality Laboratory's micro mill (Figure 11.4). When 20 or more grams of grain can be spared, test milling using the Brabender Quadrumat Junior mill is common. Milling performance is best assessed after appropriate conditioning of the grain (e.g. to 14% or 16% moisture for soft or hard wheats, respectively). Test milling has the obvious advantages of not only indicating potential milling performance (primarily yield), but also flour is produced for further testing. Milling problems often observed include poor bolting (flour sieving) and poor separation of bran from endosperm.

Dough properties

Prediction of dough-mixing properties is probably the most important and most difficult of the quality attributes that might be evaluated early in the breeding process. As indicated in Table 11.1, dough quality is an important part of processing for all aspects of wheat use worldwide, the only exception being animal feed. Unlike the other difficult-to-assess attributes (e.g. milling yield) where 'good quality' refers to only one end of the spectrum, desirable dough properties may cover the full range of possibilities, depending on the target product (Table 11.1).

Direct dough testing is obviously the most desirable alternative, provided adequate sample can be spared for the purpose – preferably as white flour, although dough-test systems for finely ground wholemeal have been devised to predict the dough properties expected for the corresponding flour (Gras and Wrigley, 1991). The Mixograph is the main dough-test instrument that has been miniaturized to suit the early-generation needs of breeders. Initially developed for 35 g flour, the Mixograph principle of planetary-moving pins has been adapted to suit 10 g flour (Finney and Shogren, 1972) or 5 g flour (Finney, 1989) in mechanically recording versions, or for 2 g flour in a direct-drive version (Figure 11.3) with computer-based data acquisition and interpretation (Rath *et al.*, 1990).

This last innovation has not only permitted the testing of much smaller samples, but it has also simplified the breeder's task of evaluating the results. Traditionally, this task has involved subjective examination of the mix trace overall, plus manual measurement of various parameters from the trace, again using subjective judgement to determine the positions of the peak and the dimensions of band widths. Computer-based recording, and profile smoothing have made these tasks automatic and objective, as indicated in Figure 11.3. Similar computer-based systems for data acquisition and interpretation have been developed by the retrofitting of recording equipment to the traditional-style mechanical Mixograph (Lang *et al.*, 1992). Small-scale versions have also been manufactured for other types of dough-testing equipment, particularly the Brabender Farinograph (for 50 g as opposed to the standard 300 g of flour). Although dough properties might be expected to be the result of a large number of genes, limited studies indicate that some aspects are highly heritable and that selection for those aspects is thus justified at early generation (Gras and O'Brien, 1992).

Predicting dough properties chemically

As dough properties are primarily determined by gluten-protein quality and composition, it should be possible to predict dough quality by chemical analysis of the appropriate aspects of protein composition. Such an analysis might be performed at any of several levels, as indicated in Table 11.3. Ultimately, total protein content (for grain or flour) is a good indicator in itself, but achieving appropriate protein quality (*versus* quantity) is a major aim of breeding.

Conventional methods of protein analysis require relatively dilute solutions of the proteins. Efficient (or representative) extraction of all protein components has been a major barrier to effective analysis of protein composition. On the other hand, degree of difficulty in extracting flour proteins has proved to be a useful indicator of dough strength, a larger proportion of difficult-to-extract protein being related to stronger doughs. This observation has been translated into screening tests for use in breeding (e.g. that of O'Brien and Orth, 1974).

Sonication of flour in a detergent-containing buffer (SDS-phosphate) has been used effectively to provide a representative extract of the range of flour proteins, whilst breaking a minimum of covalent bonds (but presumably disrupting most non-covalent bonds). Fractionation by size-exclusion high-performance liquid chromatography (SE-HPLC) permits analysis of the proportions of aggregated glutenin macropolymer (strength-related, over about 100 000 Da) and of the smaller monomeric gliadin polypeptides (Table 11.3) (Batey *et al.*, 1991). Alternatively, analysis of the size distribution of the aggregated glutenin up into the millions of molecular weight can be obtained by multi-layer SDS gel electrophoresis (Khan and Huckle, 1992; Wrigley *et al.*, 1993). Either of these analytical techniques may be used to predict dough strength (e.g. as Rmax in the Brabender Extensograph) for very small grain samples, including the half-grain approach described above (Section 11.2.1).

A labour-saving alternative for handling large numbers of samples involves the use of specific antibodies that detect quality-related protein sequences, providing correlation with dough strength (Andrews *et al.*, 1993). The advantage of this approach is that the screening techniques of medical diagnostics may be 'borrowed', such as the microtitre plate with provision for 96 samples per plate and automatic reading and recording of the results (Skerritt *et al.*, 1994).

Tracking glutenin alleles

As Table 11.4 shows, analysis of glutenin-subunit composition takes the breeder a step closer to genetic constitution. However, disulfide bonds are broken in the process, destroying some information about subunit aggregation that would indicate the effects of growth environment as it may modify genetic potential. Allelic constitution for the HMW glutenin subunits (the *Glu-1* loci) is generally determined by SDS gel electrophoresis of a grain or flour extract including SDS and an SS-breaking

Table 11.4 The various levels of organization at which gluten quality might be examined and manipulated

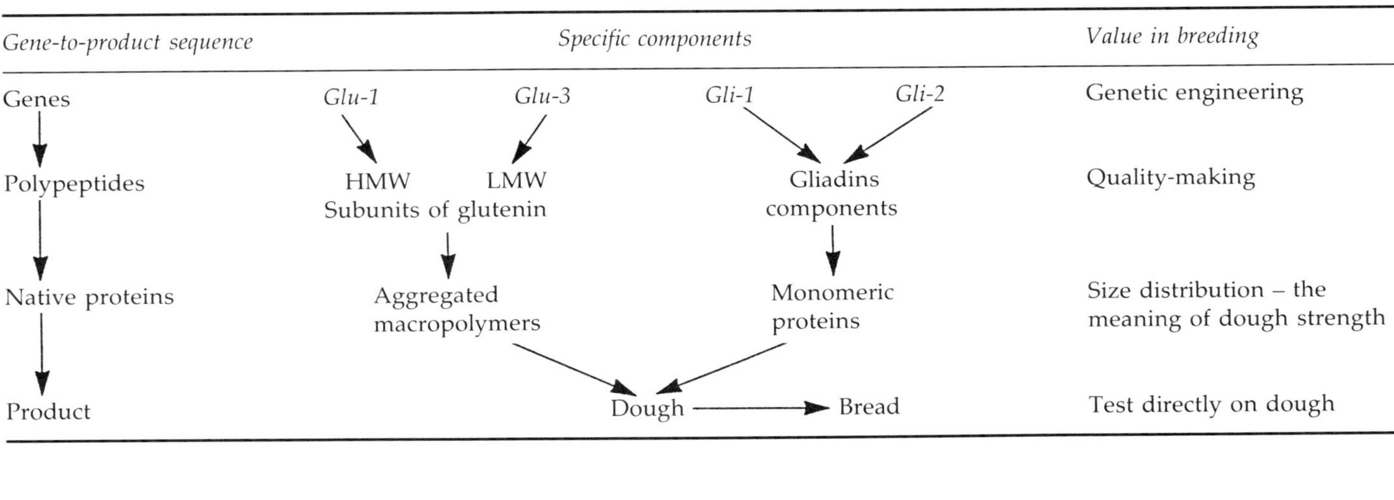

Gene-to-product sequence	*Specific components*				*Value in breeding*
Genes	*Glu-1*	*Glu-3*	*Gli-1*	*Gli-2*	Genetic engineering
Polypeptides	HMW	LMW	Gliadins components		Quality-making
	Subunits of glutenin				
Native proteins	Aggregated macropolymers		Monomeric proteins		Size distribution – the meaning of dough strength
Product		Dough →	Bread		Test directly on dough

Table 11.5 HMW glutenin subunits and dough quality. *Glu-1* dough quality scores assigned to HMW glutenin subunits and corresponding alleles, according to Payne (1987). A high score (maximum of 10) indicates a prediction of strong dough properties

	Glu-A1		*Glu-B1*		*Glu-D1*	
Score	*Allele*	*Subunit*	*Allele*	*Subunit*	*Allele*	*Subunit*
4					d	5 + 10
3	a	1	i	17 + 18		
3	b	2*	b	7 + 8		
3			f	13 + 16		
2					a	2 + 12
2					b	3 + 12
1	c	Null	a	7	c	4 + 12
1			d	6 + 8		
1			e	20		

2* is designation of subunit.

agent such as 2-mercaptoethanol (Payne, 1987), together with authentic samples of known *Glu-1* constitution. This process in turn leads to calculation of the *Glu-1* quality score for baking-quality prediction (Payne 1987; MacRitchie *et al.*, 1990; see Table 11.5 for *Glu-1* scores). This calculation involves adding the score values for each of the alleles in the three genomes. For example, a genotype having subunits 1, 17+18 and 2+12, has alleles a, i and a for loci *Glu-A1*, *Glu-B1* and *Glu-D1*, giving scores of 3, 3, and 2, which add up to 8 out of a maximum of 10 (for a, i, d). The relationships of these scores to quality rankings have been listed for national wheat sets by Payne (1987) and by MacRitchie *et al.* (1990). Catalogues of glutenin alleles have been prepared for various sets of wheats, e.g. by Cornish *et al.* (1993) and Morgunov *et al.* (1993). These will be included, together with general genetic data, in the International Wheat Information System being generated by Fox and Skovmand at CIMMYT, Mexico (e-mail contacts: pfox/bskovmand@alphac.cimmyt.mx).

Although this correlative approach with *Glu-1* alleles has proved reliable for some sets of genotypes, satisfactory prediction of dough properties also requires analysis of the LMW subunits (at the *Glu-3* loci) by performing a pre-extraction step before the extraction of glutenin subunits so that these subunits may also be seen further down the SDS gel. Preliminary quality rankings for the *Glu-3* alleles have been reported by Gupta *et al.* (1991, 1994).

Correlative tests to predict baking quality

Table 11.3 also lists several long-established tests which conform to the requirements of being relatively low in labour costs, being capable of processing large numbers of samples, and providing at least an indication of those lines that may be too poor in quality to warrant further propagation. Currently, the most popular of these is the SDS sedimentation test (Table 11.3), involving the suspension of flour in a detergent-containing solution, allowing the insoluble material to settle for a set time, and measuring the volume of the opaque sediment. This volume obviously includes the insoluble starch but also, particularly, the swollen protein material. Also in use for many breeding programmes are the Zeleny Sedimentation and the Pelshenke tests (Table 11.3).

Starch quality

Secondary to protein quality for most wheat-processing requirements are the properties of the starch, particularly its pasting properties (viscosity of the hot gelatinized starch), the gel strength on cooling ('setback') and the granule size distribution (see Chapter 9). Traditionally, the Brabender Amylograph has been used to determine the first two of these attributes, but its requirement for a large sample size has excluded it from use early in breeding. The newer Rapid ViscoAnalyser (RVA) provides similar information on only a few grams of starch or flour much more quickly (Walker *et al.*, 1988), and the results have been shown to relate to noodle-making quality (Konik *et al.*, 1993). A Standard Method for rapid pasting determination is being drafted by the International Association for Cereal Science and Technology.

Research suggests that much of the variation in peak hot paste viscosity of prime starches is attributable to amylose and amylopectin contents (Zeng and Morris, personal communication), which in turn are the result of different gene doses of the waxy (*Wx*) loci (Nakamura *et al.*, 1993; Miura and Tanii, 1994). Analysis of the 'waxy' proteins can be conducted on the starch isolated from half seeds (Miura and Tanii, 1994) as can friabilin (Bettge *et al.*, 1995).

Resistance to sprouting

In breeding programmes that cater for regions likely to experience rain at harvest, there is the further need to select for genotypes resistant to the tendency to develop excessive levels of amylases and proteases as a result of wetting after harvest maturity is reached. Breeding lines may be evaluated as intact spikes subjected to 'simulated rain' (Morris and Paulsen, 1985), or as grains wetted in petri dishes (Morris and Paulsen, 1987).

11.2.4 Intermediate and advanced generations

Many of the above characteristics and test procedures may still be appropriate later in the breeding process, although various quality attributes may already be established, e.g. grain hardness and colour. At these later stages, much more sample is available and there are relatively fewer lines, both these factors permitting more intensive and definitive testing. On the other hand, at this later stage, promising lines are being grown at various sites to determine their adaptability to diverse growing conditions.

11.2.5 Assessing the role of the environment in wheat quality

As noted in Section 11.1.8, the breeder must assess both the genetic potential of breeding lines as well as the interactive role of the genotype with the environment. The first difficulty in this endeavour is the relatively large effect simply ascribable to variation in climate, sites, or years, i.e. environments. The second difficulty is in assessing the interaction between the genotype (G) and the environment (E): (G × E). Typically, in the early stages of wheat quality testing, grain samples come from unreplicated, single-site yield trials. As such, it is not possible to assess G × E. The issue of assessing the environmental effect is dealt with by including 'check' varieties of known quality (Morris and Raykowski, 1995). These check varieties are assumed to reflect the influences of the specific growing environment. It is further assumed that the experimental lines responded in a fashion similar to the checks.

Generally, these assumptions are reasonable. Bassett *et al.* (1989) showed how most of the variation in wheat quality could be ascribed to environments (especially years) and genotypes. Although the G × E variance component was small, it was, nevertheless, significant. To assess G × E, multiple sites and/or years must be statistically analysed. This additional testing is problematic for the cereal technologist. Simply adding two or three replicated sites doubles or trebles the testing workload. Fortunately, as noted in the prior section, the number of breeding lines is usually much reduced at the intermediate to advanced stages of testing. As an example, at the Western Wheat Quality Laboratory, breeding lines reach the cultivar release stage with around 7–20 site-years of quality data over 5–8 years.

11.2.6 Cultivar-release systems

Figure 11.2 illustrates the sequence of events from the outset of the breeding process to the delivery of wheat foods to the consumer. An important part of this long process is the formal conversion of a 'breeding line' to a commercially-available cultivar. In the US, where most cultivars are developed by university or federal (USDA) scientists ('public' breeders), breeding lines are generally proposed for cultivar release by the originating breeder to a variety-release review committee.

Often this process may take 2 years, the first being for 'preliminary' approval, the second being for formal release. An advantage of the 2-year system is that, after the first approval, seed of the genotype is provided to what is referred to as the foundation-seed programme to begin amplification of seed stocks. Under the usual scheme, four classes of seed are recognized. The first, provided by the breeder to the foundation seed programme, is called **breeder seed**. This seed is considered to be the genetic starting point for the cultivar. The foundation-seed programme uses this seed to produce the second class, **foundation seed.** Any problems associated with uniformity or genetic heterogeneity are dealt with by rogueing seed fields of 'off-type' plants. Foundation seed is, in turn, used to generate **registered seed**, which in turn is used to generate **certified seed**. Obviously, at each stage, the quantity of seed increases manyfold. Certified seed is commercially available to growers without restriction. Certainly, not all wheat is planted using Certified seed. Although regions of the US vary, the self-pollinating, homozygous nature of wheat allows the planting of much 'bin run' seed. With proper care, there is often no reason that farmers may not propagate their own seed supply indefinitely. For a more detailed discussion of seed certification in the US, refer to Poehlman (1979).

Two other activities are also usually occurring at this same stage of cultivar release. The first is the formal registration of the variety with the Crop Science Society of America. The Society's journal, *Crop Science,* publishes the official description of the cultivar along with other salient information (supplied by the breeder). At the same time, a sample of the grain is sent to the USDA National Small Grains Collection as a means of germ plasm preservation and distribution. These activities are encouraged, but are not mandated by law.

The second activity is the collaborative evaluation of the cultivar by industrial users (millers and bakers). In the US, the Wheat Quality Council (WQC) coordinates this activity. The WQC is an independent organization supported primarily by producers, merchandizers and users of wheat. Normally, 'ready to be released' or 'just released' varieties are included in the collaborative testing. The main aim of the organization is to feed end-user desires for quality back to the breeding and cereal technologist community, thus providing an active role in guiding the cultivar-development process.

11.2.7 Durum wheats

Many of the attributes and test systems for bread wheats also apply to durums, except of course that the final processing test system is pasta production, for which additional characteristics such as semolina colour are needed. Table 11.1 sets out the basic requirements of high protein content, very hard grain and strong dough properties (Fabriani and Lintas, 1988).

Milling quality

Many of the quality attributes listed in Table 11.3 are appropriate to durum wheat, but in some cases with distinctive interpretations. For example, grain colour selection is for an amber hue, but still there is the requirement for plump, vitreous, uniformly sized grains, that mill well to give a high yield of semolina. Fragmentation into coarse semolina (as distinct from fine flour and bran) is a consequence of the extreme hardness required in the durum endosperm, testable in durum breeding by the range of wheat hardness tests. An actual milling test is required at later generations, when sufficient grain sample can be spared (Dick and Youngs, 1988).

Semolina colour is also an important attribute to be assessed in the test-milling process, a uniform yellow colour being desirable, free from specks of bran or 'black point'. This coloration, presumed to be due to specific xanthophylls, is highly heritable, controlled by additive gene effects (Johnson *et al.*, 1983), probably on chromosomes 2A and 2B (Joppa and Williams, 1988). It is thus feasible to commence selection for semolina colour at early generation, even if at the earliest stages this involves subjective assessment. Where sample is adequate for testing, the Standard AACC Method (14–50) would be used to determine pigment content.

Small-scale assessment of dough strength is just as applicable in durum breeding as it is for other wheats (Section 11.2.3), except for minor differences in the interpretation of mixing traces. The SDS sedimentation test has been applied successfully to the prediction of durum quality in some breeding programmes, even in a very small-scale version requiring only one gram of wholemeal per test (Dick and Quick, 1983). Dick and Youngs (1988) have reported a correlation (r^2) of 0.713 to spaghetti quality for a combination of SDS sedimentation (1-g version), Mixogram score and wheat protein content for a set of 48 samples. Combined selection for quality based on SDS-sedimentation and protein composition is recommended by Clarke *et al.* (1993) and by Kaan *et al.* (1993).

Prediction of dough strength based on gluten polypeptides has focused on the gliadins and LMW subunits of glutenin more than on the

HMW subunits. This can be partly because most durums have the null allele for *Glu-A1* (HMW subunits); selection for other alleles at this locus should improve dough strength, according to Kaan *et al.* (1993). Initial correlations related to gliadins 45 (strong-dough properties) and 42 (weak dough), 45 and 42 being alternative (allelic) bands designated for their relative mobilities on acidic gel electrophoresis (Damideaux *et al.*, 1978). These gliadin bands are genetically linked to groups of other gliadins, and also to groups of LMW subunits of glutenin, designated LMW-2 (linked to gliadin 45) and LMW-1 (associated with gliadin 42) (duCros *et al.*, 1982). Either these gliadins or the LMW subunits may be used to select for dough strength, although there is some evidence that the LMW subunits are more likely actually to contribute to dough strength (Pogna *et al.*, 1988). In a recent report of the use of these marker proteins in four durum crosses, there was a highly significant correlation ($r = 0.79$ significant at 0.1%) between good cooking quality and the presence of gliadin 45 (or LMW-2) (Federmann *et al.*, 1994). In addition, selection for HMW-subunit composition (subunits 6 + 8 or 20 associated with *Glu1*) has been shown to give useful quality-based segregation in breeding (Kovacs *et al.*, 1993), either based on SDS gel electrophoresis or on immunological detection of the appropriate protein composition. The provision of monoclonal antibodies that can distinguish between gamma-gliadin 42 and 45 offers an immunological screening method that is likely to be more cost-effective in breeding than electrophoretic analysis (Howes *et al.*, 1989; Clark *et al.*, 1993).

Pasta processing quality

A small-scale procedure, devised by Matsuo *et al.* (1972) for spaghetti-processing quality, requires 50 g semolina which is mixed into a dough in the Farinograph. The dough is extruded through a four-hole die prior to drying and cooking of the product. Evaluation involves determination of colour and brightness, as well as testing for tenderness, compressibility and resilience after cooking under specified conditions. Alternatively, a small disk of pasta may be formed for cooking and texture analysis (Cubadda, 1988).

For later generations, when at least a kilogram of grain is available, a fuller milling test is performed (e.g. by the method of Matsuo and Dexter, 1980) providing sufficient semolina for fuller-scale processing into spaghetti or other pasta product.

11.2.8 Computer-based evaluation of variety trials

The great volume of results generated in a breeding programme necessitates the use of a computer-based management system to aid in recording and interpreting results efficiently. Such systems have been developed by Morris and Raykowski (1993, 1995). Burridge *et al.* (1989) have also described a computer programme with the similar aim of recording and collating quality data, thus to quickly highlight the strengths or weaknesses of specific samples. The further difficulty of determining what weighting to place on individual quality attributes has been addressed in a companion paper (Fletcher *et al.*, 1989).

11.3 BARLEY

A century of barley improvement by cross-breeding has produced varieties in many countries that are environmentally adapted and suited to processing requirements, mainly for malting (but also the range of uses described in Chapter 4). Recent developments in breeding methodologies and in quality-selection efficiency promise to lead to further genetic improvements, both in agronomic traits and in processing quality. New breeding techniques include transformation as a means of introducing novel genes, doubled haploid and mutation breeding, and possibilities for hybrid barley. In the process of selection for grain quality, new approaches promise to provide earlier and more specific selection for quality attributes, with new gene-probing techniques (Chapter 12), antibodies specific for quality malting proteins, a better understanding of the chemistry of grain quality, and with improvements in micromalting and brewing methods.

11.3.1 Quality selection at early generation

Many of the quality attributes listed in Table 11.6 superficially resemble those for wheat, but the reasons for selecting specific attributes differ considerably.

Visual inspection

As an example, examination of grain shape and colour indicate head type (two- or six-row) and aleurone-layer type (white or blue), six-row and/or blue aleurone being used as markers of non-malting feed barleys in some countries. In general, barleys with blue aleurone are unsuitable for malting and food use, so rejection of such grains may be an essential preliminary step if a blue aleurone parent has been used in a cross. As the colour of the underlying aleurone layer may be obscured to some extent by the outer husks, a light pearling treatment (abrasion) facilitates distinction (Wrigley and Batey, 1995). Examination of the grain may also involve segregation for naked caryopsis where a breeding objective for food or feed use may be to breed for free-threshing grain.

Table 11.6 Quality testing of lines of barley in the breeding process

Quality attribute	*Test material*	*Method or equipment*	*References*
At early generation (F3–F5)			
Head type (2- or or 6-row)	Whole grains or heads	Visual	Wrigley (1995)
Grain colour (white/blue)	Whole grains	Visual, barley pearler NIR	Wrigley (1995) Henry (1985)
Hulled/naked caryopsis	Whole grain	Visual	Wrigley (1995)
Grain size	Grain	Visual, Thousand-kernel weight, slotted screens. Image analysis	Wrigley (1995); Sapirstein, 1995
Protein content	Grain, meal	NIT, NIR, Kjeldahl	Anon (1983)
β-glucan content	Grain, meal	NIT, NIR, test kit	Anon (1983); McCleary and Codd (1991)
Malting potential	Meal; Grain	NIR; NIT	Henry (1985); Ronalds *et al.* (1992)
Diastatic power	Meal	α- and β-Amlyases	McCleary and Codd (1989); McCleary and Shameer (1987)
At intermediate generation (F5–F7)			
The above tests, plus:			
Malting extract	Whole grain	Micromalting	Glennie-Holmes *et al.* (1990), Chapter 4
At late generation (>F8)			
Many of the earlier tests, plus:			
Malt extract	Whole grain	Lab and pilot-scale simulation of the industrial processes	Chapter 4
Beer production	Malt		

Grain-size distribution

Uniformity of grain size is an important attribute for malting quality, as it is likely to ensure uniformity of germination rate. Selection for large grain size is an associated objective. Both these attributes are traditionally checked by passing grain samples through a series of slotted screens, first estimating non-grain material (too large or too small) and also estimating size distribution, e.g. percentage of grains retained by 2.2, 2.5, and by 2.8-mm wide slots. Expectations for large, uniform grain sizes have been rising; a current objective might be for >90% of grains to be retained by a 2.5 mm screen.

Whilst such sieving methods are retained in general industry practice, there is reason to use this method (at least to some extent) in breeding. On the other hand, more specific information can be obtained more efficiently by image analysis (digital analysis of a video-camera image) (Sapirstein, 1995) or with an electronic balance (to record grain weights one by one). The total weight of a known (large) number of grains (as Thousand Kernel Weight) is also a useful measure of grain size, but it does not indicate size distribution.

Chemical attributes

As is shown in Table 11.6, near infrared spectroscopy is becoming used for whole-grain or ground grain analysis, not only for routine aspects of composition, such as moisture and protein content, but also for β-glucan content and to predict malting quality (generally as coarse-ground hot water extract of malt) (Henry, 1985; Glennie-Holmes, 1991; Ronalds *et al.*, 1992). For all these estimates, calibration of the NIR or NIT equipment is critical, since near infrared technology is not a primary source of data. As this technology is still in the early stages of development for attributes such as malting quality, there is not a wide range of published information on its use. Some informal reports indicate that calibrations are most reliable if they are based on samples analysed for the same season and growth region (or breeding programme) as those to be analysed.

Screening for β-glucan content, an important application of NIR spectroscopy is appropriate to malting and feed barley, both of which have a requirement of low β-glucan content, but also for human food use of barley for which high β-glucan is desirable. In cases where high β-glucan is undesirable, it is also appropriate to screen for a high activity of β-glucanase.

High diastatic power is required for malting barley. It is measured as the combined result of a range of hydrolases, or as the individual enzymic activities of one or two key enzymes, such as α-amylase and β-amylase, as is explained in detail in Chapter 4. Test kits are now available for the efficient determination of these enzymes (McCleary and Codd, 1989; McCleary and Sheehan, 1987; McCleary and Shameer, 1987), but there is also the possibility of predicting relative activities by detecting the enzymes' presence with specific antibodies. This immunological approach has the advantage that immunodiagnostic equipment can be used, such as microtitre plates, with 96 wells for rapid evaluation in an automatic plate reader (Skerritt *et al.*, 1994).

11.3.2 Micromalting and brewing

Whereas selection for the range of quality attributes described above is valuable, simulation of the whole malting process on a small scale is an ideal means of screening for malt quality (Glennie-Holmes *et al.*, 1990). This type of testing is now possible on as little as 15 g barley grain (preferably replicated), but it is generally performed on 40 g grain and is not applied generally until about F5. This approach to quality evaluation is described in Chapter 4, for processing situations in which there is no sample scarcity. Although there has been considerable success in the evaluation of malting quality for the small sample sizes mentioned above, many breeders prefer to defer this level of testing until sample sizes of over 100 g (replicated) are available. This reservation applies more obviously to the further selection test of evaluating the brewing quality of the malt (Chapter 4), a step normally performed in the late stages of the breeding selection process.

11.4 RYE AND TRITICALE

Although rye and triticale are very close relatives of wheat, they do not have the same potential for bread-making quality. Nevertheless, many of the quality requirements of rye are relevant to its use in baking, often in combination with wheat, as has been described in Chapter 7. For this reason, there is particular attention in breeding rye to select for resistance to sprout damage and thus to minimize the likely deleterious effects of the hydrolases of sprouting on baking quality. This is a particularly important requirement in the case of rye that is to be used for the production of crispbreads. Interactive with this aspect of rye quality is the suitability of the starch for such processing. There is potential for the combination of these properties to be assessed in the breeding system with the Rapid Visco Analyser (Walker *et al.*, 1988), since it can be used with much smaller samples (about 4 g) than the Brabender Amylograph, as described below for rice (Section 11.6.4). This combination of RVA tests would involve the standard methods being developed by the International Association for Cereal Science and Technology (ICC) for determining **stirring number**, equivalent to **falling number** and the ICC rapid pasting method for starch properties.

High heritabilities have been reported for resistance to sprouting (assessed as falling number) (Rattunde *et al.*, 1991; Ohnmacht *et al.*, 1990; Loock *et al.*, 1990). Despite earlier difficulties, there appear to be promising sources of genotypic variation for dormancy in rye (Loock *et al.*, 1990) and breeding successes in this respect are starting to be realized (Weipert, 1990). Possibilities for the use of molecular markers should further assist the task of generating improved genetic resistance to sprouting in rye and triticale (Gale *et al.*, 1990). The past two decades have seen advances in the potential of hybrid rye as a means of greatly increasing grain yields, with the development of **top cross hybrid varieties**, and also as a means of improving quality aspects including resistance to sprouting and feed value (Geiger, 1982; Weipert, 1990).

The tendency of rye to harbour ergot is a continuing problem in its use for human food (as well as for animal feed). It does not appear that breeding can overcome this disadvantage, as no genetic sources of resistance to ergot infection are known (Lorenz, 1991). As the feed value of rye is limited by such antinutritional factors as 5-alkyl resorcinols (Verdeal and Lorenz, 1977), phytic acid and trypsin inhibitors, there have been efforts to reduce the levels of these compounds by selection in breeding, with some success (Lorenz, 1991). The relatively high levels of non-starch polysaccharides (e.g. pentosans) in rye constitute some nutritional advantages for human food, but they also limit the extent to which rye is incorporated in feed rations. Breeding has the potential also to reduce the levels of this family of compounds, but the main approach that has been taken in this case has been to incorporate suitable enzymes (e.g. pentosanases) in the feed to reduce their growth-limiting effects (Chapter 7).

The wheat–rye hybrid, triticale, shares some of the above disadvantages of rye, with similar expectations of improving grain quality through breeding (Bushuk and Larter, 1980). The predominance of the wheat genome in triticale has always provided great hope of its value for human food. Nevertheless, triticale remains primarily a feed grain, despite attempts to improve its gluten quality by the provision of dough-strength-associated glutenin alleles (Gustafson *et al.*, 1991, and references therein). On the other hand, breeding has improved another quality deficiency, namely, the shrivelling of the grain during the later stages of grain filling, although the biochemical causes of this phenomenon have not been fully elucidated (Gustafson *et al.*, 1991; Pena *et al.*, 1982).

11.5 OATS

As in the breeding of other cereals, considerable effort with oat improvement is concentrated on yield-related issues such as disease resistance, tolerance to stresses, and resistance to lodging (Burrows, 1986). In addition, the oat breeder has the problem of being faced with diverse quality objectives, depending on whether the crop is to be used for human food processing or for animal feed. If the latter, there are the further alternatives of favouring grazing, forage or grain use, and even feed-use requirements of grain may vary, depending on whether it is destined for ruminants or for monogastrics (for which a high metabolizable energy would be required). This difficulty is being resolved in some countries by breeding separately for grain or for grazing (Peterson *et al.*, 1995). Ideally, the farmer would prefer to be able to graze the growing crop, and also harvest a good yield of plump grain. Such an objective is feasible for the breeding of general-purpose oats, mainly for use on-farm, but probably not for milling use.

The more exacting specifications of oat grain for processing into human food, as set out in Chapter 6, include a high groat content (less loss in the process of removing the hulls) and a high test weight (the result of uniformly plump kernels, with a small proportion of 'thins' – small kernels from the upper parts of the floret). Some of these goals may be better met by the breeding of naked oat types (Burrows, 1986), with an obvious (and apparent) yield penalty due to the absence of the hulls in the harvested crop. Generally, selection for these grain-quality aspects does not start until F5 or later generations, when some of the promising lines may have been discarded for other reasons (Chapter 6). On the other hand, Burrows (1986) has described the successful selection for improved chemical composition by analysing individual half-grains, retaining the germ ends for propagation as appropriate from the analyses.

Selection for genetic potential to produce grain of specific chemical composition is a viable means of suiting specific market requirements, although such selection objectives suffer from the difficulty of being subject to environmental variation. For example, a higher lipid content is likely to provide an increase in the feed value, as metabolizable energy, especially for monogastric animals, but it may also add to the processing problems associated with off-flavours due to grain damage and lipase action. Furthermore, there has been success in breeding for high protein content, as is instanced in the Canadian variety Hinoat, with up to 23% grain protein and large protein bodies embedded in the endosperm structure (Fulcher, 1986), but the associated yield penalties jeopardize the chances of extensive cultivation of such varieties.

New opportunities to select more efficiently for aspects of chemical composition include the use of near infrared analysis to determine protein and oil content. In describing such a system, Krishnan *et al.* (1994) reported that for their range of samples, growing location did not greatly contribute to protein or oil content variability. On the other hand, soil nitrogen and climatic variability were blamed for the considerable variabilities in the contents for protein and β-glucan in a study reported by Brunner and Freed (1994).

In addition to selection for variation in the proportions of chemical constituents, there is potential value in selecting for qualitative differences, particularly in starch properties, as these are likely to affect aspects of processing for food uses. Considerable variation in starch properties has been demonstrated between oat genotypes (Paton, 1986, and the Rapid Visco Analyser is being used to select efficiently for starch properties in oat (Anon, 1994).

11.6 RICE

Appearance, milling yield and cooking quality are the predominant attributes used to select for quality in rice breeding (Khush *et al.*, 1979). Small-scale evaluation of such aspects of quality is more readily attained than quality attributes for some other grain processes such as baking or brewing quality. Nevertheless, it has been important for rice breeders to identify aspects of chemical composition that would act as markers or indicators of quality attributes. For rice, starch quality is of greater importance than protein composition, in contrast to wheat quality. Furthermore, there is much greater genetic variation in starch properties available in rice than in wheat. As a result, breeding for rice quality differs considerably from this exercise for other cereals.

11.6.1 Grain appearance

As indicated in Chapter 2, preferences for rice appearance differ considerably from one market to another (Juliano, 1985). Yet appearance is a basic quality characteristic and the main basis of selection by the consumer, since rice is unusual among the food grains for being bought and eaten as the whole-grain product. The nature of segregation for this character obviously depends on the chosen target market, e.g. long or short grain. On the other hand, there are fundamental aspects of appearance that are common to all rice types and that can be readily assessed by appearance. These additional aspects include uniformity of grain weight, grain translucency (absence of chalkiness, apart from the special case of waxy rice), and grain colour (absence of pink stripes). Such characteristics can be more readily assessed by illumination of the milled grain with transmitted light, from below through a translucent plate (Ikehashi and Khush, 1979). Juliano (1985) has suggested categories for classifying according to size and shape.

As Figure 11.1 indicates, these characteristics can readily be checked at the earliest stages of grain-quality selection, initially at F2, when this examination might be restricted to brown (unmilled) rice. Although this would normally be performed subjectively by experienced workers, there is the possibility of adapting image analysis to have this task performed on a more objective and automatic basis. Grain appearance is an important character for selection at all stages of the breeding process, not only at early generation when it can be applied more readily than other tests. At later generation, when more grain is available, appearance can also be assessed after milling, providing a much better basis for quality evaluation.

11.6.2 Milling yield

As soon as adequate supplies of grain are available (about 50 g of paddy), assessments of hulling performance and milling yield are performed, generally at F3/F4 and later (Figure 11.1). Descriptions and photos of hulling and milling equipment for various sample sizes are provided by Webb (1975) and by Van Ruiten (1985). A small barley pearler can be used at the early stages for this test, with the aim of selecting lines giving a high yield of whole milled grains, and rejecting those with a tendency to grain breakage. Part of this testing is visual assessment of bran remaining in grooves on the surface of the milled grain.

11.6.3 Cooking and eating qualities

Like grain size and shape, preferences vary for eating quality, so decisions must be made about the target market. As might also be expected, cooking methods differ and so the cooking test procedure must be suited to the market goal. Cooking methods, reviewed by Juliano (1985), generally require 100 g of milled grain, and are thus restricted to the later stages of the breeding process. Specific aspects evaluated include grain elongation during cooking, volume expansion, water absorption, stickiness of the cooked rice and solids remaining in the cooking water. For specific classes of rice there will be additional qualities such as aroma and suitability for parboiling to be evaluated. Such aspects, especially eating qualities, may be evaluated by a sensory panel, especially close to the stage of release ('industry evaluation' in Figure 11.1). There have been many attempts to test eating qualities by objective means (reviewed by Juliano, 1985), and particular breeding programmes and breeders have their favoured procedures for doing so. One of these uses the Instron tester, equipped with a Kramer Shear Cell, in which the cooked rice is subjected to shear testing to simulate mouth feel, as described in more detail in Chapter 2 (Blakeney, 1979; Juliano, 1985). This test requires about 50 g milled rice, relegating it also to the mid- to late-stages of breeding.

11.6.4 Starch characteristics

Many of these eating qualities are closely related to the characteristics of the starch. Therefore, breeders have sought to test for relevant aspects of starch structure, composition or function, thereby to select early for suitable eating quality. The two major aspects of starch composition of value in this respect are amylose content and gelatinization characteristics (Webb, 1975). Both these characteristics are highly heritable, so it is valid to select for them early in the breeding process with the reasonable expectation that they will be 'fixed' in succeeding generations (Adair *et al.*, 1973).

Unlike many other cereal grains, there is wide genetic diversity in amylose content available in rice – from waxy (1–2% amylose) to high amylose contents (>30%). Waxy rice expands least during cooking (taking up little water and appearing glossy) to provide highest bulk density and least resistance to disintegration during cooking (Juliano, 1979). High amylose content is incompletely dominant to low amylose content, and is controlled by one major gene and some modifying genes (Chang and Somrith, 1979). The roles of the multiple alleles, including cytoplasmic effects, have recently been reported by Pooni *et al.* (1994). Because of the value of amylose to indicate such important attributes, a colorimetric screening test for amylose content (Juliano, 1985) is routinely applied to predict and select for the desired aspect of cooking quality. Recent study of near infrared spectroscopy (Delwiche *et al.*, 1995) suggests that breeders may soon have the advantages of NIR methods to screen efficiently for amylose content in rice, making this aspect of quality selection much simpler.

On the other hand, there are other grain attributes that contribute to cooking quality, such as grain shape and surface area, as well as the particular cooking method chosen (Juliano, 1979, 1985). A modification of the colorimetric amylose test involves the additional step of relating amylose content to solubility in boiling water (Bhattacharya *et al.*, 1978). However, consideration of gelatinization properties is probably of more value in the breeding context.

For breeding purposes, indications of starch gelatinization properties have been obtained for individual grains of milled rice by the alkaline digestibility test of Little *et al.* (1958), involving assessment of the 'spreading' of the grains during prolonged soaking (e.g. 23 h at 30°C in dilute alkali solution (1.7% KOH)). The results relate closely to gelatinization temperature, ready spreading being an indication of a low gelatinization temperature (<70°C). Alternatively, gelatinization temperature may be determined directly with a polarizing microscope by observing the birefringence end-point temperature, the point at which >90% of the starch granules have lost their characteristic birefringent appearance. Another breeders' small-scale test that generally correlates with gelatinization temperature is the determination of gel consistency (Juliano, 1985). This involves dispersing 100 mg rice flour in 2.2 ml dilute alkali (containing about 10% ethanol and dye) in a glass tube, heating the tube for 8 minutes in a boiling water bath, and measuring the height of the gel layer after cooling (Perez, 1979; Juliano, 1985).

When sufficient grain is available, it is preferable to perform an assessment of starch pasting and set-back properties in the Brabender Amylograph (requiring a relatively large sample, >50 g of rice flour and about 2 h per analysis) or using the Rapid ViscoAnalyser (needing only 3 g of rice flour and about 10 minutes for each analysis) (Welsh *et al.*, 1991). A new method (AACC Method 61–02) for the RVA test of rice has recently been approved by the American Association of Cereal Chemists. Chapter 2 has more details about these methods.

11.6.5 Overall quality characteristics

Depending on the quality objectives, there may be further aspects of quality to select for. One of these, for example, is the possibility of breeding fragrant rice, grain that when cooked has a characteristic aroma/taste attributed to 2-acetyl-1-pyrroline (Buttery *et al.*, 1983). The identification of this compound as a major sensory contributor opens the possibility of chemical analysis for aromatic lines at early generation. Traditionally, the breeder's test for this attribute is to boil the milled rice in dilute alkali (1.7% KOH) to facilitate detection of the aroma. The further (and simpler) possibility of tasting water from the soaking of individual grains has been described by Reinke *et al.* (1991). The use of this single-grain method also opens the possibility of developing genotypes in which all grains are aromatic, not only a certain proportion as is the case for much trade in fragrant rice.

Other quality types that may be part of breeding objectives include glutinous rice and genotypes suited to parboiling.These and all other preference types may be the subject of selection procedures throughout the breeding process (Figure 11.1) with respect to component attributes, but the overall aggregation of these components of quality into the processing and organoleptic assessments are the telling tests that are of overriding importance in the late stages of breeding, particularly in consultation with industry. Whereas in the early stages, the accent is on attributes with high heritability, at later stages more account must be taken of modification of genetic potential by expected variations in environmental conditions (growing, drying and storage conditions), depending on the target growth region.These environmental factors have considerable potential to modify milling quality, as has been described by Kunze (1985) for the USA and by Srinivas and Bhashyam (1985) for Asia.

11.7 MAIZE

As noted in Chapter 3, maize may be classified into five utilization groups, based on endosperm characteristics. This classification is reasonable because it is largely the composition and structure of the endosperm that impart much of the utility of maize as a cereal grain. Another way of examining maize utilization is to consider the end-user: animal (feed) or human (food). And finally, the method of processing prior to use is worthy of consideration: grinding, or otherwise reducing particle size for direct animal or human consumption, namely wet milling and dry milling.

The USA is the largest producer of maize (Glover and Mertz, 1987). The majority of this maize is yellow dent (a combination of horny and floury endosperm) and goes directly into animal rations, primarily as a source of energy. Consequently, maize breeding has generally focused on maximum production of grain per unit land area. Traits such as resistance to diseases, environmental stresses and agronomic performance (ear height, maturity, lodging resistance, etc.) have been largely viewed in the context of yield and efficiency of harvest. Quality considerations include rate of dry-down at maturity, test weight (bulk density), and breakage susceptibility (Hameed *et al.* 1994). Moisture measurement and breakage susceptibility are described in Chapter 3. The concept of maximum grain production per unit land area may be modified to reflect maximum utilizable energy per unit land area or per unit grain mass. Following this logic, the breeder may genetically alter the composition of maize. For example, the oil content of maize hybrids can be dramatically increased (Alexander and Creech, 1977). The oil content of ordinary yellow dent maize (about 4.5%) can be easily increased to 7–10%. Individual lines, though commercially impractical, may reach 20% oil (Alexander and Creech, 1977). About 85% of kernel oil resides in the germ portion and is composed mostly of triglycerides. Most commercial oil is extracted from germ isolated during wet milling by a combination of expelling and hexane extraction. Breeders may assess oil content non-destructively by wide-line NMR (Bauman *et al.*, 1963) or NIR (Hymowitz *et al.*, 1974). On the other hand, oil quality (that is, the types and composition of triglycerides in maize oil) has received little study or practical breeding effort (Alexander and Creech, 1977; Glover and Mertz, 1987).

Although commonly viewed as an energy source, maize also provides significant protein to consumers (both animal and human). Although maize contains 9–11% protein, with some lines reaching 26% (Glover and Mertz, 1987), its nutritional value is limited by a proportionally low level of lysine and tryptophan. This imbalance was largely overcome by the discovery by Mertz of the opaque-2 mutation (Glover and Mertz, 1987). Opaque-2 maize endosperm has reduced levels of zein, the main prolamine storage protein. Consequently, its overall protein composition is adequate to fulfil the nutritive requirements of children and adults (Glover and Mertz, 1987). An initial breeding problem with the opaque-2 mutation was that it caused the conversion of corny to floury endosperm. Floury endosperm caused problems with harvesting, handling, processing, and sometimes consumer acceptance. The problem was largely overcome by adding modifier genes to convert the floury endosperm back to hard.

A second genetic system similar to opaque-2 is floury-2. The modified hard endosperm opaque-2 and floury-2 genetic systems have been dubbed 'Quality Protein Maize' (QPM); the subject is reviewed by Mertz (1992). Although QPM holds promise as a means of making maize a more nutritious and valuable cereal grain, hybrids have not met with commercial acceptance in the US largely due to the yield penalty associated with the QPM hybrids, the relatively inexpensive alternative source of lysine and tryptophan in the form of soybean (*Glycine max*) meal for swine, and the general over-abundance of meat protein in the American diet. Apparently, the yield penalty of QPM hybrids has been largely eliminated in some regions (Pixley and Bjarnason, 1993). Breeding QPM inbreds can be accomplished by assaying tryptophan content (Villegas, 1975), total free amino acids by ninhydrin (Sung and Lambert, 1983), or direct measurement of lysine (Obi, 1982).

Yellow dent maize is also the main starting material for wet and dry milling in the US (see Chapter 3). Both processes aim to separate the compositional and/or botanical parts of the maize kernel. The main goal of the dry milling process is the production of large pieces of corny endosperm free of germ and pericarp. These 'flaking grits' are the largest of several fractions ranging in size down to maize flour. These fractions are used in cooked grits, corn bread and muffins, corn flakes, tortillas and snack foods. For grits, white corn devoid of the yellow carotenoid pigments is preferred. Oil is expelled from the isolated germ.

The main goal of the wet milling process is the co-production of germ (for oil) and starch. Isolated maize starch is used 'as is' in many food and industrial applications, or is converted to dextran syrups, to high fructose corn syrup, to dextrose, or is chemically modified. Although both milling processes desire different levels of corny versus floury endosperm, little breeding attention has been directed to the issue. Endosperm composition may be visually examined or may be evaluated for density, as described in Chapter 3.

Two endosperm mutations involving starch biosynthesis are important to the wet milling industry. The first is the elimination of amylose, normally comprising about 27% of starch, to yield starch of nearly pure amylopectin. Breeding for this waxy starch is accomplished by staining starch or pollen with 2% potassium iodide (Glover and Mertz, 1987). Glover and Mertz (1987) indicate that about 10% of the maize used in the US wet milling industry is waxy maize. The second important endosperm mutant is amylose-extender which produces starch composed of about 55–60% amylose. Further modification results in starch with about 85% amylose (Watson, 1977; Glover and Mertz, 1987). High amylose hybrids are grown exclusively for wet milling.

Vegetable corns incorporate other mutations of starch biosynthesis. Most common are the sugary and shrunken-2 genes used in sweetcorn, a popular fresh, frozen or canned vegetable. The shrunken, shrunken-2, brittle, and brittle-2 genes have been used to develop 'extra-sweet' or 'super sweet' sweetcorn hybrids. Alexander and Creech (1977) and Glover and Mertz (1987) provide descriptions of these as well as other endosperm mutations used in vegetable corn breeding.

11.8 FUTURE PROSPECTS FOR QUALITY IMPROVEMENT

Continuing improvements in grain quality can be expected as market pressures and higher premiums for quality create demands for specific quality types. Conventional breeding technologies have the potential to continue this process. Novel breeding technologies can both improve efficiencies of realizing these goals and create new genetic diversity from which to select appropriate genotypes. To the extent that relevant molecular markers can be identified, the use of gene or antibody probing methods can sharpen the tools of selection to make more exact identification of suitable lines possible at earlier stages of breeding, with greater assurance in distinguishing genetic from environmental influences. Achieving these dreams will also depend on a better understanding of the biochemical mechanisms for grain-quality attributes, and the implications for any one attribute of modifying other aspects of quality.

The new day that is now dawning in breeding technology offers that promise of switching genes off and on, and introducing genes from completely different species. This is the theme of the next chapter, telling more about this frontier which, when crossed, will probably achieve a range of breakthroughs comparable only to the improvements introduced a century ago by the transition from selection only to cross-breeding followed by selection.

REFERENCES

Adair, C.R., Bollich, D.H., Bowman, D.H. *et al.* (1973) Rice breeding and testing methods in the United States, in *Rice in the United States: Varieties and Production*. USDA Handbook No 289, pp. 19–64.

Alexander, D.E. and Creech, R.G. (1977) Breeding special industrial and nutritional types, in *Corn and Corn Improvement* (ed. G.F. Sprague), American Society of Agronomy, Madison, WI, pp. 363–90.

Andrews, J.L., Blundell, M.L. and Skerritt, J.H. (1993) A simple antibody-based test for dough strength. III Further simplification and collaborative evaluation for wheat quality screening. *Cereal Chemistry*, **70**, 241–6.

Anon. (1983) *Approved Methods of the American Association of Cereal Chemists*, 8th edn, The Association, St Paul, MN, USA.
Anon. (1994) *1992–1994 Biennial Report. Agricultural Research Institute, Wagga Wagga/Temora*. NSW Agriculture, Orange, NSW, Australia.
Bason, M.L., Peden, G., Zounis, S. *et al.* (1993) Detection of red-grained wheat by Tristimulus colorimetry and digital image analysis, in *Proceedings of 43rd Australian Cereal Chemistry Conference* (ed. C.W. Wrigley), Royal Australian Chemical Institute, Melbourne, pp. 29–34.
Bason, M.L., Zounis, S., Ronalds, J.A. and Wrigley, C.W. (1994) Segregating red and white wheat visually and with a Tristimulus Meter. *Australian Journal of Agricultural Research*, **46**, 89–98.
Bassett, L.M., Allan, R.E. and Rubenthaler, G.L. (1989) Genotype × environment interactions on soft white winter wheat quality. *Agronomy Journal*, **81**, 955–60.
Batey, I.L., Gupta, R.B. and MacRitchie, F. (1991) Use of size-exclusion high performance liquid chromatography in the study of wheat flour proteins: an improved chromatographic procedure. *Cereal Chemistry*, **68**, 207–9.
Bauman, L.F., Conway, T.F. and Watson, S.A. (1963) Heritability of variations in oil content of individual corn kernels. *Science*, **139**, 498–9.
Beninati, N.F. and Busch, R.H. (1992) Grain protein inheritance and nitrogen uptake and redistribution in a spring wheat cross. *Crop Science*, **32**, 1471–5.
Bettge, A.D., Morris, C.F. and Greenblatt, G.A. (1995) Assessing genotypic softness in single wheat kernels using starch granule-associated friabilin as a biochemical marker. *Euphytica*, **86**, 65–72.
Bhattacharya, K.R., Sowbhagya, C.M. and Indudhara-Swamy, Y.M. (1978) Importance of insoluble amylose as a determinant of rice quality. *Journal of Science Food Agriculture*, **29**, 359–64.
Blakeney, A.B. (1979) Instron measurement of cooked rice texture, in *Proceedings of Workshop on Chemical Aspects of Rice Grain Quality*. International Rice Research Institute, Los Banos, Philippines, pp. 343–53.
Brennan, J.P. and O'Brien, L. (1991) An economic investigation of early-generation quality testing in a wheat breeding programme. *Plant Breeding*, **106**, 132–40.
Brown, G.L., Curtis, P.S. and Osborne, B.G. (1993) Factors affecting the measurement of hardness by near infrared reflectance spectroscopy of ground wheat. *Journal of Near Infrared Spectroscopy*, **1**, 147–52.
Brunner, B.R. and Freed, R.D. (1994) Oat grain beta-glucan content as affected by nitrogen level, location, and year. *Crop Science*, 34, 473–6.
Buller, A.H.R. (1919) *Essays on Wheat*. Macmillan, New York.
Burridge, P.M., Palmer, G.A., Fletcher, R.J. and Hollamby, G.J. (1989) Computer assisted wheat quality evaluation, in *Proceedings of the 39th Australian Cereal Chemistry Conference*. (eds T. Westcott, Y. Williams and R. Ryker), Royal Australian Chemical Institute, Melbourne, Australia, pp. 29–33.
Burrows, V.D. (1986) Breeding oats for food and feed: conventional and new techniques and materials, in *Oats: Chemistry and Technology* (ed. F.H. Webster). American Association of Cereal Chemists, St Paul, MN, USA, pp. 13–46.
Bushuk, W. and Larter, E.N. (1980) Triticale: production, chemistry, and technology. *Advances in Cereal Science and Technology*, **3**, 115–57.
Buttery, R.G., Ling, L.C., Juliano, B.O. and Turnbaugh, J.G. (1983) Cooked rice aroma and 2-acetyl-1-pyrroline. *Journal of Agricultural and Food Chemistry*, **31**, 823–6.
Carver, B.F. (1994) Genetic implications of kernel NIR hardness on milling and

flour quality in bread wheat. *Journal of the Science of Food and Agriculture*, **65**, 125–32.

Chang, T.T. and Somrith, B. (1979) Genetic studies on the grain quality of rice, in *Proceedings of Workshop on Chemical Aspects of Rice Grain Quality*. International Rice Research Institute, Los Banos, Laguna, Philippines, pp. 49–58.

Chesterfield, R.S. (1971) A modified barley pearler for measuring hardness of Australian wheat. *Journal of the Australian Institute of Agricultural Science*, **37**, 148.

Clarke, J.M., Howes, N.K., McLeod, J.G. and DePauw, R.M. (1993) Selection for gluten strength in three durum wheat crosses. *Crop Science*, **33**, 956–8.

Coffman, F.A., Murphy, H.C. and Chapman, W.H. (1961) Oat breeding, in *Oats and Oat Improvement* (ed. F.A. Coffman), American Society Agronomy, Madison, Wisconsin, USA, pp. 263–29.

Cornish, G.B., Burridge, P.M., Palmer, G.A. and Wrigley, C.W. (1993) Mapping the origins of some HMW and LMW glutenin subunit alleles in Australian germplasm, in *Proceedings of the 43rd Australian Cereal Chemistry Conference* (ed. C.W. Wrigley), Royal Australian Chemical Institute, Melbourne, pp. 255–60.

Cubadda, R. (1988) Evaluation of durum wheat, semolina, and pasta in Europe, in *Durum Wheat: Chemistry and Technology* (eds G. Fabriani and C. Lintas), American Association of Cereal Chemists, St Paul, MN, USA, pp. 217–28.

Cutler, G.H. and Brinson, G.A. (1935) The granulation of whole meal and a method of expressing it numerically. *Cereal Chemistry*, **12**, 120.

Damideaux, R., Autran, J.C., Grignac, P. and Feillet, P. (1978) Mise en évidence de relations applicables en sélection entre l'éctrophorégramme des gliadines et les propriétés viscoélastiques du gluten de *Triticum durum* Desf. *Comptes Rendus de l'Académie des Sciences*, **287**, Serie D, 701–4.

Delwiche, S.R. (1993) Measurement of single-kernel wheat hardness using near-infrared transmittance. *Transactions of the American Society of Agricultural Engineers*, **36**, 1431–37.

Delwiche, S.R. and Norris, K.H. (1993) Classification of hard red wheat by near-infrared diffuse reflectance spectroscopy. *Cereal Chemistry*, **70**, 29–35.

Delwiche, S.R. and Weaver, G. (1994) Bread quality of wheat flour by near-infrared spectrophotometry: feasibility of modeling. *Journal of Food Science*, **59**, 410–5.

Delwiche, S.R., Bean, M.M., Miller, R.E., Webb, B.D. and Williams, P.C. (1995) Apparent amylose content of milled rice by near-infrared reflectance spectrophotometry. *Cereal Chemistry*, **72**, 182–87.

Dick, J.W. and Quick, J.S. (1983) A modified screening test for rapid estimation of gluten strength in early-generation durum wheat breeding lines. *Cereal Chemistry*, **60**, 315–8.

Dick, J.W. and Youngs, V.C. (1988) Evaluation of durum wheat semolina, and pasta in the United States, in *Durum Wheat: Chemistry and Technology* (eds G. Fabriani and C. Lintas) American Association of Cereal Chemists, St Paul, MN, USA, pp. 237–48.

D'Ovidio, R. and Anderson, O.D. (1994) PCR analysis to distinguish between alleles of a member of a multigene family correlated with wheat bread-making quality. *Theoretical and Applied Genetics*, **88**, 759–63.

du Cros, D.L., Wrigley, C.W. and Hare, R.A. (1982) Prediction of durum-wheat quality from gliadin-protein composition. *Australian Journal of Agricultural Research*, **33**, 429–42.

Fabriani, G. and Lintas, C. (eds) (1988) *Durum Wheat: Chemistry and Technology*. American Association of Cereal Chemists, St Paul, MN, USA.

Federmann, G.R., Goecke, E.U., Steiner, A.M. and Ruckenbauer, P. (1994)

Biochemical markers for selection towards better cooking quality in F2-seeds of durum wheat (*Triticum durum* Desf.). *Plant Varieties and Seeds*, **7**, 71–7.

Finney, K.F. (1989) A five-gram mixograph to determine and predict functional properties of wheat flours. *Cereal Chemistry*, **66**, 527–30.

Finney, K. F. and Shogren, M.D. (1972) A ten-gram mixograph for determining and predicting functional properties of wheat flours. *Bakers Digest*, **46**, 32–43, 77.

Fischer, R. A., O'Brien, L. and Quail, K.J. (1989) Early generation selection in wheat. II Grain quality. *Australian Journal of Agricultural Research*, **40**, 1135–42.

Fletcher, R.J., Hollamby, G.J., Burridge, P.M. and Palmer, G.A. (1989) Quality evaluation in wheat breeding. 2. Investigations into a selection index, in *Proceedings of the 39th Australian Cereal Chemistry Conference* (eds T. Westcott, Y. Williams, and R. Ryker), Royal Australian Chemical Institute, Melbourne, Australia, pp. 34–51.

Fulcher, R.G. (1986) Morphological and chemical organization of the oat kernel. in: *Oats: Chemistry and Technology* (ed. F.H. Webster), American Association of Cereal Chemists, St Paul, MN, USA. pp. 47–74.

Gale, M.D., Flintham, J.E. and Mares, D.J. (1990) Applications of molecular and biochemical markers in breeding for low alpha-amylase wheats, in *Fifth International Symposium on Pre-Harvest Sprouting in Cereals* (eds K. Ringlund, E. Mosleth and D.J. Mares), Westview Press, Boulder, CO, USA, pp. 167–75.

Garcia-Casado, G., Sanchez-Monge, R., Lopez-Otin, C. and Salcedo, G. (1994) Rye chromosome arm 3RS encodes a homodimeric inhibitor of insect alpha-amylase. *Theoretical and Applied Genetics*, **89**, 60–3.

Geiger, H.H. (1982) Breeding methods in diploid rye (*Secale cereale* L.). *Tagungsbericht (Akademie der Landwirtschafswissenschaften der DDR), Berlin*, **198**, 305–22.

Glennie-Homes, M. (1991) The use of unmodified barley for malting potential selection. *Journal of the Institute of Brewing*, **97**, 381–87.

Glennie-Holmes, M., Moon, R. and Cornish, P.B. (1990) A computer-controlled micromalter for barley breeding programmes. *Journal of the Institute of Brewing*, **96**, 11–16.

Glover, D.V. and Mertz, E.T. (1987) Corn, in *Nutritional Quality of Cereal Grains: Genetic and Agronomic Improvement*, American Society of Agronomy, Madison, Wisconsin, pp. 183–336.

Gras, P.W. and O'Brien, L. (1992) Application of a two-gram Mixograph to early generation selection for dough strength. *Cereal Chemistry*, **69**, 254–7.

Gras, P.W. and Wrigley, C.W. (1991) Analysis of wholemeal to predict the mixing properties of flour, in *Cereals International* (eds D.J. Martin and C.W. Wrigley), Royal Australian Chemical Institute, Melbourne, pp. 317–8.

Greenblatt, G.A., Bettge, A.D. and Morris, C.F. (1995) Relationship between endosperm texture and the occurrence of friabilin and bound polar lipids on wheat starch. *Cereal Chemistry*, **72**, 172–6.

Greenwell, P. (1992) Biochemical studies of endosperm texture in wheat. *Chorleywood Digest*, **118**, 74–6.

Greenwell, P. and Schofield, J.D. (1986) A starch granule protein associated with endosperm softness in wheat. *Cereal Chemistry*, **63**, 379–80.

Gupta, R.B. and Shepherd, K.W. (1990) Two-step one-dimensional SDS-PAGE analysis of LMW subunits of glutenin. I. Variation and genetic control of the subunits in hexaploid wheats. *Theoretical and Applied Genetics*, **80**, 65–74.

Gupta, R.B., Bekes, F. and Wrigley, C.W. (1991) Prediction of physical dough properties from glutenin subunit composition in bread wheats. *Cereal Chemistry*, **68**, 328–33.

Gupta, R.B., Paul, J.G., Bekes, F. *et al.* (1994) Allelic variation in glutenin subunit and gliadin loci, *Glu-1*, *Glu-3* and *Gli-1* of common wheats. I. Its additive and interaction effects on dough properties. *Journal of Cereal Science*, **19**, 9–17.

Gustafson, J.P., Bushuk, W., and Dera, A.R. (1991) Triticale: production and utilization, in *Handbook of Cereal Science and Technology* (eds K.J. Lorenz and K. Kulp), Marcel Dekker, New York, pp. 373–99.

Hameed, A., Pollack, L.M. and Hinz, P.N. (1994) Evaluation of Cateto maize accessions for grain yield and physical grain quality traits. *Crop Science*, **34**, 265–9.

Henry, R.J. (1985) Evaluation of barley and malt quality using near-infrared reflectance techniques. *Journal of the Institute of Brewing*, **91**, 393–6.

Hoseney, R.C. (1986) *Principles of Cereal Science and Technology*, American Association of Cereal Chemists, St Paul, MN, USA.

Howes, N.K., Kovacs, M.I., Leisle, D. *et al.* (1989) Screening of durum wheats for pasta-making quality with monoclonal antibodies for gliadin 45. *Genome*, **32**, 1096–9.

Hymowitz, T., Dudley, J.W., Collins, F.I. and Brown, C.M. (1974) Estimations of protein and oil concentration in corn, soybean, and oat seed by near-infrared light reflectance. *Crop Science*, **14**, 713–5.

Ikehashi, H. and Khush, G.S. (1979) Methodology of assessing appearance of the rice grain, including chalkiness and whiteness, in *Proceedings of Workshop on Chemical Aspects of Rice Grain Quality*. International Rice Research Institute, Los Banos, Laguna, Philippines, pp. 223–9.

Johnson, R.A., Quick, J.S. and Hammond, J.J. (1983) Inheritance of semolina color in six durum wheat crosses. *Crop Science*, **23**, 607–10.

Jolly, C.J., Rahman, S., Kortt, A.A. and Higgins, T.J.V. (1993) Characterisation of the wheat Mr 15 000 'grain softness protein' and analysis of the relationship between its accumulation in the whole seed and grain softness. *Theoretical and Applied Genetics*, **86**, 589–97.

Joppa, L.R. and Williams, N.D. (1988) Genetics and breeding of durum wheat in the United States, in *Durum Wheat: Chemistry and Technology* (eds G. Fabriani and C. Lintas), American Association of Cereal Chemists, St Paul, MN, USA, pp. 47–68.

Juliano, B.O. (1979) The chemical basis of rice grain quality, in *Proceedings of Workshop on Chemical Aspects of Rice Grain Quality*. Intern. Rice Research Inst., Los Banos, Laguna, Philippines, pp. 69–90.

Juliano, B.O. (1985) Criteria and tests for rice grain qualities, in *Rice: Chemistry and Technology*, 2nd edn (ed. B.O. Juliano), American Assoc. Cereal Chemists, St Paul, MN, USA, pp. 443–524.

Kaan, F., Branlard, G., Chihab, B. *et al.* (1993) Relations between genes coding for grain storage protein and two pasta cooking quality criteria among world durum wheat (*Triticum durum* Desf.) genetic resources. *Journal of Genetics and Breeding*, **47**, 151–6.

Khan, K. and Huckle, L. (1992) Use of multistacking gels in sodium dodecyl sulfate-polyacrylamide gel electrophoresis to reveal polydispersity, aggregation and disaggregation of the glutenin protein fraction. *Cereal Chemistry*, **69**, 686–7.

Khush, G.S., Paule, C.M. and de la Cruz, N.M. (1979) Rice grain quality evaluation and improvement at IRRI, in *Proceedings of Workshop on Chemical Aspects of Rice Grain Quality*, International Rice Research Institute, Los Banos, Philippines, pp. 21–31.

Kimber, G. (1971) The inheritance of red grain colour in wheat. *Zeitschrift für Pflanzenzüchtung*, **66**, 151–7.

Kitterman, J.S. and Rubenthaler, G.L. (1971) Assessing the quality of early generation wheat selections with the micro AWRC test. *Cereal Science Today*, **16**(9), 313–316, 28.
Konik, C.M., Miskelly, D.M. and Gras, P.W. (1993) Starch swelling power, grain hardness and protein: relationship to sensory properties of Japanese noodles. *Starch*, **45**, 139–44.
Kovacs, M.I.P., Howes, N.K., Leisle, D. and Skerritt, J.H. (1993) The effect of high-molecular weight glutenin subunit composition on tests used to predict durum wheat quality. *Journal of Cereal Science*, **18**, 43–51.
Konzak, C.F. (1977) Genetic control of the content, amino acid composition, and processing properties of proteins in wheat. *Advances in Genetics*, **19**, 408–582.
Krishnan, P.G., Park, W.J., Kephart, K.D. *et al.* (1994) Measurement of protein and oil content of oat cultivars using near-infrared reflectance spectroscopy. *Cereal Foods World*, **39**, 105–8.
Kruger, J.E. and Bietz, J.A. (1994) *HPLC of Cereal and Legume Proteins*. American Association of Cereal Chemists, St Paul, MN, USA.
Kunze, O.R. (1985) Effect of environment and variety on milling quality of rice, in *Rice Grain Quality and Marketing*, International Rice Research Institute, Manila, Philippines, pp. 37–47.
Lang, C.E., Neises, E.K. and Walker, C.E. (1992) Effects of additives on flour-water dough mixograms. *Cereal Chemistry*, **69**, 587–91.
Little, R.R., Hilder, G.B. and Dawson, E.H. (1958) Differential effect of dilute alkali on 25 varieties of milled white rice. *Cereal Chemistry*, **35**, 111–26.
Loock, A., Geiger, H.H. and Wehmann, F. (1990) Recurrent selection for Falling Number in rye, in *Fifth International Symposium on Pre-Harvest Sprouting in Cereals* (eds K. Ringlund, E. Mosleth, and D.J. Mares), Westview Press, Boulder, CO, USA, pp. 287–95.
Lorenz, K.J. (1991) Rye, in *Handbook of Cereal Science and Technology* (eds K.J. Lorenz and K. Kulp), Marcel Dekker, New York, pp. 331–71.
Luckett, D.J. and Darvey, N.L. (1992) Utilisation of microspore culture in wheat and barley improvement. *Australian Journal of Botany*, **40**, 807–28.
Lupton, F.G.H. and Derera, N.F. (1981) Soft wheat breeding in Europe and Australia, in *Soft Wheat: Production, Breeding, Milling and Uses* (eds W.T. Yamazaki and C.T. Greenwood), American Association of Cereal Chemists, St Paul, Minnesota, USA, pp. 99–127.
McCleary, B.V. and Codd, R. (1989) Measurement of beta-amylase in cereal flours and commercial enzyme preparations. *Journal of Cereal Science*, **9**, 17–33.
McCleary, B.V. and Codd, R. (1991) Measurement of (1-3)(1-4)-beta-D-glucan in barley and oats; a streamlined enzymic procedure. *Journal of the Science of Food and Agriculture*, **55**, 303–12.
McCleary, B.V. and Shameer, I. (1987) Assay of malt beta-glucanase using azo barley glucan: an improved precipitant. *Journal of the Institute of Brewing*, **93**, 87–90.
McCleary, B.V. and Sheehan, H. (1987) Measurement of cereal alpha-amylase: a new assay procedure. *Journal of Cereal Science*, **6**, 237–51.
McCluggage, M.E. (1943) Factors influencing the pearling test for kernel hardness in wheat. *Cereal Chemistry*, **20**, 686.
MacRitchie, F, duCros, D.L. and Wrigley, C.W. (1990) Flour polypeptides related to wheat quality. *Advances in Cereal Science and Technology*, **10**, 79–145.
Martin, C.R., Rousser, R. and Brabec, D.L. (1993) Development of a single-kernel wheat characterization system. *Transactions of the American Society of Agricultural Engineers*, **36**, 1399–04.

Massie, D., Slaughter, D., Abbott, J. and Hruschka, W. (1993) Acoustic single-kernel wheat hardness. *Transactions of the American Society of Agricultural Engineers*, **36**, 1393–8.

Matsuo, R.R. and Dexter, J.E. (1980) Comparison of experimentally milled durum wheat semolina to semolina produced by some Canadian commercial mills. *Cereal Chemistry*, **57**, 117–22.

Matsuo, R.R., Bradley, J.W. and Irvine, G.N. (1972) Effect of protein content on the cooking quality of spaghetti. *Cereal Chemistry*, **49**, 707–11.

Mertz, E.T. (ed.) (1992) *Quality Protein Maize*, American Association of Cereal Chemists, Saint Paul, MN, USA.

Miura, H. and Tanii, S. (1994) Endosperm starch properties in several wheat cultivars preferred for Japanese noodles. *Euphytica*, **72**, 171–5.

Morgunov, A.I., Pena, R.J., Crossa J. and Rajaram S. (1993) Worldwide distribution of *Glu-1* alleles in bread wheat. *Journal of Genetics and Breeding*, **47**, 53–60.

Morris, C.F. and Paulsen, G.M. (1985) Preharvest sprouting of hard winter wheat as affected by nitrogen nutrition. *Crop Science*, **25**, 1028–31.

Morris, C.F. and Paulsen, G.M. (1987) Development of preharvest sprouting-resistant germplasm from 'Clarks Cream' hard white winter wheat. *Cereal Research Communications*, **15**, 229–35.

Morris, C.F. and Raykowski, J.A. (1993) WAS: computer software for wheat quality data management. *Agronomy Journal*, **85**, 1257–61.

Morris, C.F. and Raykowski, J.A. (1995) A computer-aided approach to the evaluation of wheat grain and flour quality. *Computers and Electronics in Agriculture*, **11**, 229–37.

Nakamura, T., Yamamori, M., Hirano, H. and Hidaka, S. (1993) Decrease of waxy (Wx) protein in two common wheat cultivars with low amylose content. *Plant Breeding*, **111**, 99–105.

Obi, I.U. (1982) Application of the 2,4,6, trinitrobenzene-1-sulfonic acid (TNBS) method for determination of available lysine in maize seed. *Agricultural and Biological Chemistry*, **46**, 15–20.

O'Brien, L. and Orth, R.A. (1974) Effect of geographic location of growth on wheat milling yield, Farinograph properties, flour protein and residue protein. *Australian Journal of Agricultural Research*, **28**, 5–9.

O'Brien, L. and Ronalds, J.A. (1984) Yield and quality interrelationships amongst random F3 lines and their implications for wheat breeding. *Australian Journal of Agricultural Research*, **35**, 443–51.

O'Brien, L. and Ronalds, J.A. (1986) The effect on yield distribution of early generation selection for quality. *Australian Journal of Agricultural Research*, **37**, 211–8.

Ohnmacht, B., Geiger, H.H. and Dambroth, M. (1990) Genetic variation for pre-harvest sprouting in rye as a function of climatic conditions and ripening stage, in *Fifth International Symposium on Pre-Harvest Sprouting in Cereals* (eds K. Ringlund, E. Mosleth, and D.J. Mares), Westview Press, Boulder, CO, USA, pp. 256–62.

Paton, D. (1986) Oat starch: physical, chemical, and structural properties, in *Oats: Chemistry and Technology* (ed. F.H. Webster), American Association of Cereal Chemists, St Paul, MN, USA, pp. 93–120.

Patterson, F.L. and Allan, R.E. (1981) Soft wheat breeding in the United States. In *Soft Wheat: Production, Breeding, Milling and Uses* (eds W.T. Yamazaki and C.T. Greenwood), American Association of Cereal Chemists, St Paul, MN, USA, pp. 33–98.

Payne, P.I. (1987) Genetics of wheat storage proteins and the effect of allelic

variation on bread-making quality. *Annual Review of Plant Physiology*, **38**, 141–53.
Pena, R.J., Nagarajan, P. and Bates, L.S. (1982) Grain shrivelling in secondary hexaploid triticales. II. Morphology of mature and developing grains related to grain shrivelling. *Cereal Chemistry*, **59**, 459–68.
Perez, C.M. (1979) Gel consistency and viscosity of rice, in: *Proceedings of Workshop on Chemical Aspects of Rice Grain Quality*, International Rice Research Institute, Los Banos, Laguna, Philippines, pp. 293–302.
Peterson, D.M., Mattsson, B. and Heneen, W.K. (1995) Identification of oat varieties, in *Identification of Food-Grain Varieties* (ed. C.W. Wrigley), American Association of Cereal Chemists, St Paul, MN, USA, pp. 239–51.
Pixley, K.V. and Bjarnason, M.S. (1993) Combining ability for yield and protein quality among modified-endosperm opaque-2 tropical maize inbreds. *Crop Science*, **33**, 1229–34.
Poehlman, J.M. (1979) *Breeding Field Crops*, The AVI Publishing Co., Westport, Connecticut, USA, p. 483.
Pogna, C.W., Autran, J.C. and Bushuk, W. (1988). Evidence for a direct casual effect of low molecular weight subunits of glutenins on gluten viscoelasticity in durum wheats. *Journal of Cereal Science*, **7**, 211–4.
Pooni, H.S., Kumar, I. and Khush, G.S. (1994) Genetic control of amylose content in a diallel set of rice crosses. *Heredity*, **71**, 603–13.
Pomeranz, Y. (1987) *Modern Cereal Science and Technology*, VCH Publishers, New York, NY, USA, p. 486.
Rath, C.R., Gras, P.W., Wrigley, C.W., and Walker, C.E. (1990) Evaluation of dough properties from two grams of flour using the Mixograph principle. *Cereal Foods World*, **35**, 572–4.
Rattunde, H.F.W., Miedaner, T. and Geiger, H.H. (1991) Biometrical analysis of alternative plot types for selection in rye. *Euphytica*, **57**, 141–50.
Reddy, P. and Appels, R. (1993) Analysis of genomic DNA segment carrying heat high-molecular-weight (HMW) glutenin B × 17 subunit and its use as an RFLP marker. *Theoretical and Applied Genetics*, **85**, 616–24.
Reinke, R.F., Welsh, L.A., Reece, J.E., Lewin, L.G. and Blakeney, A.B. (1991) Procedures for quality selection of aromatic rice varieties. *International Rice Research Newsletter*, **16(5)**, 10–11.
Ronalds, J.A., and Blakeney, A.B. (1995) Determination of grain colour by near infrared reflectance and near infrared transmittance spectroscopy, in *Leaping Ahead with Near Infrared Spectroscopy* (eds G.D. Batten, P.C. Flinn, L.A. Walsh and A.B. Blakeney), Royal Australian Chemical Institute, Melbourne, pp. 148–53.
Ronalds, J.A., Bekes, F., Pannozo, J. and Khan, A. (1991) Prediction of wheat quality using NIR spectroscopy, in *Proceedings of 40th Australian Cereal Chemistry Conference* (eds T. Westcott and Y. Williams), Royal Australian Chemical Institute, Melbourne, pp. 147–50.
Ronalds, J.A., May, M., Glennie Homes, M. and Taylor, T. (1992) Prediction of malt quality from NIT analyses of barley, in *Proceedings of 42nd Australian Cereal Chemistry Conference* (ed. V.J. Humphrey-Taylor), Royal Australian Chemical Institute, Melbourne, pp. 130–1.
Sapirstein, H.D. (1995) Variety identification by digital image analysis, in *Identification of Food-Grain Varieties* (ed. C.W. Wrigley), American Association of Cereal Chemists, St Paul, MN, USA, pp. 91–130.
Seeborg, E.F. and Barmore, M.A. (1957) A new five-gram milling-quality test and its use in wheat breeding. *Cereal Chemistry*, **34**, 299–303.
Shoup, N.H., Pell, K.L., Seeborg, E.F. and Barmore, M.A. (1957) A new micro

mill for preliminary milling-quality tests of wheat. *Cereal Chemistry*, **34**, 296–8.
Simmonds, N.W. (1979) *Principles of Crop Improvement*, Longmans, London.
Simmonds, D.H. (1989) *Wheat and Wheat Quality in Australia*, CSIRO, Australia.
Skerritt, J.H., Andrews, J.L., Blundell, M. *et al.* (1994) Applications and limitations of immunochemical analysis of biopolymer quality in cereals. *Food and Agricultural Immunolology*, **6**, 173–84.
Srinivas T. and Bhashyam, M.K. (1985) Effect of variety and environment on milling quality of rice, in *Rice Grain Quality and Marketing*. International Rice Research Institute, Manila, Philippines, pp. 49–59.
Stein, I.S., Sears, R.G., Hoseney, R.C. *et al.* (1992) Chromosomal location of genes influencing grain protein concentration and mixogram properties in 'Plainsman V' winter wheat. *Crop Science*, **32**, 573–80.
Stoskopf, N.C. (1985) *Cereal Grain Crops*, Reston Publ. Co., Reston, Virginia, USA, p. 516.
Sung, T.M. and Lambert, R.J. (1983) Ninhydrin color test for screening modified endosperm opaque-2 maize. *Cereal Chemistry*, **60**, 84–5.
Symes, K.J. (1961) Classification of Australian wheat varieties based on the granularity of their wholemeal. *Australian Journal of Experimental Agricultural and Animal Husbandry*, **1**, 18–23.
Van Ruiten, H.T.L. (1985) Rice milling: an overview, in *Rice: Chemistry and Technology*, 2nd edn (ed. B.O. Juliano), American Association of Cereal Chemists, St Paul, MN. USA, pp. 349–88.
Verdeal, K. and Lorenz, K. (1977) Alkyl resorcinols in wheat, rye, and triticale. *Cereal Chemistry*, **54**, 475–83.
Villegas, E. (1975) An integral system for chemical screening quality protein maize, in *High-Quality Protein Maize. CIMMYT-Purdue International Symposium on Protein Quality in Maize*, El Batan, Mexico, pp. 330–6.
Walker, C.E., Ross, A.S., Wrigley, C.W. and McMaster, G.J. (1988) Accelerated characterization of starch-paste viscosity and set-back with the Rapid Visco-Analyzer. *Cereal Foods Worlds*, **33**, 491–4.
Watson, S. and Ramstad, R.E. (eds) (1987) *Corn: Chemistry and Technology*, American Association of Cereal Chemists, Saint Paul, MN, USA.
Webb, B.D. (1975) Cooking, processing and milling qualities of rice, in *Six Decades of Rice Research in Texas* (ed. J.E. Miller) Texas Agricultural Experiment Station, Texas, USA, pp. 97–106.
Weipert, D. (1990) Sprouting resistance and processing value of new conventional and hybrid rye varieties, in *Fifth International Symposium on Pre-Harvest Sprouting in Cereals* (eds K. Ringlund, E. Mosleth, and D.J. Mares), Westview Press, Boulder, CO, USA, pp. 270–7.
Welsh, L.A., Blakeney, A.B. and Bannon, D.R. (1991) Rapid viscometric analysis of rice flour. *International Rice Research Newsletter*, **16(5)**, 11–2.
Werner, E.W., Wiktorowicz, J.E. and Kasarda, D.D. (1994) Wheat varietal identification by capillary electrophoresis of gliadins and high molecular weight glutenin subunits. *Cereal Chemistry*, **71**, 397–402.
Whan, B.R. (1974) A small-scale milling technique for establishment of flour yield in wheat. *Australian Journal of Experimental Agriculture and Animal Husbandry*, **14**, 658–62.
Wrigley, C.W. (1994) Developing better strategies to improve grain quality for wheat. *Australian Journal of Agricultural Research*, **45**, 1–17.
Wrigley, C.W. (ed.) (1995) *Identification of Food Grain Varieties*, American Association of Cereal Chemists, St Paul, MN. USA.
Wrigley, C.W. and Batey, I.L. (1995) Efficient strategies for variety identification, in *Identification of Food-Grain Varieties* (ed. C.W. Wrigley), American Association of Cereal Chemists, St Paul, MN, USA, pp. 19–33.

Wrigley, C.W., Gupta, R.B. and Bekes, F. (1993) Our obsession with high resolution in gel electrophoresis; does it necessarily give the right answer? *Electrophoresis*, **14**, 1257–8.
Yamazaki, W.T. and Donelson, J.R. (1983) Kernel hardness of some US wheats. *Cereal Chemistry*, **60**, 344–50.

12

Molecular approaches to cereal quality improvement

O. Anderson

12.1 HUMAN ENGINEERING OF THE CEREAL GENOMES

Humans have participated in the alteration of cereal genomes for as long as 10 000 years. As humans moved from foraging of wild seed to the saving of seed for deliberate plantings in the next season, an unintended genetic selection occurred. For cereals such as wheat the selected characteristics included: the retention of seeds after maturity, free threshing kernels, and increased numbers of grains per stalk. As the early farmers noticed the continuity of characteristics from one seed generation to the next, they deliberately selected seeds with noticeably favourable characteristics, such as grain size, grinding properties, colour, and flavour. The result was an accelerated and directed evolution of new cereal types with altered gene pool components. A distinctive feature of these initial phases of human engineering of the cereals is that humans improved cereals through selection from the natural genetic variation.

A second phase in cereal engineering began with humans taking an active role in the faster and directed production of increased genetic variation. In the simplest cases different lines were crossed mixing the genetic potentials of the parent lines. Offspring were selected which were judged to possess favoured new combinations of characteristics. More complex and highly sophisticated strategies developed in the past few score years allow the crossing not just within the same species, but to more distantly-related grasses. As a result, modern breeders have the capability of moving sections of chromosomes between grasses in a manner that in nature would never, or seldom, occur.

This second phase in cereal engineering continues to the present, includes most current efforts to improve cereal varieties, and will continue to predominate in the near future. However, a new phase has begun. Molecular technology and advanced tissue culture techniques are developing the tools to modify, at will, single genes and metabolic pathways in the laboratory. The power of these methodologies will result in novel directions in cereal improvement and utilization. Many of these new directions are not yet conceived, but it is possible to outline some likely paths by which molecular biology can develop new cereal varieties with altered characteristics. This chapter will summarize some basic tools and approaches of modern biotechnology most applicable, at present, to cereal quality engineering. A wheat gene system will be used as an example of how molecular tools may be used, and are being used, in attempts to understand and manipulate cereal grain quality.

12.1.1 A practical definition of cereal quality

For present purposes, let us define quality as the summation of characteristics that humans desire for specific uses. Thus, a good quality bread wheat might possess dough viscoelastic characteristics relatively superior to other lines. Similarly, a high quality rice may have desirable digestibility properties and a high quality barley may have favourable malting characteristics. Agronomic traits also determine cultivar desirability and are amenable to a biotechnological approach, but for this chapter we will concentrate on properties related to defined end-uses of the grain.

12.1.2 How biotechnology can contribute to cereal quality

Let us define molecular biology/biotechnology as the laboratory manipulation of DNA and the attempt to use this DNA to alter cereal genomes and thus specific properties of cereal grains. Although the experimental details will vary for each cereal and specific gene system, Figure 12.1 outlines common critical steps. A researcher must possess DNA of a quality-related gene; the DNA must be used to transform a cereal from which can be regenerated fertile plants; and field and product testing must be carried out to assess the effect of the added DNA. The complexity of successfully carrying out all the steps in Figure 12.1 will initially restrict manipulations to simple gene systems; i.e., storage proteins, starch synthesis enzymes, etc. More complex gene systems and development pathways will become accessible to manipulation as our understanding increases, and as the technology improves. Let us first consider some features of molecular techniques that are advantageous over traditional breeding methods.

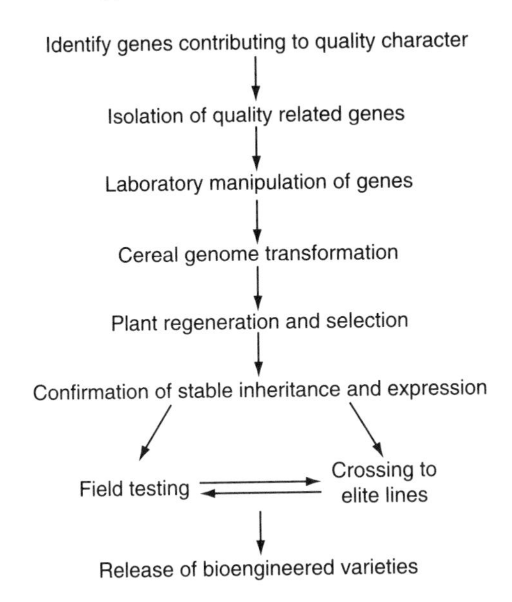

Figure 12.1 A general scheme for cereal quality molecular engineering.

Increase in the specificity of gene additions

Currently, without the use of biotechnology, the unit of gene transfer is not individual genes, but chromosomes. The size of the transfer unit ranges from an entire chromosome to a small chromosome segment (Figure 12.2). However, even the smallest chromosome segment will contain large numbers of genes, therefore allowing the potential transfer of deleterious genes along with those of interest. Well-known examples are the translocations into wheat of the long arm of the rye 1 chromosome. While these translocations confer favourable agronomic performance to wheat (Zeller, 1973), they also tend to decrease wheat bread-making quality, probably due to the simultaneous addition of the rye secalins gene proteins that also reside on this chromosome arm (Graybosch *et al.*, 1993). The molecular biological approach avoids these problems. The experimenter transfers a specific gene(s), although the recipient site of the transfer is uncontrolled, as is the number of copies of the transferred gene (transgene). The example in Figure 12.2D shows three possible transfer events, with a wheat chromosome receiving 1, 3, and 7 copies of the transgene at three random sites on the chromosome.

Increase in speed and efficiency of gene additions

Transfer of a single gene into an adapted cultivar *via* traditional breeding requires years of back-crossing to the original variety. Using

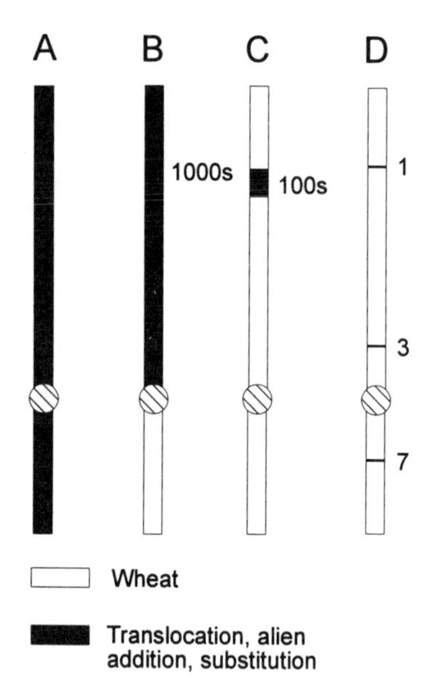

Figure 12.2 Transfer of donor chromosome segments into wheat. Traditional gene transfer procedures result in the transfer of large numbers of genes to the recipient. (A) Complete donor chromosome transferred to wheat. (B) Substitution of donor chromosome long arm. Results in the transfer of thousands of donor genes into the wheat recipient. (C) Translocation of the chromosome segment. Results in the transfer of tens to hundreds of genes. (D) Molecular techniques can accomplish the transfer of single genes, in single or multiple copies (random insertion of 1, 3, and 7 copies are shown for three separate insertion events). Striped circle, centromere.

biotechnology, it is already possible, in some cases, to begin assessing a single gene's effects within six months of starting the gene transfer process. Currently, a genotype specificity limits some cereal lines from efficient molecular biological gene transfer (discussed below).

Availability of genes not otherwise accessible

Traditional breeding practices are limited to the gene pools of those grasses that are sexually compatible with the cereals. For some cereal crops, such as maize, there are few such species available, while other cereals, wheat being a significant example, are able to cross with a wide range of wild grasses (Baum *et al.*, 1992). Several methods have been developed with wheat to enhance chromosome transfer, including the use of ionizing radiation (Sears, 1993) and mutants of the *Ph1* gene

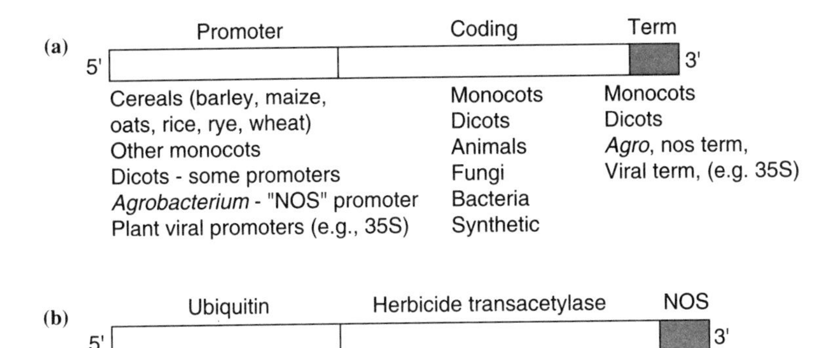

Figure 12.3 Range of gene constructions possible for cereal bioengineering. (a) Any gene can be used that possesses the following characteristics: 1. A promoter that is active in the cereal and that contains all the DNA elements necessary for developmentally correct gene expression. 2. A coding region in the correct reading frame. Introns may be present if the splice sites are correctly recognized. 3. A terminal sequence for correct transcriptional and translational termination. Examples of potential DNA sources are given below the construction segments. (b) A gene construct used to transform wheat (Weeks *et al.*, 1993). A maize ubiquitin promoter (Christensen *et al.*, 1992) and *Agrobacterium* nos termination sequences were linked to a bacterial enzyme that confers resistance to the herbicide active ingredient phosphinothricin.

which relax the stringency of meiotic chromosome pairing, allowing chromosome recombination in wide crosses (Feldman, 1993). However, even with wheat, there are both practical and absolute limits to the range of possible chromosome segment transfers. The ability to transform the cereals in the laboratory eliminates these boundaries. Thus, a wheat researcher can consider using not only monocot genes not previously available, such as from rice or sorghum, but all genes, even those from dicots, fungi, animals, or bacteria (Figure 12.3a). For example, one wheat transformation experiment has used a gene composed of segments of DNA from maize and two different bacteria (Figure 12.3b; Weeks *et al.*, 1993). The only requirement for expressing any gene is that the coding region is controlled by a promoter which is active in the transformed cereal. Whether or not the gene functions appropriately depends on the specific gene, the host plant system, sufficient planning in constructing the DNA, and an element of luck.

Modification of existing genes

As great as is the potential to access genes of other organisms, the true fruits of molecular approaches to improving cereals will come from the laboratory modifications of native genes to create new genes that have never existed in nature. Modifications of genes and even the synthesis of entirely new genes from nucleotides are reasonably straightforward tasks for a molecular biology laboratory (straightforward does not necessarily indicate short times required). If the molecular basis of a gene's contribution to cereal quality is understood, it should be possible to modify the gene to produce proteins of altered characteristics both to enhance known quality parameters and create novel utilizations.

Down-regulation of gene expression

Once a gene has been identified and isolated, strategies can be attempted to decrease or eliminate expression of that gene. One of the simplest (in principle) such strategies is **antisense** and entails designing the transcription of an RNA strand complementary to the mRNA transcript of the target gene. The complementary RNA will then hybridize with the normal transcript and reduce expression. This approach has been used in *Petunia* to inhibit flower pigmentation with an antisense gene to chalcone synthase mRNA (van der Krol *et al.*, 1990), and a similar strategy was used to confer increased resistance to tobacco golden mosaic virus of tobacco plants (Day *et al.*, 1991). An application of the technique to rice succeeded in reducing levels of granule-bound starch synthase mRNA and protein in the grain, resulting in lower levels of amylose (Shimada *et al.*, 1993). The antisense approach is not always successful, but a better understanding of the mechanistic basis of the technique will be obtained as it is used on more gene systems.

A second strategy of reducing or eliminating specific gene expression has resulted from an unexpected result occasionally observed after transformation experiments. In some cases, if a plant is transformed with a gene construction containing part or all of a gene of the host plant, both the transgene and its related endogenous gene may become inactive (see Section 12.2.2: Transgene activity). It is likely this unexplained result has uncovered an until now unknown aspect of gene expression control, and once understood, may offer another approach to specific gene inactivation.

Up-regulation of gene expression

In some cases it will be necessary to express novel genes at high levels, or native genes at higher levels than normal. The most direct methods of increasing levels of gene expression is by increasing gene copy number and pyramiding new loci created by transformation. A longer-term strategy involves the manipulation of promoter strengths, and will require considerably more understanding of the basic molecular mechanisms of gene expression (see Section 12.3.7).

12.1.3 Target gene systems

Potential target gene systems are any that affect end-use quality. This number is certainly very large, but there are a few examples of genes that affect major quality characteristics that are currently under study.

Storage proteins

Storage proteins are one of the main nutrition stores for later development. Some cereal protein families possess functional characteristics that make them major determinants of aspects of cereal quality desired by human consumers (Chapter 8). A group of these proteins from wheat is used at the end of the chapter to illustrate how bioengineering can be applied to storage protein genes.

Starch metabolism

Endosperm starch synthesis is a natural target for genetic engineering for both novel consumption and industrial utilizations. Starch is the main storage carbohydrate in plants and is the main component of cereal seeds (Chapter 9). The source of starch in plants is the ADP-glucose pathway consisting of three enzymes: ADP-glucose pyrophosphate synthase, granule-bound starch synthase, and starch branching enzyme (Preiss, 1992). Each of these genes have been cloned in the cereals (Anderson *et al.*, 1991; Ainsworth *et al.*, 1993; Fisher *et al.*, 1993). Active research programmes are attempting to manipulate the expression of these enzymes to change both the levels of starch synthesis and the structure of the starch molecules (Shimada *et al.*, 1993; Visser and Jacobsen, 1993). The importance of cereal starches and the relative simplicity of the synthetic pathway assure that modified starch will be one of the first bioengineered characteristics to appear in cereal cultivars.

Nutrition

One of the envisioned applications of plant biotechnology is to improve the nutritional quality of crops both for direct human consumption and for animal feeds. The area that has received the greatest attention is the potential to increase levels of essential amino acids. One approach is to increase the levels of certain storage proteins with a naturally high level of essential amino acids. The feasibility of this goal has been shown in dicots where a high methionine protein was expressed to 8% of the total seed protein in tobacco (Altenbach *et al.*, 1989) and 4% in canola (Altenbach *et al.*, 1992). Another approach is to modify the amino acid synthetic pathways in plant tissues. Karchi *et al.* (1993) used a feedback insensitive aspartate kinase gene from the bacteria *Escherichia coli* to increase levels of free methionine and threonine in tobacco seeds.

A second target for improving cereal nutritional value is digestibility. This is particularly important in rice where prolamines in Type I protein bodies are relatively indigestible (Ogawa *et al.*, 1987; Tanaka *et al.*, 1990). In this example the goal will be to either redistribute protein synthesis to Type II protein bodies, or to modify the structure and composition of the Type I protein body. Every percentage increase in digestibility produces the same result as a percentage increase in protein content.

Other contributions to nutrition may include factors such as mineral utilization. Pen *et al.* (1993) expressed a yeast phytase gene in tobacco seeds. Tests demonstrated that phosphorus release from feed and animal growth rates increased with supplementation from phytase expressing seeds.

Physical properties

The physical behaviour of the grain and its derivatives during processing and utilization is a critical component of quality. Some of these physical properties are well known and associated with specific grain components, such as the effect of starch structure on the stickiness of rice during cooking and the storage protein contribution to wheat dough viscoelasticity. Other properties are less well understood. One such example is wheat grain hardness, where there is progress in associating specific genes with milling and utilization characteristics (Chapter 8).

Pathology and pests

Pathogens can decrease both the yield and intrinsic quality properties of all cereal crops. To avoid both preharvest and postharvest damage, biotechnological strategies may be possible *via* the addition or modification of the appropriate genes (Lamb *et al.*, 1992; Cornellissen and Melchers, 1993). Among the many known genes that have the potential to contribute to insect resistance are: *Bacillus thuringiensis* toxin (Augustine *et al.*, 1989), chitinases (Schlumbaum *et al.*, 1986), wheat germ agglutinin (Raikhel and Wilkins, 1987), and wheat lysozyme (Audy *et al.*, 1988). Specific protection of the seed is possible as shown by Shade *et al.* (1994) who used the α-amylase protein that confers

Bruchid beetle resistance to the common bean. When the α-amylase gene was transferred to pea under control of a strong seed-specific promoter, the transgenic peas were resistant to two different genera of Bruchid beetle. Another example is the increased fungal resistance of tobacco expressing transgenic chitinase and glucanase genes (Zhu *et al.*, 1994).

Abiotic stress

Grain quality is severely affected by a number of abiotic variables; i.e. low temperatures, high temperatures, soil composition, drought, and late-season rain. Although environmental variable are uncontrollable, it may be possible to alter the plant's response to adverse conditions to mitigate detrimental effects on cereal quality. For example, variation in wheat dough properties are associated with high temperatures during grain fill (Blumenthal *et al.*, 1993). At the higher temperatures there is a change in the distribution of synthesized proteins and starch synthesis decreases. Thus, the bioengineering strategy is to either make the genes of these two grain components less responsive to heat stress, or add gene constructions that will activate during stress and compensate for down-regulated genes.

12.2 CURRENT AND FUTURE TECHNOLOGIES

The ability to directly engineer the cereal genomes to improve quality is dependent on two fundamental technologies. Researchers must be able to identify and isolate the appropriate genes, and researchers must be able to insert native and modified genes into the genome of interest such that the genes function appropriately and stably. This section selects some current methods and future trends in these areas. The technical foundations of molecular biology and plant tissue culture will continue to advance indefinitely. Thus, a step in Figure 12.1 that is a technical bottleneck with a particular cereal line and gene system can become a routine procedure within a few years.

12.2.1 Gene isolation

Researchers need physical access to those genes determining quality to attempt cereal quality molecular engineering. This requires the possession of specific probes to screen recombinant genomic libraries for the first isolate of the gene of interest (which can then be used to isolate additional closely-related alleles and genes). The majority of probes are nucleic acids, and the most common strategy is to isolate first a cDNA clone made from a mRNA population and then use the cDNA clone to isolate the entire gene. Alternatively, the initial probe can be an antibody to the gene product. An antibody-based strategy first probes a cDNA expression library for the DNA clone encoding the protein recognized by the antibody, and then the cDNA clone is used to probe the genomic library. A limitation with this basic approach is that an initial probe, nucleic acid or antibody, must be found. Fortunately, many of the genes important for cereal quality lend themselves quite readily to this approach. The storage proteins are the main products of endosperm tissue which means that most of the mRNA, and derived cDNA clones, originate from these storage protein genes.

Unfortunately these relatively direct gene isolation strategies may not be applicable to quality-related genes expressed at low levels. In some of these cases further research will succeed in gene isolation, but clearly more powerful gene isolation techniques are needed. Several such strategies are being developed. It is impossible to predict too far into the future of gene isolation techniques, but three current areas of development will probably become useful in cereal-quality gene isolation.

Map-based cloning

Recombinant DNA technology is allowing the development of genetic maps faster and at higher resolutions than maps based on phenotypic markers (Tanksley *et al.*, 1989). Random DNA clones (cDNA or genomic) from the species of interest are screened to find those which are single-copy and which identify polymorphic regions of the target genome. A genetic polymorphism is a difference in the size of the restriction fragment recognized by that probe. Sets of such probes are used to follow their homologous loci in segregating populations of the progeny of a cross of parents that differ in many such regions. In practice, restriction-fragment-length-polymorphism (RFLP) analysis is carried out by cutting the target genomic DNA with restriction enzymes, then separating the DNA on agarose gels, transferring to a membrane and hybridizing with the selected probes. The DNA fragment sizes identifying parental origin are then determined. Linked sites on a chromosome from one parent segregate together unless a recombination at meiosis occurs to separate them in the progeny DNA. In general, recombination occurs more frequently between two sites the farther apart they are on the chromosome. When the distribution of sites in a population is statistically analysed, a linkage map of the sites can be generated (Lander *et al.*, 1987; Liu and Knapp, 1990). Such genetic maps are characterized by a set of DNA clones which identify specific loci and positions on the chromosomes.

RFLP-based genetic maps have immediate uses in breeding programmes. If a desired trait is shown to be linked to a specific RFLP, that site can be used as a marker for the trait (Tanksley *et al.*, 1989). If a single gene controls the trait, the analysis is straightforward. However, many important traits are controlled by sets of genes each contributing to the phenotype. These quantitative loci (QTLs) complicate the analysis, but there are statistical tools under development to handle data from multiple interacting loci (Hayes *et al.*, 1993; Tanksley, 1993).

A second use for genetic maps is to identify RFLP sites adjacent to a gene of interest and use that RFLP and its associated clones to anchor a physical search for the new gene. A typical strategy is to isolate genomic clones containing the RFLP sites, then systematically isolate overlapping clones (referred to as 'walking' down the chromosome) until locating the desired gene. In *Arabidopsis*, a 1.7 million base pair contiguous region of a chromosome has been cloned in an overlapping ordered set of yeast artificial chromosomes (Putterill *et al.*, 1993). From these clones a gene (*CO*) involved in determining flowering time is being isolated. In tomato, the *pto* gene for bacterial resistance has been isolated with this approach (Martin *et al.*, 1993), and the isolation of *jointless*, a developmental regulatory gene involved in abscission zones, is underway (Wing *et al.*, 1994).

These examples were successful due to the well-developed genetic maps of *Arabidopsis* and tomato. In the cereals, the rice genome is small enough, and the genetic maps sufficiently well-developed, to make this approach feasible with current technology. The cereals with large genomes, such as maize, barley, and wheat, may require further technical developments. Such developments are certain to come, but RFLP mapping of the cereal genomes has uncovered two important results. The genomes of the major cereals are similar enough to use many of the same cDNA probes across species. The assumption is that these probes are identifying homologous loci in the different species. In addition, a comparison of the resulting maps shows conservation of gene order among the cereals (Ahn *et al.*, 1993). If mapping at finer resolutions confirms this conservation of gene order, it suggests an elegant solution to the problem of attempting to chromosome walk in the larger cereal genomes. A researcher can identify markers linked to a gene of interest in wheat and then isolate a homoeologous gene by chromosome walking in rice (with its 30-fold smaller genome).

Gene tagging via *insertional mutagenesis*

A classic method of identifying a gene and its phenotypic effects is by an insertion or deletion event which disrupts normal functioning of the gene. There are three such methods potentially available to cereal researchers, although all three require more development. The first method utilizes transposons – mobile genetic elements found in most organisms (reviewed in Peterson, 1987) and which may be a major source of naturally occurring insertion mutations (Wessler and Varagona, 1985). When a transposon moves from one chromosome site to another, it acts both as a mutagen and as a molecular tag identifying the new position. The transposon systems of maize, such as *Activator/ Dissociation* and *Mutator*, are classic examples and have been used to identify a number of maize genes (Walbot, 1992). This research strategy has been extended by transforming these elements into other plants without well characterized natural transposons (Baker *et al.*, 1986). Several laboratories have transformed rice with maize transposons and have shown the expected mobility (e.g., Jing-liu *et al.*, 1991). Similarly, it should be possible to establish transposon tagging systems in other diploid cereals such as barley. The polyploid cereals, such as wheat, are more difficult because recessive mutations can be masked. However, even with wheat it should be possible to use insertional mutagenesis if effects of single genes are dominant or semi-dominant; e.g., the meiotic pairing gene *ph1* (Feldman, 1993).

A second method of gene tagging uses random insertion of DNA into cereal genomes and selection for gene disruption. Most monocot transformation experiments use physical application of DNA to the monocot cells (Section 12.2.2). The resulting insertion of the applied DNA into the cereal genome will occasionally affect a gene and lead to phenotypic changes, but this approach is not commonly used to isolate genes. A more systematic approach, which is also a transformation-based approach, is being developed with dicots using *Agrobacterium* (Zambryski, 1988). During the transformation process, a piece of the bacterial plasmid DNA (T-DNA) is transferred from the bacteria into the plant genome. As with transposons, this inserted DNA can serve to randomly mutagenize the target genome and serve as a molecular marker for the mutation site. This strategy has been used successfully to isolate numerous dicot genes (Walbot, 1992) and may prove useful with monocots (see Section 12.2.2: DNA delivery systems).

Sequencing projects

Both the scientific and popular press have given considerable attention to the human genome megaprojects, the goals of which include the complete sequence of all the human chromosomes. At the same time there are smaller, and considerably less well-funded, plans to perform large-scale sequencing on other species, including the plants *Arabidopsis* and rice. Two complementary approaches are being used. The longer-term approach plans the physical isolation of clones spanning a genome, followed by sequencing the entire length of each chromosome. The second strategy, large-scale cDNA sequencing (Adams *et al.*, 1991), will yield data more quickly on the coding regions of active genes, but will not cover the intergenic regions.

Although these projects might seem removed from questions of cereal quality, the technologies being developed will eventually impact all areas of biological research, including continuing efforts to isolate quality-related cereal genes. Inevitably some genes will prove too difficult to identify and isolate until combinations of these resources are routinely available to cereal scientists; e.g. high-density genetic maps, improvements in gene isolation technology, and gene clone banks.

12.2.2 Cereal transformation

Plant science is undergoing a revolution in the last decade of this century due to the power of the coupling of recombinant DNA technology and plant transformation. These capabilities allow specific gene alterations and observations of their subsequent effects on plant development, physiology, and biochemistry. Unfortunately for cereal scientists, the monocots have proven more intransigent to transformations than the dicots. While this has been a serious hindrance to the cereals taking advantage of the newest technologies, the situation is improving and there is a reasonable expectation that cereal transformation will become as routine as in some dicots. Rice protoplasts were the first cereal tissue to be transformed and regenerated into fertile plants (Toriyama *et al.*, 1988; Zhang and Wu, 1988; Shimamoto *et al.*, 1989). Subsequently other cereals and tissues were shown to be transformable by bombardment with DNA-coated microprojectiles. The first successes using bombardment produced maize regenerants resistant to the herbicides chlorsulfuron (Fromm *et al.*, 1990) and phosphinothricin (Gordon-Kamm *et al.*, 1990). Similarly, oats (Somers *et al.*, 1992), wheat (Vasil *et al.*, 1992; Weeks *et al.*, 1993), sorghum (Casas *et al.*, 1993), rice (Li *et al.*, 1993), barley (Wan and Lemaux, 1994), and rye (Castillo *et al.*, 1994) have been transformed *via* bombardment.

Although transformation of all the major cereals is now possible, there is no standard protocol. Successes are reported with different selection procedures, different DNA delivery systems, and different target tissues. A consensus procedure from current reports would use bombardment to transform embryo-derived callus to phosphinothricin resistance. A dramatic example of the outcome of this procedure applied to wheat is shown in Figure 12.4; spraying with enough herbicide to kill the control plant allows the resistant plant to continue development and produce viable seeds.

Research plans are dependent on the number of transformed plant lines a researcher can expect to produce with available resources. If

Figure 12.4 Engineering a novel wheat phenotype. The DNA construct in Figure 12.3b was used to generate herbicide resistant wheat lines. Transformed and control plants were sprayed at the boot stage with 2% BASTA (phosphinothricin). The control plant (left) stopped development and died. The transformed plant (right) continued heading and produced fertile seed. (Photograph by T. Weeks.)

transformation procedures are too difficult, then only the gene constructions with the highest priority and expectation of success justify the transformation effort. Efficiency of transformation is the summation of all affecting factors (i.e., method of DNA delivery, frequency of integration of DNA into the cereal genome, stability of the transgenic locus, screening procedures, individual skills of the researcher, genotype dependence of the cereal line to be transformed, etc.), some of which will now be discussed in detail.

Transgene activity

Genetic engineering *via* molecular biology and plant transformation is of practical value only if the inserted genes function equally well in the *n*th generation as they do in the first generation. This requirement is not yet met by all transgenic events. Several different mechanisms are known, or suspected, to be involved in transgene inactivation. One such effect is co-suppression, the occasional inactivation, after transformation, of both the transgene and a homologous endogenous gene (Van der Krol *et al.*, 1990; Dorlhac de Borne *et al.*, 1994). The molecular basis of co-suppression is not known, but it may be post-transcriptional and involve normal elements of gene control (Finnegan and McElroy, 1994).

A second avenue to gene inactivation is *via* unwanted methylation of the inserted DNA (Matzke *et al.*, 1989; Meyer *et al.*, 1993; Ingelbrecht *et al.*, 1994). As with co-suppression, the molecular basis of the role of methylation in gene inactivation is not understood. In both cases the degree of effect varies with different transformants and some transformants are apparently not affected. While it is desirable to discover methods to avoid these effects, they are not serious blocks to cereal engineering if the researcher can produce enough transgenic lines to select those showing stable gene expression.

A third problem in predicting levels of gene activity is due to the 'position effect' that accompanies the random insertion of DNA into a complex genome. Levels of transgene expression vary over a wide range and are independent of the number of inserted gene copies. For instance, the level of activity of single copy insertions can vary over two orders of magnitude and may be more than those of a transformant with 100 copies of the same gene. It is theorized that not all regions of a chromosome are equally competent for active gene expression, possibly related to the organization of chromatin in active and inactive domains (Orkin, 1990). The position effect causes two problems in genetic engineering efforts. First, it decreases the apparent number of successful transformation events because if the insertion occurs at a chromosome site incompetent for gene expression, the event will not be detectable. The second problem is that the absolute levels of transgene expression are often lower than desired, requiring screening of larger numbers of transformed lines to find those expressing at sufficiently high levels.

Research on animal systems points to a possible solution to the position effect problem. DNA elements have been isolated that, when linked to certain gene constructions, promote position independence upon transformation (Chung *et al.*, 1993). Recently, similar results have been reported in transgenic tobacco (Schöffl *et al.*, 1993; Mlynárová *et al.*, 1994).

DNA delivery systems

The most successful method of DNA delivery to monocots is *via* biolistics, the acceleration of DNA-coated particles through plant cell walls (Klein *et al.*, 1988). Several different devices have been developed;

the most common ones using gunpowder or compressed gas to accelerate the particles.

Agrobacterium has proven useful in dicot systems, but not in monocots, although reports indicate it may be possible to adapt the system for monocot transformation (Conner and Dommisee, 1992; Chan *et al.*, 1993; discussed in Potrykus, 1991). Additional methods may also prove useful; e.g., electroporation of whole tissue (D'Halluin *et al.*, 1992), and tissue micro-injection (Holm *et al.*, 1994). It will be important to develop different DNA delivery protocols both to give researchers flexibility and to prevent dependence on a single method or device.

Target tissue

Most cereal transformations use immature embryos, but protoplasts continue to be efficient targets for rice. In addition, there are reports that using meristematic leaf or shoot tissue (Iglesias *et al.*, 1994), immature inflorescence tissue (Barcelo *et al.*, 1994), and transformation by direct injection of fertilized plant eggs may be possible (Holm *et al.*, 1994). The development of efficient microspore-based cereal transformation protocols would offer the advantage of the production of homozygous lines in a single generation (Wan and Lemaux, 1994). As with DNA delivery systems, it will be important for researchers to have available several procedures to allow maximum flexibility in research design.

Genotype independence

Ideally, a cereal researcher would be able to transform all cultivars and breeding lines with equal efficiency. This avoids the necessity of carrying out a series of crosses to move the transgene into an agronomically important cereal line. Currently, there is no convincing evidence of a reproducible genotype-independent method of transformation and regeneration in any cereal. This represents the most serious limitation of this technology, although, as with the problem of transgene inactivation, efficient enough transformation protocols will yield sufficient transgenic lines for all but the most recalcitrant cereal varieties.

12.3 THE WHEAT HIGH-MOLECULAR-WEIGHT (HMW) GLUTENINS AS AN EXAMPLE GENE SYSTEM FOR CEREAL QUALITY ENGINEERING

Let us now consider an example cereal gene system and how molecular tools are being, and could be, used to understand and manipulate cereal quality parameters. The system under consideration is the high-molecular-weight glutenin (HMW-glutenin) subunit genes of wheat

(Chapter 8; reviewed by Shewry *et al.*, 1992; Lazzeri and Shewry, 1993). These unusual polypeptides have been the object of intense study because of their known correlation with the viscoelastic properties unique to wheat dough, and a wide range of molecular tools are being used to identify the molecular basis of their functionality. The HMW-glutenin genes are also important since they will be among the first genes used in wheat transformation, and there is the potential for their use in other cereals to alter quality and utilization. Since any attempt to engineer proteins requires the isolation of the encoding genes, the HMW-glutenin system is well situated. Representatives of all members of this small gene family have been isolated and sequenced, including all six genes from the high-quality hard red winter wheat cultivar Cheyenne (Forde *et al.*, 1985; Halford *et al.*, 1987; Anderson and Greene, 1989; Anderson *et al.*, 1989). In addition, poor-quality related alleles from the D-genome have been isolated and sequenced (Thompson *et al.*, 1985; Sugiyama *et al.*, 1985).

One of the earliest and simplest goals of engineering the glutenin genes is to create new HMW-glutenin loci *via* transformation with isolated genes. Several reports correlate quality with the copy number of active HMW-glutenin genes; e.g., experiments with near isogenic wheat lines (Lawrence *et al.*, 1988), and an estimation of the percentage of total endosperm protein produced by each HMW-glutenin gene (Halford *et al.*, 1992). Therefore, there is an expectation that increasing the number of such genes will have measurable effects on dough parameters. These new loci will allow pyramiding more HMW-glutenin genes, as needed, to alter the quality characteristics of existing wheat lines. However, the use of native HMW-glutenin genes is only the first step. The real potential is to engineer enhanced, or totally different, quality characteristics. Before this is possible, there must be a more basic understanding of these unique polypeptides.

The molecular basis for HMW-glutenin functionality is not understood, but models concentrate on two prominent characteristics of their amino acid sequence: a central repetitive domain composed of 50–75 short peptide motifs and the localization of the cysteine residues within the terminal, non-repetitive regions. The cysteine residues are critical for formation of the disulfide cross-linked subunit network fundamental to dough quality, and the repetitive domain contributes to the visco-elasticity by an as yet unknown mechanism (Belton *et al.*, 1994). The remaining discussion will summarize some of the molecular approaches to understanding: (i) the basis of HMW-glutenin subunit functioning; (ii) how the expression of these genes is controlled; and (iii) some newer biotechnology tools that can contribute to bioengineering the HMW-glutenin subunits.

12.3.1 A caveat on the alteration of functional proteins

Any protein can be altered (by changing its gene sequence) and expressed. A potential problem is that alteration of the protein for one or more enhanced or new characteristics may cause a loss of functionality. This is less of a problem with most cereal storage proteins since they have only minor roles outside of serving as nitrogen stores for early development. Thus, altering the amino acid composition of a maize zein, an oat avenin, or a wheat gliadin is not expected to have significant functional effects. However, altering a wheat HMW-glutenin will probably, to some degree, affect its functionality. In addition, adding a new gene for endosperm protein expression could affect the levels of expression of other endogenous proteins, since all active endosperm-specific genes draw on the same pool of biosynthetic resources. Therefore, the addition of new endosperm-specific genes must be balanced with the possible negative effects on quality.

12.3.2 Synthetic genes

One of the powers of modern molecular biology is the freedom to make changes in gene structure. If desired, the researcher can synthesize, from nucleotides, an entirely novel gene or gene segment. This power is tempered by the fact that seldom does the researcher know all the properties of a protein needed for it to function in a useful manner in the target organism. Nevertheless, the ability to perform gene synthesis is potentially a powerful research tool, and will eventually be used to create new genes for cereal transformation.

There are two basic methods of constructing a gene. The simplest is to synthesize the two complementary gene strands and then anneal them to form the double-stranded gene. Unfortunately the current technology limits the length of any synthetic nucleotide to about 100 base pairs, but eventually the complete synthesis of entire genes of thousands of base pairs will be possible.

Currently, synthetic genes must be assembled from smaller oligonucleotides, either by synthesizing the entire double strand as overlapping oligonucleotides that can be joined by DNA ligase (Figure 12.5, left) or by synthesizing partially overlapping oligonucleotides that serve as templates for double strand formation by DNA polymerase action (Figure 12.5, right). Such techniques allow the assembly of novel HMW-glutenin construction.

12.3.3 Cysteine residue changes

Since the disulfide linkages among glutenin subunits are known to be crucial for viscoelasticity, the number and placement of cysteine

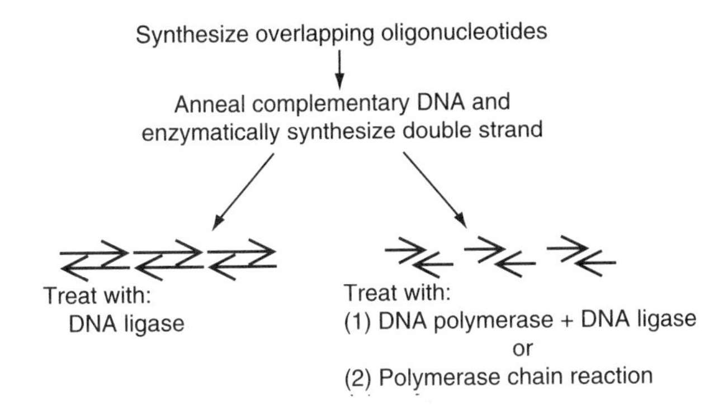

Figure 12.5 Use of synthetic oligonucleotides in gene construction. A novel gene can be constructed by synthesizing overlapping oligonucleotides. The resulting double-stranded DNA can be cloned and manipulated by standard protocols.

residues within the HMW-glutenin sequence is a major focus of attention. The simplest approach to understand their contribution to functionality is to systematically remove individual cysteines and observe the effects. One such study is the report by Shani *et al.* (1994) in which glutenin polymer formation is assessed in tobacco leaves expressing HMW-glutenin subunits. When the 3′ terminal cysteine is removed, the size distribution of the polymers in tobacco is reduced, indicating the 3′ cysteine contributes to but is not essential in the formation of HMW-glutenin polymers. Similar experiments will be performed in wheat now that transformation is possible, and a clearer understanding of the role of individual cysteines will result.

More difficult will be decisions on where to add cysteines to form a more complex cross-linked glutenin matrix. Additional cysteines may form intramolecular disulfide bonds rather than intermolecular linkages, contrary to what is intended. Analyses using current models of protein structure should be carried out to estimate, as well as possible, appropriate addition sites for intermolecular cross-links. However, our understanding of protein folding and interactions is limited, especially for such large polypeptides. For both the subtraction and addition of cysteines, it may be best to proceed both with systematic alterations and, in parallel, make glutenin constructs with more extensive changes in order to test the limitations of cysteine number and placement alterations.

Another problem is that molecular biological techniques can create hundreds of cysteine variations, but how can meaningful functionality tests be performed? Transformation, whether in wheat or heterologous systems such as tobacco, will likely remain too resource-intensive for the next few years to test more than a small subset of the possible cysteine variations. What is needed is simpler and faster physical-chemical assays for HMW-glutenin functionality. The micro-dough-mixing system of Bekes *et al.* (1994b) shows promise in this regard, but other new tools are also needed.

12.3.4 Repetitive domain changes

Modelling of the HMW-glutenin repetitive motifs has suggested that they form secondary structures of repeated β-turns that participate in a β-spiral structure similar to that theorized for elastin (Tatham *et al.*, 1984). This theory could elegantly explain a relationship between the HMW-glutenins and dough viscoelasticity. However, further physical studies do not support the similarity to elastin (Belton *et al.*, 1994), and there is no direct evidence to associate the repetitive domain with elasticity. Molecular biology offers the only tools available that can clarify the association by testing the effects of modifying the repeat structure on dough properties. Again, as with the altering of cysteine residue placement, the question is: what changes to make? One strategy is to alter the length of the repeat domain. Figure 12.6 shows a series of HMW-glutenins with varying length repeat domains. These constructs were made using some of the few available restriction sites in the repeat region of the Dx5 gene and cutting-and-splicing Dx5 gene fragments. Changes in individual repeats or repeat patterns are not practical since the repetitive region of these genes is lacking enough useful DNA restriction sites. This problem may be partly overcome by schemes to construct, in the laboratory, synthetic repeat domains which have exactly regular motifs. The same methods should also be able to produce mixtures of two or three repeat motifs. These synthetic repeat domains would be simpler than native domains, but the simplified structure may give fundamental information on the contribution of individual repeat motifs.

12.3.5 Heterologous expression of HMW-glutenins

One strategy in researching complex problems is to use model systems to isolate and study individual components. This is the rationale in using heterologous expression systems to understand how the primary structure of storage proteins relates to their post-translational processing and physical properties. The interpretations of research results using such systems must always be cautious since the proteins are synthesized outside their normal cellular environments and without the interactions with other endosperm components. Nevertheless, the power of such approaches allows research otherwise not possible.

Bacterial systems have been used for expression of several cereal

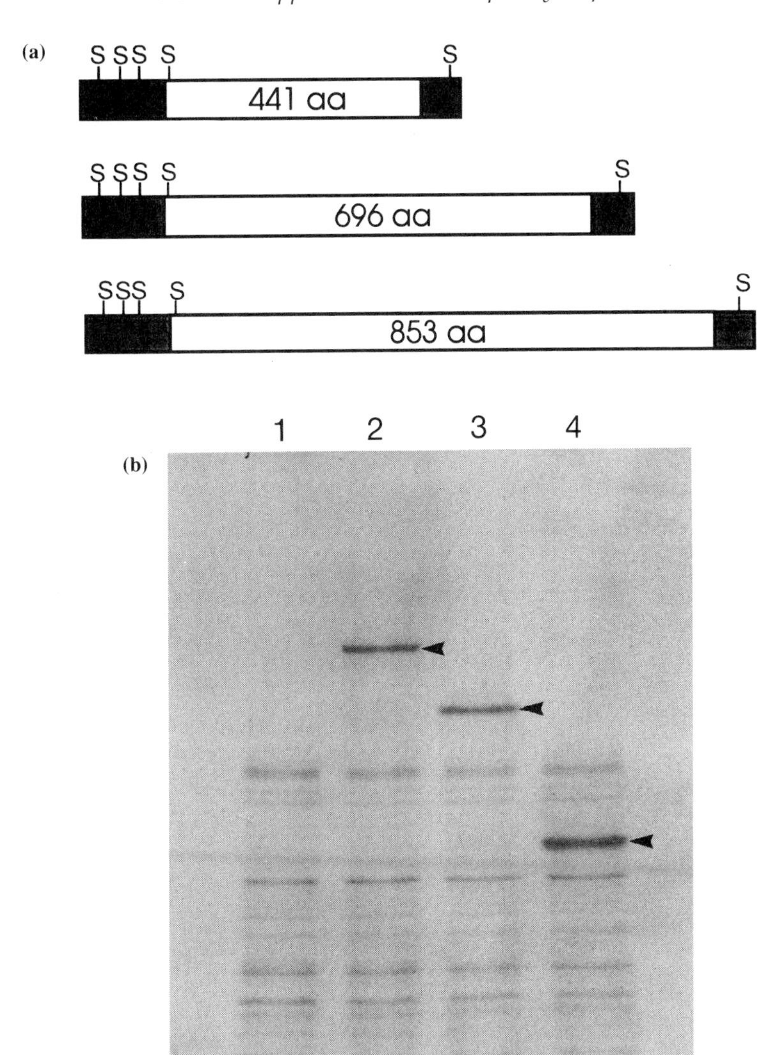

Figure 12.6 Diagram of HMW-glutenin genes with new repeat lengths. pET3A-Dx5 is a bacterial expression construction of the natural Dx5 HMW-glutenin subunit gene. It was modified (D'Ovidio and Anderson, unpublished) using unique restriction sites in the repeat domain (696 amino acids long). (a) The modified genes are identical to the natural Dx5 gene to the two terminal regions, but differed in the length of the repeat domain; i.e., 441, 696, and 853 amino acids long. (b) Modified Dx5 constructions were expressed in bacteria and a bacterial extract analysed by polyacrylmide gel electrophoresis. Lane 1: control protein extract from bacteria not possessing a HMW-glutenin expression clone. Lane 2: extract from bacteria expressing the Dx5 gene variant with a 853 amino acid repeat domain. Lane 3: a 696 amino acid repeat domain. Lane 4: a 441 amino acid repeat domain. HMW-glutenin subunit bands are indicated by arrowheads.

storage proteins. Although these are not eukaryotic systems, and their post-translational processing mechanisms are different from eukaryotes, they offer two advantages. The first is that large amounts of individual proteins can be expressed and isolated. Second, genes can be modified and expressed relatively quickly. If fidelity of processing and secondary structure formation are not critical, the bacterial systems are preferred. Galili (1989) and Shani *et al.* (1992) showed that bacterially expressed HMW-glutenins are capable of formation *in vitro* of intra- and inter-molecular disulfide bonds. Bekes *et al.* (1994a) report that the addition of such proteins can alter dough properties. The latter experiments were made possible by the development of a microdough mixing protocol based on the incorporation of exogenous proteins into the glutenin matrix during a partial reduction/reoxidation cycle prior to mixing (Bekes *et al.*, 1994b).

Eukaryotic microbial systems allow synthesis of the cereal storage proteins in an environment more likely to allow appropriate protein processing. Wheat α-gliadins (Neill *et al.*, 1987; Blechl *et al.*, 1992) and γ-gliadins (Scheets and Hedgcoth, 1989; Pratt *et al.*, 1991) have been expressed in yeast and some fidelity of processing is found; e.g. Blechl *et al.* (1992) showed correct signal peptide processing. However, the levels of expression are significantly less than in bacterial systems. In contrast, the baculovirus heterologous expression system used by Thompson *et al.* (1994) yielded low-molecular-weight (LMW) glutenin at levels 25–30% of the total extracted protein. This glutenin preparation also showed at least some physical characteristics of native LMW-glutenins and the baculovirus system is a promising addition to heterologous expression tools used to study the cereal proteins.

Another model eukaryotic system for studying aspects of mRNA activity and post-translational processing is the injection of mRNA into amphibian oocytes. This exotic sounding system has been used to study the synthesis of wheat gliadins and their packaging into protein bodies (Altschuler and Galili, 1994). In addition, oocyte experiments have shown that *in vitro* mutagenesis of a gliadin-DNA coding region can yield information on protein folding and deposition.

These different prokaryotic and eukaryotic expression systems have been and will continue to be used to study the cereal proteins. Also needed are plant expression systems with a more normal environment for plant genes. Maize zein genes have been expressed in transformed tobacco (Schernthaner *et al.*, 1988) and petunia (Ueng *et al.*, 1988), but normal zein protein synthesis and processing in seeds was not observed. Shani *et al.* (1994) report that, in tobacco leaves, polymers of HMW-glutenins can form utilizing only N-terminal cysteines. These same experiments found evidence that protein body formation and glutenin cross-linking are separate processes.

Even though cereal transformation is becoming a useable tool, the heterologous expression systems will continue to be important because of their speed in expressing modified genes, and because the different heterologous systems each offer advantages in studying specific problems.

12.3.6 Labelling HMW-glutenins to follow subunit processing and interactions

One of the major future contributions of molecular biology to understanding cereal protein functionality and post-translational processing is the ability to tag the expressed protein. To follow the processing path of a specific protein, the gene could be modified to encode a known short peptide epitope fused to the protein under study. For example, commercial antibodies (Novagen Inc.) are available that recognize the 11 amino acid N-terminus of the bacteriophage T7 gene 10; these have been used in fusion protein isolation (Tsai *et al.*, 1992). Another example is the addition of a six-residue histidine segment (His–Tag) to the polypeptide. The His–Tag allows affinity purification of the tagged polypeptide (Sauer and Stolz, 1994). Using such strategies, a tagged, modified HMW-glutenin can be followed from initial synthesis in the endosperm through post-translational processing and protein body deposition. It seems reasonable to predict that many up-to-now insoluble problems in seed development and protein interactions will yield to such protein tagging approaches.

12.3.7 Promoter modifications

The correct developmental expression of genes is fundamental to efforts to bioengineer new characteristics. The cereal storage proteins are synthesized only in a specific developmental window during endosperm development. How is this precise control achieved? Very little is known of the details, but it can be assumed that some combinations of available *trans*-acting protein factors and appropriate chromatin conformation in the regions of the storage protein genes coincide to achieve this specificity. Both these conditions involve DNA elements adjacent to the storage protein coding regions. Some of these elements are immediately adjacent; e.g., the transcription initiation elements. Other elements may be hundreds to thousands of bases upstream and/or downstream of the coding sequence. In some animal gene systems, controlling elements have been found tens of thousands of bases distal to the coding region. The complete dissection of a complex system at such distances requires intense effort and considerable resources.

However, there is optimism that in the cereal genes most important

DNA elements lie within a few thousand bases or less. Transient expression systems have been used to study a number of cereal storage protein gene promoters, including those for zeins (Thompson *et al.*, 1990), wheat gliadins (Aryan *et al.*, 1991) and HMW-glutenins (Blechl *et al.*, 1994). The results indicate that all the DNA elements necessary for gene expression under transient conditions are found within the first 600 bp (and much less in some reports).

Studies of gene expression in whole plants are more difficult than transient assays, but offer the advantage of a more normal biochemical and developmental environment. Such reports include zein promoter constructs transformed into tobacco (Schernthaner *et al.*, 1988) and petunia (Quattrocchio *et al.*, 1990), and rice glutelin (Takaiwa *et al.*, 1991) and prolamine (Zhou and Fan, 1993) promoter constructs transformed into tobacco. These results support the transient assay reports; i.e., the tissue-specific and developmental control elements are found within 1000 bp proximal to the coding region (similar results are known for dicot seed-specific genes expressed in dicot systems (Itoh *et al.*, 1993), although the exact sequences responsible have not been identified.

HMW-glutenin promoter constructions have been transformed into tobacco and these experiments showed that the first 277–375 bp proximal to the coding region were sufficient for tissue specificity (Colot *et al.*, 1987; Halford *et al.*, 1989; Thomas and Flavell, 1990) . Further evidence for the importance of the first few hundred bases is in identification of specific sites for protein binding within the first 300 base pairs upstream of the glutenin coding sequence (Hammond-Kosack *et al.*, 1993).

The analysis of known HMW-glutenin promoter DNA sequence finds that promoter DNA sequences are highly conserved for the first 500–1000 base pairs and diverge more distal to the coding region (Anderson, unpublished). This observation gives further support for the restriction of important gene controlling elements within the first approximately one thousand base pairs upstream of a coding region, assuming that the important DNA sequences, such as gene coding regions and transcription control elements, are the more conserved sequences.

The goal of the above promoter studies is to understand enough of the molecular basis of the control of seed specific genes to allow manipulation of the timing and levels of gene expression. However, there are other possible strategies that may be used to manipulate seed gene expression. One such approach is to use a seed-specific promoter to express a highly active RNA-polymerase instead of its usual protein. The polymerase would, in turn, recognize its own promoters fused to a target protein DNA coding region. Such a system can potentially amplify the expression level of the target protein and make the target protein less susceptible to physiological control mechanisms. Just such a strategy is used in bacteria for over-expressing proteins with a bacteriophage T7 RNA-polymerase (Studier *et al.*, 1991). The same polymerase has been shown to be active in mammalian cell cultures in production of human proteins and successful antisense RNA and ribozyme transcripts that inhibit specific genes (Lieber *et al.*, 1993).

For a cereal seed, this 'promoter cascade' might initiate with an endosperm-specific promoter expressing the foreign RNA polymerase which specifically transcribes the desired DNA to amplify the initial activation (Figure 12.7). McBride *et al.* (1994) have established that this strategy functions in a plant system. In their report, synthesis of bacteriophage T7 RNA polymerase was induced in tobacco by a light-sensitive promoter. The polymerase was then targeted to the plant cell plastids where it transcribed from a bacteriophage T7 promoter attached *β*-glucuronidase (GUS) reporter coding region. The GUS protein accumulated to 20–30% of the total soluble plastid protein. In theory, the same strategy should work with any tissue and gene system, including cereal endosperm-specific genes. In addition, although our understanding of eukaryotic RNA polymerases is limited at this time, it may be possible to use plant viral promoter/polymerase systems in the near future.

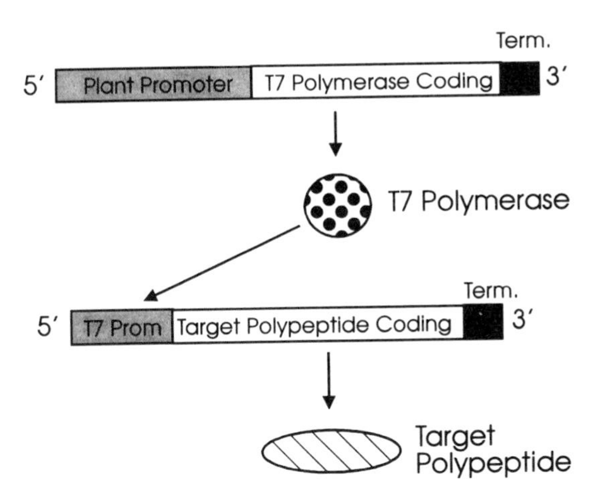

Figure 12.7 Use of a bacteriophage RNA polymerase/promoter system to express genes in plants. The system has the potential of amplifying the expression level of the target polypeptide. The T7 RNA polymerase is synthesized only in tissues appropriate for the specific plant promoter fused to the polymerase gene. The T7 polymerase must then be transported to the site of the T7 promoter/target coding construct. The highly active, specific T7 RNA polymerase will recognize only the appropriate promoter.

12.3.8 An interrelated schema for engineering the HMW-glutenins

The approaches discussed above can be used to understand both the molecular basis of HMW-glutenin functionality and propose some of the ways molecular biology can be used in glutenin bioengineering. Figure 12.8 shows how these approaches are interrelated to each other and to non-molecular biological techniques critical to a complete programme of glutenin engineering. A molecular biologist can modify a HMW-glutenin gene and express the polypeptide in a heterologous system. Protein chemists and quality technologists might use this protein in physico-chemical and quality testing. Their results could be used to make further modifications or confirm a gene modification is promising enough to use in wheat transformation experiments. Once the altered HMW-glutenin is transformed into wheat, quality testing would again be performed, and wheat agronomists and breeders would evaluate field performance. Those engineered HMW-glutenin loci showing enhanced quality may then begin appearing in new cultivars. Molecular biological approaches must interact with plant physiology and biochemistry, protein chemistry, plant breeding, and cereal processing technology to fulfil its potential contribution to improving cereal quality and utilization.

12.4 CONCLUSION

The direct manipulation of the cereal genomes may allow the transcendence of quality–factor interrelationships as we now understand them.

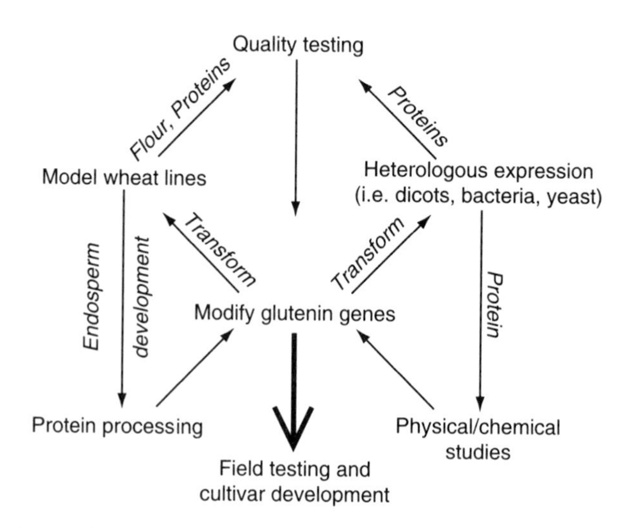

Figure 12.8 Interrelationships among a set of research approaches for molecular engineering of the wheat HMW-glutenin subunits.

Consider as an example the relative contributions of the individual wheat storage proteins to wheat quality. As important as is their contribution, the individual proteins are not the sole determinants of quality. Also important are the relative amounts of different protein classes, the size and composition of the starch granules, the lipids, etc. Let us consider a hypothetical case where two components which interact each contribute 20% of quality on an arbitrary scale.

The researcher working to engineer component 1 is not limited to optimizing within the 20% of component 1, but may be able to change component 1 such that component 2 is not as significant; i.e. in our hypothetical case component 1 can be made to contribute 30% and component 2 only 10%. Similarly, consider the relative contributions of genotype, environment, and genotype–environment interactions to quality (Peterson *et al.*, 1992).

While the environment will remain beyond our control, there is optimism that cereals can be eventually engineered to be less qualitatively sensitive to environmental variations. If an undesirable protein is produced in the grain by drought stress, endosperm-specific promoters may be used in an antisense strategy to prevent expression in the grain. If the ratio of two proteins (or protein classes) is significant, and one class is differentially turned off under a stress condition, the coding regions of the down-regulated genes can be fused to the other promoter class which is unaffected by the stress. Similarly, it may be possible to prevent other environmental, biotic, and abiotic factors from causing variations in cereal crop qualities.

The impossible task is to set timetables for these developments. In some cases positive results from engineering will occur in the near term, while more difficult problems may wait for future generations of researchers. The critical point is that there are now no fundamental technical obstacles to manipulating the cereal genomes except our limited knowledge of what manipulations to undertake. It is important to continue fundamental research in all aspects of cereal quality, from the basic chemistry of individual components to the understanding and isolation of the relevant genes. The advances in molecular biology and the ability to transform all the cereals means that every aspect of cereal quality in this volume will advance in the coming years. In closing this chapter, the easiest prediction to make is that exciting decades lie ahead for all cereal scientists, and it is certain that the potential is greater than we can envisage.

REFERENCES

Adams, M.D., Kelley, J.M., Gocayne, J.D., Dubnick, M., Polymeropoulos, M.H., Xiao, H., Merril, C.R., Wu, A., Olde, B., Moreno, R.F., Kerlavage,

A.R., McCombie, W.R. and Ventner, J.C. (1991) Complementary DNA sequencing: expressed sequence tags and human genome project. *Science*, **252**, 1651–6.

Ahn, S., Anderson, F.A., Sorrells, M.E. and Tanksley, S.D. (1993) Homologous relationships of rice, wheat and maize chromosomes. *Molecular and General Genetics*, **241**, 483–90.

Ainsworth, C., Tarvis, M. and Clark, J. (1993) Isolation and analysis of a cDNA clone encoding the small subunit of ADP-glucose pyrophosphorylase from wheat. *Plant Molecular Biology*, **23**, 23–33.

Altenbach, S.B., Pearson, K.W., Meeker, G., Starcaci, L.C. and Sun, S.S. (1989) Enhancement of the methionine content of seed proteins by the expression of a chimeric gene encoding a methionine-rich protein in transgenic plants. *Plant Molecular Biology*, **13**, 513–22.

Altenbach, S.B., Kuo, C.-C., Staraci, L.C., Pearson, K.W., Wainwright, C., Georgescu, A. and Townsend, J. (1992) Accumulation of a Brazil nut albumin in seeds of transgenic canola results in enhanced levels of seed protein methionine. *Plant Molecular Biology*, **18**, 235–45.

Altschuler, Y. and Galili, G. (1994) Role of conserved cysteines of a wheat gliadin in its transport and assembly into protein bodies in *Xenopus* oocytes. *Journal of Biological Chemistry*, **269**, 6677–82.

Anderson, J.M., Larsen, R., Laudencia, D., Kim, W.T., Morrow, D., Okita, T.W., and Preiss, J. (1991) Molecular characterization of the gene encoding a rice endosperm-specific ADPglucose pyrophosphorylase subunit and its developmental pattern of transcription. *Gene*, **97**, 199–205.

Anderson, O.D. and Greene, F.C. (1989) The characterization and comparative analysis of high M_r glutenin genes from genomes A and B of a hexaploid bread wheat. *Theoretical and Applied Genetics*, **77**, 689–700.

Anderson, O.D., Yip, R.E., Halford, N.G., Forde, J., Shewry, P.R., Malpica-Romero, J.-M. and Greene, F.C. (1989) Nucleotide sequences of two high-molecular-weight glutenin subunit genes from the D-genome of a hexaploid bread wheat, *Triticum aestivum* L. cv Cheyenne. *Nucleic Acids Research*, **17**, 461–2.

Aryan, A.P., An, G. and Okita, T.W. (1991) Structural and functional analysis of promoter from gliadin, and endosperm-specific storage protein gene of *Triticum aestivum* L. Molecular Genereal. *Genetics*, **225**, 65–71.

Audy, P., Trudel, J., and Asselin, A. (1988) Purification and characterization of a lysozyme from wheat germ. *Plant Science*, **58**, 43–50.

Augustine, J.J., Delannay, X., Dodson, R.B., Fischoff, D.A., Fuchs, R.L., Greenplate, J.T., LaVallee, B.T., Layton, J.G., Marrone, P.G., and Proksh, R.K. (1989). Field performance of transgenic tomato plants expressing the *Bacillus thuringiensis* var. Kurstaki insect control protein. *Bio/Technology*, **7**, 1265–9.

Baker, B., Schell, J., Lörz, H. and Federoff, N. (1986) Transposition of the maize controlling element 'Activator' in tobacco. *Proceedings of the National Academy of Sciences*, USA, 83, 4844–8.

Barcelo, P., Hagel, C., Becker, D., Martin, A. and Lörz, H. (1994) Transgenic cereal (tritordeum) plants obtained at high efficiency by microprojectile bombardment of inflorescence tissue. *Plant Journal*, **5**, 583–92.

Baum, M., Lagudah, E.S. and Appels, R. (1992) Wide crosses in cereals. *Annual Review of Plant Physiology*, **43**, 117–43.

Bekes, F., Anderson, O., Gras, P.W., Gupta, R.B., Tam, A., Wrigley, C.W. and Appels, R. (1994a) The contriubtions to mixing properties of 1D HMW glutenin subunits expressed in a bacterial system, in *Improvement of Cereal*

Quality by Genetic Engineering (eds Henry, R. and Ronalds, J.A.) Plenum Press, New York.
Bekes, F., Gras, P.W., Gupta, R.B., Hickman, D.R. and Tatham, A.S. (1994b) Effects of a high M_r Glutenin Subunit (1Bx20) on the Dough Mixing Properties of Wheat Flour. *Journal of Cereal Science*, **19**, 3–7.
Belton, P.S., Colquhoun, I.J., Field, J.M., Grant, A., Shewry, P.R. and Tatham, A.S. (1994) ^1H and ^2H NMR relaxation studies of a high M_r subunit of wheat glutenin and comparison with elastin. *Journal of Cereal Science*, **19**, 115–21.
Blechl, A.E., Thrasher, K.S., Vensel, W.H. and Greene, F.C. (1992) Purification and characterization of wheat α-gliadin synthesized in the yeast, *Saccharomyces cerevisiae*. *Gene*, **116**, 119–27.
Blechl, A.E., Lorens, G.F., Greene, F.C., Mackey, B.E. and Anderson, O.D. (1994) A transient assay for promoter activity of wheat seed storage protein genes and other genes expressed in developing endosperm. *Plant Science*, **102**, 69–80.
Blumenthal, C.S., Barlow, E.W.R. and Wrigley, C.W. (1993) Growth environment and wheat quality: the effect of heat stress on dough properties and gluten proteins. *Journal of Cereal Science*, **18**, 3–21.
Casas, A.M., Kononowicz, K., Zehr, U.B., Tomes, D.T., Axtell, J.D., Butler, L.G., Bressan, R.A. and Hasegawa, P.M. (1993) Transgenic sorghum plants via microprojectile bombardment. *Proceedings of the National Academy of Sciences, USA*, **90**, 11212–16.
Castillo, A.M., Vasil, V. and Vasil, I.K. (1994) Rapid production of fertile transgenic plants of rye (*Secale cereale* L.). *Bio/Technology*, **12**, 1366–71.
Chan, M.-T., Chang, H.-H., Ho, S.-L., Tong, W.-F. and Yu, S.-M. (1993) Agrobacterium-mediated production of transgenic rice plants expressing a chimeric α-amylase promoter/β-glucuronidase gene. *Plant Molecular Biology*, **22**, 491–506.
Christensen A.H., Sharrock R.A., Quail P.H. (1992) Maize polyubiquitin genes: structure, thermal perturbation of expression and transcript splicing and promoter activity following transfer to protoplasts by electroporation. *Plant Molecular Biology*, **18**, 675–89.
Chung, J.H., Whiteley, M. and Felsenfeld, G. (1993) A 5′ element of the chicken β-globin domain serves as an insulator in human erythroid cells and protects against position effect in *Dosophila*. *Cell*, **74**, 505–14.
Colot, V., Robert, L.S., Kavanagh, T.A., Bevan, M.W. and Thompson, R.D. (1987) Localization of sequences in wheat endosperm protein genes which confer tissue-specific expression in tobacco. *EMBO Journal*, **6**, 3559–64.
Conner, A. J. and Dommisse, E.M. (1992) Monocotyledonous plants as hosts for *Agrobacterium*. *International Journal of Plant Science*, **153**, 550–5.
Cornelissen and Melchers (1993) Strategies for control of fungal diseases with transgenic plants. *Plant Physiology*, **101**, 709–12.
Day, A.G., Bejarano, E.R., Buck, K.W., Burrell, M. and Lichtenstein, C.P. (1991) Expression of an antisense viral gene in transgenic tobacco confers resistance to the DNA virus tomato golden mosaic virus. *Proceedings of the National Academy of Sciences, USA*, **488**, 6721–5.
Dorlhac de Borne, F., Vincentz, M., Chupeau, Y. and Vaucheret, H. (1994) Co-suppression of nitrate reductase host genes and transgenes in transgenic tobacco plants. *Molecular and General Genetics*, **243**, 613–21.
Feldman, M. (1993) Cytogenetic activity and mode of action of the pairing homologous (*Ph1*-) gene of wheat. *Crop Science*, **33**, 894–7.
Finnegan, J. and McElroy, D. (1994) Transgene inactivation, plants fight back! *Bio/Technology*, **12**, 883–8.

Fisher, D.K., Boyer, C.D. and Hannah, L.C. (1993) Starch branching enzyme II from maize endosperm. *Plant Physiology*, **102**, 1045–6.

Forde, J., Malpica, J.-M., Halford, N.G., Shewry, P.R., Anderson, O.D. and Greene, F.C. (1985). The nucleotide sequence of a HMW glutenin subunit gene located on chromosome 1A of wheat (*Tricticum aestivum* L.). *Nucleic Acids Research*, **13**, 6817–32.

Fromm, M.E., Morrish, F., Armstrong, C., Williams, R., Thomas, J. and Klein, T.M. (1990) Inheritance and expression of chimeric genes in the progeny of transgenic maize plants. *Bio/Technology*, **8**, 833–9.

Galili, G. (1989) Heterologous expression of a wheat HMW glutenin gene in *Escherichia coli*. Proceeding of the National Academy of Sciences, USA, **86**, 7756–60.

Gordon-Kamm, W.J., Spencer, T.M., Mangano, M.L., Adams, T.R., Daines, R.J., Start, W.G., O'Brien, J.V., Chambers, S.A., Adams, W.R., Jr, Willetts, N.G., Rice, T.B., Mackey, C.J., Krueger, R.W., Kausch, A.P. and Lemaux, P.G. (1990) Transformation of maize cells and regeneration of fertile transgenic plants. *Plant Cell*, **2**, 603–18.

Graybosch, R.A., Peterson, C.J., Hansen, L.E., Worrall, D., Shelton, D.R. and Lukaszewski, A. (1993) Comparative flour quality and protein characeristics of 1BL/1RS and 1AL/1RS wheat-rye translocation lines. *Journal of Cereal Science*, **17**, 95–106.

Halford, N.G., Forde, J., Anderson, O.D., Greene, F.C. and Shewry, P.R. (1987). The nucleotide and deduced amino acid sequences of an HMW glutenin subunit gene from chromosome 1B of bread wheat (*Triticum aestivum* L.) and comparison with those of genes from chromosomes 1A and 1D. *Theoretical and Applied Genetics*, **75**, 117–26.

Halford, N.G., Forde, J., Shewry, P.R. and Kreis, M. (1989) Functional analysis of the upstream regions of a silent and an expressed member of a family of wheat seed protein genes in transgenic tobacco. *Plant Science*, **62**, 207–16.

Halford, N.G., Field, J.M., Blair, H., Urwin, P., Moore, K., Robert, L., Thompson, R. Flavell, R.B., Tatham, A.S. and Shewry, P.R. (1992) Analysis of HMW glutenin subunits encoded by chromosome 1A of bread wheat (*Triticum aestivum* L.) indicates quantitative effects on grain quality. *Theoretical and Applied Genetics*, **83**, 373–8.

Hammond-Kosack, M.C.U., Holdsworth, M.J. and Bevan, M.W. (1993) In vivo footprinting of a low molecular weight glutenin gene (LMWG-D1) in wheat endosperm. *EMBO Journal*, **12**, 545–54.

Hayes, P.M., Liu, B.H., Knapp, S.J., Chen, F., Jones, B., Blake, T., Franckowiak, J., Rasmusson, D., Sorrells, M., Ullrich, S.E., Wesenberg, D. and Kleinhofs, A. (1993) Quantitative trait locus effects and environmental interaction in a sample of North American barley germplasm. *Theoretical and Applied Genetics*, **87**, 392–401.

Holm, P.B., Knudsen, S., Mouritzen, P., Negri, D., Olsen, F.L. and Roué, C. (1994) Regeneration of fertile barley plants from mechanically isolated protoplasts of the fertilized egg cell. *The Plant Cell*, **6**, 531–43.

Iglesias, V.A., Gisel, A., Bilang, R., Leduc, N., Potrykus, I. and Sautter, C. (1994) Transient expression of visible marker genes in meristem cells of wheat embryos after ballistic micro-targeting. *Planta*, **192**, 84–91.

Ingelbrecht, I., Van Houdt, H., Van Montagu, M. and Depicker, A. (1994) Posttranscriptional silencing of reporter transgenes in tobacco correlates with DNA methylation. *Proceedings of the National Academy of Sciences, USA*, **91**, 10502–6.

Itoh, Y., Kitamura, Y., Arahira, M. and Fukazawa, C. (1993) *Cis*-acting regulatory regions of the soybean seed storage 11S globulin gene and their interactions with seed embryo factors. *Plant Molecular Biology*, **21**, 973–84.

Jing-liu, Z., Xiao-ming, L., Rui-zhu, C., Rui-xin, H. and Meng-min, H. (1991) Transposition of maize transposable element Activator in rice. *Plant Science*, **73**, 191–8.

Karchi, H., Shaul, O. and Galili, G. (1993) Seed-specific expression of a bacterial desensitized aspartate kinase increases the production of seed threonine and methionine in transgenic tobacco. *The Plant Journal*, **3**, 721–7.

Klein, T.M., Harper, E.C., Svab, Z., Sanford, J.C., Fromm, M.E. and Maliga, P. (1988) Stable genetic transformation of intact *Nicotiana* cells by the particle bombardment process. *Proceedings of the National Academy of Sciences, USA*, **85**, 8502–5.

Lamb, C.J., Ryals, J.A., Ward, E.R. and Dixon, R.A. (1992) Emerging strategies for enhancing crop resistance to microbial pathogens. *Bio/Technology*, **10**, 1436–45.

Lander, E.S., Green, P., Abrahamson, J., Barlow, A., Daly, M., Lincoln, S. and Newburg, L. (1987) MAPMAKER: an interactive computer package for constructing primary genetic linkage maps of experimental and natural populations. *Genomics*, **1**, 174–81.

Lawrence, G.J., Macritchie, F. and Wrigley, C. W. (1988). Dough and baking quality of wheat lines deficient in glutenin subunits controlled by the *Glu-A1*, *Glu-B1* and *Glu-D1* loci. *Journal of Cereal Science*, **7**, 109–12.

Lazzeri, P. and Shewry, P.R. (1993) Biotechnology of Cereals. *Biotechnology and Genetic Engineering Reviews*, **11**, 79–146.

Li, L., Qu, R., de Kochko, A., Fauquet, C. and Beachy, R.N. (1993) An improved rice transformation system using the biolistics method. *Plant Cell Reports*, **12**, 250–5.

Lieber, A., Sandig, V., Sommer, W., Bahring, S. and Strauss, M. (1993) Stable high-level gene expression in mammalian cells by T7 phage RNA polymerase. *Methods in Enzymology*, **217**, 47–66.

Liu, B.H. and Knapp, S.J. (1990) GMENDEL: a program for Mendelian segregation and linkage analysis of individual or multiple progeny populations using log-likelihood ratios. *Journal of Heredity*, **8**, 407.

McBride, K.E., Schaaf, D.J., Daley, M. and Stalker, D.M. (1994) Controlled expression of plastid transgenes in plants based on a nuclear DNA-encoded and plastid-targeted T7 RNA polymerase. *Proceedings of the National Academy of Sciences, USA*, **91**, 7301–5.

Mlynárová, L., Loonen, A., Heldens, J., Jansen, R.C., Keizer, P. (1994) Reduced position effect in mature transgenic plants conferred by the chicken lysozyme matrix-associated region. *The Plant Cell*, **6**, 417–26.

Martin, G.B., Brommonschenkel, S.H., Chunwongse, J., Frary, A., Ganal, M.W., Spivey, R., Wu, T., Earle, E.D. and Tanksley, S.D. (1993) Map-based cloning of a protein kinase gene conferring disease resistance in tomato. *Science*, **262**, 1432–6.

Matzke, M.A., Priming, M., Trnovshy, J. and Matzke, A.J.M. (1989) Reversible methylation and inactivation of marker genes in sequentially transformed tobacco plants. *EMBO Journal*, **8**, 643–9.

Meyer, P., Heidmann, I. and Niedenhof, I. (1993) Differences in DNA-methylation are associated with a paramutation phenomenon in transgenic petunia. *Plant Journal*, **4**, 89–100.

Neill, J. D., Litts, J. C., Anderson, O. D., Greene, F. C. and Stiles, J. I. (1987). Expression of a wheat alpha-gliadin gene in *Saccharomyces cerevisiae*. *Gene*, **55**, 303–17.

Ogawa, M., Kumamaru, T., Satoh, H., Iwata, N., Omura, T., Kasai, Z., Tanaka, K. (1987). Purification of protein body-1 of rice seed and its polypeptide composition. *Plant Cell Physiology*, **28**, 1517–27.

Orkin, S.H. (1990) Globin gene regulation and switching: circa 1990. *Cell*, **63**, 665–72.

Pen, J., Verwoerd,T.C.,van Paridon, P.A., Beudeker,R.F., van den Elzen, P.J.M., Geerse, K., van der Klis, J.D.,Versteegh, H.A.J., van Ooyen, A.J.J., and Hoekema, A. (1993). Phytase-containing transgenic seeds as a novel feed additive for improved phosphorus utilization. *Bio/Technology*, **11**, 811–14.

Peterson, P.A. (1987) Mobile elements in plants. *CRC Critical Reviews Plant Science*, **6**, 105–208.

Peterson, C.J., Graybosch, R.A., Baenzinger, P.S. and Grombacher, A.W. (1992) Genotype and environmental effects on quality characteristics of hard red winter wheat. *Crop Science*, **32**, 98–103.

Potrykus, I. (1991) Gene transfer to plants: assessment of Published Approaches and Results. *Annual Review of Plant Physiology and Plant Molecular Biology*, **42**, 205–25.

Pratt, K.A., Madgwick, P.J., Shewry, P.R. (1991) Expression of a wheat gliadin protein in yeast (*Saccharomyces cerevisiae*). *Journal of Cereal Science*, **414**, 223–9.

Preiss, J. (1992) Biology and molecular biology of starch synthesis and its regulation. *Oxford Surveys Plant Molecular and Cell Biology*, **7**, 69–114.

Putterill, J., Robson, F., Lee, K. and Coupland, G. (1993) Chromosome walking with YAC clones in Arabidopsis, isolation of 1700 kb of contiguous DNA on chromosome 5, including a 300 kb region containing the flowering-time gene CO. *Molecular and General Genetics*, **239**, 145–7.

Quattrocchio, F., Tolk, M.A., Coraggio, I., Mol, J.N.M., Viotti, A. and Koes, R.E. (1990) The maize zein gene zE19 contains two distinct promoters which are independently activated in endosperm and anthers of transgenic *Petunia* plants. *Plant Molecular Biology*, **15**, 81–93.

Raikel, N.V. and Wilkins, T.A. (1987) Isolation and characterization of a cDNA clone encoding wheat germ agglutinin. *Proceedings of the National Academy of Sciences, USA*, **84**, 6745–9.

Sauer, N. and Stolz, J. (1994) SUC1 and SUC2: two sucrose transporters from *Arabidopsis thaliana*; expression and characterization in baker's yeast and identification of the histidine-tagged protein. *The Plant Journal*, **6**, 67–77.

Scheets, K. and Hedgcoth, C. (1989) Expression of wheat γ-gliadin in *Saccharomyces cerevisiae* from a yeast *ADH1* promoter. *Journal of Agricultural and Food Chemistry*, **37**, 829–33.

Schernthaner, J.P., Matzke, M.A. and Matzke, A.J.M. (1988) Endosperm-specific activity of a zein gene promoter in transgenic tobacco plants. *EMBO Journal*, **7**, 1249–55.

Schlumbaum, A. Mauch, F. Vogeli, U. and Boller, T. (1986) Plant chitinases are potent inhibitors of fungal growth. *Nature*, **324**, 365–7.

Schöffl, F., Schröder, G., Kliem, M. and Rieping, M. (1993) An SAR sequence containing 395 bp DNA fragment mediates enhanced, gene-dosage-correlated expression of a chimaeric heat shock gene in tranasgenic tobacco plants. *Transgenic Research*, **2**, 93–100.

Sears, E.R. (1993) Use of radiation to transfer alien chromosome segments to wheat. *Crop Science*, **33**, 897–901.

Shade, R.E., Schroeder, H.E., Pueyo, J.J., Tabe, L.M., Murdock, L.L., Higgins, T.J.V. and Chrispeels, M.J. (1994) Transgenic Pea Seeds Expressing the α-amylase inhibitor of the common bean are resistant to Bruchid beetles. *Bio/Technology*, **12**, 793–6.

Shani, N., Steffen-Campbell, J.D., Anderson, O.D., Greene, F.C. and Galili, G. (1992) Role of the Amino- and Carboxy-Terminal Regions in the Folding and Oligomerization of Wheat High Molecular Weight Glutenin Subunits. *Plant Physiology*, **98**, 433–41.

Shani, N., Rosenberg, N., Kasarda, D.D. and Galili, G. (1994) Mechanisms of assembly of wheat high molecular weight glutenins inferred from expression of wild-type and mutant subunits in transgenic tobacco. *Journal of Biological Chemistry*, **269**, 8924–30.

Shewry, P.R., Halford, N.G. and Tatham, A.S. (1992) High molecular weight subunits of wheat glutenin. *Journal of Cereal Science*, **15**, 105–20.

Shimada, H., Tada, Y., Kawasaki, T. and Fujimura, T. (1993) Antisense regulation of the rice waxy gene expression using a PCR-amplified fragment of the rice genome reduces the amylose content in grain starch. *Theoretical and Applied Genetics*, **86**, 665–72.

Shimamoto, K., Terada, R., Izawa, T., Fujimoto, H. (1989) Fertile transgenic rice plants regenerated from transformed protoplasts. *Nature*, **338**, 274–6.

Somers, D.A., Rines, H.W., Gu, W., Kaeppler, H.F. and Bushnell, W.R. (1992) Fertile, transgenic oat plants. *Bio/Technology*, **10**, 1589–94.

Studier, F.W., Rosenberg, A.H., Dunn, J.J. and Dubendorff, J.W. (1991) Use of T7 RNA polymerase to direct expression of cloned genes. *Methods in Enzymology*, **185**, 60–89.

Sugiyama, T., Rafalski, A., Peterson, D. and Soll, D. (1985) A wheat glutenin subunit gene reveals a highly repeated strcuture. *Nucleic Acids Research*, **13**, 8729–36.

Takaiwa, F., Oono, K. and Kato, A. (1991) Analysis of the 5′ flanking region responsible for the endosperm-specific expression of a rice glutelin chimeric gene in transgenic tobacco. *Plant Molecular Biology*, **16**, 49–58.

Tanaka, K. (1990) Improvement of digestibility and nutritive quality of rice storage protein using genetic engineering. *Abstracts of the Fourth Annual Meeting of the Rockefeller Foundation's International Program on Rice Biotechnology*. The Rockefeller Foundation, New York.

Tanksley, S.D. (1993) Mapping polygenes. *Annual Review of Genetics*, **27**, 205–33.

Tanksley, S.D., Young, N.D., Paterson, A.H. and Bonierbale, M.W. (1989) RFLP mapping in plant breeding: new tools for an old science. *Bio/Technology*, **7**, 257–64.

Tatham, A., Shewry, P.R. and Miflin, B.J. (1984) Wheat gluten elasticity, a similar molecular basis to elastin? *FEBS Letters*, **177**, 205–8.

Thomas, M.S. and Flavell, R.B. (1990) Identification of an enhancer element for the endosperm-specific expression of high molecular weight glutenin. *The Plant Cell*, **2**, 1171–80.

Thompson, G.A., Boston, R.S., Lyznik, L.A., Hodges, T.K. and Larkins, B.A. (1990) Analysis of promoter activity from an α-zein gene 5′ flanking sequence in transient expression assays. *Plant Molecular Biology*, **15**, 755–64.

Thompson, R.D., Bartels, D. and Harberd, N.P. (1985) Nucleotide sequence of a gene from chromosome 1D of wheat encoding a HMW glutenin subunit. *Nucleic Acids Research*, **13**, 6833–46.

Thompson, S., Bishop, D.H.L., Madgwick, P., Tatham, A.S. and Shewry, P.R. (1994) High-level expression of a wheat LMW glutenin subunit using a baculovirus system. *Journal of Agricultural and Food Chemistry*, **42**, 426–31.

Toriyama, K., Arimoto, Y., Uchimiya, H., Hinata, K. (1988) Transgenic rice plants after direct gene transfer into protoplasts. *Bio/Technology*, **6**, 1072–4.

Tsai, D.E., Kenan, D.J. and Keene, J.D. (1992) *In vitro* selection of an RNA epitope immunologically cross-reactive with a peptide. *Proceedings of the National Academy of Sciences, USA*, **89**, 8864–8.

Ueng, P., Galili, G., Sapanara, V., Goldsbrough, P.B., Dube, P., Beachy, R.N. and Larkins, B.A. (1988) Expression of a maize storage protein gene in petunia plants is not restricted to seeds. *Plant Physiology*, **86**, 1281–5.

Van der Krol, A.R., Mur, L.A., de Lange, P., Mol, N.M. and Stuitje, A.R. (1990) Inhibition of flower pigmentation by antisense CHS genes: promoter and minimal sequence requirements for the antisense effect. *Plant Molecular Biology*, **14**, 457–66.

Vasil, V., Castillo, A., Fromm, M., and Vasil, I. (1992) Herbicide resitant fertile transgenic wheat plants obtained by micro-projectile bombardment of regenerable embryogenic callus. *Bio/Technology*, **10**, 667–74

Visser, R.G.F. and Jacobsen, E. (1993) Towards modifying plants for altered starch content and composition. *Trends in Biotechnology*, **11**, 63–8.

Walbot, V. (1992) Strategies for mutagenesis and gene cloning using transposon tagging and T-DNA insertional mutagenesis. *Annual Review of Plant Physiology and Plant Molecular Biology*, **43**, 49–82.

Weeks, J.T., Anderson, O.D. and Blechl, A.E. (1993) Rapid production of multiple independent lines of fertile transgenic wheat (*Triticum aestivum*). *Plant Physiology*, **102**, 1077–84.

Wessler, S.R. and Varagona, M.J. (1985) Molecular basis of mutations at the waxy locus of maize: Correlation with the fine structure genetic map. *Proceedings of the National Academy of Sciences, USA*, **82**, 4177–81.

Wing, R.A., Zhang, H.-B. and Tanskley, S.D. (1994) Map-based cloning in crop plants. Tomato as a model system: I. Genetic and physical mapping of *jointless*. *Molecular and General Genetics*, **242**, 681–8.

Wan, ? and Lemaux, P.G. (1994) Generation of large numbers of independently transformed fertile barley plants. *Plant Physiology*, **104**, 37–48.

Zambryski, P. (1988) Basic processes underlying agrobacterium-mediated DNA transfer to plant cells. *Annual Review of Genetics*, **22**, 1–30.

Zeller, F.J. (1973) 1B/1R wheat-rye chromosome substitutions and translocations. *Proceedings of the 4th International Wheat Genetics Symposium*, Sears, E.R. and Sears, L.M.(eds), University of Columbia, Missouri pp. 209–221.

Zhang, W., Wu, R. (1988) Efficient regeneration of transgenic plants from rice protoplasts and correctly regulated expression of the foreign gene in the plants. *Theoretical and Applied Genetics*, **76**, 835–40.

Zhou, X. and Fan, Y.-L. (1993) The endosperm-specific expression of a rice prolamin chimeric gene in transgenic tobacco plants. *Transgenic Research*, **2**, 141–6.

Zhu, Q., Maher, E.A., Masoud, S., Dixon, R.A. and Lamb, C.J. (1994) Enhanced protection against fungal attack by constitutive co-expression of chitinase and glucanase genes in transgenic tobacco. *Bio/Technology*, **12**, 807–12.

Part Four

Production of Quality Cereals

13 Agronomy and cereal quality

P.S. Kettlewell

13.1 INTRODUCTION

In addition to the genetic differences in cereal grain quality discussed in the previous two chapters there are several agronomic factors (e.g. nitrogen supply) which have marked effects on quality. The effects of these factors are, however, often strongly modified by the environment (e.g. rainfall) and indeed some, such as harvest date, are mediated largely through environmental influences. This chapter will therefore consider the role of environment as well as agronomy.

The chapter will broadly follow the sequence of growing a cereal crop from sowing to harvest. Environmental factors will be considered at points in the sequence where they have their main effects. It is impossible in a chapter of this size to comprehensively cover all the agronomic and environmental factors which influence quality of every cereal species. The chapter will therefore be selective and aims to illustrate primarily the large and important effects. Wheat will be the main crop referred to, but examples will also be drawn from barley, maize and rice.

The quality characters considered will be in three broad categories:

1. Physical grain characteristics – kernel size, test (specific) weight, and to a lesser extent kernel vitreousness, hardness and colour. There is a vast literature on agronomic and environmental effects on mean kernel weight, but this is largely ignored since mean kernel weight (or 1000 kernel weight) is rarely used commercially for indicating quality. In any case, there is not always a correlation between mean kernel weight and test weight (Bayles, 1977).
2. Grain protein concentration and protein quality (i.e. composition) and associated characters such as malt extract, and bread dough strength and loaf volume.

3. Grain α-amylase activity and related measurements such as Hagberg falling number.

For many agronomic factors, e.g. sowing date, there is a plethora of published results from field experiments conducted in various countries where only the yield and grain quality have been measured. Many of these experiments demonstrate conflicting conclusions. The reasons for these conflicts are often unknown since usually no growth or physiological data have been collected to help understand the mechanisms involved. Frequently, environmental data is also inadequate to understand the interactions between agronomy and environment. Thus, in this chapter, broad generalizations have had to be made based either on the majority of empirical observations from the field or from the few detailed physiological studies, often carried out in controlled environments. The physiological basis of agronomic effects on grain quality deserves greater attention, and a few careful physiological investigations can yield far more useful information than a myriad of empirical field experiments.

There are a few simple principles of physiology of kernel growth which determine the response of quality to agronomic and environmental factors. Once these principles are learnt it becomes easier to understand how quality responds to agronomy and environment. Thus the next section outlines the physiological determinants of grain quality.

13.2 PHYSIOLOGY OF CEREAL QUALITY

13.2.1 Kernel size and test weight

Although crop growth is very responsive to several agronomic factors, kernel size and test weight are relatively stable. This is because many of these factors influence the growth of the crop early in development when compensation can take place between yield components which develop before kernel growth starts. The number of kernels per unit ground area will usually adjust according to the supply of photosynthetic assimilate by mechanisms such as tiller death and floret death; thus many crops have a similar supply of assimilate per kernel leading to stability of kernel size (Biscoe and Gallagher, 1977). The exceptions are when crops undergo development before anthesis (flowering) in more favourable conditions than occur post-anthesis so that the kernel number is adjusted to a larger supply of assimilate than can be provided. Even in these situations small reductions in assimilate supply may be compensated for by increased remobilization of stem carbohydrate reserves (Schnyder, 1993). If a reduction in assimilate supply after anthesis is too great to be compensated for by this mechanism there are two ways in which kernel growth can be affected depending on the time at which assimilate supply is restricted; either kernel size will be reduced, or grain will be shrivelled.

The maximum potential size and weight of wheat kernels is determined largely by the number of endosperm cells (Jenner *et al.*, 1991). The endosperm undergoes cell division in the first 15–20 days after anthesis, and any factor which reduces photosynthesis and assimilate supply in this period will reduce cell number and therefore final kernel size. Maximum grain size is achieved at about the end of this period when much of the grain consists of water.

After the cell division period, accumulation of kernel weight continues through deposition of dry matter in the endosperm cells. If there is inadequate assimilate in this period to fill the endosperm cells then shrivelled kernels may be formed. Grain shrivelling may also occur from respiratory loss of dry matter before harvest, especially if pre-harvest sprouting occurs. Test weight depends on both kernel density and packing efficiency and is often lower if grain is shrivelled (Bayles, 1977).

13.2.2 Grain protein concentration

The concentration of protein in grain is clearly a balance between accumulation of protein and the other components of the grain, principally starch. The bulk of the starch originates from photosynthesis during kernel growth; the remainder (about 10% in wheat and barley) is from retranslocation from stem reserves (Schnyder, 1993). In contrast much of the protein in the grain derives from nitrogen taken up during the vegetative phase and subsequently remobilized and translocated from senescing leaves and other green plant parts. Up to half the grain protein may, however, be taken up from the soil during kernel growth in some circumstances. Any restriction to uptake of nitrogen after anthesis will increase the proportion of grain protein derived from retranslocation, e.g. low soil fertility, lack of water for uptake (Evans *et al.*, 1975).

The final grain protein concentration will depend on the relative effects of agronomic and environmental factors on supply and accumulation of starch and protein in the grain. Factors favouring protein accumulation will increase grain protein concentration and *vice versa*. Thus greater availability of nitrogen to the crop, either from soil or from accelerated leaf senescence, will increase grain protein concentration (Evans *et al.*, 1975). Since much of the nitrogen in the grain is from retranslocation from green parts of the plant, and since a large proportion of that nitrogen is a component of the photosynthetic system (largely in the enzyme rubisco), then extensive retranslocation must be associated with reduced photosynthetic activity (Gregory *et al.*, 1981). Therefore less photosynthate will be available to contribute to yield, but grain protein concentration will be high. Doubtless, this intrinsic interrelationship between nitrogen retranslocation and photosynthesis is a major reason for the typical negative correlation between grain yield and grain protein concentration (Simmonds, 1995). In some situations, however, this explanation is unlikely to apply and the negative correlation may simply be a result of dilution of a given amount of nitrogen by the more abundant starch in higher yielding crops (Kibite and Evans, 1984).

13.2.3 α-Amylase activity

High α-amylase activity in harvested grain can occur if dormancy breaks before harvest and the grain is moist enough to germinate, either from frequent wetting by rain or if the crops lodges severely and the heads are in close contact with moist soil. The time of dormancy break varies greatly between cultivars, and is strongly influenced by temperature during grain ripening. The higher the temperature the shorter the period of dormancy and *vice versa* (Belderok, 1968).

In some cultivars of wheat α-amylase can also form during grain ripening in the absence of germination (Flintham and Gale, 1988). This may be related to some extent to the rate at which the grain dries. Slower drying can lead to greater enzyme activity although other environmental factors may also be involved.

13.3 SEEDING

13.3.1 Cultivar

Choice of the cultivar for seeding is a major influence on quality and the subsequent use or sale of grain, as has been illustrated clearly in preceding chapters. Quality characters differ considerably, however, in their heritability. Some characters, e.g. hardness, are highly heritable with relatively little environmental influence, whereas others, e.g. protein concentration, are mainly determined by environmental and agronomic factors. The following list is the view of the author, rather than exact heritabilities, on the relative importance of cultivar choice on the main quality characters of wheat. This list is given in decreasing order of influence of cultivar (or increasing order of influence of the growing environment): hardness, protein quality, α-amylase activity, test weight, protein concentration, kernel size, dockage.

There are strong interactions between cultivar and several other aspects of agronomy. Space does not permit an extensive consideration of these interactions, but some examples of the most marked interactions are mentioned in subsequent sections.

13.3.2 Seeding date

Delaying seeding past the normal seeding date may reduce kernel size slightly (Lauer and Partridge, 1990), probably by reducing the duration of the grain filling period (Darwinkel *et al.*, 1977). Root and stem-base diseases such as take-all (*Gaeumannomyces graminis*) in wheat may also be more severe in early-seeded crops leading to small, shrivelled grain of low test weight. Seeding earlier than the normal date can, in some circumstances, reduce test weight by predisposing the crop to lodging (Stevens, 1986).

Crop nitrogen uptake is less influenced by seeding date than is grain yield, and since grain yield is often reduced by delaying seeding then grain protein concentration tends to increase. Conversely, early seeding may lead to lower protein concentration, along with greater yield (Figure 13.1). Quality characters associated with protein concentration

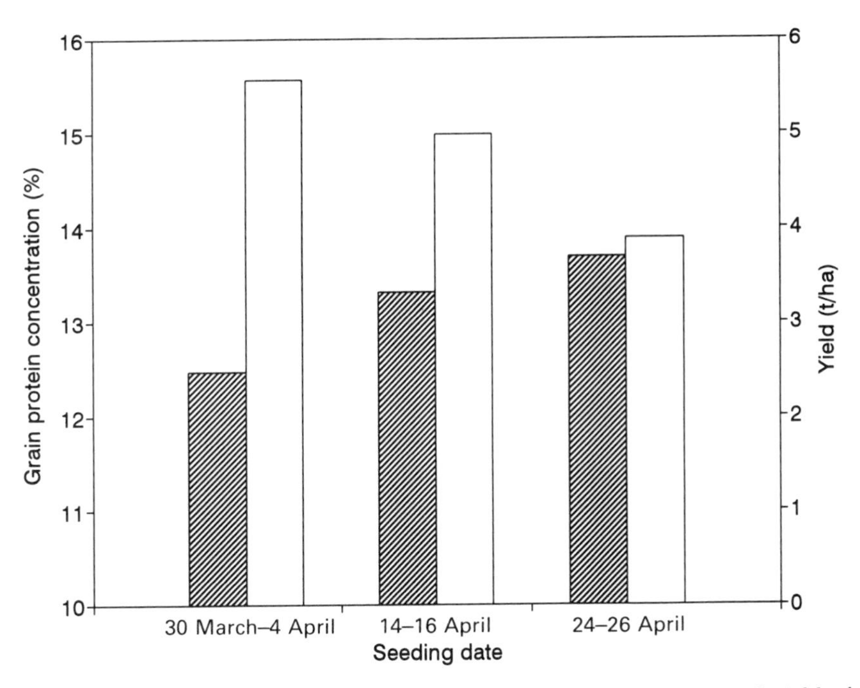

Figure 13.1 Effect of seeding date on grain protein concentration and yield of spring wheat grown in Poland. Hatched bars are grain protein concentration; unhatched bars are yield. Data are means of two cultivars grown in 1984, 1985 and 1986 (Fatyga, 1991).

such as malt extract and bread loaf volume will also change with seeding date. Thus early seeding can increase malt extract and reduce loaf volume. These effects are more marked in dry environments where water stress is more severe in late-seeded crops and appears to follow the general relationship of decreasing protein concentration as yield increases (Section 13.2.2). In moister regions or irrigated cropping systems there is less effect of seeding date (Lauer and Partridge, 1990).

α-Amylase activity can be influenced by seeding date in two ways. First, lodging induced by early seeding is likely to increase pre-harvest sprouting once dormancy has broken (see Section 13.7.4). Second, later seeding can delay the harvest such that the weather may have deteriorated sufficiently to stimulate pre-harvest sprouting. An example of the latter effect contributed to poor quality of many wheat crops in Zimbabwe in 1993. The previous year many farmers had grown the longer-season crop maize in preference to the more usual shorter-season crop soybeans. This led to late seeding of the following wheat crops which, together with other factors, delayed harvest from the usual October into November when the rainy season had begun. Thus many crops sprouted before harvest and suffered from high α-amylase activity as shown by a low Hagberg falling number (P.D. Wells, personal communication).

13.3.3 Seeding rate

There is often little effect of seeding rate on kernel size or test weight, although there is a tendency for very low seed rates to lead to larger kernels and very high seed rates to lead to smaller kernels (Cromack and Clark, 1987). Disease, drought and lodging will exaggerate these tendencies. This is because dense crop canopies arising from high seeding rates, respectively, encourage spread of some foliar or head diseases (e.g. *Septoria* spp. on wheat), use more water from soil reserves and have weaker stems.

Grain protein concentration is also little affected by seed rate, except when crops are droughted. High seeding rates can favour greater water use in the vegetative phase, and lower water use during grain growth which can give rise to higher protein concentration in the grain (Tompkins *et al.*, 1991). The response of grain protein concentration to seeding rate is more marked in maize than in small-grain cereals. This is related to the restricted tillering ability of maize. Thus maize grain yield is very dependent on plant density, whereas in the profuse-tillering small-grain cereals yield varies little across a wide range of seed rates if moisture is plentiful. Therefore the usual negative correlation between yield and protein concentration is found in maize sown at different densities: the higher the plant density the greater the yield and the lower the protein concentration (Ahmadi, 1990). High plant densities of maize also lead to increased kernel breakage susceptibility (Bauer and Carter, 1986) which may be a consequence of lower protein concentration (Ahmadi, 1990).

There is little effect of seed rate on α-amylase activity except at very high seed rates. Lodging is encouraged in the resulting dense crops, and consequently sprouting risk will be greater (see Section 13.7.4).

13.1 NUTRIENT SUPPLY

Crops obtain their nutrients from the soil, from manures and fertilizer and from rainfall. Consideration of the soil as a nutrient source has to take account of numerous factors such as: soil type, soil texture, previous cropping, soil pH, weather. All these influence nutrient reserves, availability and uptake and subsequently quality. Utilization of fertilizer is also strongly influenced by soil and weather. Rain is usually a small, but sometimes significant, influence on quality through its nutrient content.

Nutrients often interact in their effects on quality; a response to one nutrient is often dependent on adequate supply of other nutrients. The experiments referred to in this section mostly have non-limiting supplies of nutrients except the one under investigation, so that interpretation is simplified.

Grain quality is most responsive to nitrogen of all the nutrients, and this section concentrates on nitrogen with brief mention of sulfur, potassium, phosphorus, copper and molybdenum.

13.4.1 Nitrogen

Kernel size and test weight

Despite considerable responses of yield to nitrogen, kernel size and test weight vary much less (see Section 13.2.1). There is, however, a tendency for kernel size to decline with applied nitrogen (Batey and Reynish, 1976). The physiological basis of this effect remains unclear. Drought will exacerbate the effect because crops given more nitrogen deplete soil water faster, but reduced kernel size can occur in crops well-supplied with water. Lodging induced by nitrogen may also contribute to the effect (Section 13.7.4), but unlodged crops also suffer.

The response of test weight to nitrogen is variable – in some years the response is positive, in some negative (Dyke and Stewart, 1992). The reasons for this variability remain unexplained. Although shrivelling can be worse in grain from crops given more nitrogen, test weight may not be affected (Bayles *et al.*, 1978).

Protein concentration

Nitrogen is a major component of protein, and nitrogen supply to cereal crops is the principal factor influencing grain protein concentration. Any factor which increases the supply of nitrogen to the crop will increase protein concentration, except at very low levels of fertility. Crop rotation is important, and the greater soil nitrate available after fallow or after crops which leave nitrogen residues increases grain protein concentration compared with a continuous cereal (McGuire *et al.*, 1988). This comparison is, however, complicated by fertilizer nitrogen. If comparisons are made at the optimum rate of nitrogen fertilizer for yield then the continuous cereal will give a higher protein concentration and lower yield than where the cereal is grown after a non-cereal crop (Vaidyanathan *et al.*, 1987). This is possibly a consequence of the reduction in disease in cereal crops after fallow and non-cereal crops, which in turn leads to greater yield and a dilution of grain protein. This may therefore be a further expression of the negative protein concentration:yield relationship referred to previously (Section 13.2.2).

The response of grain protein concentration to soil nitrogen is sigmoid rather than linear (Figure 13.2). Initial increments of nitrogen have little effect or decrease protein concentration, because yield is increased at a

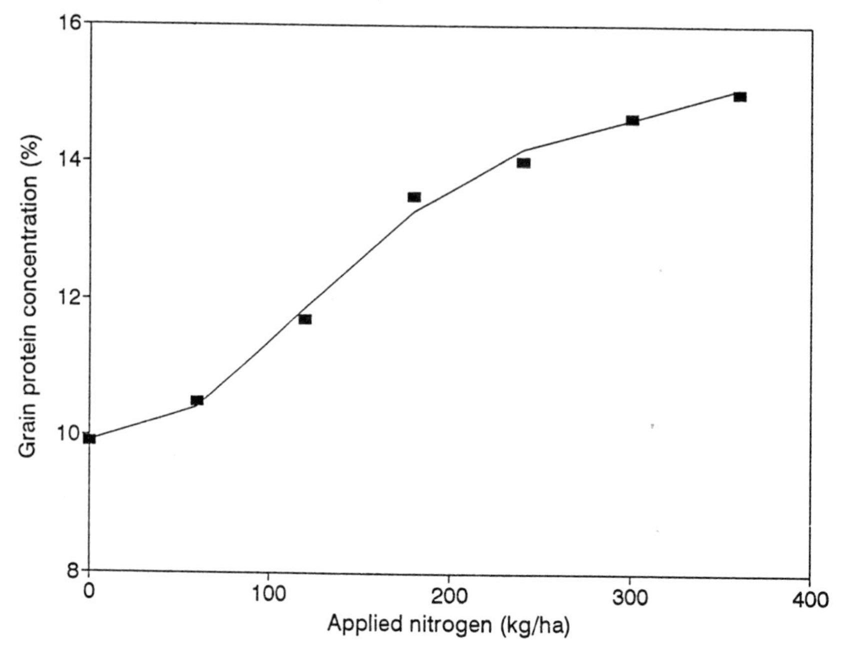

Figure 13.2 The response of grain protein concentration (% of dry matter) of winter wheat to application rate of nitrogen fertilizer in Shropshire, UK in 1986. Data are means of four cultivars (Gooding, 1988).

greater rate than nitrogen uptake. Thereafter, protein concentration increases linearly with increasing nitrogen supply until the optimum for yield is reached. After this point increases in protein concentration are smaller, but can continue. The response to nitrogen is strongly influenced by water supply and this is discussed in Section 13.5.

The timing of nitrogen supply also has considerable influence on protein concentration. In essence, the earlier the nitrogen is supplied the greater the yield and the lower the protein concentration and *vice versa*. Applications of fertilizer at the beginning of kernel growth have little or no effect on yield, but have the greatest effect on protein concentration. The use of nitrogen loss inhibitors can extend the period over which nitrogen is available, as well as reducing losses, and therefore can alter the effective timing of nitrogen leading to greater protein concentration (Tsai *et al.*, 1992). In many climates there is inadequate natural soil moisture for such late applications to be taken up and irrigation or foliar application is essential to ensure uptake by the plant. The considerable effect of timing of nitrogen application is clearly illustrated by data from Finney *et al.* (1957) on foliar urea application to wheat (Figure 13.3).

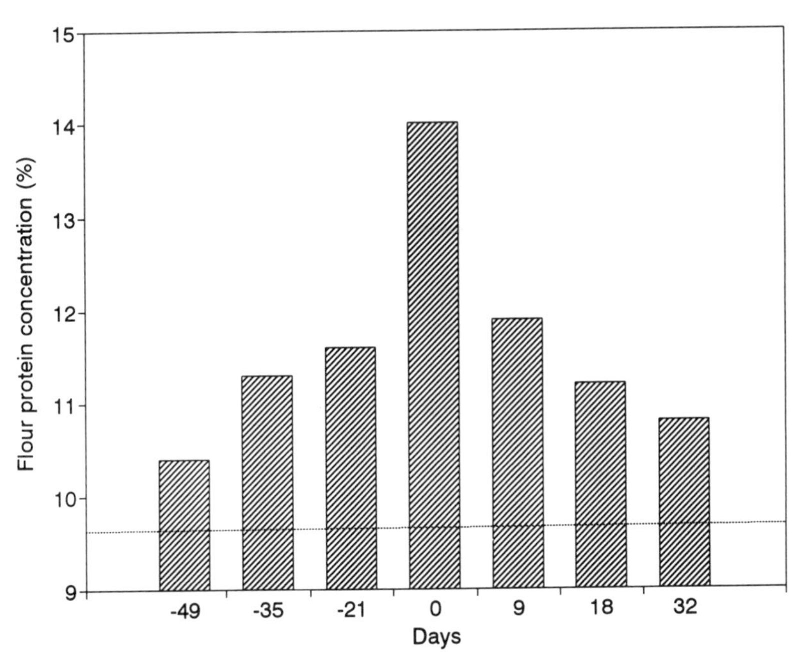

Figure 13.3 The response of flour protein concentration to 56 kg N/ha applied as urea solution at different times, either before or after flowering (zero days represents the date of flowering) to Pawnee wheat in Kansas, USA, in 1950. The broken line is the protein concentration with no urea applied. Data from Finney *et al.* (1957).

Early applications at seeding or stem extension can increase yield since they are able to influence the size of the leaf canopy and thereby the supply of photosynthate for increasing head number and grain number per head, i.e. the two components of the sink. Late applications in early grain growth cannot increase the size of the leaf canopy because tillering and leaf expansion have ceased. Duration of the canopy can be greater, but the sink size is frequently the limiting factor for yield. The effects of timing on protein concentration may, therefore, result from dilution of the same quantity of nitrogen in a smaller or larger mass of grain.

Since protein concentration of flour is a major determinant of baking quality, bread loaf volume and other bread quality characteristics also increase as nitrogen supply to wheat crops increases, and malt extract of barley decreases.

Vitreous kernels

The effects of nitrogen on grain protein concentration are parallelled by the effects on the proportion of vitreous kernels, extent of vitreousness and hardness of grains. In maize, for example, kernels become more vitreous as the concentration of the major grain storage protein (zein) rises in response to increasing nitrogen fertilizer application. Susceptibility to breakage during handling and processing is lower for the harder and more vitreous high protein kernels (Tsai *et al.*, 1992). Similarly, in rice the higher grain protein concentration produced by delaying nitrogen fertilizer application from panicle initiation to heading leads to fewer non-vitreous (chalky) kernels and greater recovery of whole kernels (head rice) after milling through reduced breakage (Seetanun and De Datta, 1973). These effects are probably economically most important for durum wheat, since the grain price for durum is very dependent on percentage of non-vitreous kernels (yellow-berry).

Protein quality

The supply of nitrogen influences the balance of protein fractions which accumulate in grain. For wheat, barley and maize the prolamin protein fractions increase most in response to increasing supply of nitrogen. This protein has a low concentration of lysine (an essential amino acid for non-ruminants) and therefore the nutritional value of the protein for non-ruminants declines. The total protein concentration increases with nitrogen supply as discussed earlier, and there is a net increase in lysine concentration in the grain dry matter (Byers *et al.*, 1978). Thus the effect of nitrogen supply on nutritional quality of cereal grains is of relatively little significance, and this is shown in feeding experiments (Fuller *et al.*, 1989). Nevertheless, the proportion of different amino acids has to be taken into account in formulating animal feed. A supplementary source of protein richer in lysine will be needed with grain grown at a high nitrogen supply than with grain grown where less nitrogen is available in order to achieve the same amino acid balance.

For food processing quality the greater total protein concentration from enhanced supply of nitrogen is usually of more consequence than the change in protein quality. Nevertheless, there are some circumstances in which baking quality of wheat flour may not be as good as expected from the high grain protein concentrations produced by nitrogen fertilization (Tipples *et al.*, 1977). In these situations, the proportion of ω-gliadins in the protein have been found to increase (Timms *et al.*, 1981). These effects of nitrogen supply are closely linked with sulfur supply and are discussed further in Section 13.4.2

α-Amylase activity

Nitrogen supply can influence α-amylase activity derived both from sprouting and from pre-maturity formation in the absence of sprouting. Nitrogen can exacerbate lodging (Section 13.7.4) and in doing so can predispose crops to pre-harvest sprouting (Brun, 1982). Grain from crops adequately supplied with nitrogen may also sprout more rapidly once dormancy has broken irrespective of any effect of lodging (Morris and Paulsen, 1985). Nitrogen can also reduce α-amylase activity formed before maturity, although the way in which this reduction occurs is not clear (Kettlewell and Cooper, 1993). The magnitude of any reduction in α-amylase from nitrogen is small compared with differences between seasons and is usually associated with elevated yield and protein concentration.

13.4.2 Sulfur

Sulfur in the soil behaves in a similar way to nitrogen in many respects, but one difference is that deposition to the soil either directly from the atmosphere or dissolved in rain, and also absorption by plants in the gaseous form, can be substantial and a major source of this nutrient in industrialized regions of some countries (Terman, 1978). Reduction of atmospheric pollution can alter crop sulfur status and quality unless other sulfur sources are substituted.

Few studies of sulfur nutrition have included measurement of kernel size or test weight, but thousand kernel weight has been assessed. One study involving 38 field experiments in France showed that thousand kernel weight could be increased or decreased by sulfur fertilizers, but that decreases occurred more often (Thevenet and Taureau, 1987). These decreases were related to corresponding increases in number of heads per unit area without a change in kernel number per head and may reflect yield component compensation. It should be borne in mind that changes in thousand kernel weight do not necessarily imply corresponding changes in kernel size or test weight (Bayles, 1977). Ramig *et al.* (1975) indicated that effects of sulfur on test weight were small despite a large yield response.

Sulfur supply has relatively little effect on protein concentration, but considerable influence on protein quality. In wheat the ω-gliadins decrease in proportion as sulfur supply increases. These proteins are low in the sulfur-containing amino acids cysteine and methionine. Bread loaf volume increases with sulfur supply (Figure 13.4) and this is associated with greater dough extensibility. Such changes are attributed to an increase in sulfur for forming disulfide bridges involved in dough extensibility (Randall and Wrigley, 1986).

The responses of quality to sulfur supply are closed related to responses to nitrogen since it is the ratio of sulfur to nitrogen in the grain which determines the protein quality. Hence it is possible to induce the effects of sulfur deficiency by excessive application of nitrogen fertilizer. The lack of response of baking quality to very high protein concentrations described in Section 13.4.1 is related to an increase in N:S ratio in the grain.

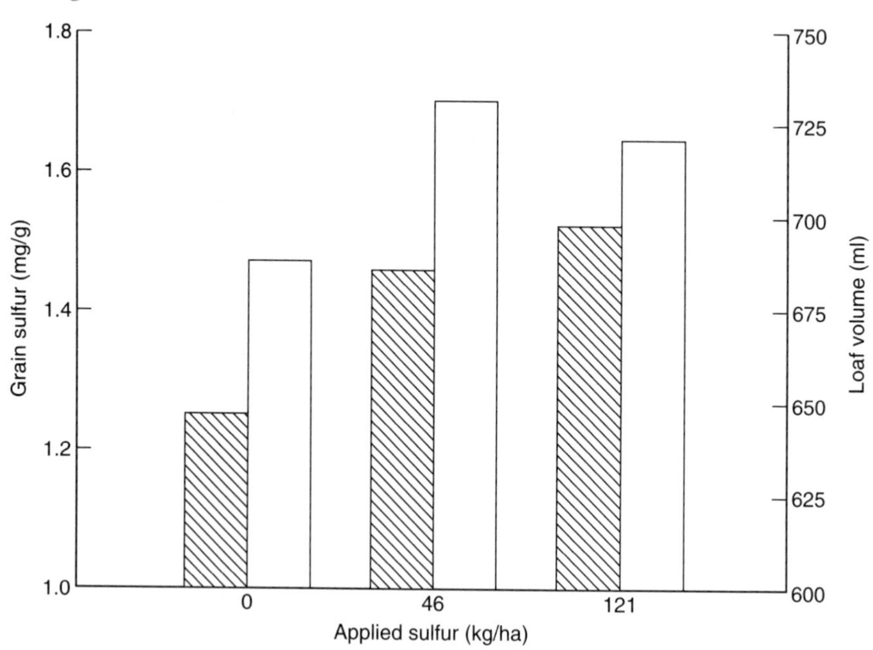

Figure 13.4 Effect of sulfur fertilizer on grain sulfur concentration and bread loaf volume (RMT method) of winter wheat in Germany in 1990. Hatched bars are grain sulfur concentration; unhatched bars are loaf volume. Data are means of three sites (Schnug *et al.*, 1993).

13.4.3 Other nutrients

Potassium

Potassium can influence several aspects of grain quality. Kernels are larger and test weight greater as potassium supply increase (Usherwood, 1985). Protein concentration may increase through greater uptake of nitrogen from soil and also greater retranslocation from leaves. If the effect of potassium on yield is large, however, no effect or even a small decrease in protein concentration can occur. Cereal stems are thicker and stronger when adequate potassium is supplied (Usherwood, 1985), so that the risk of lodging and pre-harvest sprouting is reduced.

Phosphorus

Kernel size increases with phosphorus supply, but other effects on quality may be detrimental. There is not usually an increase in protein concentration, and there may be a decrease if the yield response is large (Finck, 1992). Crops well-supplied with phosphorus may be at greater risk of lodging and pre-harvest sprouting as a result of decreased lignin synthesis in the stem.

Copper

Copper is essential for satisfactory grain filling, and grain from deficient wheat crops has low test weight (Reith, 1968). Protein synthesis depends on an adequate copper supply and thus deficient wheat grain also has low protein concentration (Alloway and Tills, 1984). Copper deficient crops are also more susceptible to lodging, with possible consequences for pre-harvest sprouting.

Molybdenum

A deficiency of molybdenum in maize prevents the onset of dormancy, so that pre-maturity sprouting occurs; this is known as vivipary. A similar phenomenon has also been shown to occur in molybdenum-deficient wheat, manifested as reduced dormancy (Cairns and Kritzinger, 1992).

13.5 WATER

Since water is fundamental to crop growth it is not surprising that it can influence many aspects of grain quality. Crop water supply in dryland agriculture is dependent on the soil available water capacity (determined by soil depth and texture) as well as rainfall. In areas with very low rainfall previous agronomic practices, e.g. fallowing, will have a considerable influence on soil water reserves for the current crop.

Water deficits before anthesis usually have only small effects on kernel size and test weight because of yield component compensation, but during grain growth have marked effects on both kernel size and test weight (Figure 13.5). During the early period of grain growth, when endosperm cell division is taking place, water stress can reduce endosperm cell number which will limit kernel size.

After cell division is complete, when starch and protein are being deposited in the endosperm, water deficits can lead to grain shrivelling (Brocklehurst *et al.*, 1978). This occurs through inhibition of photosynthesis, largely by inducing stomatal closure, and thereby assimilate supply to the grains is restricted. Small water deficits may have little effect because remobilization of stem carbohydrate reserves can help maintain assimilate supply. One way in which a pre-anthesis deficit might exacerbate loss of grain quality is by reducing the accumulation of stem carbohydrate reserves, so that the plant is less able to compensate for post-anthesis water stress.

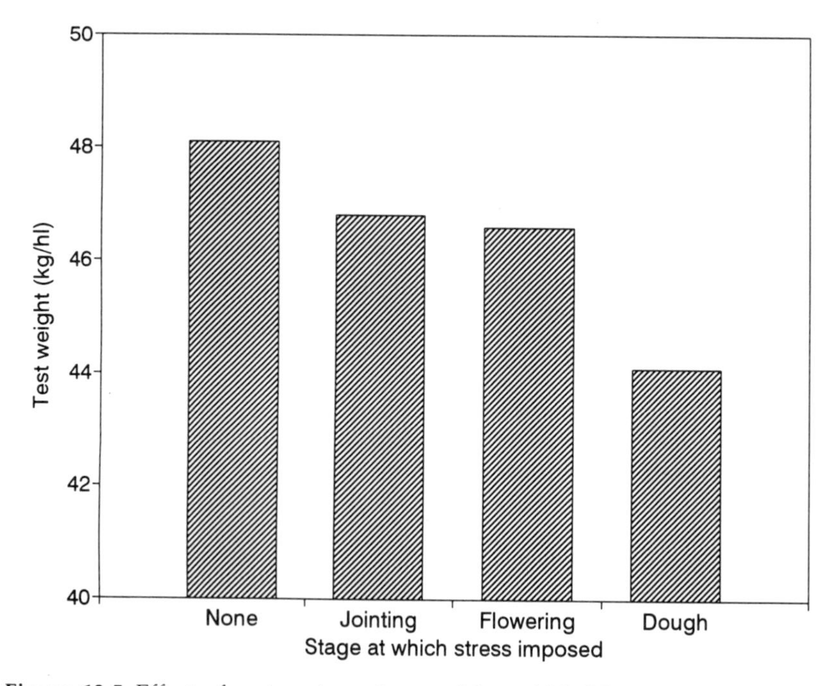

Figure 13.5 Effect of water stress imposed by withholding irrigation on test weight of Maricopa spring wheat in Arizona, USA. Data are means of 1966 and 1967 (Day and Intalap, 1970).

Water deficits also have a considerable effect on grain protein concentration. Generally, grain protein concentration increases with increasing water deficit because carbohydrate deposition in the grain is reduced, but protein deposition is less affected. Yield is lower and although nitrogen uptake from soil is reduced, remobilization of nitrogen from vegetative parts of the plant is enhanced as senescence accelerates in response to drought. Indeed the combined effects of nitrogen and water supplies can account for a large proportion of the variation in grain protein in dryland wheat production. Smika and Greb (1973) studied relationships between grain protein concentration and soil nitrogen and water supply in wheat crops grown over a period of up to 18 years without nitrogen fertilizer at two locations in Nebraska and one in Colorado. They found that 86% of the variation in protein concentration could be accounted for by soil nitrate-nitrogen and available soil water, both determined at seeding.

Although maize protein concentration increases with a drought, oil concentration decreases. Another effect of reduced water supply in maize is a lower kernel breakage susceptibility, assumed to result from the greater protein concentration (Bauer and Carter, 1986). In small-grain cereals pre-anthesis drought can stimulate production of late tillers which ripen late (Talukder *et al.*, 1987). This can lead to either immature kernels at harvest, or increased risk of pre-harvest sprouting if the farmer waits for the late tillers to ripen.

Conversely, excess rainfall can reduce grain protein concentration by leaching soil nitrate and preventing uptake. Waterlogging early in a cereal crop's development restricts root growth to upper soil horizons and can indirectly amplify subsequent effects of drought by limiting extraction of soil water. Soil compaction has a similar effect.

Excess water supply as rain also has dramatic consequences for quality by inducing sprouting and high α-amylase activity in cereal grains which have broken dormancy before harvest. The reduction in Hagberg falling number has been quantified and incorporated into a predictive model by Karvonen and Peltonen (1991). These effects of rain are also dependent on head type (lax or club) and presence of awns. Club heads and awns are both associated with faster water uptake and more rapid sprouting (King and Richards, 1984).

13.6 TEMPERATURE

Temperature is negatively correlated with rainfall in many environments and effects of low rainfall often cannot be clearly separated from effects of high temperature. Nevertheless, temperature does have independent effects on grain quality. Temperature is the fundamental driving force behind kernel development, and stages of kernel maturity can be quantified and predicted using accumulated temperature. Thus the higher the mean temperature during kernel growth, the shorter the duration. The rate of kernel growth is also faster at high temperatures, but the increase in rate does not adequately compensate for the reduction in duration (Sofield *et al.*, 1977a). The net effect is smaller kernels in response to high temperature (MacNichol *et al.*, 1993).

Nitrogen accumulation in grain is much less responsive to temperature than carbohydrate accumulation with the consequence of greater protein concentration in the grain as temperature rises (Sofield *et al.*, 1977b). There is an optimum temperature, however, beyond which protein concentration may decrease. Smika and Greb (1973) found the optimum temperature to be 32°C during the period 15 to 20 days before maturity in Nebraska and Colorado (Figure 13.6). Benzian and Lane (1986) also showed that the best prediction of wheat grain protein concentration from weather data in the cool, moist environment of the UK was from air temperature during early grain filling. In their study solar radiation during grain filling reduced protein concentration, presumably by enhancing photosynthesis and carbohydrate deposition in grain. Global

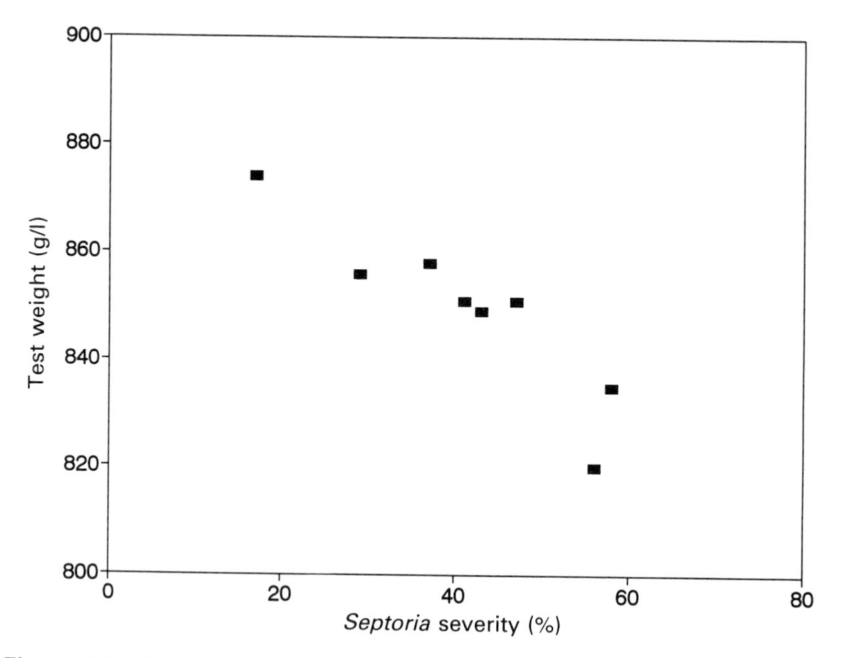

Figure 13.6 Relationship between wheat grain protein concentration and average maximum air temperature between 15 and 20 days before maturity. Data from 21 years in Nebraska, USA and 18 years in Colorado, USA (Smika and Greb, 1973).

warming from rising carbon dioxide concentrations in the atmosphere might be expected to enhance protein concentration, but the elevated carbon dioxide can increase yield and reduce grain protein (Thompson and Woodward, 1994). Thus the effect of the carbon dioxide might counteract any response to the associated temperature rise. Very high temperatures (above 35°C) during kernel growth are detrimental to protein quality of wheat by reducing glutenin, but not gliadin synthesis. This results in weaker dough and may lead to reduced loaf volume (Blumenthal *et al.*, 1993).

The duration of grain dormancy is shorter with high temperature during early kernel growth. Thus the problem of dormant barley for malting is much worse following cool weather during early kernel development. This effect of temperature can be successfully used to predict the end of dormancy and the risk of pre-harvest sprouting damage in wheat (Couvreur *et al.*, 1993). The extent of sprouting once dormancy has broken will depend on the duration of wetness of the head, and temperature will clearly affect the rate at which the head dries. The development of pre-maturity α-amylase in wheat may be dependent on grain drying rate (Section 13.2.3) and temperature may play a role in determining grain drying rate.

Temperatures during kernel growth which are low enough to kill the plant will have detrimental consequences for grain quality. If maize plants are killed before the kernels are mature the test weight and oil content will be reduced. If the kernels are also frozen then 'soft corn' results with a high moisture content and a low test weight. Frost can also be damaging to quality of small grain cereals. In wheat, temperatures below −3°C reduce test weight, protein concentration and loaf volume, but only at grain moisture above about 45% (Preston *et al.*, 1991).

13.7 CROP PROTECTION

13.7.1 Weeds

Weed seeds can clearly be a major contributor to dockage in grain, but their influence on other aspects of grain quality is less marked. This is because most of the competition between weeds and crop takes place before grain growth and most of the effect is through reductions in head number and kernel number per head. Weeds which exert most of their competitiveness late in the life of the cereal crop, such as *Galium aparine* in small grain cereals, can reduce kernel size and specific weight through a restriction of photosynthesis by shading or depletion of soil water. Weeds compete for soil nitrogen and may reduce grain protein concentration as a result, and shading from weeds may weaken stems and induce lodging, leading to reduced quality.

Cereal plants growing from seed shed by a previous crop can lead to contamination of the grain from the current crop. This can reduce several aspects of quality, but particularly Hagberg falling number since high α-amylase grain can have a larger effect on Hagberg than expected from the proportion of grain involved. Thus if the previous crop was a cereal species or cultivar with high α-amylase activity then there is a risk of reduced falling number in the next crop (Garstang, 1993). Furthermore, self-sown cereal plants often mature earlier than the crop in which they are a weed and are at greater risk of pre-harvest sprouting in wet weather. Even if the self-sown plants are of the same cultivar as the current crop, there is consequently still a risk of reduced falling number. Clearly, there are several factors which will have considerable impact on the incidence of self-sown plants: the extent of grain shedding before harvest of the previous crop; the losses from the combine-harvester; the type of and interval between stubble cultivations after the previous crop together with the quantity of rain, which will determine germination of shed grain for subsequent killing; the use of herbicides on stubble; the depth and completeness of burial of shed grain, both germinated and dormant, by ploughing.

Herbicides usually do not have direct effects on grain quality, unless they are phytotoxic. One exception is pre-harvest application to small grain cereals to kill weeds after grain filling has finished. These herbicide applications (e.g. of glyphosate) enhance the drying rate of wheat grain (Darwent *et al.*, 1994) and can thus allow harvesting to take place earlier without incurring extra drying costs. This could be of benefit in situations where there is a risk of deteriorating weather leading to pre-harvest sprouting. These herbicide applications will also kill late-formed tillers which will normally be green when the main shoots and early-formed tillers are mature and ready to harvest. Late tillers may develop after drought is followed by rain or irrigation, and in crops which lodge early and expose the stem bases to extra light. They lead to green kernels in the harvested grain, and pre-harvest herbicide applications can improve the appearance of the grain by killing green kernels before the time of harvest.

13.7.2 Disease

Diseases can have serious consequences for grain quality, and control by resistant cultivars, cultural means or fungicides is important in preserving grain quality. There are also instances of effects of fungicides on quality in the absence of disease. There may, however, be harmful effects of disease control in some circumstances.

Diseases may be caused by bacteria, viruses or fungi, but fungal diseases probably have the major impact on quality and will be the main examples used. This section is sub-divided according to which part of the plant shows the main symptoms of disease: seedling, root and stem, foliage or head.

Seedling diseases

Seedling diseases which severely reduce crop establishment (e.g. *Fusarium* spp.) could indirectly increase kernel size in the same way as lowering the seeding rate (Section 13.3.3). Otherwise these diseases are unlikely to affect quality. Some of the chemicals used as fungicidal seed dressings to control these diseases also possess growth regulator activity which can have repercussions on quality irrespective of any disease control. Triadimenol and fuberidazole mixture abolishes subcrown internode extension leading to deeper positioning of the stem apex in the soil and potentially greater insulation from frost (Davies and Kettlewell, 1987). This may in turn prevent a reduction in plant number from frost damage (Webb and Stephens, 1936) and possible changes in kernel size. These chemicals also reduce lodging and may lessen the risk of pre-harvest sprouting in consequence.

Root and stem diseases

Root diseases, such as take-all (*Gaeumannomyces graminis*), indirectly influence quality by restricting water and nutrient uptake. The former probably has greater impact on quality since the main symptoms shown in grain from take-all infected plants are reduced size, test weight and shrivelling (Manners and Myers, 1981).

Similarly, stem diseases, such as eyespot (*Pseudocercosporella herpotrichoides*), interfere with water and nutrient transport and the grain symptoms are similar to those of root diseases. A further effect on quality from diseases like eyespot arises from stem weakening and enhanced lodging risk. This may reduce test weight and stimulate pre-harvest sprouting. Any agronomic factor which influences infection by these diseases will also indirectly affect grain quality. Thus early seeding and seeding following a previous cereal crop can exacerbate these diseases and reduce quality. Conversely, fungicidal control or, for take-all, control by ammonium fertilizers can prevent excessive loss of quality. Nitrogen fertilizer also can help offset effects of take-all, by reducing the need for a large root system and encouraging new roots to grow (Huber, 1981).

Foliage diseases

The responses of grain quality to foliage diseases are dependent on the stage of development of the crop when the disease occurs. The effects of foliage disease are largely analogous to those of water deficit, because the primary result is reduced supply of photosynthate. There is unlikely to be much effect on kernel size and test weight of disease during tillering and early stem extension because of yield component compensation. During the later stages of stem extension it is possible that accumulation of stem carbohydrate reserves could be reduced (Carver and Griffiths, 1981). This may lead to grain shrivelling, especially if the crop has greatly reduced photosynthesis during grain filling, from disease or water deficit.

Late-season foliage diseases, such as rust (*Puccinia* spp.) or mildew (*Erysiphe graminis*), with reduce kernel size if the attack occurs during the endosperm cell division phase of grain growth, and will induce shrivelling if attack occurs after cell division has ceased. Test weight will often suffer, as illustrated in Figure 13.7. Grain protein concentration is less sensitive to disease than kernel size and test weight, and both

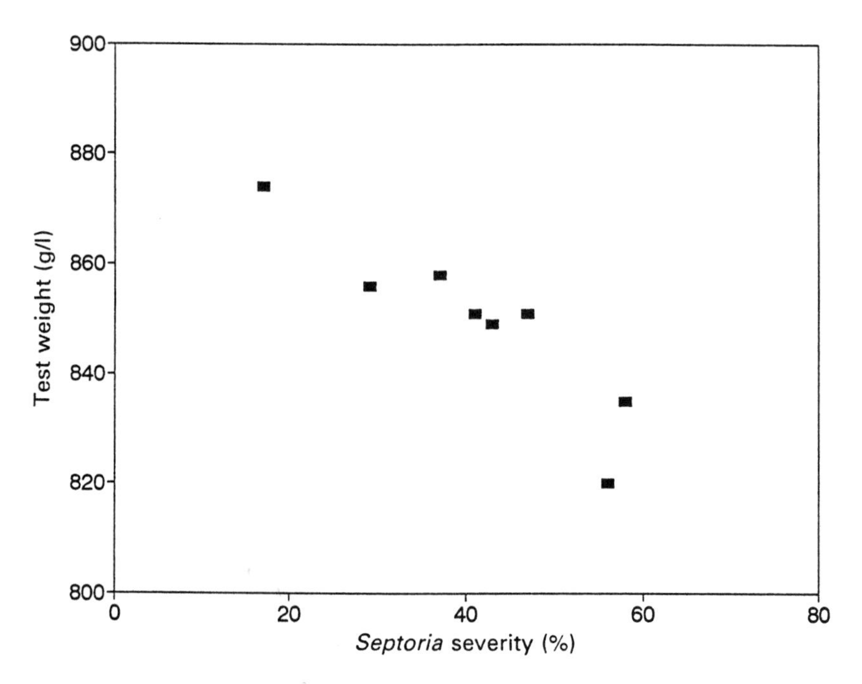

Figure 13.7 Relationship between test weight and severity of *Septoria tritici* on foliage of wheat at grain dough stage. Data are from cultivar Florida 302 and are means of two sites in Arkansas, USA in 1989 and 1990 (Milus, 1994).

reductions and increases have occurred depending on the disease involved (Gooding *et al.*, 1994). *Septoria* spp. have a detrimental effect on grain carbohydrate accumulation but may have little effect on nitrogen retranslocation and may therefore increase grain protein concentration. In contrast, mildew (*E. graminis*) may reduce protein concentration by reducing nitrogen retranslocation from leaves and inducing greater loss of nitrogen as ammonia.

Wheat protein quality can be reduced by foliage disease and this may be reflected in poorer bread quality. In contrast, pre-maturity α-amylase activity in wheat can be reduced by disease, which leads to higher Hagberg falling number. This effect only appears to occur in cool, wet weather during the later stages of grain growth, when grains dry slowly. The response to disease may be mediated through faster grain drying correlated with faster crop senescence. Such an effect ought to be of benefit to bread quality, but changes in protein quality may have an overriding influence so that bread quality still suffers (Gooding *et al.*, 1994).

Control of foliage diseases by cultural or chemical means is clearly important in preventing reduced quality. In areas where the economics of cereal production allow fungicides to be used, there is little justification, however, for altering fungicide use from that which gives an economic yield response, since most of the beneficial effects of disease control on quality accompany yield increases.

Head diseases

Some seedborne diseases only become manifest after heading, such as bunt (*Tilletia caries*) of wheat. They destroy the grain on infected plants and may also contaminate sound grain from uninfected plants leading to undesirable taint or colour. Photosynthesis in the head is a substantial contributor to grain carbohydrate in small grain cereals and thus head diseases which attack the glumes, such as *Septoria* spp., will reduce kernel size and specific weight in the same way as foliage diseases. Other head diseases, such as *Cladosporium* spp. on wheat, may have their main effects through discolouring grain which leads to poor flour colour (Baker *et al.*, 1958). In these circumstances, late fungicide use for preventing loss of quality might be justified in the absence of a yield response.

13.7.3 Pests

There are numerous pests of cereal crops which kill plants or tillers early in development, for example, black cutworm (*Agrotis ipsilon*) of maize, wheat bulb fly (*Delia coarctata*). These may increase kernel size in the same way as reducing the seeding rate (Section 13.3.3), but will otherwise have little effect on quality.

Pests of roots (e.g. cereal cyst nematode, *Heterodora avenae*) can limit water uptake and lead to shrivelled grain in the same way as root pathogens. Some such pests may do sufficient root damage to induce lodging with consequent risk of reduced test weight, e.g. northern corn rootworm (*Diabrotica longicornis*) on maize. Some pests damage the stem, such as European corn borer (*Ostrinia nubilalis*) on maize, or feed on sap, e.g. grain aphid (*Sitobion avenae*) on wheat. These pests reduce kernel size and test weight by preventing assimilates from reaching the developing kernels and may also reduce protein quality. Figure 13.8 illustrates the consequence of aphid infestation on SDS sedimentation volume of wheat grain – a measure of the total quantity of proteins suitable for bread-making. Since grain protein concentration in this study was unchanged by aphids, the effect on SDS volume must reflect a reduction in protein quality.

Pests which damage foliage are only likely to affect quality if they attack late in development, as for foliage diseases. Pests which directly attack developing kernels will clearly cause considerable damage to

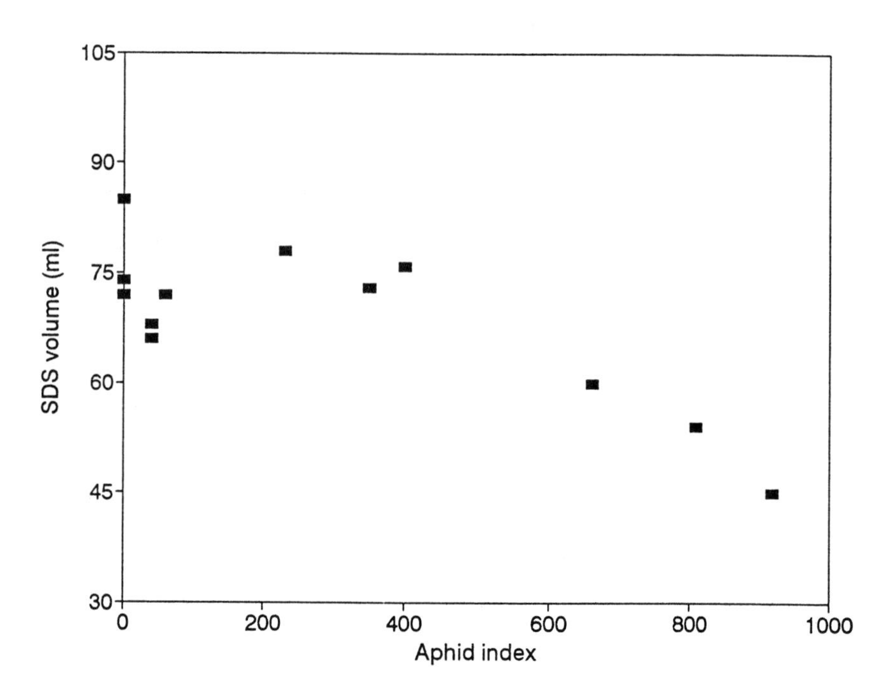

Figure 13.8 Relationship between SDS sedimentation volume of Maris Widgeon wheat and cumulative aphid index (*Sitobion avenae*). Data from Hampshire, UK in 1978 (Lee *et al.*, 1981).

quality. For example, the larvae of orange blossom midge of wheat (*Sitodiplosis mosellana*) feed on immature kernels and not only cause shrivelling, but also increase α-amylase activity (Helenius and Kurppa, 1989).

13.7.4 Lodging

Lodging is induced by wind, rain, irrigation or hail, but cereal crops can be predisposed to lodge by excessive seeding rates, early seeding, soil loosening by frost, nitrogen fertilizer, weeds, root or stem diseases and pests (Pinthus, 1973). Cultivars differ greatly in their resistance to lodging and this can be a major factor indirectly influencing quality. The effects on grain quality of lodging mediated by most of these factors have been mentioned in previous sections.

The two main effects of lodging on grain quality are reduction in kernel size (Figure 13.9) and pre-harvest sprouting. Lodging reduces

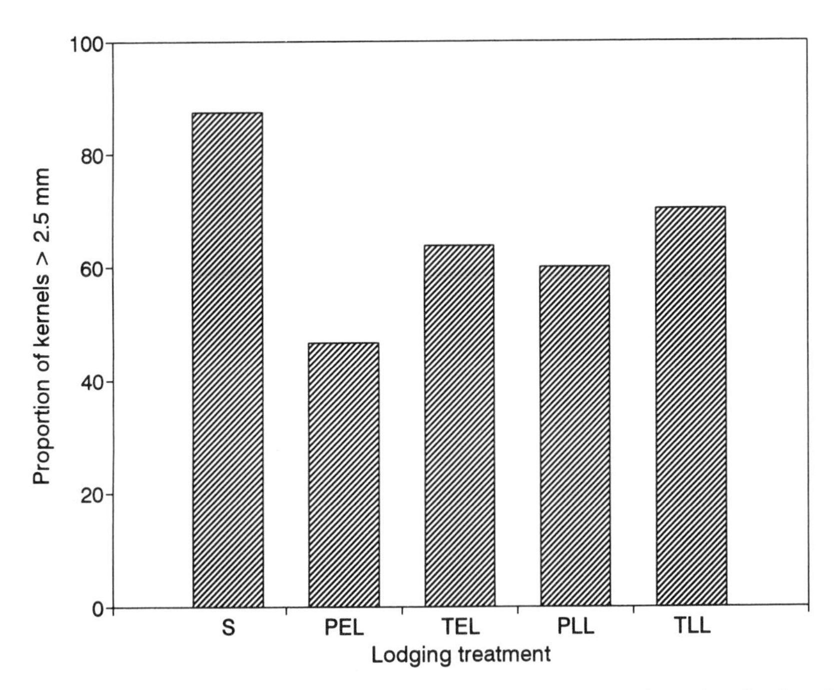

Figure 13.9 Effect of artificial lodging on kernel size of spring barley in Cambridge, UK. Data are means of eight cultivars (Stanca *et al.*, 1979). S, supported; PEL, permanent early lodging (from anthesis to maturity); TEL, temporary early lodging (from anthesis until 7 days later); PLL, permanent late lodging (from 20 days after anthesis until maturity); TLL, temporary late lodging (from 20 days after anthesis until 7 days later).

photosynthesis by shading, and if lodging occurs during early kernel growth kernel size may be smaller through lower endosperm cell number. Lodging occurring later in kernel growth will lead to shrivelling and reduced test weight. Lodging may also lead to green kernels at harvest by allowing more light to reach the stem base which can induce tiller production to restart. Protein concentration frequently increases in lodged crops because lodging reduces carbohydrate accumulation, but not nitrogen retranslocation (Stanca *et al.*, 1979).

The risk of pre-harvest sprouting can be greater in lodged crops as a result of contact between moist soil and the heads. Thus once dormancy has broken the grain will readily germinate if the soil is moist enough. Also rain on the heads will evaporate more slowly than from heads on a standing crop, where air can circulate freely. α-Amylase activity thus rises and Hagberg falling number drops as lodging becomes more severe.

Clearly, reduction of lodging with chemical plant growth regulators, such as chlormequat or ethephon, will be of benefit to quality. These chemicals may, however, have small, detrimental effects themselves and ideally should only be used where there is a clear risk of lodging. Chlormequat can reduce kernel size and test weight slightly (Stevens, 1986), and ethephon may stimulate late tillering leading to green kernels in harvested grain.

13.8 HARVEST

13.8.1 Time of harvest

Cereal crops can be combine-harvested and successfully threshed at very high moisture contents – above 50%. At this stage yield will be lost because maximum dry weight will not have been achieved, and in any case most farmers will want to avoid harvesting grain at a moisture which necessitates costly artificial drying. One exception is grain destined for preservation by anaerobic storage or acid treatment (see Chapter 14), but such grain is only suitable for animal feed. Even if grain harvested at high moisture is then dried, test weight will be reduced if the grain is immature at harvest. Another disadvantage of immature kernels from an early, high moisture harvest in rice is that the immature kernels are unlikely to mill well and will lead to a lower recovery of head rice (Figure 13.10). This is aggravated by increasing quantities of nitrogen fertilizer. Similarly, maize harvested at high moisture is soft and easily punctured.

Conversely, harvesting at too low a moisture can be detrimental to quality because of brittleness of grain leading to cracking and breakage. Thus there is an optimum moisture range for combine-harvesting which

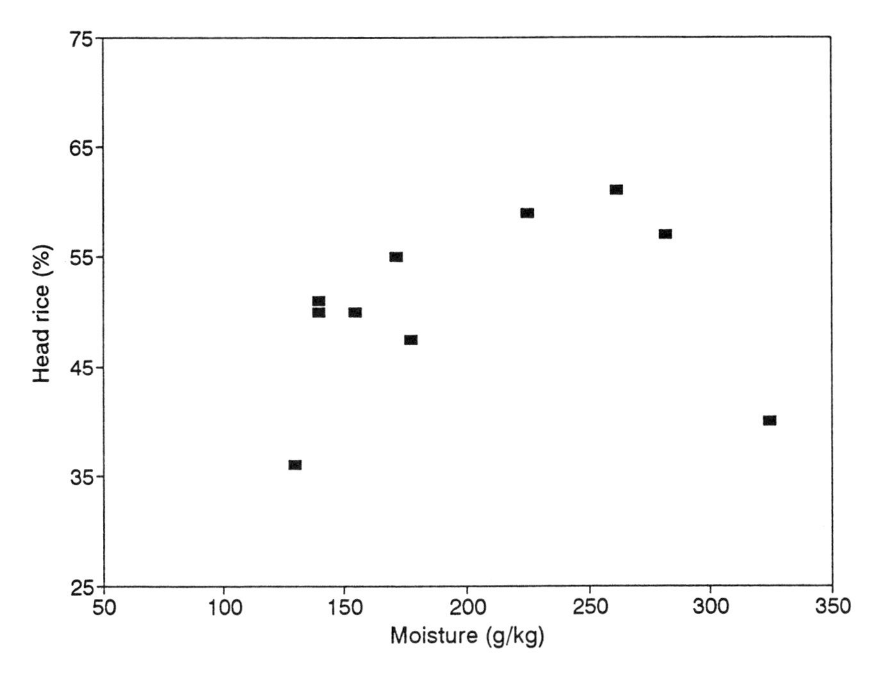

Figure 13.10 Relationship between harvest grain moisture and head rice recovery after milling for different application rates of nitrogen. Data for cultivar L202 in California, USA in 1987 and 1988 (Jongkaewwattana *et al.*, 1993).

maximizes recovery of undamaged kernels. For example, with rice (in California) this is 20–27% depending on cultivar (Jongkaewwattana *et al.*, 1993).

If crops are left in the field after they have dried to a suitable moisture for harvest they may be at risk of weather damage. Repeated wetting from rain reduces test weight, stimulates pre-harvest sprouting and encourages mould growth and discoloration. The longer the crop is left the worse the damage will be. There are, however, interactions of delayed harvest with cultivar dormancy, head type, lodging, nitrogen fertilizer and harvest method on sprouting damage. There are considerable genetic differences in sprouting resistance which mean that some cultivars, e.g. white-grained wheats, can tolerate little delay in harvest if wet weather occurs. Conversely, red-grained wheats can be left for much longer periods after maturity; several weeks for some cultivars despite intermittent wet weather. Earlier sections have covered the influence of head type (13.5), lodging (13.7.4) and nitrogen (13.4.1), all of which exacerbate the sprouting response to delayed harvest. Harvest method is considered in the next section.

13.8.2 Harvest technique

If crops are cut and left to dry in the field, i.e. windrowed or swathed, they are at greater risk of sprouting damage if wet weather occurs. This is because, like lodged crops, they are slower to dry again if wetted by rain compared with standing crops (Clarke, 1983). An alternative technique to achieve faster drying and promote uniformity of maturity is to spray with a herbicide after maximum kernel weight is achieved. This technique has the benefit of allowing the crop to redry faster should rain occur (Darwent *et al.*, 1994).

The combine-harvester settings are important in determining mechanical damage and, in particular, the higher the cylinder speed and the narrower the cylinder concave adjustment the greater the risk of kernel cracking and breakage. Rotary combine harvesters can reduce kernel breakage at low moisture compared with conventional combine harvesters (Herum, 1987).

13.9 POST-HARVEST

It is possible to improve grain quality after harvest by either cleaning or gravity separation. Removal of small, shrivelled kernels with an appropriate size of screen in a grain cleaner often increases test weight of the remaining grain (McLean, 1987). Sometimes, however, removal of small weed seeds and other components of dockage can be detrimental to test weight since they may have been filling the interstices between kernels and enhancing the bulk density. The exact circumstances in which test weight can be increased by cleaning are not clearly defined.

Gravity separators, which remove less dense kernels, can not only increase test weight of the remaining grain, but also protein concentration and Hagberg falling number of some grain lots (Vaidyanathan, 1987). Grain cleaners with efficient aspiration may achieve a similar, but smaller, response. The responses are not easily predictable in advance, but for falling number it is assumed that sprouted kernels are less dense through respiratory loss of carbohydrate. Removal of these kernels should therefore leave grain of lower α-amylase activity. One possible reason for lack of success in raising falling number by gravity separation of some grain lots may be when pre-maturity α-amylase forms in the absence of sprouting. No density change would be anticipated and thus high amylase kernels would not be removed with a gravity separator. A trial and error approach to gravity separation has to be adopted, which can be especially costly when mobile separators travel to farms, as occurs in the UK, only to discover that the grain is unresponsive. Some simple, rapid means of identifying unresponsive grain would be of considerable value.

REFERENCES

Ahmadi, M. (1990) *Cultural practices and endosperm effects on corn kernel chemical and physical properties.* PhD dissertation, Ohio State University, Columbus, USA.

Alloway, B.J. and Tills, A.R. (1984) Copper deficiency in world crops. *Outlook on Agriculture*, **13**, 32–42.

Baker, G.J., Greer, E.N., Hinton, J.J.C., Jones, C.R. and Stevens, D.J. (1958) The effect on flour colour of *Cladosporium* growth on wheat. *Cereal Chemistry*, **35**, 260–75.

Batey, T. and Reynish, D.J. (1976) The influence of nitrogen fertilizer on grain quality in winter wheat. *Journal of the Science of Food and Agriculture*, **27**, 983–90.

Bayles, R.A. (1977) Poorly filled grain in the cereal crop. 1 The assessment of poor grain filling. *Journal of the National Institute of Agricultural Botany*, **14**, 232–40.

Bayles, R.A., Evers, A.D. and Thorne, G.N. (1978) The relationship of grain shrivelling to the milling and baking quality of three winter wheat cultivars grown with different rates of nitrogen fertilizer. *Journal of Agricultural Science, Cambridge*, **90**, 445–6.

Bauer, P.J. and Carter, P.R. (1986) Effect of seeding date, plant density, moisture availability, and soil nitrogen fertility on maize kernel breakage susceptibility. *Crop Science*, **26**, 1220–6.

Belderok, B. (1968) Seed dormancy problems in cereals. *Field Crop Abstracts*, **21,** 203–11.

Benzian, B. and Lane, P.W, (1986) Protein concentration of grain in relation to some weather and soil factors during 17 years of English winter wheat experiments. *Journal of the Science of Food and Agriculture*, **37**, 435–44.

Biscoe, P.V. and Gallagher, J.N. (1977) Weather, dry matter production and yield, in *Environmental effects on crop physiology* (eds J.J. Landsberg and C.V. Cutting), Academic Press, London, UK, pp. 75–100.

Blumenthal, C.S., Barlow, E.W.R. and Wrigley, C.W. (1993) Growth environment and wheat quality: the effect of heat stress on dough properties and gluten proteins. *Journal of Cereal Science*, **18,** 3–21.

Brocklehurst, P.A., Moss, J.P. and Williams, W. (1978) Effects of irradiance and water supply on grain development in wheat. *Annals of Applied Biology*, **90**, 265–76.

Brun, L. (1982) Combined variety and nitrogen fertilizer trial with cereals, 1975–1979. *Forskning og Forsok i Landbruket*, **33**, 133–42.

Byers, M., Kirkman, M.A. and Miflin, B.J. (1978) Factors affecting the quality and yield of seed protein, in *Plant Proteins* (ed. G. Norton), Butterworths, London, UK, pp. 227–43.

Cairns, A.L.P. and Kritzinger, J.H. (1992) The effect of molybdenum on seed dormancy in wheat. *Plant and Soil*, **145**, 295–7.

Carver, T.L.W. and Griffiths, E. (1981) Relationship between powdery mildew infection, green leaf area and grain yield of barley. *Annals of Applied Biology*, **99**, 255–66.

Clarke, J.M. (1983) Wheat weathering damage in windrower/combine vs direct-combine harvesting systems, in *Third International Symposium on Pre-Harvest Sprouting in Cereals* (eds J.E. Kruger and D.E. LaBerge), Westview Press, Boulder, USA, pp. 287–93.

Couvreur, F., Dessacs, G. and Martin, G. (1993) Influence des conditions de maturation et des techniques culturales sur la qualité des semences de blé, in *Fourth International Workshop on Seeds – Basic and Applied Aspects of Seed Biology,*

Vol. 3 (eds D. Come and F. Corbineau), Université Pierre et Marie Curie, Paris, France, pp. 905–11.

Cromack, H.T.H. and Clark, A.N.S., (1987) Winter wheat and winter barley – the effect of seed rate and sowing date on grain quality, in *Cereal Quality*, Aspects of Applied Biology 15, Association of Applied Biologists, Warwick, UK, pp. 171–9.

Darwent, A.L., Kirkland, K.J., Townley-Smith, L., Harker, K.N., Cessna, A.J., Lukow, O.M. and Lefkovitch, L.P. (1994) Effect of preharvest applications of glyphosate on the drying, yield and quality of wheat. *Canadian Journal of Plant Science*, **74**, 221–30.

Darwinkel, A., ten Hag B.A. and Kuizenga, J. (1977) Effect of sowing date and seed rate on crop development and grain production of winter wheat. *Netherlands Journal of Agricultural Science*, **25**, 83–94.

Davies, W.P. and Kettlewell, P.S. (1987) Morphological influences of seed-applied triadimenol, flutriafol and other compounds on spring barley, in *Application to Seeds and Soil*, BCPC Monograph No. 39 (ed. T. Martin), British Crop Protection Council, Thornton Heath, UK, pp. 91–7.

Day, A.D. and Intalap, S. (1970) Some effects of soil moisture stress on the growth of wheat (*Triticum aestivum* L. en Thell.). *Agronomy Journal*, **62**, 27–9.

Dyke, G.V. and Stewart, B.A. (1992) Factors affecting the grain yield, milling and breadmaking quality of wheat, 1969–1972 I. Grain yield, milling quality and flour protein content in relation to variety and nitrogen fertilizer. *Plant Varieties and Seeds*, **5**, 115–28.

Evans, L.T., Wardlaw, I.F. and Fischer, R.A. (1975) Wheat, in *Crop Physiology* (ed. L.T. Evans), Cambridge University Press, Cambridge, UK, pp. 101–49.

Fatyga, J. (1991) Effect of sowing date and nitrogen fertilizer application on yield and quality of spring wheat grain. *Roczniki Nauk Rolniczych. Seria A, Produkcja Roslinna*, **109**, 71–84.

Finck, A., (1992) Phosphate fertilization and crop quality, in *Phosphorus, Life and Environment*, World Phosphate Institute, Casablanca, Morocco, pp. 231–51.

Finney, K.F., Meyer, J.W., Smith F.W. and Fryer, H.C. (1957) Effect of foliar spraying of Pawnee wheat with urea solutions on yield, protein content and protein quality. *Agronomy Journal*, **49**, 341–7.

Flintham, J.E. and Gale, M.D. (1988) Genetics of pre-harvest sprouting and associated traits in wheat: review. *Plant Varieties and Seeds*, **1**, 87–97.

Fuller, M.F., Cadenhead, A., Brown, D.S., Brewer, A.C., Carver, M. and Robinson, R. (1989) Varietal differences in the nutritive value of cereal grains for pigs. *Journal of Agricultural Science, Cambridge*, **113**, 149–63.

Garstang, J.R. (1993) The effects of volunteers on cereal quality and profitability, in *Volunteer Crops as Weeds*, Aspects of Applied Biology 35 (eds R.J. Froud-Williams, C.M. Knott and P.J.W. Lutman), Association of Applied Biologists, Warwick, UK, pp. 67–74.

Gooding, M.J. (1988) *Interactions between late-season foliar applications of urea and fungicide on foliar disease, yield and breadmaking quality of winter wheat*, PhD thesis (CNAA), Harper Adams Agricultural College, Newport, UK.

Gooding, M.J, Smith, S.P, Davies, W.P. and Kettlewell, P.S. (1994) Effects of late-season applications of propioconazole and tridemorph on disease, senescence, grain development and the breadmaking quality of winter wheat. *Crop Protection*, **13**, 362–70.

Gregory, P.J., Marshall, B., and Biscoe, P.V. (1981) Nutrient relations of winter wheat 3. Nitrogen uptake, photosynthesis of flag leaves and translocation of nitrogen to grain. *Journal of Agricultural Science, Cambridge*, **96**, 539–47.

Helenius, J. and Kurppa, S. (1989) Quality losses in wheat caused by the orange wheat blossom midge *Sitodiplosis mosellana*. *Annals of Applied Biology*, **114**, 409–17.

Herum, F.L. (1987) Harvesting and postharvest management, in *Corn: Chemistry and Technology* (eds S.A. Watson and P.E. Ramstad), American Association of Cereal Chemists, St. Paul, USA, pp. 83–123.

Huber, D.M. (1981) The role of nutrients and chemicals, in *Biology and Control of Take-all* (eds M.J.C. Asher and P.J. Shipton), Academic Press, London, UK, pp. 317–41.

Jenner, C.F., Ugalde T.D. and Aspinall, D. (1991) The physiology of starch and protein deposition in the endosperm of wheat. *Australian Journal of Plant Physiology*, **18**, 211–26.

Jongkaewwattana, S., Geng, S., Brandon, D.M. and Hill, J.E. (1993) Effect of nitrogen and harvest grain moisture on head rice yield. *Agronomy Journal*, **85**, 1143–6.

Karvonen, T. and Peltonen, J. (1991) A dynamic model for predicting the falling number of wheat grains in Scandinavian climatic conditions. *Acta Agriculturae Scandinavica*, **41**, 65–73.

Kettlewell, P.S. and Cooper, J.M. (1993) Field studies on α-amylase activity of wheat grain in the absence of sprouting: relationship with grain drying rate and with nitrogen fertilization, in *Pre-harvest Sprouting in Cereals 1992* (eds M.K. Walker-Simmons and J.L. Ried), American Association of Cereal Chemists, St. Paul, MN, USA, pp. 354–61.

Kibite, S. and Evans, L.E. (1984) Causes of negative correlations between grain yield and grain protein concentration in common wheat. *Euphytica*, **33**, 801–10.

King, R.W. and Richards, R.A. (1984) Water uptake and pre-harvest sprouting damage in wheat: ear characteristics. *Australian Journal of Agricultural Research*, **35**, 327–36.

Lauer, J.G. and Partridge, J.R. (1990). Planting date and nitrogen rate effects on spring malting barley. *Agronomy Journal*, **82**, 1083–8.

Lee, G., Stevens, P.J., Stokes, S. and Wratten, S.D. (1981) Duration of cereal aphid populations and the effects on wheat yield and breadmaking quality. *Annals of Applied Biology*, **98**, 169–78.

McGuire, C.F., Stallknecht, G.F. and Larson, R.A. (1988) Impact of cropping systems on end-use quality of hard red spring wheat. *Field Crops Research*, **18**, 203–6.

McLean, K.A. (1987). Post-harvest manipulation and measurement of grain quality. In *Cereal Quality*, Aspects of Applied Biology 15, Association of Applied Biologists, Warwick, UK, pp. 483–94.

MacNicol, P.K., Jacobsen, J.V., Keys, M.M. and Stuart, I.M. (1993) Effects of heat and water stress on malt quality and grain parameters of Schooner barley grown in cabinets. *Journal of Cereal Science*, **18**, 61–8.

Manners, J.G. and Myers, A. (1981) Effects on host growth and physiology, in *Biology and Control of Take-all* (eds M.J.C. Asher and P.J. Shipton), Academic Press, London, UK, pp. 237–48.

Milus, E.A. (1994) Effect of foliar fungicides on disease control, yield and test weight of soft red winter wheat. *Crop Protection*, **13**, 291–95.

Morris, C.F. and Paulsen, G.M. (1985) Pre-harvest sprouting of hard winter wheat as affected by nitrogen nutrition. *Crop Science*, **25**, 1028–30.

Pinthus, M.J. (1973) Lodging in wheat, barley, and oats: the phenomenon, its causes, and preventive measures. *Advances in Agronomy*, **25**, 209–63.

Preston, K.R., Kilborn, R.H., Morgan, B.C. and Babb, J.C. (1991) Effects of frost and immaturity on the quality of a Canadian hard red spring wheat. *Cereal Chemistry*, **68**, 133–8.

Ramig, R.E., Rasmussen, P.E., Allmaras, R.R. and Smith, C.M. (1975) Nitrogen-sulfur relations in soft white winter wheat I. Yield response to fertilizer and residual sulfur. *Agronomy Journal*, **67**, 219–24.
Randall, P.J. and Wrigley C.W. (1986) Effects of sulphur supply on the yield, composition, and quality of grain from cereals, oilseeds, and legumes. *Advances in Cereal Science and Technology*, **8**, 171–206.
Reith, J.W.S. (1968), Copper deficiency in crops in north-east Scotland. *Journal of Agricultural Science, Cambridge*, **70**, 39–45.
Schnug, E., Haneklaus, S. and Murphy, D. (1993) Impact of sulphur supply on the baking quality of wheat, in *Cereal Quality III*, Aspects of Applied Biology 36 (eds P.S. Kettlewell, J.R. Garstang, C.M. Duffus, N. Magan, W.T.B. Thomas and N.D. Paveley), Association of Applied Biologists, Warwick, UK, pp. 337–45.
Schnyder, H. (1993) The role of carbohydrate storage and redistribution in the source-sink relations of wheat and barley during grain filling – a review. *New Phytologist*, **23**, 233–45.
Seetanun, W. and De Datta, S.K. (1973) Grain yield, milling quality and seed viability of rice as influenced by time of nitrogen application and time of harvest. *Agronomy Journal*, **65**, 390–4.
Simmonds, N.W. (1995) The relation between yield and protein in cereal grain. *Journal of the Science of Food and Agriculture*, **67**, 309–15.
Smika, D.E. and Greb, B.W. (1973) Protein content of winter wheat grain as related to soil and climatic factors in the semiarid Central Great Plains. *Agronomy Journal*, **65**, 433–6.
Sofield, I., Evans, L.T., Cook, M.G. and Wardlaw, I.F. (1977a) Factors influencing the rate and duration of grain filling in wheat. *Australian Journal of Plant Physiology*, **4**, 785–97.
Sofield, I., Wardlaw, I.F., Evans, L.T. and Zee, S.Y. (1977b) Nitrogen, phosphorus and water contents during grain development and maturation in wheat. *Australian Journal of Plant Physiology*, **4**, 799–810.
Stanca, A.M., Jenkins, G. and Hanson, P.R. (1979) Varietal responses in spring barley to natural and artificial lodging and to a growth regulator. *Journal of Agricultural Science, Cambridge*, **93**, 449–56.
Stevens, D.B. (1986) Agronomic influences on marketability, in *Combinable Crops – Prospects for the Future*, Proceedings of the 19th NIAB Crop Conference, National Institute of Agricultural Botany, Cambridge, UK, pp. 27–35.
Talukder, M.S.V., Mogensen, V.O. and Jensen H.E. (1987) Grain yield of spring wheat in relation to water stress I. Effect of early drought on development of late tillers. *Cereal Research Communications*, **15**, 101–7.
Terman, G.L. (1978) *Atmospheric sulphur – the agronomic aspects*. Technical Bulletin No. 23, The Sulphur Institute, Washington DC, USA.
Thevenet, G. and Taureau, J.C. (1987) L'alimentation soufrée du blé tendre d'hiver. Effets de la fumure soufrée sur le rendement. Etude des possibilités de diagnostic et de pronostic, in *Elemental Sulphur in Agriculture*, Volume 2, Syndicat Francais du Soufre, Marseille, France, pp. 469–82.
Thompson, G.B. and Woodward, F.I. (1994) Some influences of CO_2 enrichment, nitrogen nutrition and competition on grain yield and quality in spring wheat and barley. *Journal of Experimental Botany*, **45**, 937–42.
Timms, M.F., Bottomley, R.C., Ellis, R.S. and Schofield, D. (1981) The baking quality and protein characteristics of a winter wheat grown at different levels of nitrogen fertilisation. *Journal of the Science of Food and Agriculture*, **32**, 684–98.
Tipples, K.H., Dubetz, S. and Irvine, G.N. (1977) Effects of high rates of

nitrogen on Neepawa wheat grown under irrigation. II. Milling and baking quality. *Canadian Journal of Plant Science*, **57**, 337–50.

Tompkins, D.K., Fowler, D.B. and Wright, A.T. (1991) Water use by no-till winter wheat: influence of seed rate and row spacing. *Agronomy Journal*, **83**, 766–9.

Tsai, C.Y., Dweikat, I., Huber, D.M. and Warren, H.L. (1992) Interrelationship of nitrogen nutrition with maize (*Zea mays*) grain yield, nitrogen use efficiency and grain quality. *Journal of the Science of Food and Agriculture*, **58**, 1–8.

Usherwood, N.R. (1985) The role of potassium in crop quality, in *Potassium in Agriculture* (ed. R.D. Munson), American Society of Agronomy, Madison, USA, pp. 489–513.

Vaidyanathan, L.V. (1987) Precision and reliability of measuring Hagberg falling number of wheat including variability associated with crop husbandry and grain handling, in *Cereal Quality*, Aspects of Applied Biology 15, Association of Applied Biologists, Warwick, UK, pp. 495–513.

Vaidyanathan, L.V., Sylvester-Bradley, R., Bloom, T.M. and Murray, A.W.A. (1987) Effects of previous cropping and applied nitrogen on grain nitrogen content in winter wheat, in *Cereal Quality*, Aspects of Applied Biology 15, Association of Applied Biologists, Warwick, UK, pp. 227–37.

Webb, R.B. and Stephens, D.E. (1936) Crown and root development in wheat varieties. *Journal of Agricultural Research*, **52**, 569–83.

Part Five

Post-harvest Management of Cereal Quality

14

Quality of stored cereals

J.T. Mills

14.1 INTRODUCTION

Cereal grains are subject to quality loss during storage and transportation, often resulting in considerable diminution in grade and value. Quality loss is of concern to many persons in the grain industry including producers, elevator managers, shippers, regulators, exporters, and purchasers, both domestic and foreign. Considerable research has been carried out to study, detect, and prevent loss in quality of stored grains and their products during storage. The work has been undertaken from diverse viewpoints, involving the disciplines of biochemistry, mycology, entomology, toxicology, food science, ecology and agricultural engineering. It is known that quality loss in stored grains is caused mainly by deterioration, a natural process which breaks down organic matter through either physical/chemical processes or biological processes where contained nutrients and energy are used by other life forms. The effects of deterioration can be considerably diminished through careful stored grain management based on a knowledge of the principles governing deterioration and by knowledge of the storage behaviour of the particular grains involved. This chapter discusses the concept of quality in stored cereals, then describes the factors affecting stored cereal quality, the management of stored cereals to minimize quality loss, the storage characteristics of world cereal crops, and offers suggestions for future work.

14.2 STORED CEREAL QUALITY

Our present understanding of the concepts of grain quality has resulted from research in several key areas. These include: grain respiration and heating (Pomeranz, 1992); the role of moulds in quality loss (Sauer *et al.*, 1992); methods for detecting and measuring grain deterioration (Muir *et al.*, 1985); grain moisture content/equilibrium relative humidity relationships (Gough and King, 1980; Henderson, 1987); the estimation of grain infestability by insects (Sinha, 1972); aeration and drying of grains (Brooker *et al.*, 1992); and grain quality standards (Anon., 1978; Anon., 1992). Quality as related to stored grains means different things, however, to different people. For example, a seed merchant is interested in seeds with high levels of varietal purity, viability, soundness, and low admixture levels, whereas a poultry farmer seeks a product which is nutritionally sound and toxin-free.

Several general methods are available for assessing the quality of stored cereal grains including visual or light microscope examination to detect the presence of other grains, chemical residues, or a dull appearance which may indicate storage moulds (Mills, 1992). Other tests include the 1000-kernel weight test for estimating the degree of plumpness (Gedye *et al.*, 1981), and the dockage test for determining the amount of weed seeds and fine material (Anon., 1992). Many additional tests are available including those for seed viability (Wallace and Sinha, 1962); free fatty acid levels (FFA), a measure of seed biochemical deterioration (Pomeranz, 1992); mycotoxins (Abramson, 1991; Wilson and Abramson 1992); and presence of field and storage moulds (Wicklow, 1995), and insects and mites (Sinha, 1964; Pedersen, 1992).

To ensure that grains are traded fairly, grade standards of predetermined quality have been established for cereals in many countries (Anon., 1992; Manis, 1992). Grade standards for grain, according to Hill (1988), facilitate marketing, identify economic factors important to end-users, and reflect storability. In addition, a grade provides an incentive for grain quality maintenance and improvement (Brooker *et al.*, 1992). The use of grade standards to define contracts between sellers and purchasers ensures that the purchaser will receive a product of an agreed upon level of quality. As a safeguard, cargoes on arrival are often sampled by internationally bonded companies to establish the grade and act as impartial witness in disputes.

The most effective control of quality is achieved with a complete grading system. It must be sufficiently complex to take account of variety, soundness, admixtures, and moisture level. Variety has little effect on storage *per se*; within a given class of grain all varieties tend to store equally well. By contrast, the remaining three quality factors do affect storage. Grain stores better when it is sound, clean, and especially when it is dry. Where there is no grading system, all wheat may be bulked, the good varieties with the poor, the sound grain with the less sound, the dry with the damp. With the blending that occurs on farms, at collection points, and in terminals, the majority of large shipments will level out at 'fair average quality' (Anderson, 1973).

14.3 FACTORS AFFECTING STORED CEREAL QUALITY

14.3.1 Introduction

Cereal grains are alive, generally in a resting state, and subject to change from abiotic and biotic environmental factors. Abiotic influences include climate; moisture; temperature; storage period; atmospheric composition; grain engineering properties; the methods and frequency of handling, and the type of storage structure. Biotic influences include the grains themselves; associated debris; moulds; insects and mites, and rodents and birds. Loss of stored grain quality occurs through interactions among abiotic and biotic variables (Sinha, 1973). The moisture and temperature of the grains are generally the factors which control the organisms, and ultimately grain quality deterioration (Mills, 1986).

14.3.2 Abiotic factors

Climate

The climates of the world have been divided into four major classes and several subdivisions (Trewartha, 1968). Climate is a major regulator of grain storage ecosystems. Generalizations can be made regarding the relationships of microorganisms, insects, and mites to stored-cereal grains in each climatic region. For example, deteriorating stored grain, especially wheat, in regions with temperate climates has a relatively dominant mite component and fewer species of insects and microorganisms. In contrast, on similar substrates in humid tropical climates, one can expect to find insects primarily, microbes to a lesser extent, and mites least of all. In hot dry climates, insects are dominant. In subtropical climates, the three groups of organisms are often equally abundant, but their relative abundance may be influenced by the type of grain and by the harvesting, drying, and storage practices in a given country within the subtropical region. Bacteria could cause most of the damage to stored grain in humid tropical and subtropical climates, but it is a common practice for farmers or grain handlers to use either sunlight or artificial heat to dry grain down to a moisture level that limits damage to that caused by insects and moulds. Although several major pest insect species occur in most food-producing areas of the world, some species are particularly destructive to grain in certain climatic zones but are unimportant in others. To cause outbreaks in any climatic zone, a species must be able to reproduce quickly during favourable periods of the year and survive during unfavourable periods (Sinha, 1991).

Moisture

During storage, moisture within grain kernels reaches an equilibrium with the air around them and produces a relative humidity level that may favour the development and growth of deteriorative micro-organisms. The lower limit of moisture level for mould growth is near the upper limit for the designation of dry (straight grade) seed. The maximum moisture levels under the Canadian Grain Act at which cereals can be sold as straight grade are listed in Table 14.1.

Because seed with these moisture levels can be sold without penalty, such values are often assumed to represent safe levels (Moysey and Norum, 1975). In practice, however, the safe moisture levels are below those given in Table 14.1. This is because some seed lots may have a higher number of fungal propagules, or a higher level of damage than others, have immature, high moisture weed seeds or other debris, or may have suffered the effects of high-temperature drying, reducing viability and making the grain more susceptible to invasion by moulds and subsequent deterioration (Mills, 1989).

Deteriorative organisms affecting stored cereals require different levels of relative humidity for normal development. The level for bacteria is generally above 90%, for storage moulds above 70%, for storage mites above 60%, and for storage insects in the range 30–50%. Relative humidity and moisture level are both dependent on temperature. For example, if the temperature of an air sample having a relative humidity level of 50% is increased 5°C from 25°C to 30°C its relative humidity level will decrease to 38%; if the temperature of the air sample is decreased 5°C from 25°C to 20°C, then the relative humidity level will increase to 69% (Mills, 1989; Brooker *et al.*, 1992).

The relationship between grain equilibrium moisture, relative humidity and temperature is shown in Table 14.2 for selected cereals

Table 14.1 Maximum moisture levels for straight grade seeds (Anon., 1992)

Type of cereal grain	*Maximum moisture level (g/kg)*
Barley	148*
Corn	155
Oats	140
Rye	140
Triticale	140
Wheat (includes soft red winter, hard red winter, hard red spring, durum)	145

*Reduced to 140 g/kg for malting barley in 1994.

Table 14.2 Equilibrium moisture levels (g/kg) for cereal grains (after Hall, 1980; Sinha and White, 1982)

Cereal grain	*Temperature (°C)*	*Relative humidity (%)*						
		40	*50*	*60*	*70*	*80*	*90*	*100*
Barley	25	97	108	121	**135**	158	195	268
Corn, shelled, YDent	25	98	112	129	**140**	156	196	238
Oats	25	91	103	118	**130**	149	185	241
Rice, whole grain	25	109	122	133	**141**	152	191	
	38	98	111	123	**133**	148	191	
Rye	25	99	109	122	**135**	157	206	267
Sorghum	25	98	110	120	**138**	158	188	219
Triticale	22	120	125	130	**151**	181	220	
Wheat								
soft red winter	25	97	109	119	**136**	157	197	256
hard red winter	25	97	109	125	**139**	158	197	250
hard red spring	25	98	111	125	**139**	159	197	250
durum	25	94	105	115	**131**	154	193	267

Grain moisture in equilibrium with 70% RH as shown in bold as most stored grain spoilage begins at about that value due to mould activity.

(Hall, 1980; Sinha and White, 1982). Grain moistures in equilibrium with 70% RH are shown in bold type in Table 14.2 as most stored-grain spoilage begins about this level as a result of storage mould activity. The relationship between the moisture level of a particular grain and its equilibrium relative humidity at the particular temperature can be expressed by an equilibrium moisture curve. The values plotted for each curve correspond to particular temperatures (Anon., 1993). Grain which is gaining moisture by adsorption has a lower equilibrium moisture level than grain losing water by desorption. Desorption curves are used more often than adsorption curves because grain drying processes are more common than moisture sorption processes during cereal-grain management (Muir, 1973). A concise explanation of the theory of moisture in stored products is given by Mackay (1967).

Water (or moisture) content is commonly used to express the amount of water in a substance but it gives little idea of the availability of the water for microbial growth since the relationship between water content and water availability differs between different types of seed. Water availability in grain is better indicated by equilibrium relative humidity (ERH), water activity (A_w) or water potential (ψ). ERH is the relative humidity of the intergranular atmosphere in equilibrium with the water in the substrate; A_w is the ratio of the vapour pressure of water over a substrate to that over pure water at the same temperature and pressure; and ψ is the sum of osmotic, matric and turgor potentials. ERH and A_w are numerically the same, except that ERH is expressed as a percentage and A_w as a decimal fraction of one (Lacey and Magan, 1991).

Temperature

Important facts regarding the temperature of stored grains are: (i) high temperatures of grain harvested and binned on a hot day are retained within unaerated grain bulks for many months due to the insulation properties of the grain (Yaciuk *et al.*, 1975); (ii) temperature and moisture influence enzymatic and biological activities and, therefore, the rate of spoilage; (iii) temperature differences within bulk commodities favour mould development through moisture migration resulting from convection currents caused by sinking colder dense air, followed by rising warmer air and subsequent moisture absorption near the cool top surface in winter (Muir, 1973).

The growth of moulds and insects in stored grain produces metabolic heat, thus a rise in grain temperature may indicate deterioration. There are, however, difficulties in using and interpreting temperature results (Mills, 1990). For example, when a small pocket of grain spoils, the temperature at the centre of the pocket may reach 65°C, whereas only 50 cm away the grain may be 10°C. To detect such pockets, temperatures must be measured at many points or at least where spoilage is most probable. However, bulk temperatures above outside air temperatures do not necessarily indicate the occurrence of spoilage, as straight grade crops can be harvested and placed into storage in excellent condition at warm temperatures. If crops are stored dry in large, unventilated storages, temperatures near the centre can remain relatively high throughout cold winter months without any spoilage occurring. In the tropics and subtropics, the prevailing temperature conditions are characterized by maximum shade temperatures of 38°C to 43°C and minimum temperatures of 30°C to 35°C in the hot, dry areas, and by maximum shade temperatures ranging from 21°C to 35°C in warm, wet areas. These conditions are important for the following reasons:

1. High temperatures up to 43°C speed up the life processes of all organisms.
2. Natural drying can be carried out during long periods of high temperature when relative humidity is low.
3. Commodities containing a certain amount of heat are put into store and the heat is retained during the storage period. Problems due to condensation on the stored commodity will, therefore, occur in areas where there are wide fluctuations in temperature (such as differences

between day and night temperatures or between high and low altitude areas (Hall, 1970)).

Storage period

Safe storage of a commodity depends largely on its level of moisture, temperature, and period of storage. Grains in a bin, pile, or hold of a ship, after a certain period, are influenced by air movements within the grain sometimes resulting in moisture condensation and mould spoilage (Mills and Wallace, 1979; Sinha and Wallace, 1977; Tuite and Foster, 1979). The length of the intended storage period is of prime importance as the maximum moisture level needs to be modified for safe storage. Thus, for the safe storage of barley for 1 year the maximum moisture level is 130 g/kg whereas for 5 years it is 110 g/kg (Hall, 1980).

Atmospheric composition

The composition of the intergranular atmosphere may also influence grain quality. The air composition within sealed storages may be manipulated so that it reduces the impact of, or is lethal to pests (Agboola, 1993). For example, high moisture grains are stored anaerobically in silos to stop mould spoilage, and carbon dioxide (CO_2) and nitrogen (N_2) are added to stored grains in sealed bins to kill insect pests (see Section 14.4.5: Controlled atmospheres). In unsealed storages, respiration by seedborne moulds, insects and mites produces CO_2 which can be detected using a simple syringe device and a colour indicator based on a chemical reaction. By measuring the CO_2 concentration, the presence of pests or initiation of spoilage can be detected before serious grain damage occurs (Mills, 1989)

Engineering properties of grain

A grain bulk has five engineering properties: porosity, flow, layering, sorption, and thermophysical mass-exchange properties whose interplay influences deterioration of grain. Because a grain bulk is composed mainly of living grain, most of its physical properties are derived from and influenced by interactions of biotic and abiotic factors. It is impossible to separate a single physical property of a grain bulk without considering one or more chemical and biotic factors. For example, porosity is due both to the nature of the grain kernel itself and the presence of intergranular spaces within grain bulks. The extent of porosity depends on the size and shape of grains, elasticity, surface state, dockage level, weight, compaction, storage period, and the distribution of moisture in the bulk. These physical characteristics, in turn, influence the movement of air, heat and moisture. Together they affect storage stability of the grain (Sinha, 1973). Recent articles on the physical properties of grains are by Cenkowski and Zhang (1995) and Hoseney and Faubion (1992); tables listing the physical dimensions of grain kernels are given by Brooker *et al.* (1992).

Types of structure

Well-constructed and weatherproofed structures are essential to preserve crop quality and prevent infestations during storage. Structures on high, well-drained land protect the crop from heavy rainfall and spring floods. Steel granaries, when empty, provide fewer places for insects to breed than empty wooden structures. Steel bins are erected on steel-reinforced concrete slabs to prevent cracks and moisture transfer through the floor. When yields are above average, crops are sometimes stored in machinery sheds or barns. To minimize the effect of residual insect populations, cracks in the floors and side walls of machine sheds are caulked. In double-sided barns, the bottom boards are removed to clean out infested grain dust and to apply pesticides. For temporary crop storage, plywood sheets can be used to construct circular cribs. Otherwise the crop is piled on the ground until storage space becomes available or until it is sold (Mills, 1990). Storage units can generally be classified as horizontal or vertical and further classified as round, rectangular, cylindrical or box-shaped. Storages may be further classified according to the construction material (Hall, 1980).

The sophisticated storage structures used in developed countries may not be practical under the conditions of the tropics, due to the variation in climatic conditions and the cost of construction, operation and maintenance. Hence, storage structures must be based on locally available materials which are simple, cheap and durable. The type of structure depends on the storage system (i.e. for handling bulk or bagged materials), on the level of storage (i.e. for on-farm or rural storage for smaller quantities and commercial storage for larger quantities), and whether above-ground or underground (Salunkhe *et al.*, 1985). Descriptions of storages in developing countries are provided by Hall (1980) and Salunkhe *et al.* (1985); for grain handling systems by Hall and Maddex (1980); and for structural design of grain storages, by Manbeck *et al.* (1995).

14.3.3 Biotic factors

Grains

Living seeds respire, producing heat, moisture and carbon dioxide. Stored dry seeds are largely inactive and have very low respiration

(White *et al.*, 1982). Freshly harvested immature seeds or seeds with a high moisture level, however, have a much higher respiration resulting from biological activities of the seeds or mould infection. Most of the respiration is by moulds which are located on and within the seed coat (Hall, 1970). The original condition of grain is a major factor affecting their future storage quality. During the processes of growth and maturation, plant material in the field goes through a number of chemical reactions catalysed by enzymes. Freshly harvested grains entering storages are often immature and may have increased enzymatic activity, resulting in high temperature rates and in heat production (Mills, 1989). Respiration is a self-accelerating process. The moisture produced can increase the moisture of the grain which in turn can cause an increase in respiration rate; also, the heat produced can raise the grain temperature, in turn increasing the respiration rate (Hall, 1970). These changes often lead to reduced grain quality, therefore freshly harvested or high moisture grains should be carefully monitored during initial storage and remedial action such as aeration, drying or other measures applied.

Seed characteristics

The embryo and endosperm of stored cereal seeds are protected to a considerable extent from deteriorative agents by several seed coats and coverings varying in effectiveness with the type of seed. For example, in rice, the hull functions as a barrier to fungal penetration and prevents the grain from becoming rancid by protecting the bran layer from mechanical damage during harvesting and handling. The outer pericarp consists of a very hard impermeable tissue and protects the kernel from oxidative and enzymatic deterioration. Varieties of rough paddy rice with a tight hull and no gaps between the palea and lemma coverings are known to be more resistant to insect infestation. In corn, the endosperm has an outer hard layer, and grain hardness has been positively correlated with resistance to pest invasion. In sorghum, the thick-walled epidermis and tannins in the testa layer of the seed coat also serve to protect the seed against invasion. In wheat and barley, mechanical injury to the seed coat at the end of the seeds during harvesting and subsequent handling increases the likelihood of deterioration in storage (Salunkhe *et al.*, 1985).

Post-harvest maturity

Complex biochemical changes occur in the days and weeks following the storage of freshly harvested grain (Sinha, 1973). According to Trysvyatskii (1966), if these changes do not follow an orderly sequence of events in the 60 days following harvest, storage stability of the grain in terms of seed viability, fat synthesis, and other technological characteristics become seriously impaired. The final period of chemical synthesis, which begins in the maturing heads of cereal plants in the field, can be completed successfully only after grain has ripened in the field and been harvested with minimum injury or in storage within the range of critical moisture level (below 140 g/kg) and temperature (15–45°C). To mature properly the processes of synthesis, i.e. synthesis of starch from sugars, proteins from amino acids, fats from fatty acids and glycerine, must predominate in the kernels over hydrolytic processes, i.e. enzyme activity and respiration.

The milling quality of newly harvested wheat is of interest to the miller because changes in flour quality and milling characteristics may occur during wheat storage (Pomeranz, 1992). Changes that occur in milling and baking performance have been defined in small-scale milling experiments with hard red winter wheat during the sweating period (Posner and Deyoe, 1986). Wide fluctuations in milling characteristics were observed resulting from the storage of freshly harvested wheat. Post-harvest maturation is of significance in countries in which wheat is harvested at relatively high moisture levels. The maturation seems to be of less consequence in wheat harvested with a rather low moisture level (Pomeranz, 1992).

Debris

Debris is waste material, often originating from the field, associated with grains within and immediately outside storages. It includes cereal stalks, chaff, shrunken or underdeveloped seeds, weed seeds, broken kernels, fine particulates and dust. The presence of debris in stored grains can result in severe loss in quality of the grains through spoilage and heating. During loading operations, debris becomes aggregated in the bin core and under spout lines, providing a suitable habitat for certain insects (Pedersen, 1992) and moulds. The metabolic activity of these organisms and of immature or other moist material often results in deterioration and heating in such locations. To prevent loss of grain quality, debris is removed from in and around storages prior to filling, and removed from fresh grains by cleaning. Dockage is a term used in the grading of grains. In the USA and Canada, dockage in wheat, barley and rye consists of material, other than the predominant grain, that can be readily removed by the appropriate sieves and cleaning devices and underdeveloped, shrivelled, and small pieces of grain that cannot be recovered by proper rescreening or recleaning. Dockage in sorghum consists only of the material that passes through the prescribed sieve (Manis, 1992).

Fungi

Grains in storage provide food and an environment for many micro-organisms including pre-harvest and post-harvest moulds. Of these, certain post-harvest or storage moulds are the most important cause of grain deterioration (Christensen and Kaufmann, 1969; Wicklow, 1995). Storage moulds exist as spores in soil, on decaying debris, in harvesting equipment, and within storage structures and are gathered by the combine harvester and distributed among the grains. The various types of storage moulds each require a different relative humidity and temperature for their growth and development. Some species, such as *Eurotium amstelodami* Mangin, grow at low humidities and produce water during their growth, which enable more damaging moulds to grow. Such moulds include *Aspergillus candidus* Link and *Penicillium* species, both frequently associated with hot spots, areas within bulk grain that have a higher temperature than the surrounding material. Growth and development of storage moulds affect grains in storage by causing adverse quality changes, grain aggregation, heat-damage, and production of toxins and allergens (Mills, 1989). The effects and consequences of their activities on stored cereal grains are summarized in Table 14.3.

Growth of storage moulds on the surface of seeds often results in a dull rather than a bright appearance as in normal seeds. Dull appearance is sometimes regarded as a degrading factor. The presence of storage moulds on seeds is also often associated with musty odours, which are a degrading factor (Anon., 1972; Anon., 1992). Other important effects of storage moulds on seeds include reduction in germinability and discolouration of whole seeds or portions of them, including the germ. Under the right moisture conditions, storage moulds invade the germs with no visible signs of moulding, weaken the seeds, and eventually cause seed death. Some strains of the storage moulds *Aspergillus restrictus* G. Sm., *A. candidus*, and *A. flavus* Link can cause severe damage and kill the germs quickly. As fungal invasion of the germs of the seeds continues, the tissues of the germ become brown and then black (Sauer *et al.*, 1992). Discolouration caused by fungi results in lower grades both in the USA (Anon., 1972) and in Canada (Anon., 1992).

Under suitable conditions of moisture and temperature, some strains of storage moulds produce mycotoxins on stored grains. When mycotoxin-contaminated grains are eaten by susceptible animals, mycotoxicoses may result. The effects of mycotoxins on animals vary, depending on the age and species of the animal, and the type and amount of toxin in the feed. Disease effects include lack of weight gain, formation of tumours, loss in productivity, fetal abnormalities, and

Table 14.3 Effects and consequences of mould activities on stored grains (Mills, 1989)

	Effects	*Consequences*
1. Adverse quality changes	Dull appearance	
	Musty odours	Possible degrading
	Visible moulds	
	Reduced germination	Rejection for malting
	Germ damage, discolouration	Degrading
	Increased free fatty acids	Rejection for processing
2. Aggregation of product	Clogging of pipes, augers	Interruption of operations
	Sticking to bin walls	Uneven pressure effects, partial bin collapse
	Bridging of bin contents	Dangerous air space
	Aggregation and/or fusion of bin contents	Cleaning out costs, unusable facilities
3. Heating of product	Bin-burning	Damage to product and premises
		Possible degrading, rejection, extra costs
		Could lead to fire-burning explosions
4. Contamination of product by harmful substances	Mycotoxins	Livestock poisoning, feed refusal
		Rejection of shipments
		Loss of markets
		Chronic human health problems
	Respiratory/allergenic effects	Breathing problems in animals and humans
		Employees handling grain may need to be reassigned

sudden death (Mills, 1989). The relationship between mould development in grain and accumulation of mycotoxins is reviewed in Chelkowski (1991). The worldwide risks from mycotoxins are summarized by Mannon and Johnson (1985).

Insects and mites

Several hundred different species of insects are found associated with stored grains and their products worldwide, of which about 50 species are serious pests. The insects most injurious to stored cereal grains originated in areas where wheat, barley and rice were the principal grain foods and were subsequently carried by commerce to all parts of the world (Cotton, 1963). Pest species include beetles and moths with life cycle stages adapted for feeding on grains. Some insect pests, such as the rice weevil, *Sitophilus oryzae* (L.), develop and feed inside kernels consuming the entire seed, while other species such as the red flour beetle, *Tribolium castaneum* (Herbst), feed mainly on the germs thus reducing germination (Howe, 1972). Psocoptera are detritus feeders and cannot damage whole grains, and others, including the squarenosed fungus beetle, *Lathridius minutus* (L.), feed on fungal mycelium. These detritus and fungiverous feeders, however, are also pests as their faeces and body parts contaminate grains and their metabolic activity may permit storage moulds to grow. The feeding action by insects facilitates the entrance of moulds and mites into seeds (Mills, 1986).

Many insect species can thrive on low moisture grain. Insect metabolic activity within dry grain bulks of 150 g/kg moisture can result in heating up to 42°C (Cotton and Wilbur, 1982). In the vicinity of insect-induced hot spots increases in grain moisture can permit storage moulds to grow and sometimes produce temperatures of up to 62°C. Stored-product insects cannot reproduce below 17°C; however, where grain temperatures remain above 25°C for long periods, insects can cause extensive damage. The effect of insect damage is worsened in high moisture grains (Mills, 1986). The identification and detection of stored grain insects and the damage caused by them has been summarized by Pedersen (1992).

Upon initial storage, grain may be already contaminated by species of mites present on crops in the field; further contamination can occur by other species present in the empty granary (Liscombe and Watters, 1962) or carried to the granary by flying stored-product beetles (Barker, 1993). Mites are minute creatures that thrive in moist grain, feeding on broken grain, weed seeds, and moulds. Stored-product mites have specialized feeding requirements. The grain mite, *Acarus siro* (L.), feeds on grain germs and pre-harvest moulds while the long-haired mite, *Lepidoglyphus destructor* (Schrank), feeds on seedborne pre-harvest and post-harvest moulds and debris. The cannibal mite, *Cheyletus eruditus* (Schrank), preys upon other mites and yet other species act as scavengers. Species of mites within the grain mass exist as a community and develop in an ecological succession (Sinha, 1984). Under certain conditions, widespread infestations of mites can occur in stored cereals with *A. siro* multiplying to enormous numbers (Solomon, 1962). The occurrence of mites on large numbers of seeds lowers seed quality through visible presence and damage, and development of unpleasant odours. There is often a strong minty odour to the grain which, when heavily infested, makes animal feed unpalatable. Mite-infested cargoes may develop extensive infestations during transit and be subject to delays on arrival, demurrage charges, and fumigation costs which will be passed on to the vendor. Infestations can be controlled by keeping cereal grains generally below a moisture level of 100 g/kg, by turning to break up moist pockets of grain, by aeration, and by fumigation (Mills, 1986). The distribution of stored product mites in bulks of wheat and barley was studied by Armitage (1984).

Rodents and birds

The most important rodents affecting stored-cereal quality include the house mouse, *Mus musculus* L.; the Norway rat, *Rattus norvegicus* (Berkenhout); and the roof rat, *R. rattus* L. All are highly successful, cosmopolitan, commensal invaders of grain storages; their success lies in their extreme adaptability, high reproductive rate and omniverous habits (Sinha, 1991). Other rodents may be of regional importance, such as the lesser bandicoot rat, *Bandicota bengalensis* (Gray and Hardwicke), in South and Southeast Asia (Frantz and Davis, 1991). Rats eat about 10% of their weight in food each day and contaminate a great deal more with their droppings, urine and hair, thereby rendering grain unfit for human consumption (Dykstra, 1973); they can also cause considerable damage to storage structures. The bionomics and integrated pest management of commensal rodents have been recently summarized by Frantz and Davis (1991), and rodent biology and control by Harris and Baur (1992).

Birds affect stored grain quality through their activities in standing crops and in grain storages. Direct damage by red-winged blackbirds (*Agelaius phoeniceus* (L.)), red-billed queleas (*Quelea quelea* (L.)), house sparrows (*Passer domesticus* (L.)), common grackles (*Quiscalus quiscula* (L.)) and common starlings (*Sturnus vulgaris* (L.)) on corn, oats and other ripening crops makes the subsequent stored crop vulnerable to infection by insects and mites (Sinha, 1991). English sparrows, domestic pigeons (*Columba livis* Gmelin) and domestic chickens (*Gallus gallus* L.) are casual invaders of grain storages (Cotton, 1963). The impact of their consumption and contamination of stored grain by excreta and introduction of stored-product mites becomes appreciable only when proper storage practices are consistently ignored. This occurs when doors, windows and ventilators are left open, thus providing easy access to the interior of granaries (Sinha, 1991).

14.3.4 Interrelationships among abiotic and biotic factors

Once filled, a bin of grain becomes a specific entity with particular physical, chemical and biological patterns of development. The binned grain provides a source of nutrients and a suitable environment for organisms living in and on the grains. The binned grains are influenced by grain moisture level, grain condition, management practices and other factors which largely determine the types and sequence of events leading to loss of stored grain quality (Mills, 1986). Intergranular conditions affect the growth of deteriorative and other kinds of flora and fauna with relative humidity of the intergranular air largely determining which organisms develop on the seed (Sauer *et al.*, 1992). Generally, organisms adapted for life in stored grains are opportunistic and exploit the environment to maximize their survival. They rarely act alone but usually in association with other organisms (Figure 14.1) and their biological activity creates suitable environments for the sequential development of other species. Organisms living in and on stored grains frequently also develop cyclically, often because of seasonal and annual temperature variations (Sinha, 1973). Because spoilage of grain results from ecological relationships among several biotic and abiotic variables over a period of time, it can only be understood and prevented in the long run by adopting a multidisciplinary approach (Sinha *et al.*, 1969).

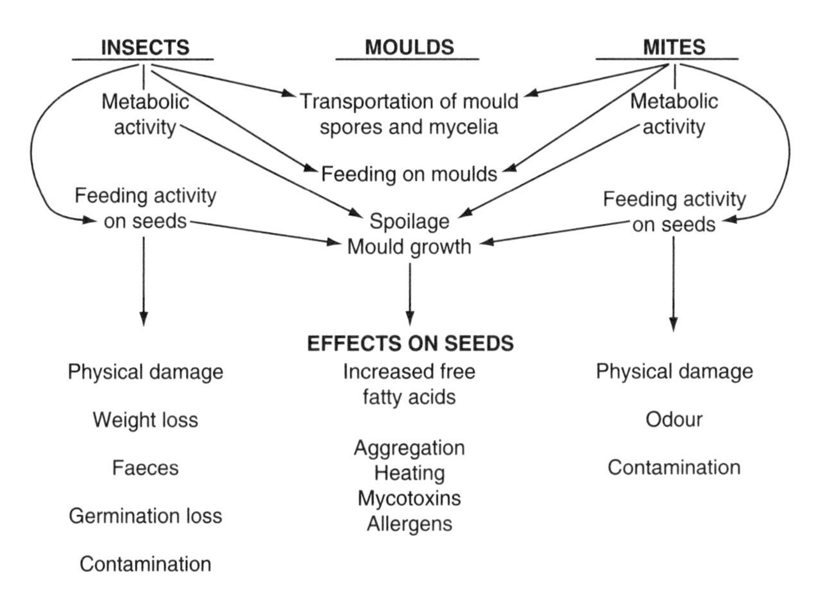

Figure 14.1 Associations/interactions among insects, mites and moulds, and the effects of these organisms on stored grains (from Mills, 1986).

14.4 MANAGEMENT OF STORED CEREALS TO MINIMIZE QUALITY LOSS

14.4.1 Introduction

Quality loss can be minimized in stored cereals through a knowledge of grain behaviour in storage, safe storage guidelines, grain and facility management, and transportation and quality control procedures.

14.4.2 Crop behaviour in storage

Each type of grain has its own physical, chemical, and biological characteristics that influence its behaviour in storage. The size, shape, structure and constituents of the grain, dimensions of the spaces between the grains, amount of oils present, hardness of the grains, and their nutritional suitability to support pests, all influence grain storability (Cenkowski and Zhang, 1995; Mills, 1992). Environmental conditions during growth and maturation, the degree of maturity at harvest, the methods of harvesting, and handling practices prior to storage also influence storability (Kreyger, 1972). Varietal differences in cereals may influence relative respiratory rates; softer types of wheat respire more rapidly than harder types at similar moisture levels and temperatures (Pomeranz, 1992).

14.4.3 Grain storage guidelines

Grain storage guidelines, which relate the present quality of grain to its future short- or long-term storability, have been developed for several cereals. The guidelines permit grain handlers to determine whether the crop is safe from spoilage, the risk of transporting particular crops from temperate to tropical climates or vice versa, and of other handling situations. The guidelines have two basic components: (i) a chart or table showing deteriorative or non-deteriorative changes occurring with time as a function of initial seed moisture and temperature or other parameters; and (ii) an assessment of the changes in quality occurring during the storage period based on an understanding of the characteristics and storage problems associated with the particular type of grain (Mills, 1992).

14.4.4 Management before storage

Decisions made while the crop is still in the field can prevent potential storage problems. Such decisions might be to air-dry small grains sufficiently in the field to ensure safe moisture levels during binning; to determine the seed moisture level and temperature of samples obtained from the combine to predict, by means of charts, the keeping quality of the crop (Kreyger, 1972; Wallace *et al.*, 1983); and to separately harvest wetter outer or lower areas of the field, placing this grain, often immature and of high respiratory activity, into smaller observation bins until it is dried or aerated (Mills, 1989).

14.4.5 Management during storage

Bin and grain preparation

To avoid future storage problems, certain bin management decisions are required. The first step is to obtain the advice of a registered professional engineer on the most suitable types of storage structures for the region and their intended use (Manbeck *et al.*, 1995). Once selected, the structures are located on a well-drained site on properly designed foundations, thus avoiding ingress of drainage water and cracked floors. The bins are emptied completely at regular intervals and examined within to detect adherent material (hang-ups) sticking to walls. The empty bins are cleaned thoroughly and then walls and floor sprayed with an appropriate insecticide, or dusted with an inert product such as diatomaceous earth, to kill any remaining insects which might infest the new grain. In addition, old grain and debris near bin doors and under aeration and drier floors is removed; likewise, any vegetation growing near the bins is removed as it could harbour grain pests. Structures are kept in a good state of repair and kept weathertight to keep out wind-driven rain or snow which, on melting, provides moisture for mould development. This may require the repair of cracks in concrete walls and floors and the sealing of gaps at the wall/concrete base interface to keep out moisture and to reduce the number of hiding places for insects (Mills, 1989).

Cleaning harvested material to remove high-risk debris, broken seeds, immature weed seeds, chaff, fine particulates, and dust improves the efficiency of aerators and bin driers by increased airflow. Cleaning units incorporated into standard grain augers now permit grain cleaning during augering. As the harvested material is placed in the bin it is sampled to determine the range of moisture levels present; the highest moisture level is the one of greatest concern. On completion of binning, one or more loads are withdrawn through the bottom of the bin to remove fines and to make an inverted cone suitable for ambient air-cooling rather than a peaked surface (Mills, 1989).

Aeration and drying

The quality of cereal grains can be maintained economically by forcing air through bulk-stored crops. Air is blown in or sucked out by means of a fan attached to a bin equipped with either a perforated duct or perforated floor. When air is blown in, the last part of the bulk to cool will be the top layer; most spoilage begins near the top centre where moisture condensation is greatest. If the air is sucked out by reversing the airflow, the last part to cool is the bottom layer. In this case, spoilage may occur at the bottom of the bin where it is much more difficult to control or monitor. There are three main options for moving air through bins as a means of preventing spoilage: aeration to cool the bulk; in-bin drying with near-ambient air; and in-bin drying with heated air (Mills, 1990).

Dehumidified air and chilled air are also used to prevent spoilage of moist grain in store. In the dehumidification process, unheated air is passed through large beds of absorbent material, such as granular silica gel, then the dry air is passed through the grain. The absorbent material eventually becomes fully charged with moisture and has to be dried at high temperatures, making the technique costly. Artificial chilling or cooling involves the passage of air over a refrigerator coil before it is blown into the grain. Artificial chilling provides cool conditions when these are most urgently needed, i.e. at harvest time when cold weather spells have not yet arrived; disadvantages of the technique are equipment and operational costs. Further details of these techniques are given in Nash (1985).

Aeration is the practice of forcing unheated air, by means of a small fan, through grain to cool it and maintain its condition and reduce the chances of spoilage and heating. This system lowers the temperature of the bulk if the grain is above ambient air temperature, maintaining a uniform air temperature throughout the grain mass which reduces or eliminates moisture migration, hot spots, mould and insect development and storage odours. Proper floor design and fan size are important in aeration systems and require consultation with a professional agricultural engineer. Localized spoilage may occur if the air flow is too low in some regions of the bin, if there is excessive debris, or if the fan is shut off before all the grain has been cooled in the fall or warmed in the spring; condensation may occur between cooler and warmer parts of the grain mass (Mills, 1989). Aeration principles and the proper operation of aeration systems are described by Friesen and Huminicki (1986).

The purpose of grain drying is to reduce the moisture in the crop resulting in an equilibrium relative humidity level at which the mould growth rates are zero or so low that during the anticipated storage period, the spoilage will be negligible. In commercial practice, two main types of drying procedure are used, the near-ambient and heated-air methods. The near-ambient method uses low air temperatures, which restrict the rate of mould growth so that drying can be completed before significant moulding occurs. Heated-air drying avoids the mould problem by using air at temperatures lethal to moulds to dry the grain rapidly. A major consideration of heated-air drying is loss of grain quality. Of particular concern during improper heated-air drying is the decrease in the milling quality of wheat and rice and the increase in breakage susceptibility in corn.

The influence of temperature on grain quality is categorized below in order of increasing temperature. The damage which occurs depends, in addition to grain temperature, on the exposure time, and the moisture and its change during this time (Bruce and Ryniecki, 1991).

1. Loss of viability. The ability of a seed to germinate and produce a viable seedling depends on complex reactions involving enzymes which help to break down starch into sugars and amino acids for the growing germ (Roberts, 1972). These enzymes begin to be inactivated during drying around 60–65°C.
2. Loss of bread-making quality. Dough must be elastic to be able to hold the CO_2 gas produced by fermentation of sugars by yeast. Its elasticity is a property of the gluten which is a complex formed by wheat protein and water. If the temperature of the wheat grain is raised above 65–70°C during drying the gluten in protein starts to become denatured, i.e. its functional properties are impaired by cross-links which are formed in the long spiral-shaped molecules of protein and which reduce the elasticity. When the gas is formed during proving of the dough, the inelastic gluten ruptures allowing the gas to escape. Therefore dough made from heat-damaged wheat does not rise well and the resulting loaf is dense and has a very crumbly texture.
3. Loss of amino acids. Ruminant animals can synthesize any amino acids not present in their feed to enable them to manufacture their body proteins. Provided the grains are not scorched during drying, their protein feeding value to ruminants is not impaired and the energy content may be even increased because starch digestibility can be improved by heating (Sullivan *et al.*, 1975). For non-ruminants, including humans, there are essential amino acids which the animal cannot synthesize and which therefore must be present in the diet. Each animal protein requires the essential amino acids to be present in a certain proportion and if one, the limiting amino acid, is lacking, the protein cannot be synthesized and the overall protein value of the feed is reduced proportionately. Wheat, barley and maize are all particularly deficient in one of the essential amino acids, lysine, so damage to the lysine content of the grains reduces the protein feed value of the grains to non-ruminants (Muhlbauer and Christ, 1974). The amino acids not synthesized to protein are metabolized for

energy. Damage to the lysine content of wheat starts to occur during drying at a grain temperature of 105–110°C.

4. Gelatinization of starch. This occurs at grain temperatures of 105°C and above. If the intact starch is required, e.g. for flour, the grain temperature must be kept below this level. However, even below this temperature there may be effects on flour colour and separability from bran. As noted in (3) above, gelatinization may improve starch digestibility for ruminants.
5. Charring or scorching. At grain temperatures during drying of 120°C and above conversion to carbon begins to occur and the grains take on a darkened appearance.

Most of the corn crop in the USA is dried on farms; less than 5% is dried at local elevators or terminals. In France the situation is reversed, as most of drying takes place at the elevator level. In the Third World, the majorty of the grain is still dried by spreading the wet grain in thin layers on the ground in the open air to be dried by the sun (Brooker *et al.*, 1992). The subject of grain drying is complex; for more in-depth information on theories of grain drying and optimal drying of wheat, rice, corn, barley, sorghum and oats see Brooker *et al.* (1992); and for the effects of drying on grain quality, see Foster (1973) and Bruce and Ryniecki (1991).

Pest control

To prevent and control infestations it is important to know where and when insect, mite, mould, rodent and bird pests occur. Most empty granaries are infested with low numbers of insects and mites (Smith and Barker, 1987). Animal feeds, trucks, and farm machinery are other sources of infestations. Most insects can fly as well as walk, which increases their ability to infest stored crops. New cereal crops should be stored only in clean empty bins as bins that contain old grain may harbour undetected infestations (Mills, 1990). Insect control methods include the use of fumigants to eliminate infestations from grain, residual pesticides to protect the grain against invasion by insects and mites, and other methods such as controlled atmospheres.

Fumigants are particularly useful in controlling pests in grain and grain products because they can diffuse through space and penetrate into protected places that are inaccessible to liquid and solid pesticides. Control of pests with fumigants is direct; fumigants exert their effect during the exposure period but diffuse away afterwards leaving little or no residue, but also no residual protection. Certain fumigants, e.g. methyl bromide, however, are no longer environmentally acceptable because of ozone depletion. Residual contact pesticides are applied to grain and to storage facilities to kill indigenous pests and to prevent infestation during the storage period (Bond, 1973). Major problems with chemical control are human safety, e.g. insecticide residues in grains (Smith, 1990), health concerns, e.g. the effect of phosphine on human genes (Garry *et al.*, 1989), and the development of insecticide resistance in target species (Champ and Dyte, 1976). Further information on control of insects in stored grains is given by Banks and Fields (1995), Bell and Armitage (1992), Harein and Davis (1992) and White and Leesch (1996); on rodent control by Harris and Baur (1992); and on bird control by Baur and Jackson (1982).

Controlled atmospheres

The atmosphere surrounding any commodity can be altered to kill pests in that commodity. When concentrations of component atmospheric gases are controlled accurately, the resulting mixture is called a controlled atmosphere (CA), whereas when constituent gas concentrations are less critically controlled, the atmosphere is regarded as a modified atmosphere (MA). CAs or MAs may be referred to by their active components as oxygen-deficient, low-oxygen, high-carbon dioxide, nitrogen or hermetic storage. An oxygen-deficient atmosphere can be provided by the addition of N_2, CO_2, helium or combustion gases. In practice, N_2 and CO_2 are the most common gases used in CA storage worldwide. CA storage employs only the major gaseous components of the Earth's atmosphere and excludes any gases or vapours normally considered to leave toxic residues (Agboola, 1993).

Advantages of using CA for crop storage include:

1. It has the potential to kill all animals including all stages of insects without the use of toxic chemicals.
2. It inhibits the growth of microorganisms, particularly aerobic fungi, thereby reducing the production of mycotoxins.
3. It reduces grain respiration.
4. It reduces oxidative degradation.
5. The structures used protect against external reinfestation from pests.
6. It has minimal adverse effect on either the commodity or the environment since CA constituents are present normally in the storage atmosphere and in the environment.
7. There is little chance of development of pest resistance to low-oxygen CA.

Disadvantages of CA storage are:

1. High initial capital costs of the technology to make structures air-tight.

2. The process of disinfestation is slower than with toxic chemicals.
3. There is inadequate information about its reliability as a technology for crop storage, especially in the developing countries.

Additional information is available on the use of CA gases in grains (Bell and Armitage, 1992; Jayas *et al.*, 1991), on quality changes in grain under CA storage (White and Jayas, 1993), and on CA storage of yellow maize, rice and sorghum in Nigeria (Agboola, 1993).

High moisture grain storage

Mould growth on high moisture grains (above 220–250 g/kg) is prevented by limiting the available oxygen supply. When high moisture grain such as corn is placed in a sealed silo, the grain undergoes fermentation, oxygen is depleted, and carbon dioxide is increased by the respiration of the grains, yeasts, and bacteria. Aerobic mould growth is halted but germination is impaired, making the seed only suitable for animal feed. On removal from the silo, mould growth recommences; therefore exposed grains must be fed rapidly to keep ahead of mould growth. The high moisture grains may be stored in glass-lined, oxygen-limiting steel silos or in other kinds of airtight bins that have a breathing system to prevent structural failure because of differential pressures and to limit the exchange of in-storage gas with outdoor air.

Another method of preventing mould growth in high moisture grains is by the application of inhibitory chemicals. When such grain is treated with the recommended dosage of a registered mould inhibitory chemical the grain usually can be removed from storage without concern for spoilage. Propionic acid is the commonest material used; the dosage rate depends on the moisture level of the grain (higher moisture levels requiring more acid), temperature, and length of storage period. Acid-treated grain does not need any particular type of storage structure, but when galvanized steel sheets are used severe corrosion can result. This problem can be corrected by prior coating with chlorinated rubber paint (Mills, 1989).

Grain that has been treated with a preservative can be subject to mould spoilage and should be inspected regularly during storage. Mould growth releases moisture, enabling the moulds to spread to the treated grains. This occurs under the following conditions:

- When an incorrect dosage rate of preservative for the particular grain moisture is used.
- When insufficient acid is used on all or part of the bulk.
- When wet spots develop through moisture migration.
- When treated grains are in contact with unprotected concrete or steel.

Such surfaces should be covered with plastic or acid-resistant paint (Anon., 1984).

14.4.6 Bag management

The choice between bag and bulk grain storage depends on local factors including type of produce, duration of storage, value of produce, climate, transport system, cost and availability of labour and sacks, and incidence of rodents and certain types of insect infestation. Advantages of bag storage include flexibility of storage, low capital cost, and being partially mechanizable. Disadvantages include slow handling, considerable spillage, high operating costs, high rodent loss potential and pest reinfestation. Nevertheless in certain areas of the world, bag storage is the method of choice, such as where road and transport facilities are poorly developed; where quantities of grain delivered and handled by traders are in small lots, and where a wide variety of products is to be handled (Hall, 1970). Buildings used for bag storage in the tropics and subtropics are often not designed for grain storage. A well-planned bag storage warehouse should have a waterproof roof, walls and floor. There must be sealable openings for controlled ventilation and complete building fumigation. The metal roof should be insulated to minimize temperature buildup, bags should be raised above the floor to provide insulation against floor temperature and moisture, and the building should be bird- and rodent-proof (Salunkhe *et al.*, 1985).

Where bagged produce in store is exposed to high humidity air, reabsorption of moisture takes place especially in the outer areas of bags. The increased moisture accelerates product biological activity resulting in deterioration and quality loss (Hall, 1970). Bagged produce in store is also subject to pest infestation. Stored-product beetles and moths lay their eggs directly into the product through the meshes of the bags or through needle holes along the seams or where the tops are sewed (Cotton, 1963). Stacked bags need to be kept free of dust to aid natural ventilation processes (Calverley and Hallam, 1982).

Information is available on the management of bags in storages (Hall, 1970); on fumigation of bagged cargoes (Bond, 1973); and on fumigation of sealed stacks in Indonesia and Australia (Sidik and van Graver, 1993).

14.4.7 Transportation

Grain-quality loss also occurs during long distance transportation (Mills, 1989). For example, spoilage has been reported in corn during transport by river barge (DeHoff *et al.*, 1984) and ocean freighter from North and South American ports (Milton and Jarrett, 1969; Paulsen *et al.*, 1991). In the United States, substantial amounts of corn are shipped from the midwest during cold weather to warmer southern states either for local consumption as animal feed or for export. Spoilage occurs onboard ship because the corn is already infected by storage fungi on arrival by barge at the Gulf Ports, picks up moisture from the humid air, and has a substantially increased level of broken corn foreign material (BCFM) due to numerous handlings en route (Tuite and Foster, 1979). The likelihood of spoilage is further enhanced by ocean shipment at moisture levels above 150 g/kg and elevated temperatures. In one case, the mean temperature of corn, one metre below the grain surface, increased from 22°C to 47°C during a voyage from the Gulf of Mexico to Japan (Paulsen *et al.*, 1991). During shipment to the tropics and subtropics, the storage life of corn rapidly decreases with increased breakage levels (Figure 14.2; Calverley and Hallam, 1982).

14.4.8 Quality control

Quality control procedures are important in every phase of grain storage including the initial purchase, throughout the storage period, and before and after shipment. Persons receiving grain at a facility have a right to refuse loads that have doubtful keeping quality or are in poor condition, thus presenting later handling, storage and quality problems (Ferket and Jones, 1989; Mills, 1989). Purchasing materials from

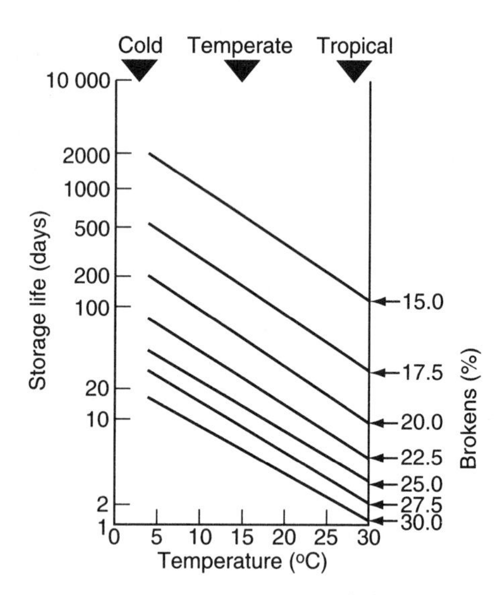

Figure 14.2 Storage life of maize in different environments (after Calverley and Hallam, 1982). 'Brokens' refers to the amount of broken corn in the samples examined.

unknown suppliers is risky. Their reputation, and if possible their facilities, should first be checked. Cereal grains may be frost-damaged, shrivelled, spoiled, or heated which are all conditions that reduce the nutritive value of grains. Grains may also be contaminated with chemicals, pests and pest by-products, other grains, and debris and stones (Mills, 1995).

During storage, the moisture levels of the stored stocks should be determined at intervals, particularly if stocks are unaerated. The storage structure and its contents should be regularly inspected for leaks, and the stored material probed regularly for signs of spoilage and heating, especially in the top centre. It is helpful to install remote CO_2 and temperature sensors to detect deteriorative changes occurring in the stocks. The probable development of spoilage moulds in a lot of grain during storage is predicted by the amount of existing spoilage moulds present and a knowledge of the history, condition and type of grain involved. Decisions are then made regarding how long to keep a particular lot, whether it should be disposed of, or which preventive measures need to be taken (Mills, 1989). When cereals are shipped in bulk or bags, proper representative samples of the stocks should be obtained prior to shipment to protect the shipper from any claims regarding quality at point of destination. This process is particularly important when shipment is by barge or ocean vessel over long distances (Calverley and Hallam, 1982; DeHoff *et al.*, 1984).

14.5 STORAGE CHARACTERISTICS OF WORLD CEREAL CROPS

14.5.1 Introduction

Cereal grains differ in their composition, kernel size, packing characteristics, moisture relationships and other parameters which influence storage quality. Some crops are more prone to spoilage, heating problems and mycotoxin development than others. The grain handler, in order to ensure optimal preservation in storage quality, needs to be aware of the peculiarities of the crops in his or her care. This section describes the main storage characteristics of selected world cereal crops; each crop is examined from the viewpoint of moisture/temperature/time relationships, storage problems and seed management.

14.5.2 Storage characteristics of major cereals

*Wheat (*Triticum aestivum *L.)*

Moisture/temperature/time relationships

Grain moisture is usually high (300 g/kg) at harvest (Salunkhe *et al.*, 1985) in some countries, necessitating swathing and drying in the field before harvest, while in other countries, grain is harvested directly from standing plants in dry conditions. Maximum moisture for safe storage is 130 g/kg for commercial wheat and 120 g/kg for seed wheat. For long-term storage of commercial wheat the maximums are 130–140 g/kg for up to 1 year and 110–120 g/kg for up to 5 years (Hall, 1980). In soft red winter, hard red winter, hard red spring, and durum wheats an intergranular relative humidity of 70% equilibrates with 135, 139, 139 and 137 g/kg, respectively, at 25°C (Table 14.2). At 10°C, 70% r.h. equilibrates with a moisture of 150 g/kg (Friesen and Huminicki, 1986). The storage time chart in Figure 14.3 predicts the storability of wheat over 6 months under selected temperature and moisture conditions (Wallace *et al.*, 1983; Wilkins, 1983). Hard red spring wheat was obtained from farm granaries, conditioned to various moisture levels, stored at several temperatures in the laboratory, and grain quality assessments made at predetermined intervals.

Storage problems

Wheat stores readily but on occasion hot spots may develop in non-aerated grains, originating from either fungal and or insect activity

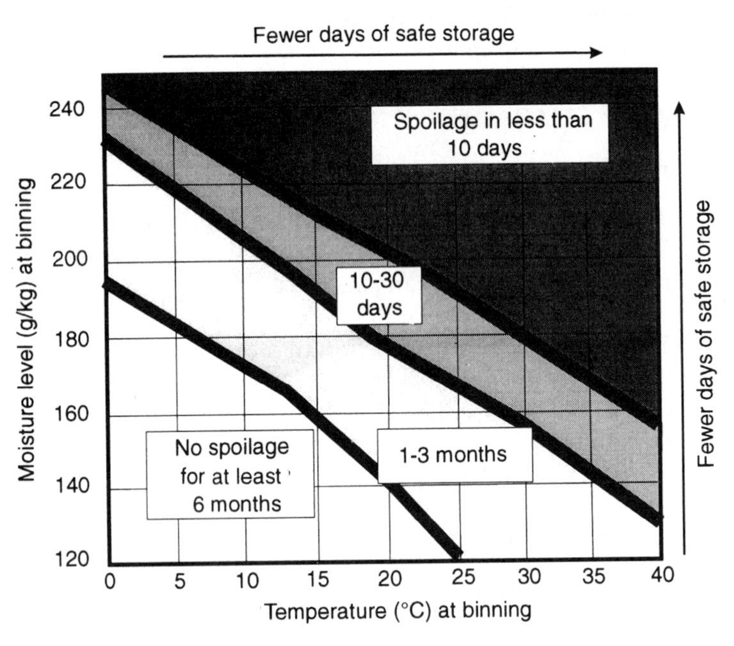

Figure 14.3 Wheat storage time chart showing moisture level and temperature zones in which spoilage occurs in less than 10 days, within 10–30 days, within 1–3 months, and no spoilage for at least 6 months (Wilkins, 1983).

(Sinha, 1973). Storey *et al.* (1983) found that 25.1% of all wheat surveyed in 8000 USA farm storages was infested in 1980.

Seed management

The maximum grain temperatures for safe drying are 60°C for seed required for seeding purposes, 65°C for commercial use, and 80–100°C for feed (Friesen, 1981). Excessive heat during drying of wheat damages the endosperm protein, impairing the suitability of the flour for bread-making (Freeman, 1980).

*Rice (*Oryza sativa *L.)*

Moisture/temperature/time relationships

Rough paddy rice is harvested at a moisture level of 160–280 g/kg (Brooker *et al.*, 1992). In the USA, a moisture level of 125 g/kg is generally considered suitable for rice storage (Kunze and Calderwood, 1980). For rough paddy rice the advised maximum level for safe storage up to 27°C is 140 g/kg and for milled rice it is 120 g/kg (Muckle and Stirling, 1971). For whole grain rice, an intergranular relative humidity of 70%, at which moulds can be expected to appear, equilibrates with a moisture level of 141 g/kg at 25°C (Hall, 1980). For rice at a moisture level of 130 g/kg, mould growth is inhibited below 21.1°C, and insect activity is considerably reduced below 15.6°C (Steffe *et al.*, 1980).

Storage problems

Most of the storage problems which occur are the result of unsuitable moisture and temperature regimens often favouring mould and/or insect activities. Post-harvest yellowing of rice is a major problem in the humid tropics, especially in Southeast Asia. Phillips *et al.* (1988) monitored post-harvest yellowing in rough bulk rice dried by aeration and stored over a period of one year. Yellowing increased from 0–0.5% up to 4.5–5.0%, at moisture levels of less than 140 g/kg. This was related to earlier mould growth before and during the drying period, particularly in the upper layers of grain which had taken longer to dry.

Seed management

To avoid drying problems when more than one crop is being dried, it is important to realize that the bulk density of paddy rice is significantly lower than that of corn and wheat, and the air velocity in rice is correspondingly larger at constant dryer output. A commercial bulk-storage system designed for long-term safe storage of rough rice must provide proper aeration to prevent self-heating and maintain the rice at a low moisture of about 135 g/kg. The operation of aeration systems for bulk storages in high humidity environments calls for constant operator attention (Steffe *et al.*, 1980). Additional information on drying rice is given in Brooker *et al.* (1992).

*Corn/maize (*Zea mays *L.)*

Moisture/temperature/time relationships

The maximum moisture level for safe storage of corn is 139 g/kg for 1 year and 110 g/kg for 5 years. In shelled corn an intergranular relative humidity of 70% equilibrates with 140 g/kg at 25°C (Table 14.2; Hall, 1980). The allowable storage time for corn, calculated from spoilage, moisture, temperature and time data, is given in Friesen and Huminicki (1986). Corn stored at a 220 g/kg moisture level will be safe at 27°C for about 5 days, at 19°C for 10 days and at 4°C for 60 days.

Storage problems

Corn is prone to breakage during handling and after drying at too high a temperature and drying rate (Tuite and Foster, 1979). As a result, broken corn foreign material is increased and constitutes a storage hazard when it collects in the central core or spout lines of grain bins. Spoilage may begin in the spout line partly because the fines, which consist mainly of fragments of corn endosperm, are more susceptible than whole kernels to invasion by storage moulds, and partly because insects and mites thrive in the fines and promote the growth of such moulds (Sauer *et al.*, 1992). Corn is prone to spoilage during transport by river barge, and by ocean freighter from North and South American ports (see Section 14.4.7).

Seed management

Corn that is too dry is subject to breakage during handling, and if too moist is prone to mould development. To avoid these problems, corn is harvested moist, and carefully dried. During the drying and storage period, deterioration such as dry matter loss needs to be minimized (below 0.5%) to reduce producer costs. Storage time charts, based on a formula for estimating accumulative deterioration, have been developed for corn (Misra *et al.*, 1980). Their use enables the producer to select the proper storage moisture drying temperature and rates to reduce costs through more efficient energy use by not overdrying.

Barley (Hordeum vulgare *L.)*

Moisture/temperature/time relationships

The maximum amount of moisture for safe storage of barley is 130 g/kg for 1 year and 110 g/kg for 5 years (Hall, 1980). Barley containing 103–121 g/kg initial moisture and having a temperature of 22–35°C can be kept in good condition with no increase in fatty acids in Manitoba, Canada farm bins for 3 years (Sinha and Wallace, 1977). Barley of 135 g/kg moisture equilibrates with a relative humidity of 70% at 25°C (Mills, 1989). Freshly harvested grain with above 140 g/kg may heat and go out of condition (Dickson, 1959). Malting barley may be more sensitive to quality deterioration than barley used for food or feed.

Storage problems

Only a moderate development of storage moulds is needed to destroy the germination ability of barley and give it a musty odour (Dickson, 1959). Spoilage of moist barley (230–400 g/kg) may occur in sealed and unsealed silos and in structures containing acid-treated grains for animal feeds. Spoilage can occur in high moisture barley treated with propionic acid when inadequate acid is used, and when condensation occurs, diluting the acid treatment (Mills, 1989).

Seed management

Barley that is used for seed or malting purposes requires frequent monitoring and special care in storage (Dickson, 1959). Inadequate drying leads to development of moulds during storage (Salunkhe *et al.*, 1985). Burrell (1970) examined the moisture content–temperature combinations at which mould and mite problems may be expected over a 32-week period under UK farm conditions. He showed that high-value malting barley needed to be dried down to a level of 120 g/kg and cooled to avoid risk of mite infestation. The maximum safe drying temperatures are 45°C for barley required for seeding or malting purposes, 55°C for commercial use, and 80–100°C for feed (Friesen, 1981).

Sorghum (Sorghum bicolor *(L.) Moench)*

Moisture/temperature/time relationships

The maximum moisture level for safe storage of grain sorghum is 130 g/kg for 1 year and 100–110 g/kg for 5 years (Hall, 1980). The maximum moisture level for safe storage of sorghum at 27°C is 135 g/kg according to Muckle and Stirling (1971), but this figure varies considerably between varieties. In sorghum, an intergranular relative humidity of 70% equilibrates with a moisture level of 138 g/kg (Table 14.2; Hall, 1980).

Storage problems

Infection of sorghum by storage moulds can result in loss of viability, reduction in market value, and aflatoxin production (Christensen and Kaufmann, 1969). Stored sorghum is more susceptible to insect attack than stored rice. Major pests of stored sorghum include the Angoumois grain moth, rice weevil, lesser grain borer and red flour beetle (Salunkhe *et al.*, 1985).

Seed management

Sorghum is readily stored if the usual management practices for cereals are employed. The maximum grain temperatures for safe drying are 43°C for grain sorghum intended for seeding purposes, 60°C for commercial use, and 82°C for feed (Hall, 1980). Additional information on sorghum storage can be found in Doggett (1970) and Sorensen and Person (1970).

*Millet (*Pennisetum americanum *L.)*

Moisture/temperature/time relationships

In millet, a moisture level of 160 g/kg equilibrates with 70% RH at 27°C (Hall, 1970).

Storage problems

Owing to a thick pericarp and a smaller grain size, millet is relatively more resistant to pest attacks than sorghum. However, insects, moulds, and rodents are known to cause considerable losses in millet during storage; the germ of pearl millet is particularly susceptible to pest attack (Salunkhe *et al.*, 1985). Five common species of stored-product insect were shown to develop and multiply well on millet cultivars (Sinha, 1972). Factors affecting rancidity in ground pearl millet were studied by Kaced *et al.* (1984); increases in fat acidity and peroxide value show that pearl millet rapidly becomes rancid after it is ground. Pomeranz (1982) also found deterioration occurred more rapidly in ground millet than in whole grain.

Seed management

Storage practices recommended for sorghum are equally applicable to millet.

*Oats (*Avena sativa *L.)*

Moisture/temperature/time relationships

Oats at harvest may contain about 150–160 g/kg or even more moisture (Salunkhe *et al.*, 1985). Oats can be safely stored at grain moistures of 120–130 g/kg for up to 6 months. Longer-term storage requires a moisture of 120 g/kg or less (Schrickel *et al.*, 1992). Oats of 130 g/kg moisture equilibrate with a relative humidity of 70% at 25°C (Mills, 1989).

Storage problems

If oats contain more than 140 g/kg moisture when binned they tend to become musty or heat-damaged due to mould activity. This markedly reduces their feed value and makes them unfit for use as food (Stanton, 1959). Stored oats have a low susceptibility to mycotoxin development when compared to other cereals (Abramson, 1991), and instances of self-heating are rarely reported (Mills, 1989). Levels of hydrolytic rancidity occurring in stored hulled and hull-less oats were investigated by Welch (1977). The level of hydrolytic rancidity was found to increase at higher moisture contents and with longer storage periods, but the level in hull-less oats only exceeded that in hulled oats if the grain was severely bruised. Insect infestation of stored oats is mostly by secondary insects that feed only on cracked, dehulled, or mouldy grain (Cotton, 1963).

Seed management

Oat grain that is combined directly and transported immediately to storage bins often contains 180–200 g/kg moisture, in which case it must be artificially dried to a safe level (Schrickel *et al.*, 1992). The maximum grain temperatures for safe drying are 50°C for oats required for seeding purposes, 60°C for commercial use, and 80–100°C for feed (Friesen, 1981). Oats of between 150–170 g/kg moisture should be cooled from 15 to 5°C, respectively, to prevent mould development during medium-term (45 weeks) storage (Kreyger, 1972). Hull-less oats probably should be stored in hoppered bins, which allow grain to flow out easily, and maintained at a moisture of 120 g/kg or less. Grain temperature should be checked frequently and monitoring of the fat acidity is also desirable (Schrickel *et al.*, 1992).

Rye (Secale cereale *L.)*

Moisture/temperature/time relationships

As rye matures early in the summer, the moisture reaches a safe storage level more quickly than wheat or other grains (Shands, 1959). To avoid spoilage, the moisture level should not be over 130 g/kg (Rozsa, 1976). Kreyger (1972) recommends a maximum moisture level of 140 g/kg for storage of rye seed. Long-term storage at this moisture level requires cooling to 15°C, or less. Mould development occurs rapidly on seeds stored at moistures above 140 g/kg with visible moulds occurring on 150 g/kg moisture seed stored at 25°C after only 1 month.

Storage problems

To avoid heating and mould growth, rye harvested at a moisture level of 150 g/kg needs immediate drying to reduce the moisture to 120–130 g/kg (Salunkhe *et al.*, 1985).

Seed management

Since rye is similar to wheat in storage requirements, satisfactory storage conditions for wheat are usually satisfactory for rye (Rozsa, 1976). The maximum grain temperatures for safe drying are 45°C for rye required for seeding purposes, 60°C for commercial use, and 80–100°C for feed (Friesen, 1981).

Triticale

Triticale is a hybrid of wheat and rye. In triticale, an intergranular relative humidity of 70% equilibrates with a moisture level of 151 g/kg at 22°C. The moisture-equilibrium relative humidity values for triticale at 22°C are higher than those for rye at 25°C or wheat at 20°C or 25°C. Triticale has a density about 20% less than that of wheat and 15% less than that of rye, and this may have some bearing on its higher moisture–relative humidity values (Sinha and White, 1982). From the foregoing, triticale is less likely to spoil than wheat when stored at the same moisture level and temperature (Mills, 1989).

14.6 FUTURE WORK

Considerable progress has been made in recent years on the improvement of quality in stored cereals. Research and development emphasis now needs to be directed towards the determination of storage parameters for existing, new or genetically modified crops; integrated pest control methods, rapid testing for pests, mycotoxins and other specific compounds; and the determination of crop susceptibility to mycotoxins and toxin resistance in existing, or new, crops and varieties. It is envisaged that government, industry and universities will be involved in the identification of these research and development (R&D) areas and their funding, and that teams of specialists working in their own facilities will be constituted to solve specific complex multi-disciplinary problems. For example, a team has been formed in Winnipeg to work on the storage parameters of hull-less barley, an increasingly important crop in western Canada. The team consists of an agricultural engineer to examine the engineering properties of the stored crop, a mycologist to work on moisture/temperature/time guidelines in relation to spoilage, an entomologist to study infestation potential, and a feed scientist to examine nutritional aspects.

Another R&D approach is the establishment of partnerships involving industries to ensure research is directed towards priority problems and rapidly transferred to clients such as flour millers. It is likely that cereal genetic manipulation and other biotechnological techniques will be utilized to improve the storage quality of grain and to solve R&D problems. It is hoped that R&D solutions will be implemented with minimal disturbance to the ecological balance for the long-term benefit of mankind, e.g., environmental hazards caused by the fumigant methyl bromide which depletes atmospheric ozone, and the use of safer, non-chemical alternatives for control of pests. With recent improvements in communications, identification and implementation times for R&D will be greatly reduced through faster accession of storage and handling information, and more widely available, easily understood vehicles for transfer of the required technology.

REFERENCES

Abramson, D. (1991) Development of molds, mycotoxins and odors in moist cereals during storage, in *Cereal Grain Mycotoxins, Fungi and Quality in Drying and Storage* (ed. J. Chelkowski), Elsevier, Amsterdam, pp. 119–47.

Agboola, S.D. (1993) Current status of controlled atmosphere storage in Nigeria, in *Proceedings of the International Conference on Controlled Atmosphere and Fumigation in Storages* (eds S. Navarro and E. Donahaye), Caspit Press, Jerusalem, pp. 51–68.

Anderson, J.A. (1973) Problems of controlling quality in grain, in *Grain Storage: Part of a System* (eds R.N. Sinha and W.E. Muir), AVI, Westport, CT, pp. 1–14.

Anon. (1972) *Grain Inspection Manual.* US Department of Agriculture, Agriculture Marketing Service, Washington DC.

Anon. (1978) *Official United States Standards for Grain.* US Department of Agriculture, Agriculture Marketing Service, Washington DC.

Anon. (1984) *Management of On-Farm Stored Grain,* College of Agriculture, University of Kentucky, Lexington, KY.

Anon. (1992) *Official Grain Grading Guide*, Canadian Grain Commission, Winnipeg, MB.

Anon. (1993) *Standards, Engineering Practices, Data*, American Society of Agricultural Engineers, St. Joseph, MI.

Armitage, D.M. (1984) The vertical distribution of mites in bulks of stored produce, in *Acarology VI*, Vol. 2 (eds D.A. Griffiths and C.E. Bowman), Ellis Horwood, Chichester, pp. 1006–13.

Banks, H.J. and Fields, P.G. (1995) Physical methods for insect control in stored-grain ecosystems, in *Stored Grain Ecosystems* (eds D.S. Jayas, N.D.G. White and W.E. Muir), Marcel Dekker, New York, NY, pp. 353–409.

Barker, P.S. (1993) Phoretic mites found on beetles associated with stored grain in Manitoba. *Canadian Entomologist*, **125**, 715–7.

Baur, F.J. and Jackson, W.B. (eds) (1982) *Bird Control in Food Plants*, American Association Cereal Chemists, St. Paul, MN.

Bell, C.H. and Armitage, D.M. (1992) Alternative storage practices, in *Storage of Cereal Grains and Their Products* (ed. D.B. Sauer), American Association of Cereal Chemists Inc., St. Paul, MN, pp. 249–311.

Bond, E.J. (1973) Chemical control of stored grain insects and mites, in *Grain Storage: Part of a System* (eds R.N. Sinha and W.E. Muir), AVI, Westport, CT, pp. 137–79.

Brooker, D.B., Bakker-Arkema, F.W. and Hall, C.W. (1992) *Drying and Storage of Grains and Oilseeds*, AVI, Van Nostrand Reinhold, New York, NY.

Bruce, D.M. and Ryniecki, A. (1991) Economic methods of cereal grain drying to prevent spoilage and loss of quality, in *Cereal Grain Mycotoxins, Fungi and Quality in Drying and Storage* (ed. J. Chelkowski), Elsevier, Amsterdam.

Burrell, N.J. (1970) *Conditions for safe grain storage*. Technical Note 16, Home Grown Cereals Authority, London.

Calverley, D.J.B. and Hallam, J.A. (1982) Problems of storing grain from temperate climates in tropical countries including a recent case history, in *Grain-Trade, Transportation and Handling* (ed. P. Findlay), Proceedings of the Graintrans 82 Conference, CS Publications, Worcester Park, London, pp. 143–50.

Cenkowski, S. and Zhang, Q. (1995) Engineering properties of grains and oilseeds, in *Stored Grain Ecosystems* (eds D.S. Jayas, N.D.G. White and W.E. Muir), Marcel Dekker, New York, NY.

Champ, B.R. and Dyte, C.E. (1976) *Report of the FAO Global Survey of Pesticide Susceptibility of Stored Grain Pests*, FAO Plant Protection Series No. 2.

Chelkowski, J. (ed.) (1991) *Cereal Grain Mycotoxins, Fungi and Quality in Drying and Storage*, Developments in Food Science 26, Elsevier, Amsterdam.

Christensen, C.M. and Kaufmann, H.H. (1969) *Grain Storage. The Role of Fungi in Quality Loss*, University of Minnesota Press, Minneapolis, MN.

Cotton, R.T. (1963) *Pests of Stored Grain and Grain Products*, Burgess Publishing Co., Minneapolis, MN.

Cotton, R.T. and Wilbur, D.A. (1982) Insects, in *Storage of Cereal Grains and Their Products* (ed. C.M. Christensen), American Association of Cereal Chemists Inc., St. Paul, MN, pp. 281–318.

DeHoff, T.W., Stroshine, R., Tuite, J. and Baker, K. (1984) Corn quality during barge shipment. *Transactions of the American Society of Agricultural Engineers*, **27**, 259–64.

Dickson, A.D. (1959) Barley, in *The Chemistry and Technology of Cereals as Food and Feed* (ed. S.A. Matz), AVI, Westport, pp. 76–95.

Doggett, H. (1970) *Sorghum*, Tropical agriculture series, Longmans, London.

Dykstra, W.W. (1973) Rodents in stored grain, in *Grain Storage: Part of a System* (eds R.N. Sinha and W.E. Muir), AVI, Westport, CT, pp. 181–8.
Ferket, P.R. and Jones, F.T. (1989) North Carolina research leads to ingredient quality findings. *Feedstuffs* (September 25, 1989), pp. 32; 39–40; 57–9; 70.
Foster, G.H. (1973) Heated-air grain drying, in *Grain Storage: Part of a System* (eds R.N. Sinha and W.E. Muir), AVI, Westport, CT, pp. 189–208.
Frantz, S.C. and Davis, D.E. (1991) Bionomics and integrated pest management of commensal rodents, in *Ecology and Management of Food-Industry Pests* (ed. J.R. Gorham), Association Official Analytical Chemists, Arlington, VA, pp. 243–313.
Freeman, J.E. (1980) Quality preservation during harvesting, conditioning and storage of grains and oilseeds, in *Crop Quality Storage and Utilization* (ed. C.S. Hoveland), American Society of Agronomy, Madison, WI, pp. 187–226.
Friesen, O.H. (1981) *Heated-Air Grain Dryers.* Agriculture Canada Publ. No. 1700, Ottawa, ON.
Friesen, O.H. and Huminicki, D.N. (1986) *Grain Aeration and Unheated Air Drying.* Manitoba Agriculture Agdex 732–1, Winnipeg, MB.
Garry, V.F., Griffith, J., Danzi, T.J., Nelson, R.L., Wharton, E.B., Krueger, L.A. and Cervenka, J. (1989) Human genotoxicity: pesticide applicators and phosphine. *Science*, **246**, 251–5.
Gough, M.C. and King, P.E. (1980) Moisture content/relative humidity equilibria of some tropical cereal grains. *Tropical Stored Products Information*, **39**, 13–7.
Gedye, D.J., Doling, D.A. and Kingswood, K.W. (1981) *A Farmers Guide to Wheat Quality*, National Agricultural Centre Cereal Unit, Kenilworth, UK.
Hall, C.W. (1980) *Drying and Storage of Agricultural Crops*, AVI, Westport, CT.
Hall, C.W. and Maddex, R.L. (1980) Systems for handling of grain, in *Drying and Storage of Agricultural Crops* (ed. C.W Hall), AVI, Westport, CT, pp. 234–57.
Hall, D.W. (1970) *Handling and Storage of Food Grains in Tropical and Subtropical Areas*, FAO Agricultural Development Paper 90, Rome.
Harein, P.K. and Davis, R. (1992) Control of stored-grain insects, in *Storage of Cereal Grains and Their Products* (ed. D.B. Sauer), American Association of Cereal Chemists, St. Paul, MN, pp. 491–534.
Harris, K.L. and Baur, F.J. (1992) Rodents, in *Storage of Cereal Grains and Their Products* (ed. D.B. Sauer), American Association of Cereal Chemists, St. Paul, MN, pp. 393–434.
Henderson, S. (1987) A mean moisture content-equilibrium moisture content relationship for nine varieties of wheat. *Journal of Stored Products Research*, **23**, 143–7.
Hill, L.D. (1988) The challenge of developing workable grades for grain. *Cereal Foods World*, **33**, 348–9.
Hoseney, R.C. and Faubion, J.M. (1992) Physical properties of cereal grains, in *Storage of Cereal Grains and Their Products* (ed. D.B. Sauer), American Association of Cereal Chemists, St. Paul, MN, pp. 1–38.
Howe, R.W. (1972) Insects attacking seeds during storage, in *Seed Biology, Vol. 3, Insects and Seed Collection, Storage, Testing and Certification* (ed. T.T. Kozlowski), Academic Press, New York, NY, pp. 247–300.
Jayas, D.S., Khangura, B. and White, N.D.G. (1991) Controlled atmosphere storage of grains. *Postharvest News and Information*, **2**, 423–7.
Kaced, I., Hoseney, R.C. and Varriano-Marston, E. (1984) Factors affecting rancidity in ground pearl millet (*Pennisetum americanum* (L.) Leeke). *Cereal Chemistry*, **61**, 187–92.

Kreyger, J. (1972) *Drying and Storing Grains, Seeds and Pulses in Temperate Climates*, Publ. 205, IBVL, Wageningen.

Kunze, O.R. and Calderwood, D.L. (1980) Systems for drying of rice, in *Drying and Storage of Agricultural Crops* (ed. Hall, C.W.), AVI, Westport, CT, pp. 209–33.

Lacey, J. and Magan, N. (1991) Fungi in cereal grains: their occurrence and water and temperature relationships, in *Cereal Grain Mycotoxins, Fungi and Quality in Drying and Storage* (ed. J. Chelkowski), Elsevier, Amsterdam, pp. 77–118.

Liscombe, E.A.R. and Watters, F.L. (1962) Insect and mite infestations in empty granaries in the Prairie Provinces. *Canadian Entomologist*, **94**, 433–41.

Mackay, P.J. (1967) Theory of moisture in stored produce. *Tropical Stored Products Information*, **13**, 9–14.

Manbeck, H.B., Britton, M.G. and Puri, V.M. (1995) Theory of moisture in stored produce, in *Stored Grain Ecosystems* (eds D.S. Jayas, N.D.G. White and W.E. Muir), Marcel Dekker, New York, NY.

Manis, J.M. (1992) Sampling, inspecting, and grading in *Storage of Cereal Grains and Their Products* (ed. D.B. Sauer), American Association of Cereal Chemists, St. Paul MN, pp. 561–88.

Mannon, J. and Johnson, E. (1985) Fungi down on the farm. *New Scientist*, **105** (1445), 12–6.

Mills, J.T. (1986) Postharvest insect-fungus associations affecting seed deterioration, in *Physiological-pathological Interactions Affecting Seed Deterioration* (ed. S.H. West), Crop Society of America Special Publ. No. 12, pp. 39–51.

Mills, J.T. (1989) *Spoilage and Heating of Stored Agricultural Products. Prevention, Detection and Control*. Agriculture Canada Publ. No. 1823E, Ottawa.

Mills, J.T. (ed.) (1990) *Protection of Farm-Stored Grains and Oilseeds from Insects, Mites and Molds*. Agriculture Canada Publ. 1851E, Ottawa.

Mills, J.T. (1992) Safe storage guidelines for grains and their products. *Postharvest News and Information*, **3**, 111N–5N.

Mills, J.T. (1995) Ecological aspects of feed mill operation, in *Stored Grain Ecosystems* (eds D.S. Jayas, N.D.G. White and W.E. Muir), Marcel Dekker, New York, NY.

Mills, J.T. and Wallace, H.A.H. (1979) Microflora and condition of cereal grains after a wet harvest. *Canadian Journal of Plant Science*, **59**, 645–51.

Misra, R.N., Keener, H.M. and Glen, T.L. (1980) Safe storage of shelled corn under Ohio weather conditions. *Ohio Report*, May–June 1980, 41–3.

Milton, R.F. and Jarrett, K.J. (1969) Storage and transport of maize 1. Temperature, humidity and microbiological spoilage. *World Crops*, **21**, 356–7.

Moysey, E.B. and Norum, E.R. (1975) Storage, drying and handling of oilseeds and pulse crops, in *Oilseed and Pulse Crops in Western Canada – A Symposium* (ed. J.T. Harapiak), Western Cooperative Fertilisers Ltd, Calgary, pp. 507–40.

Muckle, T.B. and Stirling, H.G. (1971) Review of the drying of cereals and legumes in the tropics. *Tropical Stored Products Information*, **22**, 11–30.

Muhlbauer, W. and Christ, W. (1974) The admissible drying time of maize kernels for animal nutrition with different grain temperatures. *Grundlagen der Landtechnik*, **24**, 161–4.

Muir, W.E. (1973) Temperature and moisture in grain storages, in *Grain Storage: Part of a System* (eds R.N. Sinha and W.E. Muir), AVI, Westport, CT, pp. 49–70.

Muir, W.E., Waterer, D. and Sinha, R.N. (1985) Carbon dioxide as an early indicator of stored cereal and oilseed spoilage. *Transaction of the American Society of Agricultural Engineers*, **28**, 1673–5.

Nash, M.J. (1985) *Crop Conservation and Storage in Cool Temperate Climates*, 2nd edn, Pergamon Press, Oxford.
Paulsen, M.R., Hill, L.D. and Shove, G.C. (1991) Temperature of corn during ocean vessel transport. *Transaction of the American Society of Agricultural Engineers*, **34**, 1824–9.
Pedersen, J.R. (1992) Insects: identification, damage, and detection, in *Storage of Cereal Grains and Their Products* (ed. D.B. Sauer), American Association of Cereal Chemists, St. Paul., MN, pp. 435–89.
Phillips, S., Widjaja, S., Wallbridge, A. and Cooke, R. (1988) Rice yellowing during post-harvest drying by aeration and during storage. *Journal of Stored Products Research*, **24**, 173–81.
Pomeranz, Y. (1992) Biochemical, functional, and nutritive changes during storage, in *Storage of Cereal Grains and Their Products* (ed. D.B. Sauer), American Association of Cereal Chemists, St. Paul, MN, pp. 55–141.
Posner, E.S. and Deyoe, C.W. (1986) Changes in milling properties of newly harvested hard wheat during storage. *Cereal Chemistry*, **63**, 451–6.
Roberts, E.H. (1972) *Viability of Seeds*, Chapman and Hall, London.
Rozsa, T.A. (1976) Rye milling, in *Rye: Production, Chemistry and Technology* (ed. W. Bushuk), American Association of Cereal Chemists, St. Paul, MN, pp. 111–25.
Salunkhe, D.K., Chavan, J.K. and Kadam, S.S. (1985) *Postharvest Biotechnology of Cereals*, CRC Press, Boca Raton, FL.
Sauer, D.B., Meronuck, R.A. and Christensen, C.M. (1992) Microflora, in *Storage of Cereal Grains and Their Products* (ed. D.B. Sauer), American Association of Cereal Chemists Inc., St. Paul, MN, pp. 313–40.
Schrickel, D.J., Burrows, V.D. and Ingemansen, J.A. (1992) Harvesting, storing, and feeding of oat, in *Oat Science and Technology* (eds H.G. Marshall and M.E. Sorrells), American Society of Agronomy and Crop Science Society of America, Madison, WI, pp. 223–45.
Shands, H.L. (1959) Rye, in *The Chemistry and Technology of Cereals as Food and Feed* (ed. S.A. Matz), AVI, Westport, CT, pp. 96–119.
Sidik, M. and van S. Graver, J. Jr. (1993) Sealed stacks: a progress report on the technology, in *Proceedings of the International Conference on Controlled Atmosphere and Fumigation in Grain Storages* (eds S. Navarro and E. Donahaye), Caspit Press, Jerusalem, pp. 471–5.
Sinha, R.N. (1964) Mites of stored grain in western Canada – ecology and survey. *Proceedings of the Entomological Society of Manitoba*, **20**, 19–33.
Sinha, R.N. (1972) Infestability of oilseeds, clover and millet by stored product insects. *Canadian Journal of Plant Science*, **52**, 431–40.
Sinha, R.N. (1973) Interrelations of physical, chemical and biological variables in the deterioration of stored grains, in *Grain Storage: Part of a System* (eds R.N. Sinha and W.E. Muir), AVI, Westport, CT, pp. 15–47.
Sinha, R.N. (1984) Acarine community in the stored rapeseed ecosystem, in *Acarology VI, Vol. 2*, (eds D.A. Griffiths and C.E. Bowman), Ellis Horwood, Chichester, pp. 1017–25.
Sinha, R.N. (1991) Storage ecosystems, in *Ecology and Management of Food-Industry Pests* (ed. J.R. Gorham), Association Official Analytical Chemists, Arlington, VA, pp. 17–30.
Sinha, R.N. and Wallace, H.A.H. (1977) Storage stability of farm-stored rapeseed and barley. *Canadian Journal of Plant Science*, **57**, 351–65.
Sinha, R.N. and White, N.D.G. (1982) Moisture relations of stored triticale and its susceptibility to infestation by mites and infection by microflora. *Canadian Journal of Plant Science*, **62**, 351–60.

Sinha, R.N., Wallace, H.A.H. and Chebib, F.S. (1969) Principal component analysis of interrelations among fungi, mites and insects in grain bulk ecosystems. *Ecology*, **50**, 536–47.
Smith, B.L. (ed.) (1990) *Codex Alimentarius;* abridged version, Food and Agriculture Organization/World Health Organization, Rome, Italy.
Smith, L.B. and Barker, P.S. (1987) Distribution of insects found in granary residues in the Canadian prairies. *Canadian Entomologist*, **119**, 873–80.
Solomon, M.E. (1962) Ecology of the flour mite, *Acarus siro* L. (*Tyroglyphus farinae* DeG.). *Journal of Applied Ecology*, **1**, 178–84.
Sorensen, J.W. Jr. and Person, N.K. (1970) Drying, storing and handling sorghum grain, in *Sorghum Production and Utilization* (eds J.S. Wall and W.M. Ross), AVI, Westport, CT.
Stanton, T.R. (1959) Oats, in *The Chemistry and Technology of Cereals as Food and Feed* (ed. S.A. Matz), AVI, Westport, CT, pp. 59–75.
Steffe, J.F., Singh, R.P. and Miller, G.E. (1980) Harvest, drying and storage of rough rice, in *Rice: Production and Utilization* (ed. B.S. Luh), AVI, Westport, CT, pp. 311–59.
Storey, C.L., Sauer, D.B. and Walker, D. (1983) Insect populations in wheat, corn, and oats stored on the farm. *Journal of Economic Entomology*, **76**, 1323–30.
Sullivan, J.E., Costa, P.M., Owens, F.N., Jensen, A.H., Wikoff, K.E. and Hatfield, E.E. (1975) The effect of heat on the nutrition value of corn, in *Corn Quality in World Markets* (ed. L.D. Hill), Interstate Publishers, Danville, IL.
Trewartha, G.T. (1968) *An Introduction to Climate*, McGraw-Hill, New York, NY.
Trisvyatskii, L.A. (1966) *Khranenie Zerna (Storage of Grain)* English translation by D.M. Keane. Vol. 1–3 1969. National Lending Library of Science and Technology, Boston Spa, UK.
Tuite, J. and Foster, G.H. (1979) Control of storage diseases of grain. *Annual Review of Phytopathology*, **17**, 343–66.
Wallace, H.A.H. and Sinha, R.N. (1962) Fungi associated with hot spots in farm-stored grain. *Canadian Journal of Plant Science*, **42**, 130–41.
Wallace, H.A.H., Sholberg, P.L., Sinha, R.N. and Muir, W.E. (1983) Biological, physical and chemical changes in stored wheat. *Mycopathologia*, **82**, 65–76.
Welch, R.W. (1977) The development of rancidity in husked and naked oats after storage under various conditions. *Journal of Science of Food and Agriculture*, **28**, 269–74.
White, N.D.G. and Jayas, D.S. (1993) Controlled atmosphere storage research and technology in Canada, in *Proceedings of the International Conference on Controlled Atmosphere and Fumigation in Storages* (eds S. Navarro and E. Donahaye), Caspit Press, Jerusalem, pp. 205–14.
White, N.D.G. and Leesch, J.G. (1995) Chemical control, in *Management of Insects in Stored Products* (eds Bh. Subramanyam and D.W. Hagstrum), Marcel Dekker, New York, NY.
White, N.D.G., Sinha, R.N. and Muir, W.E. (1982) Intergranular carbon dioxide as an indicator of biological activity associated with the spoilage of stored wheat. *Canadian Agricultural Engineering*, **24**, 35–42.
Wicklow, D.T. (1995) Mycology of stored grain, in *Stored Grain Ecosystems* (eds D.S. Jayas, N.D.G. White and W.E. Muir), Marcel Dekker, New York, NY.
Wilkins, D. (1983) Safe storage of wheat. *Country Guide*, **102**, 30A, 30D.
Wilson, D. M. and Abramson, D. (1992) Mycotoxins, in *Storage of Cereal Grains and Their Products* (ed. D.B. Sauer), American Association of Cereal Chemists, St. Paul, MN, pp. 341–91.
Yaciuk, G., Muir, W.E. and Sinha, R.N. (1975) A simulation model of temperatures in stored grain. *Journal of Agricultural Engineering Research*, **20**, 245–58.

Index

Page numbers appearing in *italics* refer to tables and page numbers appearing in **bold** refer to figures.